Carbon Nanostructures

Series editor

Paulo Araujo, Tuscaloosa, AL, USA

More information about this series at http://www.springer.com/series/8633

Nianjun Yang · Xin Jiang
Dai-Wen Pang
Editors

Carbon Nanoparticles and Nanostructures

Springer

Editors
Nianjun Yang
Institute of Materials Engineering
University of Siegen
Siegen
Germany

Dai-Wen Pang
Wuhan University
Wuhan
China

Xin Jiang
Institute of Materials Engineering
University of Siegen
Siegen
Germany

ISSN 2191-3005 ISSN 2191-3013 (electronic)
Carbon Nanostructures
ISBN 978-3-319-80423-1 ISBN 978-3-319-28782-9 (eBook)
DOI 10.1007/978-3-319-28782-9

Printed on acid-free paper

This Springer imprint is published by Springer Nature
The registered company is Springer International Publishing AG Switzerland

Preface

Carbon is an extraordinary element. Because of its ability to covalently bond with different orbital hybridizations, a uniquely rich array of molecular structures are formed. Carbon is thus the basis of all-known life on earth. For millennia, there were only two known substances of pure carbon atoms: graphite and diamond. In recent decades, a series of new carbon nanostructures have been discovered, including fullerenes in the mid-1980s, carbon nanotubes in the early 1990s, graphene in 2003, onions, nanoparticles, nanohorns, nanobells, nanopeapods, and nanofoams. The properties of these different carbon materials are actually determined by their carbon–carbon covalent bonding and the organization of the carbon atoms into a characteristic nano- and microstructure.

Carbon nanostructures have been thus classified from the hybridizations of the sp atomic orbitals of carbon, different from the approaches using the topological dimension of carbon, or the characteristics of carbon structures. If one takes the topological dimension of carbon as an example, carbon nanostructures cover zero-dimensional fullerenes and carbon nanoparticles, one-dimensional nanotubes and diamond nanorods, two-dimensional graphene and diamond nanoplates, and three-dimensional ultrananocrystalline diamond films. Owing to their characteristic size, shape, and spatial arrangement, carbon nanostructures and nanoparticles have shown different properties. Numerous varied applications using carbon nanostructures and nanoparticles have been realized as well in different fields.

This book contains a collection of the most important progress achieved in the fields of the preparation and applications of carbon nanostructures and nanoparticles. Ten chapters have been collected from international experts, which can be divided into three parts. The first part (Chaps. "Nanodiamonds: From Synthesis and Purification to Deposition Techniques, Hybrids Fabrication and Applications" and "One-Dimensional Carbon Nanostructures: Low-Temperature Chemical Vapor Synthesis and Applications") concerns the synthesis of carbon nanostructures and nanoparticles. Chapter "Nanodiamonds: From Synthesis and Purification to Deposition Techniques, Hybrids Fabrication and Applications" summarizes the recent advances in the production and the purification methods of diamond nanoparticles. The different strategies for seeding and patterning of surfaces are detailed. The CVD growth of carbon nanostructures at low temperatures (<450 °C) and their growth mechanisms are overviewed in Chap. "One-Dimensional Carbon Nanostructures: Low-Temperature Chemical Vapor Synthesis and Applications". The second part (Chaps. "Carbon Nanohorns and Their High Potential in Biological Applications"–"Polyglycerol-Functionalized Nanoparticles for Biomedical Imaging") is devoted to the biological, medical, imaging, and quantum sensing applications of carbon nanostructures and nanoparticles. In Chap. "Carbon Nanohorns and Their High Potential in Biological Applications", the formation, properties, and and biological applications of carbon nanohorns are introduced. Chapter "Bioimaging and Quantum Sensing Using NV Centers in Diamond Nanoparticles" is about bioimaging and quantum sensing using diamond nanoparticles with NV negatively charged nitrogen vacancy centers. Polyglycerol-functionalized nanoparticles, including detonation nanodiamond, superparamagnetic iron oxide nanoparticle, and fluorescent nanodiamond, for biomedical imaging is presented in Chap. "Polyglycerol-Functionalized Nanoparticles for Biomedical Imaging". The third part (Chaps. "Carbon Based Dots and Their Luminescent Properties and Analytical Applications"–"Catalytic Applications of Carbon Dots") focuses on analytical, electrochemical, and catalytic applications of carbon nanostructures and nanoparticles. Chapters "Carbon Based Dots and Their Luminescent Properties and Analytical Applications" and "Photoluminescent Properties of Carbon Nanodots" cover the synthesis, the luminescent properties, and the analytical applications of carbon dots, while the catalytic applications of carbon dots are highlighted in Chap. "Catalytic Applications of Carbon Dots". Diamond electrochemistry at the nanoscale using diamond nanostructures and nanoparticles is shown in Chap. "Diamond Nanostructures and Nanoparticles: Electrochemical Properties and Applications". Chapter "Carbon-Based Nanostructures for Matrix-Free Mass Spectrometry" presents recent developments in the use of carbon-based materials for matrix-free mass spectrometry.

From our point of view, all the chapters in this book coupled with their citations will be useful to both specialists and early-stage researchers. It is hoped that this book will attract a broad readership ranging from materials scientists, chemists, biologists, physicists, and engineers. We strongly believe this book will simulate more researchers to devote their effort and energy to the progress of the preparation and the applications of carbon nanostructures and nanoparticles in the forthcoming years.

Nianjun Yang
Xin Jiang
Dai-Wen Pang

Contents

Nanodiamonds: From Synthesis and Purification to Deposition Techniques, Hybrids Fabrication and Applications

J.C. Arnault

Abstract The present chapter summarizes the recent advances in the production and the purification methods of nanodiamonds. The different strategies for seeding and patterning of surfaces are detailed. First reports of hybrids based on nanodiamonds are included like core shell particles or decoration with carbon dots or metallic atoms. Finally, an overview of applications for composites and nanomedicine is provided.

Keywords Nanodiamonds · Purification · Hybrids · Seeding · Surfaces · Applications

1 Introduction

Nanodiamonds (NDs) are nowadays intensively studied. Indeed, diamond material at nanoscale combines the numerous bulk properties of diamond with specific assets due to the high specific surface area of NDs (400 m^2/g for detonation). The present chapter is built from the recent advances on nanodiamonds namely reported within the two last years. These contributions represent more than the third of references.

Sources of diamond nanoparticles are more and more diversified with the recent development of crushing procedures from bulk diamond, progress in detonation synthesis which allows the production of smaller NDs. Each source of NDs has its own characteristics (size distribution, shape, impurities level, defects). The first part of the present chapter focuses on these aspects (Sect. 2).

After their synthesis or milling, NDs often exhibit a scattered surface chemistry. The second topic deals with the purification protocols of NDs and their deaggregation methods to obtain stable colloids (Sect. 3). The different strategies employed to reach a homogeneous surface chemistry are also summarized. The recent labeling of detonation NDs with tritium using micro-wave plasma is addressed. The last

J.C. Arnault (✉)
CEA, LIST, Diamond Sensors Laboratory, 91191 Gif sur Yvette, France
e-mail: Jean-Charles.ARNAULT@cea.fr

N. Yang et al. (eds.), *Carbon Nanoparticles and Nanostructures*,
Carbon Nanostructures, DOI 10.1007/978-3-319-28782-9_1

investigations on specific surface properties of hydrogenated and surface graphitized NDs are then presented.

In a third part, the progresses in seeding of surfaces with NDs are provided. After an overview of the experimental strategies (deposition of NDs on surface and dip coating), the different approaches to achieve a patterning of NDs on surfaces are detailed (Sect. 4).

A current trend consists in building hybrids from nanoparticles. Indeed, the combination of luminescence or magnetic properties and functionalization can lead to theranostics applications. First, core shells of silica, gold or microgels based on NDs are presented. Then, decoration of NDs surface with carbon dots or metallic atoms is shown (Sect. 5).

This chapter concludes with a brief overview of the main applications (Sect. 6). Nanodiamonds are currently used for composites to strengthen mechanical and thermal properties of polymers. Their addition in oils led to enhanced friction and wear behaviours. Concerning bioapplications, drug delivery, bio-imaging and tissue engineering are addressed.

2 Nanodiamonds from Different Origins

Diamond nanoparticles or nanodiamonds (NDs) can be produced either directly using detonation synthesis (Sect. 2.1), using pulsed laser ablation [1–3] or by crushing of diamond bulk materials grown by HPHT or CVD synthesis (Sects. 2.2 and 2.3). After presentation of these techniques, the most important characteristics of diamond nanoparticles influencing their properties will be discussed (Sect. 2.4).

2.1 Detonation Synthesis

In 1963, Russian groups achieved synthesis of NDs by decomposition of explosive mixtures [4]. Indeed, according to thermodynamic calculations, free carbon produced in detonation products of condensed carbon-containing individual explosives with a negative oxygen balance should condense into diamond [5]. Most used explosives are trinitrotoluene (TNT), hexogen, octogen and their mixtures. More recently, hydrogen free detonation NDs were produced with benzotrifuroxane as an explosive [6]. Temperature and pressure conditions ensuring diamond stability are preserved during several tenths of a microseconde. This explains the nanometric size of synthetized diamond particles (Fig. 1).

Briefly, detonation synthesis can be decomposed into three main steps. First, explosives are modified under HPHT conditions. This leads to the production of free carbon among detonation products. Pressure and temperature conditions associated with the detonation wave allow the conservation of diamond and avoid its graphitization. In a second step, expansion of detonation products rapidly occurs

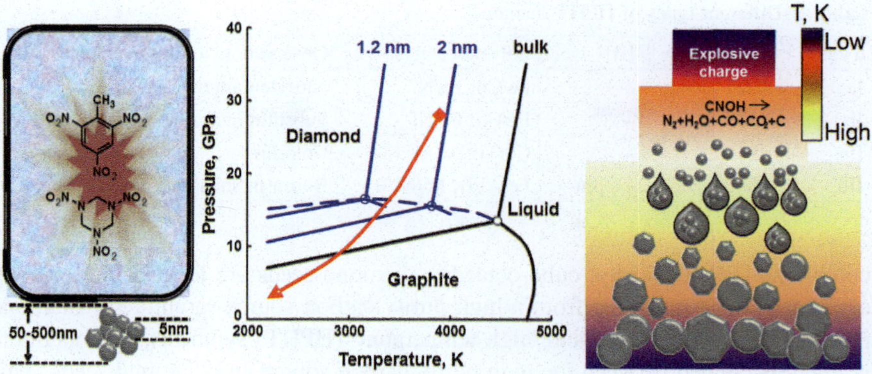

Fig. 1 Detonation synthesis of nanodiamonds: explosive mixture, phase diagram and scheme of the detonation wave propagation (with courtesy of Mochalin et al. [7])

with cooling of diamond particles down to temperature insufficient for graphitization. Then, very important heat and mass exchanges are taking place between detonation products and surrounding medium. The ND yield per explosive unit mass is related to explosive characteristics (mass, shape, density, chemical composition). It is also dependent on the environment within detonation chamber (gaseous mixture, amount of inert medium i.e. coolant). Especially, ND yield increases with the specific heat of the coolant which may be within 3000–4000 K min^{-1} to conserve diamond phase [8]. In practice, water, water-based foams or ice are used as coolants. More details about detonation synthesis could be found in the review of Dolmatov [9].

Recently, the impact of nanostructured explosives on the mean size of synthetized nanodiamond was reported. For explosives of 40 nm, NDs of 2.6 nm were obtained instead of 5 nm for micro-structured explosives (Fig. 7). A model based on the number of nucleation sites in detonation media was proposed to explain this effect [10]. This approach constitutes a serious alternative to accurately control the size of nanodiamonds during detonation process.

2.2 *Synthesis of Bulk Diamond*

2.2.1 High Pressure High Temperature (HPHT)

The first evidence of diamond formation from graphite under high pressure (13–16 GPa) and high temperature (>1700 °C) was reported in 1955 [11]. Then, the use of metal catalysts (Fe, Co, Ni) allowed the lowering of pressure/temperature

Table 1 Different types of HPHT diamond

Type	[N] (ppm)	[B]	Solvent/catalyst	Observations on dopants and color
Ia	3000	–	Fe/Co; Fe/Ni	Nitrogen aggregates, yellow
Ib	<500	–	Fe/Co; Fe/Ni	Substitutional nitrogen, yellow
IIa	<5	–	Co/Ti; Fe/Al	Colorless
IIb	<0.01	1 ppm	Fe/Al/B; Fe/Ni/B	Boron in substitution, blue

conditions [12]. Basically, cubo-octaedric diamond seeds are located in a solvent/catalyst mixture separated from a high purity carbon source (graphite or diamond powder). During high pressure/high temperature (HPHT) synthesis, a temperature gradient is created between the high purity carbon source and diamond seeds. This gradient induces the carbon diffusion toward diamond seeds causing the growth of diamond crystals [13]. During HPHT synthesis, the level of impurities like nitrogen or boron can be tuned by changing the solvent/catalyst mixture which could include nitrogen getters like Al, Ti, Zr or Hf [13]. In this way, several types of HPHT diamonds presented on Table 1 can be grown [14]. Depending on nitrogen concentration, different atomic configurations of defects have been determined [15]. Recently, HPHT monocrystalline boron doped diamonds of very high crystalline quality were reported [16].

2.2.2 Chemical Vapor Deposition (CVD)

The chemical vapor deposition of diamond was first tested in 1968 [17]. Its principle is based on the induction of a plasma from gaseous species (more often few methane vol. % diluted in dihydrogen). CH_4 and H_2 molecules could be dissociated either by a hot filament (Hot Filament CVD, HFCVD) or by microwaves (Micro-wave Plasma Assisted CVD, MPACVD). Methyl radicals ($CH_3^{\bullet}$) are species responsible for diamond growth while atomic hydrogen ($H^{\bullet}$) plays an essential role for stabilization of diamond surface and etching of sp^2-hybridized faster than sp^3-hybridized carbon [18]. Experimental conditions of CVD are metastable i.e. far from thermodynamic conditions. This explains why diamond can be obtained under low pressure (20–100 Torr) and temperature (600–1000 °C) conditions compared to HPHT. A recent MRS Bulletin was focused on CVD diamond and its applications [19].

Using the CVD technique, different diamond microstructures can be achieved, from monocrystalline to polycrystalline films (with micron-sized or nano-sized grains) (Fig. 2). Diamond deposition can be performed on large surfaces, up to 4 inches. Under specific conditions, homoepitaxial diamond films can be grown by CVD onto HPHT substrates. For growth of diamond films on heterosubstrates (Si, 3C-SiC, Ir,…), nucleation methods like bias enhanced nucleation (BEN) or seeding protocols were developed [20]. In CVD diamond, the level of impurities can be

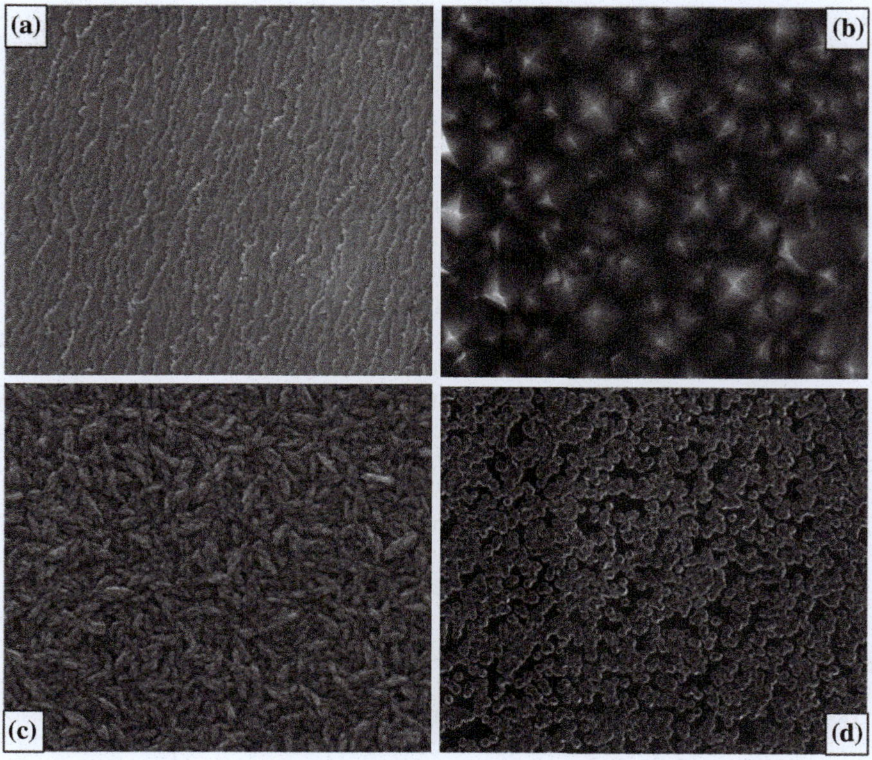

Fig. 2 Diamond films grown by micro-wave assisted CVD **a** monocrystalline (150 × 150 μm^2); **b** polycrystalline (40 × 40 μm^2); **c** ultra-nanocrystalline (2.5 × 2.5 μm^2); **d** micro-particles doped with boron (70 × 70 μm^2) (with courtesy of R. Nemanich et al. [19])

accurately controlled. High quality p-doped and n-doped diamond films are grown by CVD using boron [21, 22] and phosphorous [23] as dopants.

Using surface wave plasma, nanocrystalline diamond films have been deposited onto plastic substrates at temperature down to 100 °C [24]. Films morphology corresponds to nanometric diamond crystals embedded in an amorphous carbon matrix according to X-ray diffraction and TEM observations [25]. This surface wave plasma allows large area deposition up to 300 × 300 mm^2.

2.3 *Crushing of Diamond Bulk Material*

To better tune the crystallinity as well as the concentration of impurities within diamond particles, crushing procedures of natural diamond or diamond grown by HPHT or CVD were developed during the last years. Starting from p-doped or n-doped diamond films, these methods may permit to get doped diamond nanoparticles. The remaining challenge is the available amount of crushed material.

2.3.1 From HPHT Diamond

Diamond nanoparticles could be produced from micrometer-sized particles (200–250 μm) used for polishing applications. HPHT diamond has a high availability and it is a low cost material (a few $ per gram). Moreover, using HPHT, different types are available with controlled concentrations of dopants (Table 1).

NDs with size lower than 10 nm were obtained by a milling procedure including two main steps [26]. This procedure was applied to HPHT Ib diamond particles (size 150–190 μm) of cubo-octaedral shape. Particles were previously irradiated with 10 meV electrons to produce vacancies. The first milling was carried out using a nitrogen jet at 8 bars. A grey powder was produced with 97 % of particles having a size of 2 μm. A second milling was performed under argon using WC-Co beads followed by a cooling step. Particles with size lower than 10 nm represent 15 wt%. These nanodiamonds exhibit rounded shape as revealed by High Resolution Transmission Electron Microscopy (HRTEM) (Fig. 3a).

Another protocol was developed by Mahfouz et al. [27] including cycles of grinding, chemical purification and fractioning by ultracentrifugation [28]. HPHT diamond microparticles (size 210–250 μm) with a nitrogen concentration of 200 ppm were used as raw material. The grinding step was achieved using steel beads of 5 mm. Cycles of grinding (1 h)/cooling (30 min) were applied during 20 h. After chemical purification, diamond nanoparticles were dispersed into ultra-pure water and underwent a new grinding step. NaCl was added to the suspension. Indeed, the role of salts in deagglomeration of detonation NDs was previously emphasized [29]. Using this procedure, 15 nm NDs were isolated exhibiting facets and sharp edges as shown on Fig. 3b [27].

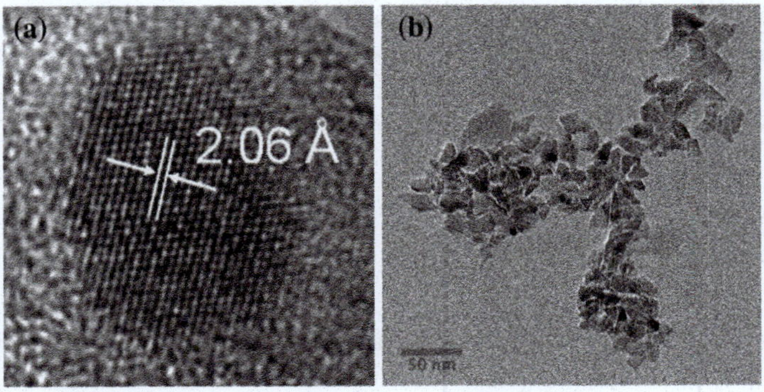

Fig. 3 HRTEM pictures of diamond nanoparticles obtained by crushing of HPHT diamond **a** with courtesy of Boudou et al. [26]; **b** with courtesy of Mahfouz et al. [27]

2.3.2 From CVD Diamond

Recent progresses of CVD synthesis allow an excellent control of incorporated impurities [19] and the possibility to get doped diamond with boron, phosphorous or nitrogen.

A free-standing diamond film is first obtained by etching of the heterosubstrate. The grinding of bulk diamond is then carried out using the bead assisted sonic disintegration (BASD) method previously developed to deagglomerate diamond nanoparticles (see Sect. 3.2) [30, 31]. The principle is based on the strong interactions of ceramic microbeads (namely 50 μm ZrO_2) with agglomerates under sonication. Shock waves produced by ultrasonic cavitation push beads toward polycrystalline diamond films leading to splitting of grains. Micrometer-sized crystals are then obtained [32]. A size selection of milled particles is performed by successive centrifugations. During this BASD procedure, the milled material underwent several chemical treatments to eliminate contaminants. Especially, the removal of zirconia traces requires a specific treatment with phosphoric acid or a strong basis. In this way, diamond particles with a size distribution centered around 70–80 nm were obtained from a polycrystalline diamond film 2 μm thick [32].

A similar milling procedure was very recently used to produce diamond particles from a boron doped diamond film grown by CVD [33]. The boron concentration was $[B] = 2 \times 10^{21}\ cm^{-3}$. Dynamic light scattering (DLS) reveals a size distribution between 30 and 60 nm for NDs (Fig. 4). The corresponding colloid has a blue color. Here, steel beads were preferred for BASD to avoid specific purification treatment. HRTEM and Electron Energy Loss Spectroscopy (EELS) investigations reveal that boron atoms are located on substitutional sites in the diamond lattice, suggesting a dopant behavior (Fig. 4).

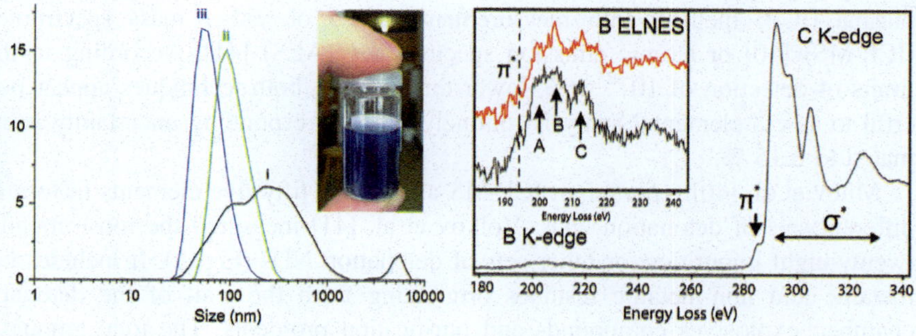

Fig. 4 *Left* Size distributions measured by Dynamic Light Scattering (DLS) for nanodiamonds produced by milling and after several centrifugations and purifications. The colloidal suspension has a *blue color* related to the presence of boron; *right* EELS spectrum exhibiting the presence of boron, B K-edge structure is in agreement with boron incorporation in substitution (with courtesy of Heyer et al. [33])

2.4 *Relevant Characteristics for NDs*

According to the literature, NDs obtained from detonation synthesis or crushed from HPHT or CVD diamond can exhibit very different physical or chemical properties. Indeed, NDs can behave variable chemical compositions (especially, impurities concentrations), morphologies (shape and size) or crystalline qualities (concentrations of crystalline defects). Such major parameters which influence NDs properties are now discussed.

2.4.1 Purity

Elemental analysis of NDs appears more and more essential to understand their intrinsic properties. Indeed, chemical impurities present in the diamond lattice or in the outer shells, especially metal residues or nitrogen, may significantly affect physical and chemical properties of NDs like fluorescence, conductivity and optical absorbance. These aspects are of first importance for biomedical or catalysis applications. The nature of impurities and their amount may also affect NDs toxicity in cells.

Initially, optical emission spectroscopy (OES) was used to estimate the amount of metallic impurities contained in natural diamond. Nevertheless, the high limit of detection (10^{-3}–10^{-4} wt%) as well as the poor quantification strongly limit its development [34]. Spectroscopic analyses of residues formed by combustion of detonation NDs were then reported [35, 36]. Drawbacks of this approach are the possible formation of new compounds during annealing and the likely loss of volatile species like nitrogen or sulfur. Non-destructive X-ray spectroscopies also behave high limits of detection 0.01–0.1 wt% [37, 38]. Instrumental neutron activation analysis (INAA) demonstrated excellent limits of detection (10^{-7} wt%) [39] but its access is more limited.

Recently, two groups reported elemental analysis based on inductively coupled plasma (ICP) method with measurements based on either mass spectrometer (ICP-MS) [40] or atomic emission spectroscopy (AES) [41]. According to their limits of detection of 10^{-8} wt%, lower than 1 ppm, both techniques appear powerful to assess elemental analysis routinely. The corresponding uncertainty is estimated to ± 15 %.

Mitev et al. [40] reported an ICP-MS analysis of fifty five elements in fourteen different sets of detonation NDs. Volkov et al. [41] measured the most abundant twenty-eight impurities for other sets of detonation NDs (Fig. 5). It includes both metallic and non-metallic residues originating from the wall of the detonation chamber, explosives compounds and purification protocols. The total amount of impurities represent up to 8 wt%. According to the explosives usually used for detonation synthesis (Fig. 1), hydrogen may be also incorporated in nanodiamonds. Its presence may have consequences on NDs properties. This aspect has been up to now poorly investigated.

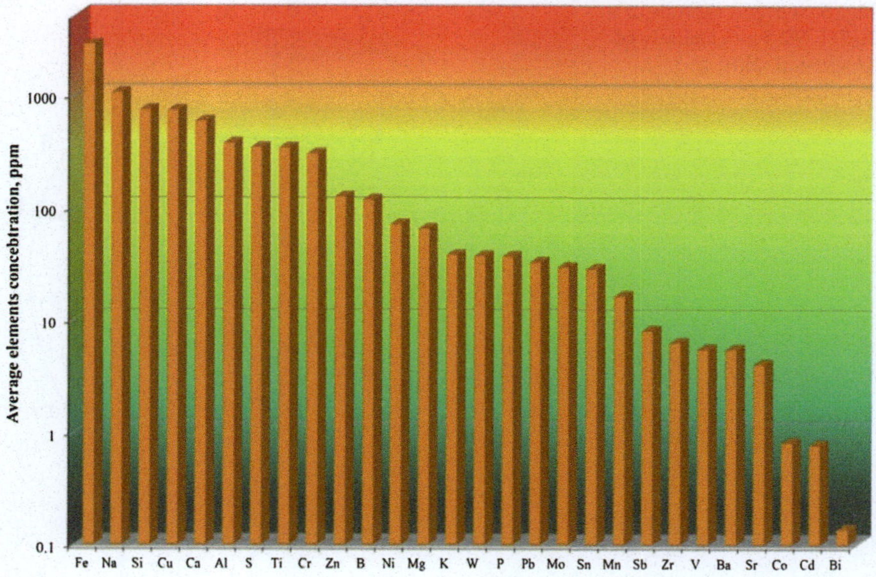

Fig. 5 Mean concentrations of twenty-eight elements measured by ICP-AES in detonation NDs (with courtesy of Volkov et al. [41])

For HPHT nanodiamonds, contaminations can originate from catalysts (Fe, Co, Ni), nitrogen getters (Ti, Al, Zr, Hf), copper added during synthesis to reduce formed carbides or purification treatments [13]. Mitev et al. [40] reported an amount of impurities of 3 wt% for one HPHT source. The concentrations of twenty three elements are reported on Fig. 6. The different purification protocols developed to improve NDs purity will be detailed in Sect. 3.

2.4.2 Shape

Nanodiamonds produced by milling of bulk HPHT material most often exhibit crystallographic facets and sharp edges resulting from fracture of diamond crystalline lattice [27, 42]. However, it is possible to obtain round-shaped NDs after milling of HPHT diamond as illustrated on Fig. 3a [26]. Moreover, starting from HPHT NDs, an oxidation treatment using melted potassium nitrate (KNO_3 at 560 °C) was powerful to round up corners and edges [43].

For detonation NDs, the shape strongly depends on detonation conditions. This fact was well illustrated for different carbon precursors, explosives combinations or coolant natures [44]. According to HRTEM investigations, resulting detonation NDs could have either well defined shapes with facets or no clear morphology. Several reported HRTEM observations underlined quasi-spherical shape for NDs synthetized by detonation [45–47]. Turner et al. [48] have even shown that facets could be present for the smallest NDs of 2 nm.

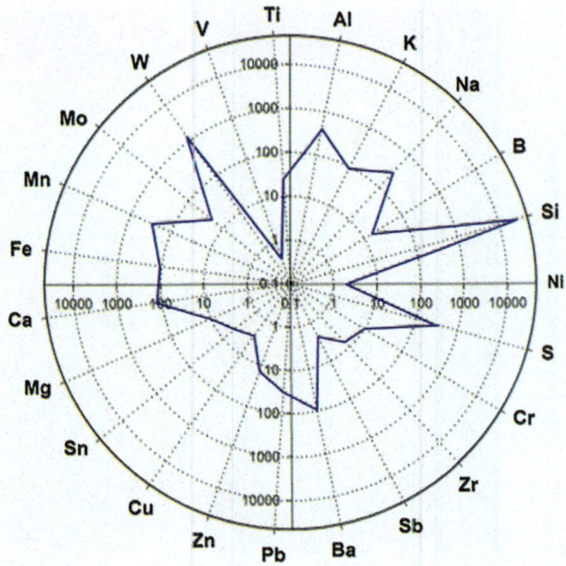

Fig. 6 ICP-MS analysis of a HPHT ND (with courtesy of Mitev et al. [40])

The presence of facets or edges at their surface may confer specific surface properties to NDs. Indeed, most observed facets correspond to low index {111}, {110} and {100} planes behaving proper atomic densities. Differences in densities of chemical bonds may induce different chemical reactivity. Another consequence concerns electrostatic effects. According to theoretical calculations, the nature of facets together with size of NDs may have an effect on protonation affinity [49] or on possible electron transfer [50]. Facets may play an important role in the aggregation of nanodiamonds in suspension. These phenomena were investigated experimentally and theoretically for detonation NDs [51]. Very recently, the effects of NDs morphology on excretion by HepG2 cells were investigated. Excretion rates of as received and round-up HPHT NDs were experimentally determined [52]. Round-up NDs can be excreted by exocytosis while as received NDs stay longer in the cytoplasm.

2.4.3 Size Distribution

Detonation nanodiamonds exhibit narrow size distributions centered between 2 and 10 nm. This nanometric size is related to the duration of the denotion wave (Sect. 2.1). Size distributions of detonation NDs produced from several explosives determined after TEM observations are shown on Fig. 7. The use of nanostructured explosives allows the production of smaller NDs [10]. On the other hand, NDs produced from HPHT diamond showed a broader dispersion in size. For example, size distributions measured on NDs from Microdiamond Switzerland reveal wide distributions (Fig. 8) [42]. This TEM study emphasizes the presence of a significant population of NDs with diameters lower than 5 nm representing 33 % in number.

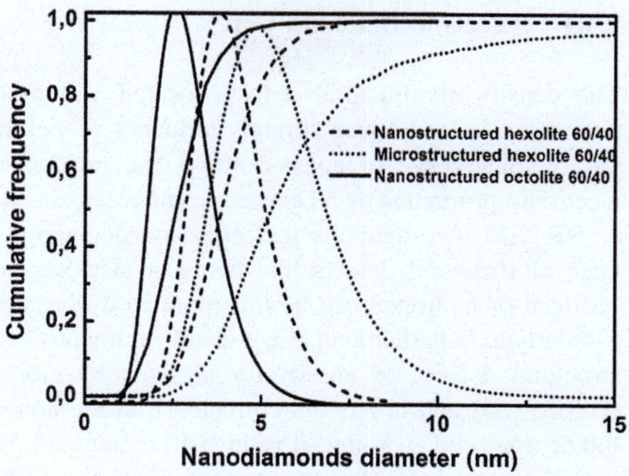

Fig. 7 Size distributions from TEM observations for detonation NDs synthetized with different explosives (with courtesy of Pichot et al. [10])

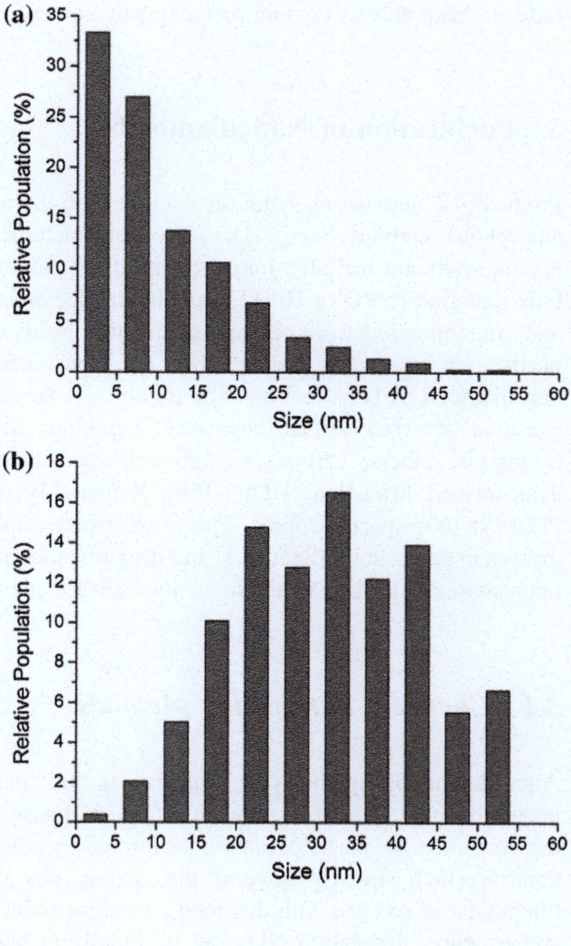

Fig. 8 Size distributions extracted from TEM observations for HPHT nanodiamonds. **a** Number-weighted; **b** volume weighted (with courtesy of Rehor et al. [42])

2.4.4 Defects in Diamond Core

The density of structural defects located within the diamond core is a major parameter. Indeed, it can strongly influence properties like thermal conductivity or photoluminescence of color centers. The incidence of defects concentration on electronic properties of NDs was recently calculated [53].

HRTEM investigations performed on detonation NDs have revealed the presence of structural defects like twins or stacking faults [48, 54]. Impurities like nitrogen or hydrogen can be incorporated during detonation synthesis (Sect. 2.1). Similarly to bulk diamond, these chemical impurities are preferentially trapped onto structural defects, as shown for nitrogen by electron energy loss spectroscopy (EELS) [55] and aberration-corrected transmission electron microscopy [56]. On the contrary, NDs produced from HPHT diamond exhibit less structural defects in their diamond core. However, recent NMR and EPR investigations underlined the presence of defects at the surface of NDs obtained by milling procedure [57]. Indeed, these defects contain paramagnetic centers.

3 Purification of Nanodiamonds

Products of detonation synthesis include non-diamond carbons like graphite or amorphous carbon (Sect. 2.1). Moreover, metallic impurities originating from reactor walls are included into detonation soot (Sect. 2.4). Crushing procedure of bulk diamond (CVD or HPHT) can also induce amorphization of diamond surface and contamination from ceramic or metallic beads (Sect. 2.3). For these reasons, purification treatments mainly based on strong acids were developed to remove non-diamond carbons and metallic impurities. These different chemical treatments led to a scattered surface chemistry involving carbon-oxygen terminations like hydroxyls, ethers, carboxyls, carbonyls or anhydrides as shown by Fourier Transformed Infra-Red (FTIR) [58], Raman [59] or Thermal Desorption Mass (TDMS) [60] spectroscopies. The present paragraph summarizes the recent progresses in purification (Sect. 3.1) and deaggregation methods (Sect. 3.2). Lastly, the main strategies leading to a homogenous surface chemistry are indicated (Sect. 3.3).

3.1 Chemical Purification Methods

After detonation synthesis or crushing of bulk material, diamond nanoparticles generally underwent oxidation either in gas-phase or in liquid-phase. The main drawback for gas-phase oxidation performed by annealing in air is the temperature control which strongly governs the etching rate of diamond [61]. Indeed, the interaction of oxygen with diamond particles produces volatile CO and CO_2. As a consequence, detonation NDs can be rapidly etched for temperature higher than

430 °C [61]. For bigger HPHT NDs (45 and 140 nm), an air oxidation was optimized at 510 °C [43]. A monitoring of carbon and oxygen *K*-edge X-ray absorption allowed a fine characterization of surface modifications for temperatures included between 475 and 575 °C [62]. Nevertheless, air oxidation at 400 °C allows the promotion of carboxylated terminations, it constitutes a way to control surface chemistry (Sect. 3.3). As an alternative, air enriched with ozone was also efficient to purify detonation soot from non-carbon species [63]. Another gas phase purification can be performed via plasma treatments. The effect of an oxygen plasma was found very similar to air annealing in terms of oxidation state [43].

The second approach is based on liquid-phase oxidation with concentrated mineral acids. Different acids or acid mixtures have been used to purify detonation NDs [9, 64] or diamond particles obtained by crushing [26, 27, 33, 42]. As an illustration, detonation NDs synthetized by Pichot et al. were cleaned using a HF/HNO_3 mixture (25/75 % in mass) at room temperature during 6 h [65]. This treatment was found more efficient than the one applied with H_2SO_4 sulfuric acid at 160 °C. A number of 0.85 COOH sites per nm^2 was estimated by Boehm titration using sodium bicarbonate ($NaHCO_3$) which selectively reacts with carboxyls and lactones [66]. Recently, a purification method assisted by micro-waves was also reported using different HNO_3, H_2SO_4 and HCl mixtures [67]. A significant reduction of non-carbon contaminants was confirmed by ICP-MS elemental analysis with a 120 factor compared to as received NDs (Sect. 2.4). Improvements of oxidation in liquid-phase were shown for detonation NDs by optimization of the temperature and the acid concentration [68].

For nanoparticles produced from bulk diamond, purification steps were applied to remove non-carbon species and metallic contaminations induced by beads (WC-Co, ZrO_2, steel). For this purpose, different hot acid mixtures were used: HF/HNO_3 [26], successive HCl and $H_2SO_4/HNO_3/HClO_4$ [27], successive H_2SO_4/HNO_3 and HCl [33, 42]. In addition to oxidation, an annealing of HPHT NDs carried out in melted KNO_3 at 560 °C [43] was very efficient to produce rounded NDs as shown by TEM (Fig. 9). These changes in morphology confer new intra-cellular properties to NDs [52] (Sect. 2.4).

3.2 Toward Colloidal Stability of NDs

The control of colloids in different media is essential for numerous applications of NDs especially their bioapplications (Sect. 6.2). After detonation synthesis, nanoparticles could be aggregated i.e. linked with strong chemical bonds. If colloidal stability is closely related to electrostatic repulsive interactions between nanoparticles mainly governed by their surface chemistry, opposite interactions could also promote agglomeration i.e. nanoparticles linked with weak bonds like Van der Waals forces. Purification treatments previously presented allow the elimination of graphite or amorphous carbon surrounding NDs which could have a role in their aggregation. Nevertheless, additional deaggregation and dispersion

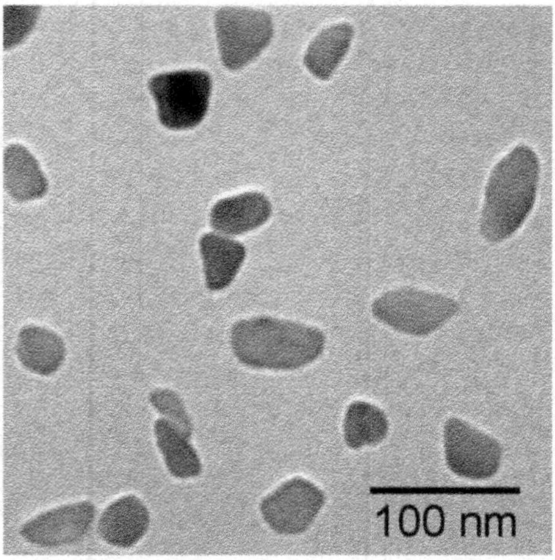

Fig. 9 Rounded HPHT NDs obtained after annealing in melted KNO_3 observed by TEM (with courtesy of Cigler et al. [43])

strategies are needed to obtain stable colloidal suspensions of well-dispersed nanodiamonds.

Deaggregation was first achieved using wet milling with micron-sized silica beads. This method was called stirred-media-milling [69]. The bead-assisted sonic disintegration (BASD) was then reported coupling milling with zirconia beads (ZrO_2) and sonication [70]. This later technique was even used to enhance the functionalization rate of NDs [71]. The main drawback of BASD is the possible contaminations from beads. Alternative chemical assisted milling methods were proposed to achieve deagglomeration of nanodiamonds using electrolytes or surfactants [72] like anionic sodium oleate [73]. Their major limitation was the smallest size of aggregates close to 20 nm. Milling in presence of sodium chloride or sucrose was successfully performed to promote deaggregation of detonation NDs [29]. Obtained suspensions exhibited a mean hydrodynamic radius of 10 nm measured by Dynamic Light Scattering (DLS). Contrary to beads, the use of salts or sugars induces minor contamination and no specific purification step is needed. The efficacy of lecithin for dispersion of NDs in water and in physiological solution was also demonstrated [74]. In some cases, an ultracentrifugation applied after purification by air annealing was sufficient to prepare stable colloidal suspensions of detonation NDs with primary size [75]. An activation with sodium hydride was more recently reported to favor repulsion of NDs in suspension during functionalization [76].

3.3 *Attempts Towards Homogeneous Surface Chemistry*

The colloidal stability of NDs in different environments and their functionalization require a fine control of their surface chemistry. Different strategies were developed

Surface terminations		Treatments	Objectives
Hydrogenated		- H_2 plasma - annealing at 500°C under H_2	- positive ZP - grafting
Oxidized	*Carboxylated*	- different acid treatments - air annealing at 400-430°C - ozone treatment at 150-200°C	- negative ZP - disaggregation - grafting of peptides
	Hydroxylated	- borane reduction - Fenton reagent - milling with beads - photochemistry	- grafting - silanization
Fluorinated		- F_2/H_2 exposure at 150-470°C - CF_4 plasma	- grafting
Aminated		- Cl-term NDs in gaseous ammonia - covalent grafting of amine derivative	- positive ZP - grafting
Surface graphitized		- annealing at 750°C under vacuum - long beads milling	- positive ZP - hybrid properties

Fig. 10 The different physical and chemical treatments to modify NDs surface chemistry. For the corresponding references see [77]

to monitor the surface terminations of NDs using physical (annealing under different atmospheres, plasma) or chemical (wet chemistry, photochemistry) treatments. These procedures are summarized on Fig. 10. More details could be found in [77].

Recently, the hydrogenation procedure based on microwave CVD plasma [78] was extended to tritium (^3H). A convincing radioactive labeling was shown for detonation NDs exposed to a tritium plasma [79]. The total measured radioactivity was 9120 µCi/mg (Fig. 11). Moreover, a significant amount of ^3H atoms (7 %) is

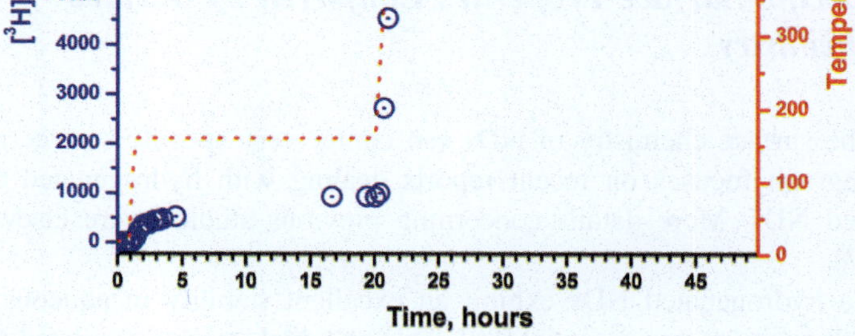

Fig. 11 Radioactive titration of tritium released from tritiated NDs after three annealings in air (with courtesy of Girard et al. [79])

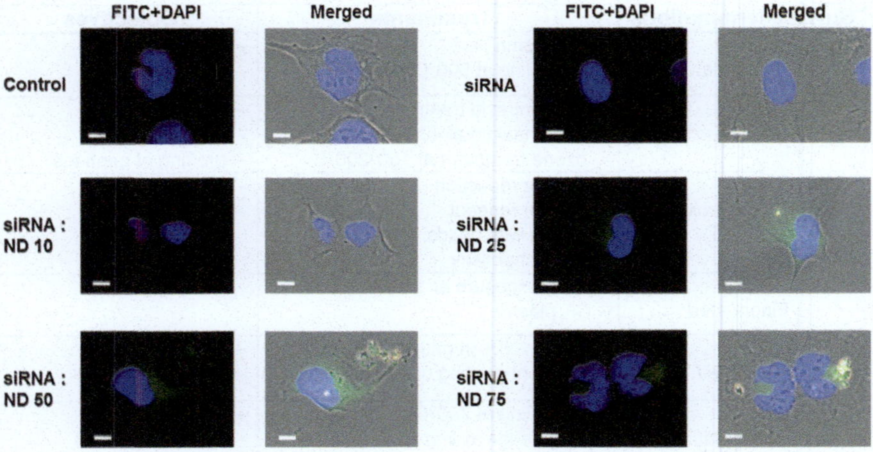

Fig. 12 A673 cells observed by epifluorescence microscopy for increasing ND-H/siRNA ratio. In *blue*: nucleus coloration with DAPI; in *green*: FITC dye attached to siRNA. Scale bar: 10 μm (with courtesy of Bertrand et al. [85])

located in diamond cores. This method is thus very efficient to provide stable radiolabeled NDs for biodistribution or pharmacokinetics studies. Indeed, ^{3}H labeled NDs can be further functionalized with biological moieties of interest.

Tailoring the surface chemistry allows the tuning of the Zeta potential (ZP) of modified NDs. Positive ZP (>30 mV) were measured for NDs after plasma hydrogenation [80], surface graphitization by annealing under vacuum [81] or amination [82–84]. On the other hand, NDs annealed under air often exhibit a negative ZP provided by carboxylate groups stable at pH higher than 5. Cationic hydrogenated detonation NDs were recently used for electrostatic adsorption of small interfering RNA (siRNA) [85]. An efficient delivery of siRNA into A673 cells was observed with an inhibition of the gene expressing Ewing sarcoma disease (Fig. 12).

3.4 Specific Surface Properties Conferred by Surface Chemistry

Tuning the surface chemistry of NDs can confer very specific surface properties. This paragraph focuses on recent reports dealing with hydrogenated or surface graphitized NDs. More details concerning previous studies were elsewhere provided [77].

Plasma hydrogenated NDs exhibit an excellent stability in aqueous solutions with a high positive Zeta potential [80]. This colloidal stability is related to the high

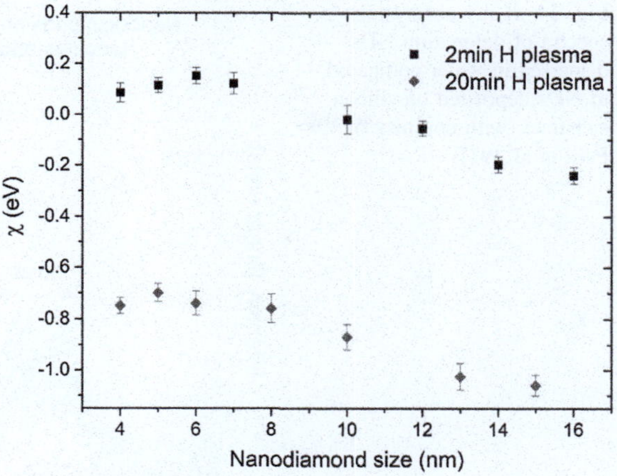

Fig. 13 Electron affinity (Y axis) versus NDs size for two hydrogenation treatments (with courtesy of Bolker et al. [86])

hydrophilicity of hydrogenated NDs measured by BET. Scanning Tunneling Spectroscopy (STS) and Kelvin Force Microscopy (KFM) experiments performed on isolated plasma hydrogenated NDs exposed to air demonstrated recently a transfer doping mechanism similar to the one observed for hydrogenated bulk diamond [86]. Namely, this surface conductivity arises from the combination of the negative electron affinity, fingerprint of the hydrogenated surface (Fig. 13), and the adsorption of molecules at the surface. These experiments reveal that this surface property of hydrogenated bulk diamond is preserved at nanoscale. Measurements of electrical conductivity conducted on detonation NDs annealed under hydrogen flux also suggest the existence of a surface conductivity [87]. The surface properties of hydrogenated nanodiamonds can explain their radiosensitization behavior observed in vitro on tumor cells under gamma irradiation [88, 89].

The surface graphitization of detonation NDs obtained by accurate annealing under vacuum allows the preparation of stable colloids in water [46, 81]. The positive zeta potential was linked to an oxygen hole doping. In fact, these graphitized NDs stable in water exhibit fullerene like reconstructions (FLRs) at their surface according to HRTEM observations [81]. The proposed model corresponds to a surface mechanism of charge transfer occurring between adsorbed oxygen molecules and electron-enriched FLRs via an electron transfer from the diamond core. This model was supported by a recent study combining ozone treatment and graphitization of detonation NDs [90]. Moreover, X-ray absorption and emission spectroscopies directly performed in a microjet demonstrate the presence of valence holes for NDs dispersed in water (Fig. 14) [91]. On the same time, π^* transitions characteristics of sp^2 carbon probed for dried NDs disappear when NDs are dispersed in water. These measurements strongly suggest a charge transfer which could explain the positive zeta potential.

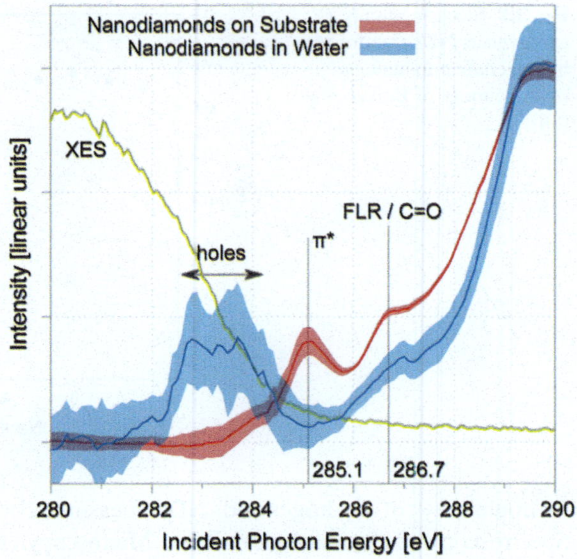

Fig. 14 X-ray absorption spectra of detonation NDs dispersed in water compared to NDs deposited on silicon substrate (with courtesy of Petit et al. [91])

4 Nanodiamonds and Surfaces

The deposition of thin diamond films with thickness lower than 100 nm is an important issue for applications. Nanodiamonds act as seeds which can be grown under CVD conditions. The different ways to control the morphology of semiconductor thin films starting from suspensions were recently reviewed [92]. The different approaches for seeding of nanodiamonds are first presented (Sect. 4.1). Then, strategies elaborated to locate NDs on surfaces for patterning like grafting or self-organization are detailed (Sect. 4.2).

4.1 Deposition of NDs

Nowadays, nanodiamonds are commonly used as seeds for CVD growth of diamond films on heterosubstrates [93]. Indeed, seeding with NDs represents an alternative to in situ nucleation methods like Bias Enhanced Nucleation (BEN) [20]. Sequential XPS analysis confirmed the stability of detonation NDs (5 nm) under CVD conditions [94]. Their thermal stability was also investigated on different substrates like silicon [95] or silicon nitride [96]. The seeding technique demonstrates its high versatility. It can be applied to various substrates [97–99], on large surfaces, up to eight inches [100, 101]. Contrary to BEN, 3D substrates like silicon tips [102, 103] or silica optical fibers [104] can be homogeneously coated [105]. The main challenge is the improvement of diamond films adhesion [20, 93].

An optimized seeding would combine a high density of seeds (up to $10^{11}/cm^2$) with a seed size as small as possible. Different experimental strategies were developed in this objective using HPHT or detonation NDs. Progresses of seeding methods were closely linked to improvements of the NDs dispersion in colloids (Sect. 3.2). A first approach was to deposit NDs on the substrate by different ways. Drop casting usually leads to a non-uniform deposition due to convection forces acting on NDs during drying. Spin coating methods were thus developed to obtain uniform deposits of NDs. This was reported on eight inches silicon wafers [106]. The spin coating technique is well suitable for flat substrates. It is more hardly applicable to other morphologies of substrates due to screening effects. An alternative method, the aerosol deposition of detonation NDs was carried out on silicon wafers under argon or nitrogen flow providing seeds density from 10^8 to $10^{11}/cm^2$ versus suspension concentration [107].

A second strategy uses substrates immersed in colloidal suspensions or dip-coating. Sonication was used to ensure NDs dispersion and enhance the deposition during dip-coating in aqueous solutions [108]. This approach was powerful to obtain isolated seeds with a density close to $10^{11}/cm^2$ on silicon wafers. A similar technique was developed for detonation NDs in dimethylsulfoxide (DMSO) suspensions [109]. The dip-coating assisted sonication was successfully applied to seed carbon nanotubes [110] or polymeric stripes [111] with NDs. Dip-coating assisted by low power sonication was performed with detonation NDs dispersed in different solvents like acetone, ethanol, isopropyl alcohol, methanol or water on gold electrodes [112]. Such seeded electrodes were further used for biosensing. Very recently, to further improve the surface covered by diamond seeds, adamantane ($C_{10}H_{16}$) molecules i.e. subnanometer unit of diamond crystal [113] were deposited on sapphire substrates by dip-coating [114]. Diamond films were grown on these seeded substrates by microwave CVD plasma at 700 °C. This later result demonstrates the stability of diamondoids under CVD conditions despite their size of 1–2 nm.

Seeding can also be driven by electrostatic forces taking place between NDs and surfaces. Playing with the surface chemistry of NDs, the solvent nature and the suspension pH, it is possible to tune their surface charge from negative to positive (Sect. 3.3). For example, positive detonation NDs which underwent an annealing under hydrogen were seeded onto silicon dioxide substrates with a high density [115]. Similarly, the seeding of negative carboxylated NDs was achieved onto positive copper substrate [116]. A polarization of the substrate was used in electrophoresis deposition of NDs [111, 117–119]. Using this technique, negatively charged NDs previously annealed under air were dispersed using an ultracentrifugation and deposited on mica [120].

Seeding of NDs can alternatively be improved by controlling the surface charge of the substrate. This was well illustrated for AlN substrates modified by gas discharge plasmas (N_2, O_2 or CF_4) [121]. Detonation NDs were dispersed in water after an annealing under hydrogen. The higher density of seeds close to $10^{11}/cm^2$ was obtained after fluorination with an enhancement of three orders of magnitude compared to other treated NDs. Another way consists in functionalization of NDs to control their charge. It was achieved by Cetyltrimethylammonium chloride (CTAC) grafting forming an ionic complex with carboxylic groups present on detonation NDs surface [122]. These hydrophobic complexes remain at the water surface and NDs can be deposited on mica substrates using the Langmuir-Blodgett technique.

To extend seeding to all substrates whatever its nature, its electronic properties (conductive to insulator) or its morphology, polymer were used either as a matrix for NDs for spin-coating or as a charged coating before deposition of NDs by dip-coating. NDs embedded into polyvinyl alcohol (PVA) matrix were deposited on 4 inches silicon by spin-coating. The PVA polymer is quickly burned away after eight seconds of CVD plasma as shown by XPS sequential analysis [123]. Consequently, no contamination of NDs arises from the polymer. This method was also applied to Si_3N_4 and Ti substrates.

Another strategy consists to cover the substrate by an organic polymer either the cationic Poly(diallyldimethylammonium chloride) (PDDAC) or the anionic poly (styrenesulfonate) (PSS). These polymers are soluble in water. Such layer by layer deposition was previously achieved for other nanoparticles [124, 125] and later applied to NDs [126–129]. Negatively charged NDs were then deposited through a layer by layer process on substrates coated with PDDAC [126]. This was reported for HPHT NDs of different sizes (15–50 nm) as well as for detonation NDs [130]. The method was applied to silicon, gold and quartz substrates. Using the same principle, detonation NDs functionalized with metalloporphyrin were deposited on surface acoustic wave (SAW) devices. Positively charged complexes are adsorbed by the layer by layer technique using PSS anionic polymer. This sensitive layer was further used for gas sensing, especially Dinitrotoluen (DNT) detection [129].

The high versatility of this dip-coating method allowed the coverage of silicon substrates of different morphologies from tips to wells [130]. Nanostructured silicon substrate with wells of different diameters (100–300 nm) and depths (400–1200 nm) was fabricated using e-beam lithography and reactive ion etching (RIE). Wells were then seeded using a PDDAC coating and negatively charged detonation NDs. These substrates constituted molds to further grow diamond pillars (Fig. 15). This seeding procedure was particularly powerful to cover the whole surface of wells with NDs [130]. During CVD growth, the diffusion of reactive species from the plasma down to deeper wells (1200 nm) was effective.

Using a prior coating of polyallylamine (PAAm), microdiamonds were decorated with HPHT nanodiamonds (10–150 nm) [131]. Several bilayers of PAAM/NDs can be deposited onto microdiamonds. The resulting porous hybrids exhibit higher surface areas and were used for high-performance liquid chromatography (HPLC) measurements.

4.2 Patterning and Seeding

Several important technological applications of diamond like resonators (MEMS, NEMS), optical (photonic crystals) or biomedical devices require patterned

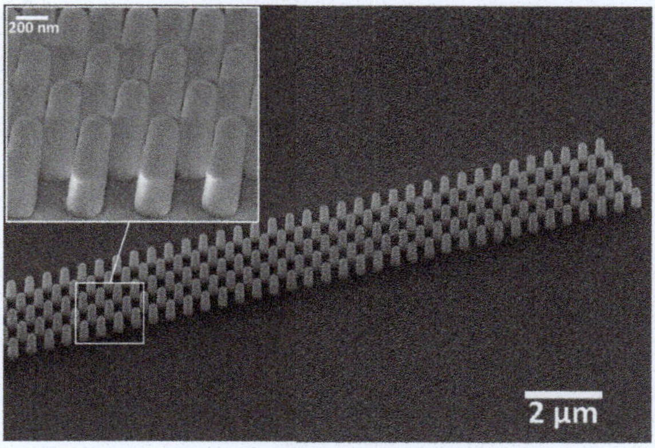

Fig. 15 Diamond pillars obtained from seeded silicon molds (with courtesy of Girard et al. [130])

diamond coatings [132]. Although the top-down approach based on selective etching of diamond films can lead to high quality patterns, these technics are often time consuming [133, 134]. Consequently, different bottom-up procedures have been developed using selective seeding of NDs.

Patterning of seeded substrates was achieved via a plasma etching using an aluminum mask [135]. An alternative way was the wet chemical elimination of NDs with a buffer oxide etchant [136].

On the contrary, patterns can be first created on the substrates before NDs deposition. A first way used photoresist lithography and metal masking followed by UV light exposure. Negatively or positively charged detonation NDs were then dip-coated on such patterns on silicon, AlN and sapphire substrates [137]. To reach sub-micrometer patterning, e-beam lithography was performed on substrates covered by poly(methyl methacrylate) (PMMA) (Fig. 16).

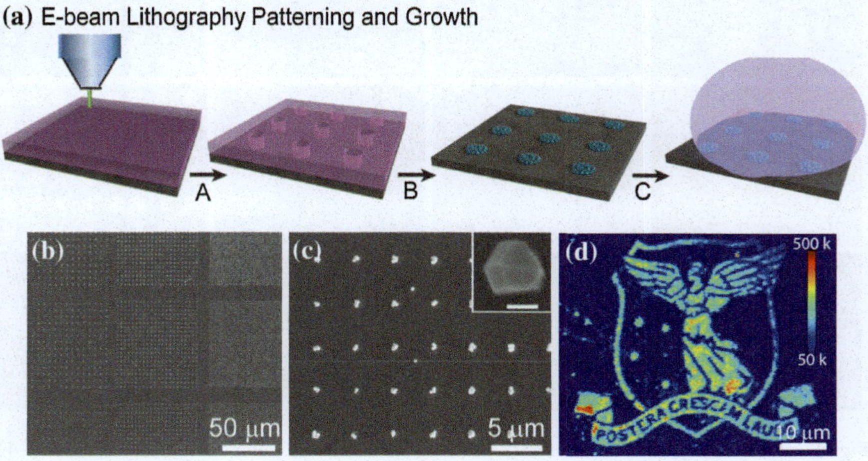

Fig. 16 Seeding on patterns fabricated by electron lithography, see details in [137] (with courtesy of Shimoni et al. [137])

Another approach to create patterns prior seeding uses self-assembled monolayers of octadecanethiol (ODT) printed by microcontact on gold substrates [138]. Carboxylated HPHT NDs suspended in water were seeded preferentially on non-coated gold by wettability effects. Indeed, stripes covered by ODT are highly hydrophobic.

An oxidized poly(dimethylsiloxane) (PDMS) prepared by photolithography was used as a stamp onto which detonation NDs were fixed by dip-coating [139]. Then, detonation NDs are transferred from PDMS to silicon substrates via a spin coated thin layer of poly(methyl methacrylate) (PMMA). Micrometer patterns were reported with seeds density up to $3 \times 10^{10}/cm^2$. More recently, PDMS moulds were also used in a microfluidic approach to seed NDs by contact printing [140]. A low cost patterning was achieved using the inkjet technique [141]. A liquid ink was developed with nanodiamonds suspended into ethylene glycol.

To get an accurate positioning of NDs at nanometer scale, a dip-pen nanolithography approach was recently reported [142]. HPHT NDs suspended in polymeric ink were prepared in DMSO, polyethylenimine (PEI) and glycerol. An AFM tip was then dip-coated within this solution and used for localized deposition of NDs on SiO_2 substrates to create $66 \times 66\ \mu m^2$ arrays. Silicon vacancy (Si-V) color centers were then incorporated into NDs by a microwave CVD plasma. The far-red corresponding luminescence was then investigated by epi-fluorescence (Fig. 17). Functionalized nanodiamonds were previously patterned on glass substrates using a hollow tip via microchannels [143].

A new strategy based on self-assembly behavior of biological systems was applied to functionalized NDs [144]. Proteins can form 1D, 2D or 3D scaffolds on

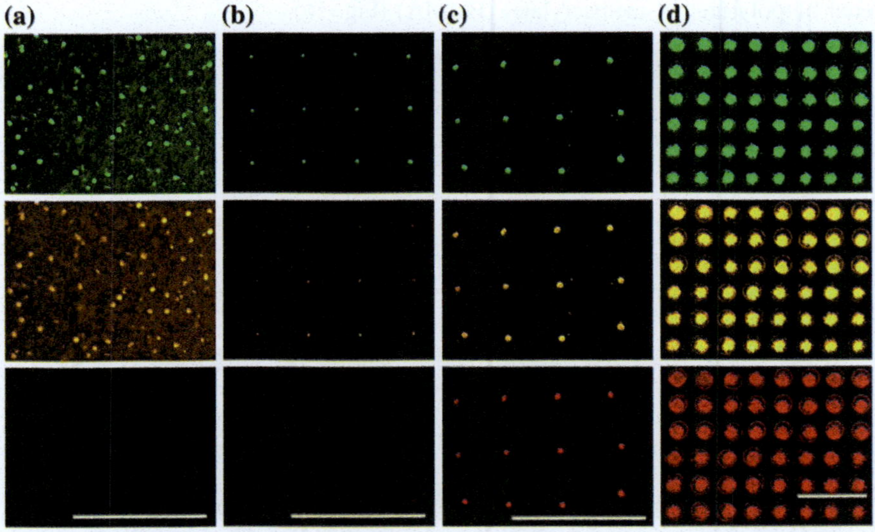

Fig. 17 Arrays of HPHT NDs with Si-V centers observed by epifluorescence, see details in [142] scale bar 20 μm (with courtesy of Singh et al. [142])

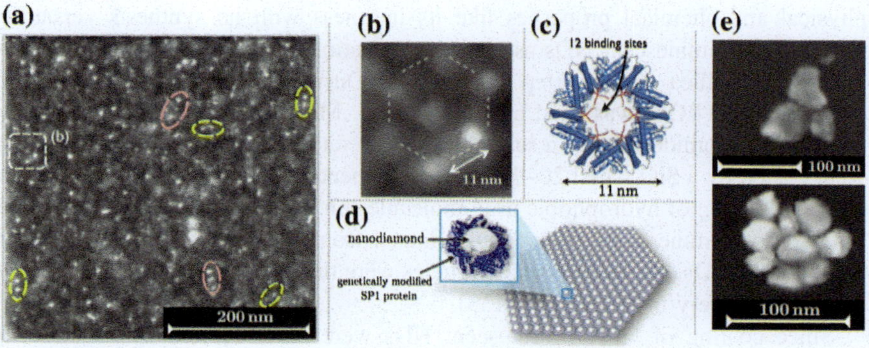

Fig. 18 Arrays of NDs built from SP1 scaffolds see details in [144] (with courtesy of Albrecht et al. [144])

which NDs can be precisely inserted (Fig. 18). SP1 protein originating from poplar trees [145] can self-assemble to an 11 nm ring-shape dodecamer. SP1 was deposited on substrates by Langmuir-Blodgett technique. Such structures were further used to study the dipolar coupling interactions between NV centers located into NDs.

5 Hybrids Built from Nanodiamonds

According to their low toxicity [146], their tunable surface charge (Sect. 3) and the existence of photoluminescent color centers in their core [147], nanodiamonds are promising candidates for biomedical applications (Sect. 6.2). Nevertheless, theranostics applications combining imaging and drug delivery require hybrid structures [148]. Core shell nanoparticles have been synthesized from NDs with different coatings like silica, gold or microgels (Sect. 5.1). Other interesting hybrids which could confer photoluminescence or magnetic properties to NDs were reported using carbon dots or metal ions (Sect. 5.2).

5.1 Core Shells

5.1.1 SiO_2/NDs

Core shell particles are promising materials to build multifunctional platforms [149]. The engineering and characterization of inorganic core shell nanoparticles was recently reviewed [150]. Among investigated coatings, porous silica (SiO_2) was successfully used for the realization of core-shells from iron oxide, quantum dots or metallic nanoparticles like gold or platinum [151]. Indeed, silica combines physical and chemical properties like its inertness with its synthesis versatility. HPHT photoluminescent NDs as well as detonation NDs were recently coated by mesoporous silica in one step [152, 153]. During this synthesis, tetraethoxy orthosilicate (TEOS) was used as silica source. Moreover, CTAB (cetyltrimethyl ammonium bromide) acts as a surfactant to generate pores into silica coating. NDs were coated by a SiO_2 shell 20–90 nm thick depending on TEOS/CTAB ratio. The efficient loading of hydrophobic drug molecules on porous silica was demonstrated [152]. Photoluminescent properties originated from NV centers are preserved for SiO_2-coated NDs. These core-shell particles were then used to combine bioimaging and drug delivery in HeLa cells [153].

Silica coating of HPHT fluorescent NDs were also reported by Rehor et al. [154]. NDs were encapsulated into a first silica shell using TEOS. The SiO_2 coating 10–20 nm thick provides a pseudo-spherical shape to nanoparticles which initially present sharp edges. Then, a second layer was deposited using a (3-aminopropyl) triethoxysilane (APS)/1,2-bis(triethoxysilyl)ethane (BTSE) mixture to provide amino-functionalized layer favorable for further grafting of polyethyleneglycol (PEG) chains. PEG grafted hybrids exhibit excellent colloidal properties in buffers and ensure furtive hybrids in cells. PEG-SiO_2/NDs were further internalized into LNCaP cells (human cancer cell line) [154]. Photoluminescence from NV centers was preserved for these hybrid nanoparticles.

5.1.2 Au/NDs

Dielectric cores coated with metallic shells constitute promising hybrid nanoparticles for plasmonic applications [155]. Indeed, multimodal imaging probes can be built combining photoluminescence of color centers and surface plasmon resonance. After their doping with SiV color centers achieved by PECVD, oxidized NDs of 100–200 nm were coated by gold [156]. This was performed in colloidal suspension by the chemical reduction of an Au (III) precursor [157, 158]. Especially, gold chloride ($AuCl_3$) was reduced to metallic gold by hydroxylamine. 70 % of the photoluminescence intensity of SiV centers was recorded for Au/NDs hybrids [156].

5.1.3 Au/SiO_2/NDs

The SiO_2/NDs hybrids previously described [154] were encapsulated in a gold shell [159]. Gold nanoparticles suspended in a colloid were first electrostatically adsorbed on SiO_2/NDs and act as seeds for the formation of a gold layer by reduction of the chloroauric acid ($HAuCl_4$) [160]. The thickness of the gold shell is 12 nm as measured by scanning transmission electron microscopy (STEM) in the high angle annular dark field configuration (HAADF) as shown on Fig. 19.

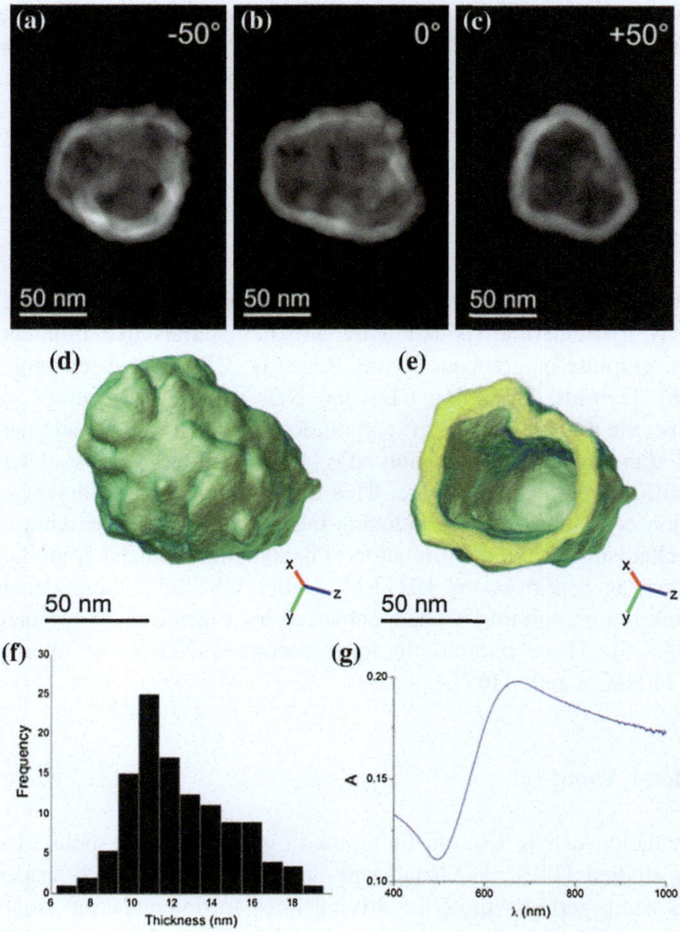

Fig. 19 Gold/SiO_2/NDs core shells studied by HAADF-STEM see details in [159] (with courtesy of Rehor et al. [159])

5.1.4 NIPAM/NDs

Hybrid smart microgels can self-assemble into 2D arrays via conformal modifications induced by temperature [161], pH [162] or ionic strength [163]. The N-isopropylacrylamide (NIPAM) monomer was polymerized by Girard et al. in the presence of carboxylated HPHT NDs [164]. During this one step method, the size of NDs/pNIPAM microspheres can be adjusted by tuning the temperature. 2D arrays of NDs/pNIPAM microgels were carried out by dip-coating on silicon surfaces. Self-assembly of loosely packed or closed packed microspheres was obtained depending on the suspension temperature during deposition. After CVD growth, an ordered array of diamond crystals was obtained. Such smart hybrids open the way of building scaffolds of NDs for catalysis, optical or biological applications.

5.2 *Decoration of Nanodiamonds*

5.2.1 Carbon Dots

Carbon dots (CDs) exhibit tunable photoluminescence properties [165]. CDs can be obtained by hydrothermal oxidation of carbon materials like fullerenes, carbon nanotubes, graphite or graphene oxide. Recently, CDs have been produced from NDs [166]. Hybrids combining CDs and NDs were even recently synthetized. Using a specific acid treatment, sp^2 carbon dots from 1 to a few nanometers can be generated at the surface of detonation NDs [167]. The detonation soot was oxidized using a sulfuric/nitric acid mixture. This acid mixture is an efficient reagent for intercalation of graphitic planes inducing their dilatation and breaking [168, 169]. These mechanisms lead to the formation of nanometric oxidized graphitic species at NDs surface as confirmed by HRTEM. Under UV light, these decorated NDs exhibit pink-red photoluminescence, enhanced by a factor of 20 compared to NDs alone (Fig. 20). These photoluminescent decorated NDs were internalized and localized in HaCat cells [167].

5.2.2 Metal Atoms

The intercalation of Cu, Co, Ni, Fe atoms in graphitic outer shells of detonation NDs was studied [170]. As metal salts are added to the NDs suspension, ion exchanges occur and constitute the driving force for intercalation. Nanodiamonds were then annealed under hydrogen or argon atmosphere. Such modified NDs were characterized by means of chemical and structural probes (X-ray diffraction, X-ray scattering). Several Cu^{2+} or Co^{2+} ions per NDs located at the surface were detected by nuclear magnetic resonance (NMR) [171, 172]. According to these studies, the number of metal ions is limited by the density of carboxylic groups present at NDs

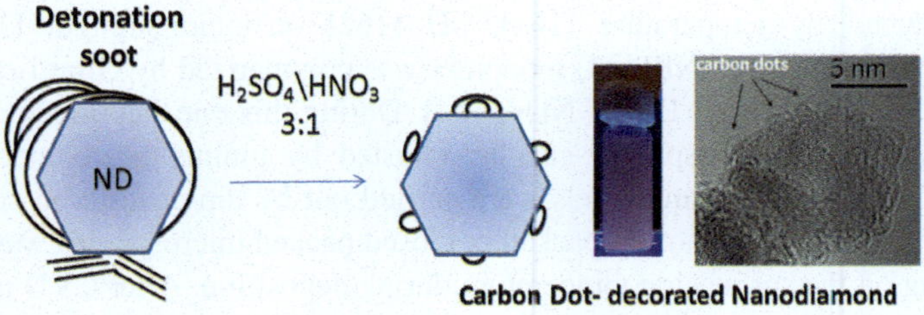

Fig. 20 NDs decorated with carbon dots (with courtesy of Shenderova et al. [167])

surface. Indeed, Cu ions may be located between two oxygen anions O^- of deprotonated carboxylic groups [172, 173]. Recent DFT (density functional theory) calculations have shown that stable binding of Cu chelate is ensured at edges of the NDs surface [174]. Nevertheless, a higher amount of metal ions is required to use these metal/NDs hybrids for Magnetic Resonance Imaging (MRI).

6 Applications of Diamond Nanoparticles: An Overview

This last part focuses on the most promising applications of nanodiamonds. A first trend concerns the elaboration of composites to enhance mechanical and thermal properties of polymers or oils (Sect. 6.1). A second active domain concerns bioapplications especially nanomedicine (Sect. 6.2).

6.1 Composites

6.1.1 Polymers

If elastomers exhibit high elasticity, their mechanical and thermal properties remain limited. As nanosized oxides, metal oxides or other carbon nanostructures, nanodiamonds are excellent candidates to strengthen polymers. Performances of rubbers reinforced by NDs were recently reviewed [175]. Tribological properties of different NDs-polymer composites like friction coefficient or wear rate were very significantly enhanced [176–180]. An original method based on electrospinning was used to produce nano and microfibers of NDs-polymer composites [181]. The surface chemistry of NDs plays an essential role in composites properties. This was in particular underlined by Neitzel et al. who used aminated NDs to reinforce epoxy (Fig. 21). The corresponding Young modulus was improved by 700 % compared to epoxy alone [182]. Other reports concerned NDs grafted with vinyl or silyl groups [183, 184].

Depending of NDs-composites, the concentration of NDs could be critical. As an illustration, the impact of NDs concentration on mechanical (strength, toughness, and elastic modulus) and thermal performances of polyethylene was recently investigated [185]. These properties are enhanced for NDs concentrations up to 1 wt%. Higher concentrations of NDs lead to a deterioration of composite properties. Another challenge concerns the control of NDs distribution in the polymer matrix. For example, polyepoxide-based nanohybrid film was elaborated by self-assembly of NDs under electric field [186]. The formation of chain-like structures allows the improvement of thermal conductivity of the composites. Applications of NDs-composites are multiple. In biomedical field, a biodegradable ND-PLLA (poly (L-lacticacid) composite was used for bone tissue engineering [187]. The same PLLA biopolymer was covalently grafted on hydroxylated NDs via a polymerization

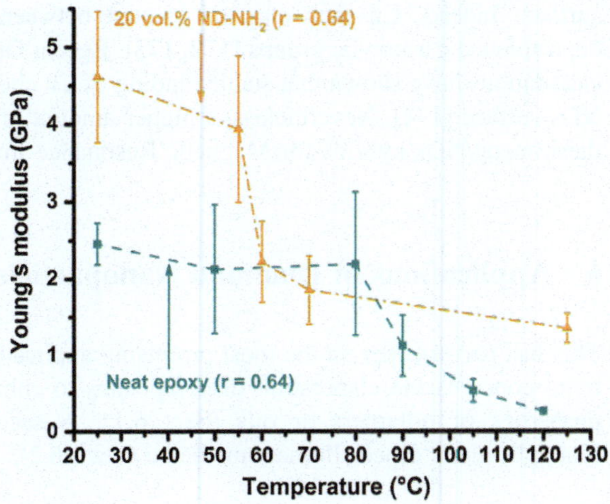

Fig. 21 Evolution of Young modulus versus temperature for epoxy reinforced with aminated NDs compared to epoxy alone (with courtesy of Neitzel et al. [182])

starting at the surface [188]. The biocompatibility of diamond-like carbon (DLC) films for fibroblasts was enhanced by addition of NDs in DLC matrix [189]. New progresses on surface functionalization of NDs are nevertheless required to further improve mechanical and thermal properties of NDs-polymer composites.

6.1.2 Lubricants

Addition of NDs in thermal fluids was investigated to improve viscoelastic and thermal properties [190]. Results showed a significant increase of the dynamic viscosity with NDs concentrations (range 0.01–0.1 wt%). An enhanced thermal conductivity, up to 70 %, is obtained for increasing NDs concentrations (Fig. 22).

Additives based on NDs or detonation soot provided better performances for lubricants, especially it induces a reduction of friction and wear coefficients [191–194]. If the effects on mechanical behavior have been well characterized, involved mechanisms are still under investigations. Indeed, they are strongly dependent on the dispersion of NDs in oils and on the material in contact with lubricants. For example, two commercial oils with NDs additives were tested on steel and aluminum alloy by tribological pin-on-disk tests [195]. For both materials, wear losses were reduced. However, involved mechanisms are rather different: both viscosity of suspension and hardness of the surface contact are active for steels while viscosity mainly governs for aluminum. Mechanisms appear even more complex in the recent report of Ivanov et al. [196] who showed a wear reduction for different NDs-commercial oils suspensions for steel/steel friction. On the other hand, an increased wear is observed for WC/steel friction. The improvement of wear loss was also attributed to NDs incorporation into the tribolayer which governs anti-friction and anti-wear properties [197].

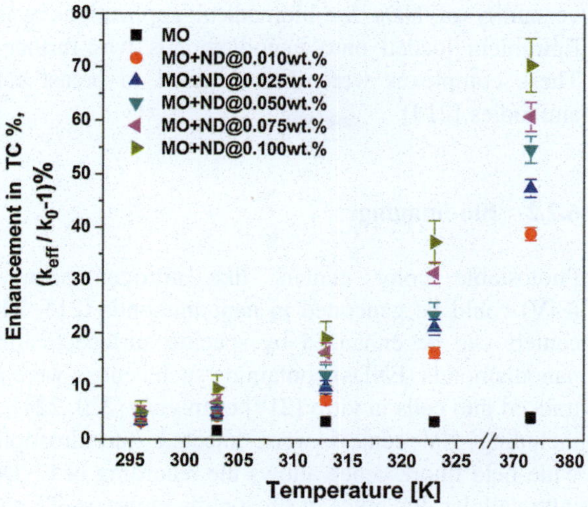

Fig. 22 Thermal conductivity TC of different nanofluids including various NDs concentrations versus temperature (with courtesy of Taha-Tijerina et al. [190])

6.2 *Bio-applications*

The main bio-applications of nanodiamonds concern bio-imaging, drug delivery [148, 198–200] and tissue engineering [201]. This part will summarize the most recent contributions from the literature concerning these applications.

6.2.1 Drug Delivery

The nanodiamond surface allows grafting of biomolecules using the rich chemistry of carbon. Indeed, controlled surface treatments permit to reach surface chemistries suitable for specific grafting routes applied to biomolecules (Sect. 3.3). These different strategies were recently reviewed [202]. Anti-cancer drugs [203], nucleic acids [87, 204] and insulin [205] can be either electrostatically adsorbed onto NDs or covalently grafted [206, 207]. Drug loading and its release by a response to a stimulus like the pH value [208–210] are current challenges. These issues are closely related to the drug binding strength and the cell environment (temperature, ionic strength, pH,…). Adsorption-desorption kinetics of doxorubicin and polymyxin B adsorbed on NDs have been investigated tuning their surface chemistry, purity or aggregation [211]. Drug adsorption and release are highly sensitive to NDs surface chemistry. However, a high drug loading can correspond to a low binding strength and inversely. Drug retention in breast cancer cells can be enhanced using NDs as carriers. Mitoxantrone (MTX) was electrostatically adsorbed onto detonation NDs. The MTX release from these MTX-NDs complexes was controlled by pH and protein concentration in the media [212]. Moreover, the therapeutic efficacy was better compared to MTX delivered alone. Similar effects were reported for doxorubicin adsorbed on NDs to treat prostate cancer [213]. The versatility of NDs for biomedical applications was illustrated by Moore et al. Epirubicin loaded onto nanodiamonds was further encapsulated by lipid films. These complexes were then targeted to breast cancer cells using biotinylated antibodies [214].

6.2.2 Bio-imaging

Photostable color centers like nitrogen-vacancy (NV) or silicon-vacancy (SiV) could be generated in nanodiamonds [215, 216]. The fluorescence of color centers can be enhanced by specific surface treatments [217, 218]. Fluorescent nanodiamonds (FNDs) containing such centers were used for bio-labeling and were tracked into cells in vitro [219] or in vivo [220, 221]. More recently, nanodiamonds including NV centers were injected into drosophila embryos [222]. In vivo wide-field fluorescence allows the recording of FNDs movements giving access to intra-cellular dynamics of drosophila embryos. To enhance the lateral resolution of the wide-field fluorescence imaging, a time-gating with applied [223]. Its efficiency was demonstrated for HeLa cells labeled with FNDs in whole blood with a lateral resolution of 0.5 μm. FND-labeled lung cancer cells were tracked in blood vessels. In vivo conditions were mimicked for a cell sample covered with a thin layer of chicken breast (Fig. 23). In the latter case, the lateral resolution is 5 μm.

Nanodiamond reveals as a versatile platform which could combine drug delivery and imaging for theranostics applications [148, 198]. This constitutes the major potential for their use in biomedical applications.

6.2.3 Tissue Engineering

Nowadays, therapies based on scaffolds elaborated by tissue engineering are strongly required [224]. Among carbon nanostructures, nanodiamonds are also considered as biomimetic materials providing a scaffold for tissue regeneration [225–227]. NDs were combined with polymer scaffolds either by adsorption,

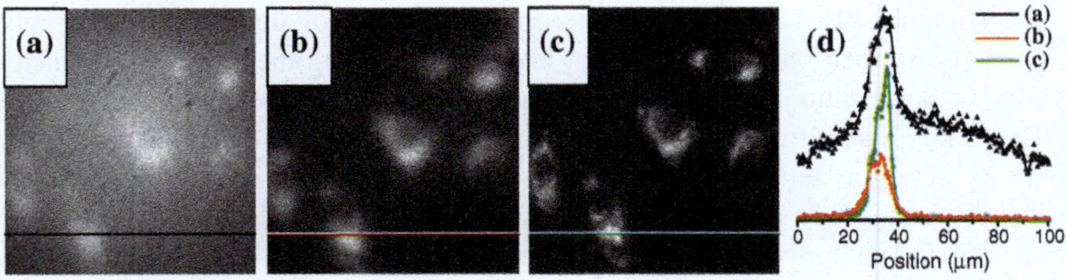

Fig. 23 In-vitro imaging by wide-field fluorescence of FND-labeled cells covered with chicken breast **a** without **b** with time gating **c** time-gated fluorescence image **d** intensity profiles along the *three lines* denoted in **a**, **b** and **c** (with courtesy of Hui et al. [223])

covalent bonding or encapsulation. The release of a protein linked to bone morphogenesis was compared in vitro and in vivo for the different hybrids. Scaffolds with covalently bonded NDs were the most promising [227]. In addition, NDs were used to elaborate electrodes for cell growth [226].

7 Conclusion and Perspectives

According to recent advances in detonation synthesis and in milling techniques, it is now possible to fabricate nanodiamonds tunable in size (down to 2 nm) containing different impurities (nitrogen, boron, phosphorous) with controlled concentrations (Sect. 2). Their surface chemistry can be tailored to monitor their surface charge (Sect. 3). Colloidal suspensions of NDs can be further used to electrostatically adsorb drugs or nucleic acids (Sect. 6.2). Adjusted surface chemistry is essential for the elaboration of composites with polymers or oils (Sect. 6.1). In addition, the versatile surface of NDs allows specific grafting of biological moieties. In parallel, new developments in seeding protocols currently permit to control accurate patterning of NDs on surfaces (Sect. 4.2). During the last three years, first core shell materials based on NDs were reported (Sect. 5). These hybrids open the way for theranostics applications combining imaging and drug delivery or hyperthermia (Sect. 6.2).

From the material side, the role of impurities in NDs on their properties remains to be investigated. This work has been largely performed for nitrogen, it would be important to better understand effects due to hydrogen which is likely to be incorporated during synthesis of NDs. A second challenge concerns the combination of surface treatments to reach hybrid surface chemistries. This was recently reported for detonation NDs combining ozone treatment and surface graphitization [90]. Seeding techniques may be applied to 3D substrates to synthetize diamond foam or porous materials [105, 228, 229]. Finally, the field related to hybrids particles based on NDs is rapidly growing up. Alternatives will be elaborated in the near future to combine imaging (color centers or MRI) with drug delivery or hyperthermia.

Dealing with applications, better performances of NDs composites with polymers should be expected by the use of new surface functionalization more adapted to elastomers [175, 182]. Improvements of NDs dispersion into polymers is also an important issue. For lubricants, mechanisms responsible for friction and wear properties have to be better identified and understood [190, 196]. Indeed, they appear strongly dependent on the material involved in the contact with lubricant.

For applications of NDs in nanomedicine, biodistribution and pharmacokinetic studies are absolutely required. Currently, very few studies were reported [230]. The tritium labeling of detonation NDs reported by Girard et al. [79] will facilitate such investigations. Moreover, in addition to previous studies [146, 231], the toxicity of NDs must be deeply investigated. More precisely, it appears essential to identify which physical and chemical parameters are conferring toxicity: size, shape, surface chemistry,... Another linked perspective is the ecotoxicity which is an essential issue for nanodiamonds applications. Nevertheless, from our knowledge, no study was currently reported on nanodiamonds. Recent results also indicated that ND can be used as an active nanoparticle. A radiosensitization effect was demonstrated for hydrogenated NDs under gamma irradiation of cancer cells [88, 89].

New fields of applications may emerge from current investigations of nanodiamonds. The most promising one is related to the emission of solvated electrons from hydrogenated diamond films under UV illumination [232, 233]. The extension of this behavior to nanoparticles may lead to applications in CO_2 reduction and catalysis.

Acknowledgments J.C. Arnault would like to thank his co-workers involved in surface modifications of nanodiamonds at CEA LIST, especially H.A. Girard, C. Gesset and T. Petit. He also acknowledges his collaborators from other laboratories for fruitful interactions.

References

1. L. Yang, P.W. May, L. Yin, J.A. Smith, K.N. Rosser, Growth of diamond nanocrystals by pulsed laser ablation of graphite in liquid. Diam. Relat. Mater. **16**, 725–729 (2007). doi:10.1016/j.diamond.2006.11.010
2. D. Adams, A.C. Chenus, G. Ledoux, C. Dujardin, C. Reynaud, O. Sublemontier, K. Masenelli-Varlot, O. Guillois, Nanodiamond synthesis by pulsed laser ablation in liquids. Diam. Relat. Mater. **18**, 177–180 (2009). doi:10.1016/j.diamond.2008.10.035
3. M.V. Baidakova, Y.A. Kukushkina, A.A. Sitnikova, M.A. Yagovkina, D.A. Kinlenko, V.V. Sokolov, M.S. Shestakov, A.Y. Vul', B. Zousman, O. Levinson, Structure of nanodiamonds prepared by laser synthesis. Phys. Solid State **55**, 1747–1753 (2013). doi:10.1134/S1063783413080027
4. K.V. Volkov, V.V. Danilenko, V.I. Elin, Synthesis of diamond from the carbon in the detonation products of explosives. Combustion Explosion and Shock waves **26**, 366–368 (1990). doi:10.1007/BF00751383
5. V.Y. Dolmatov, M.V. Veretennikova, V.A. Marchukov, V.G. Sushchev, Currently available methods of industrial nanodiamond synthesis. Phys. Solid State **46**, 611–615 (2004). doi:10.1134/1.1711434
6. S.S. Batsanov, A.N. Osavchuk, S.P. Naumov, A.E. Efimov, B.G. Mendis, D.C. Apperley, A. S. Batsanov, Synthesis and properties of hydrogen-free detonation diamond. Propellants Explos. Pyrotech. **35**, 1–8 (2010). doi:10.1002/prep.201400039
7. V.N. Mochalin, O. Shenderova, D. Ho, Y. Gogotsi, The properties and applications of nanodiamonds. Nature Nanotech. **7**, 11–23 (2012). doi:10.1038/NNANO.2011.209
8. A.L. Vereshchagin, E.A. Petrov, G.V. Sakovich et al., U.S. Patent No. 591.655, 1999
9. V.Y. Dolmatov, Detonation-synthesis nanodiamonds: synthesis, structure, properties and applications. Russ. Chem. Rev. **76**, 339–360 (2007). WOS:000247118100004
10. V. Pichot, M. Comet, B. Risse, D. Spitzer, Detonation of nanosized explosive: new mechanistic model for nanodiamond formation. Diam. Relat. Mater. **54**, 59–63 (2015). doi:10.1016/j.diamond.2014.09.013
11. F.P. Bundy, H.T. Hall, H.M. Strong, R.H. Wentorf, Man-made diamond. Nature **176**, 51 (1955). doi:10.1038/176051a0
12. H.P. Bovenkerk, F.P. Bundy, H.T. Hall, H.M. Strong, R.H. Wentorf, Preparation of diamond. Nature **184**, 1094 (1959). doi:10.1038/1841094a0

13. R.C. Burns, J.O. Hansen, R.A. Spits, M. Sibanda, C.M. Welbourn, D.L. Welch, Growth of high purity large diamond crystals. Diam. Relat. Mater. **8**, 1433 (1999). doi:10.1016/S0925-9635(99)00042-4
14. H. Kanda, M. Akaishi, S. Yamaoka, Synthesis of diamond with the highest nitrogen concentration. Diam. Relat. Mater. **8**, 1441 (1999). doi:10.1016/S0925-9635(99)00022-9
15. A. Dobrinets, V.G. Vins, A.M. Zaitev, *HPHT-Treated Diamonds*, Springer series in Material Science, vol 181 (Springer, Berlin, 2013). doi:10.1007/978-3-642-37490-6_1
16. V.S. Bormashov, S.A. Tarelkin, S.G. Buga, M.S. Kuznetsov, S.A. Terentiev, A.N. Semenova, V.D. Blank, Electrical properties of the high quality boron-doped synthetic single-crystal diamonds grown by the temperature gradient method. Diam. Relat. Mater. **35**, 19–23 (2013). doi:10.1016/j.diamond.2013.02.011
17. J.C. Angus, H.A. Will, W.S. Stanko, Growth of diamond seed crystals by vapor deposition. J. Appl. Phys. **39**, 2915 (1968). doi:10.1063/1.1656693
18. J.C. Angus, C.C. Hayman, Low pressure metastable growth of diamond and diamond like phases. Science **241**, 913–921 (1988). doi:10.1126/science.241.4868.913
19. R.J. Nemanich, J.A. Carlisle, A. Hirata, K. Haenen, CVD diamond—research, applications and challenges. MRS Bull. **39**, 490–494 (2014). doi:10.1557/mrs.2014.97
20. J.C. Arnault, H.A. Girard, Diamond nucleation and seeding techniques: two complementary strategies for growth of ultra-thin diamond films, in *Nanodiamonds, Royal Society Chemistry*, ed. by O.A Williams (2014), pp. 221–252. ISBN 978-1-84973-639-8
21. N. Fujimori, T. Imai, A. Doi, Characterization of conductive diamond films. Vacuum **36**, 99–102 (1986). doi:10.1016/0042-207X(86)90279-4
22. Y. Takano, M. Nagao, T. Takenouchi, H. Umezawa, I. Sakaguchi, M. Tachiki, H. Kawarada, Superconductivity in polycrystalline diamond thin films. Diam. Relat. Mater. **14**, 1936–1938 (2005). doi:10.1016/j.diamond.2005.08.014
23. S. Koizumi, M. Kamo, Y. Sato, H. Ozaki, T. Inuzuka, Growth and characterization of phosphorous doped 111 homoepitaxial diamond thin films. Appl. Phys. Lett. **71**, 1065–1067 (1997). doi:10.1063/1.119729
24. K. Tsugawa, M. Ishihara, J. Kim, M. Hasegawa, Y. Koga, Large-area and low-temperature nanodiamond coating by microwave plasma chemical vapor deposition. New Diamond Front. Carbon Technol. **16**, 337–346 (2006). WOS:000246559400005
25. K. Tsugawa, M. Ishihara, J. Kim, Y. Koga, M. Hasegawa, Nanocrystalline diamond film growth on plastic substrates at temperatures below 100 °C from low-temperature plasma. **82**, 125460 (2010). doi:10.1103/PhysRevB.82.125460
26. J.P. Boudou, P.A. Curmi, F. Jelezko, J. Wrachtrup, P. Aubert, M. Sennour, G. Balasubramanian, R. Reuter, A. Thorel, E. Gaffet, High yield fabrication of fluorescent nanodiamonds. Nanotechnology **20**, 235602 (2009). doi:10.1088/0957-4484/20/23/235602
27. R. Mahfouz, D.L. Floyd, W. Peng, J.T. Choy, M. Loncar, O.M. Bakr, Size-controlled fluorescent nanodiamonds: a facile method of fabrication and color-center counting. Nanoscale **5**, 11776–11782 (2013). doi:10.1039/c3nr03320a
28. W. Peng, R. Mahfouz, J. Pan, Y. Hou, P.M. Beaujuge, O.M. Bakr, Gram-scale fractionation of nanodiamonds by density gradient ultracentrifugation. Nanoscale **5**, 5017–5026 (2013). doi:10.1039/c3nr00990d
29. A. Pentecost, S. Gour, V. Mochalin, I. Knoke, Y. Gogotsi, Deaggregation of nanodiamond powders using salt- and sugar-assisted milling. ACS Appl. Mater. Interfaces **2**, 3289–3294 (2010). doi:10.1021/am100720n
30. M. Ozawa, M. Inakuma, M. Takahashi, F. Kataoka, A. Krueger, E. Osawa, Preparation and behavior of brownish, clear nanodiamond colloids. Adv. Mater. **19**, 1201–1206 (2007). doi:10.1002/adma.200601452
31. Y. Liang, M. Ozawa, A. Krueger, A general procedure to functionalize agglomerating nanoparticles demonstrated on nanodiamond. ACS Nano **3**, 2288–2296 (2009). doi:10.1021/nn900339s
32. E. Neu, C. Arend, E. Gross, F. Guldner, C. Hepp, D. Steinmetz, E. Zscherpel, S. Ghodbane, H. Sternschulte, D. Steinmüller-Nethl, Y. Liang, A. Krueger, C. Becher, Narrowband

fluorescent nanodiamonds produced from chemical vapor deposition films. Appl. Phys. Lett. **98**, 243107 (2011). doi:10.1063/1.3599608
33. S. Heyer, W. Janssen, S. Turner, Y.G. Lu, W.S. Yeap, J. Verbeeck, K. Haenen, A. Krueger, Toward deep blue nano hope diamonds: heavily boron-doped diamond nanoparticles. ACS Nano **8**, 5757 (2014). doi:10.1021/nn500573x
34. F.A. Raal, A spectrographic study of the minor element content of diamond. Am. Mineral. **42**, 354–361 (1957). WOS:A1957XF01500004
35. V.Y. Dolmatov, Detonation synthesis ultradispersed diamonds: properties and applications. Russ. Chem. Rev. **70**, 607–626 (2001). Accession Number: WOS:000184202800004
36. H. Sakurai, N. Ebihara, E. Osawa, M. Takahashi, M. Fujinami, K. Oguma, Adsorption characteristics of a nanodiamond for oxoacid anions and their application to the selective preconcentration of tungstate in water samples. Anal. Sci. **22**, 357–362 (2006). doi:10.2116/analsci.22.357
37. B.V. Spitsyn, J.L. Davidson, M.N. Gradoboev, T.B. Galushko, N.V. Serebryakova, T.A. Karpukhina, I.I. Kulakova, N.N. Melnik, Inroad to modification of detonation nanodiamond. Diam. Relat. Mater. **15**, 296–299 (2006). doi:10.1016/j.diamond.2005.07.033
38. A.P. Koscheev, Thermodesorption mass spectrometry in the light of solution of the problem of certification and unification of the surface properties of detonation nanodiamonds. Russ. J. Gen. Chem. **79**, 2033–2044 (2009). doi:10.1134/S1070363209090357
39. S. Merchel, U. Ott, S. Herrmann, B. Spettel, T. Faestermann, K. Knie, G. Korschinek, G. Rugel, A. Wallner, Presolar nanodiamonds: faster, cleaner, and limits on platinum-HL. Geochim. Cosmochim. Acta **67**, 4949–4960 (2003). doi:10.1016/S0016-7037(03)00421-6
40. D.P. Mitev, A.T. Townsend, B. Paull, P.N. Nesterenko, Screening of elemental impurities in commercial detonation nanodiamond using sector field inductively coupled plasma-mass spectrometry. J. Mater. Sci. **49**, 3573–3591 (2014). doi:10.1007/s10853-014-8036-3
41. D.S. Volkov, M.A. Proskurnin, M.V. Korobov, Elemental analysis of nanodiamonds by inductively-coupled plasma atomic emission spectroscopy. Carbon **74**, 1–13 (2014). doi:10.1016/j.carbon.2014.02.072
42. I. Rehor, P. Cigler, Precise estimation of HPHT nanodiamond size distribution based on transmission electron microscopy image analysis. Diam. Relat. Mater. **46**, 21–24 (2014). doi:10.1016/j.diamond.2014.04.002
43. J. Havlik, V. Petrakova, I. Rehor, V. Petrak, M. Gulka, J. Stursa, J. Kucka, J. Ralis, T. Rendler, S.Y. Lee, R. Reuter, J. Wrachtrup, M. Ledvina, M. Nesladek, P. Cigler, Boosting nanodiamond fluorescence: towards development of brighter probes. Nanoscale **5**, 3208–3211 (2013). doi:10.1039/c2nr32778c
44. O.A. Shenderova, I.I. Vlasov, S. Turner, G. Van Tendeloo, S.B. Orlinskii, A.A. Shiryaev, A. A. Khomich, S.N. Sulyanov, F. Jelezko, J. Wrachtrup, Nitrogen Control in Nanodiamond Produced by Detonation Shock-Wave-Assisted Synthesis. J. Phys. Chem. C **115**, 14014–14024 (2011). doi:10.1021/jp202057q
45. A.E. Aleksenskii, V.Y. Osipov, A.T. Dideikin, A.Y. Vul', G.J. Adriaenssens, V.V. Afanas'ev, Ultradisperse diamond cluster aggregation studied by atomic force microscopy. Tech. Phys. Lett. **26**, 819–821 (2000). doi:10.1134/1.1315505
46. T. Petit, J.C. Arnault, H.A. Girard, M. Sennour, P. Bergonzo, Early stages of surface graphitization on nanodiamond probed by x-ray photoelectron spectroscopy. Phys. Rev. B **84**, 233407 (2011). doi:10.1103/PhysRevB.84.233407
47. M.V. Baidakova, in *Methods of Characterization and Models of Nanodiamond Particles in Detonation Nanodiamonds: Science and Applications*, ed. by A.Y. Vul', O.A. Shenderova (Pan Stanfford Publishing Pte Ltd.). ISBN 978-981-4411-27-1
48. S. Turner, O.I. Lebedev, O. Shenderova, I.I. Vlasov, J. Verbeeck, G. Van Tendeloo, Determination of size, morphology, and nitrogen impurity location in treated detonation nanodiamond by transmission electron microscopy. Adv. Funct. Mater. **19**, 2116–2124 (2009). doi:10.1002/adfm.200801872
49. A.S. Barnard, M.C. Per, Size and shape dependent deprotonation potential and proton affinity of nanodiamond. Nanotechnology **25**, 445702 (2014). doi:10.1088/0957-4484/25/44/445702

50. L. Lai, A.S. Barnard, Tuning the electron transfer properties of entire nanodiamond ensembles. J. Phys. Chem. C **118**, 30209–30215 (2014). doi:10.1021/jp509355g
51. L.Y. Chang, E. Osawa, A.S. Barnard, Conformation of the electrostatic self-assembly of nanodiamonds. Nanoscale **3**, 958–962 (2011). doi:10.1039/c0nr00883d
52. Z. Chu, S. Zhang, B. Zhang, C. Zhang, C.Y. Fang, I. Rehor, P. Cigler, H.C. Chang, G. Lin, R. Liu, Q. Li, Unambiguous observation of shape effects on cellular fate of nanoparticles. Scientific Reports **4**, 4495 (2014). doi:10.1038/srep04495
53. A. Barnard, Modeling polydispersive ensembles of diamond nanoparticles. Nanotechnology **24**, 085703 (2013). doi:10.1088/0957-4484/24/8/085703
54. V.Y. Dolmatov, G.S. Yurev, V. Myllymaki, K.M. Korolev, Why detonation nanodiamonds are small. J. Superhard Mater. **35**, 77–82 (2013). doi:10.3103/S1063457613020020
55. I.I. Vlasov, O. Shenderova, S. Turner, O.I. Lebedev, A.A. Basov, I. Sildos, M. Rähn, A.A. Shiryaev, G. Van Tendeloo, Nitrogen and luminescent nitrogen-vacancy defects in detonation nanodiamond. Small **6**, 687–694 (2010). doi:10.1002/smll.200901587
56. S. Turner, O. Shenderova, F. Da Pieve, Y.G. Lu, E. Yücelen, J. Verbeeck, D. Lamoen, G. Van Tendeloo, Aberration-corrected microscopy and spectroscopy analysis of pristine, nitrogen containing detonation nanodiamond Phys. Status Solidi A **210**, 1976–1984 (2013). doi:10.1002/pssa.201300315
57. A.M. Panich, N.A. Sergeev, A.I. Shames, V.Y. Osipov, J.P. Boudou, S.D. Goren, Size dependence of 13C nuclear spin-lattice relaxation in micro- and nanodiamonds. J. Phys. Condens. Matter **27**(7), 072203 (2015). doi:10.1088/0953-8984/27/7/072203
58. H.A. Girard, T. Petit, S. Perruchas, J.C. Arnault, P. Bergonzo, Surface properties of hydrogenated nanodiamonds: a chemical investigation. Phys. Chem. Chem. Phys. **13**, 11511–11516 (2011). doi:10.1039/c1cp20424f
59. M. Mermoux, A. Crisci, T. Petit, H.A. Girard, J.C. Arnault, Surface modifications of detonation nanodiamonds probed by multiwavelength Raman spectroscopy. J. Phys. Chem. C **118**, 23415–23425 (2014). doi:10.1021/jp507377z
60. O. Shenderova, A. Koscheev, N. Zaripov, I. Petrov, Y. Skryabin, P. Detkov, T. Turner, G. Van Tendeloo, Surface chemistry and properties of ozone-purified detonation nanodiamonds. J. Phys. Chem. C **115**, 9827–9837 (2011). doi:10.1021/jp1102466
61. S. Osswald, G. Yushin, V. Mochalin, S.O. Kucheyev, Y. Gogotsi, Control of sp2/sp3 carbon ratio and surface chemistry of nanodiamond powders by selective oxidation in air. J. Am. Chem. Soc. **128**, 11635–11642 (2006). doi:10.1021/ja063303n
62. A. Wolcott, T. Schiros, M.E. Trusheim, E.H. Chen, D. Nordlund, R.E. Diaz, O. Gaathon, D. Englund, J.S. Owen, Surface structure of aerobically oxidized diamond nanocrystals. J. Phys. Chem. C **118**, 26695–26702 (2014). doi:10.1021/jp506992c
63. I. Petrov, O. Shenderova, V. Grishko, V. Grichko, T. Tyler, G. Cunningham, G. McGuire, Detonation nanodiamonds simultaneously purified and modified by gas treatment. Diam. Relat. Mater. **16**, 2098–2103 (2007). doi:10.1016/j.diamond.2007.05.013
64. V.G. Sushchev, V.Y. Dolmatov, V.A. Marchukov, M.V. Veretennikova, Fundamentals of chemical purification of detonation nanodiamond soot using nitric acid. J. Superhard Mater. **30**, 297–304 (2008). doi:10.3103/S1063457608050031
65. V. Pichot, M. Comet, E. Fousson, C. Baras, A. Senger, F. Le Normand, D. Spitzer, An efficient purification method for detonation nanodiamonds. Diam. Relat. Mater. **17**, 13–22 (2008). doi:10.1016/j.diamond.2007.09.011
66. L. Schmidlin, V. Pichot, M. Comet, S. Josset, P. Rabu, D. Spitzer, Identification, quantification and modification of detonation nanodiamond functional groups. Diam. Relat. Mater. **22**, 113–117 (2012). doi:10.1016/j.diamond.2011.12.009
67. D.P. Mitev, A.T. Townsend, B. Paull, P.N. Nesterenko, Microwave-assisted purification of detonation nanodiamond. Diam. Relat. Mater. **48**, 37–46 (2014). doi:10.1016/j.diamond.2014.06.007
68. V.Y. Dolmatov, A. Vehanen, V. Myllymaki, K.A. Rudometkin, A.N. Panova, K.M. Korolev, T.A. Shpadkovskaya, Deep purification of detonation nanodiamond material. J. Superhard Mater. **35**, 408–414 (2013). doi:10.3103/S1063457613060099

69. A. Krüger, F. Kataoka, M. Ozawa, T. Fujino, Y. Suzuki, A.E. Aleksenskii, A.Y. Vul', E. Osawa, Unusually tight aggregation in detonation nanodiamond: identification and disintegration. Carbon **43**, 1722–1730 (2005). doi:10.1016/j.carbon.2005.02.020
70. M. Ozawa, M. Inakuma, M. Takahashi, F. Kataoka, A. Krueger, E. Osawa, Preparation and behavior of brownish, clear nanodiamond colloids. Adv. Mater. **19**, 1201–1206 (2007). doi:10.1002/adma.200601452
71. Y. Liang, M. Ozawa, A. Krueger, A general procedure to functionalize agglomerating nanoparticles demonstrated on nanodiamond. ACS Nano **3**, 2288–2296 (2009). doi:10.1021/nn900339s
72. X.Y. Xu, Y.W. Zhu, B.C. Wang, Z.M. Yu, S.Z. Xie, Mechanochemical dispersion of nanodiamond aggregates in aqueous media. J. Mater. Sci. Technol. **21**, 109–112 (2005). WOS:000226775000026
73. Y.Y. Xu, Z.M. Yu, Y.M. Zhu, B.C. Wang, Effect of sodium oleate adsorption on the colloidal stability and zeta potential of detonation synthesized diamond particles in aqueous solutions. Diamond Relat. Mater. **14**, 206–212 (2005). doi:10.1016/j.diamond.2004.11.004
74. X.Y. Zhang, S.Q. Wang, M.Y. Liu, J.F. Hui, B. Yang, L. Tao, Y. Wei, Surfactant-dispersed nanodiamond: biocompatibility evaluation and drug delivery applications. Toxicol. Res. **2**, 335–342 (2013). doi:10.1039/c3tx50021g
75. A.E. Aleksenskiy, E.D. Eydelman, A.Y. Vul', Deagglomeration of detonation nanodiamonds. Nanosci. Nanotechnol. Lett. **3**, 68–74 (2011). doi:10.1166/nnl.2011.1122
76. Y. Sun, P. Olsen, T. Waag, A. Krueger, D. Steinmüler-Nethl, A.C. Albertsson, A. Finne-Wistrand, Disaggregation and anionic activation of nanodiamonds mediated by sodium hydride—a new route to functional aliphatic polyester-based nanodiamond materials. Part. Part. Syst. Charact. **32**, 35–42 (2015). doi:10.1002/ppsc.201400098
77. J.C. Arnault, Surface modifications of nanodiamonds and current issues for their biomedical applications, in *Novel Aspects of Diamond, Topics in Applied Physics*, vol 121, ed. by N. Yang (2014). doi:10.1007/978-3-319-09834-0_4
78. H.A. Girard, J.C. Arnault, S. Perruchas, S. Saada, T. Gacoin, J.P. Boilot, P. Bergonzo, Hydrogenation of nanodiamonds using MPCVD: a new route toward organic functionalization. Diam. Relat. Mater. **19**, 1117–1123 (2010). doi:10.1016/j.diamond.2010.03.019
79. H.A. Girard, A. El Kharbachi, S. Garcia-Argote, T. Petit, P. Bergonzo, B. Rousseau, J.C. Arnault, Tritium labeling of detonation nanodiamonds. Chem. Comm. **50**, 2916–2918 (2014). doi:10.1039/c3cc49653h
80. T. Petit, H.A. Girard, A. Trouve, I. Batonneau-Genner, P. Bergonzo, J.C. Arnault, Surface transfer doping can mediate both colloidal stability and self-assembly of nanodiamonds. Nanoscale **5**, 8958–8962 (2013). doi:10.1039/c3nr02492j
81. T. Petit, J.C. Arnault, H.A. Girard, M. Sennour, T.Y. Kang, C.L. Cheng, P. Bergonzo, Oxygen hole doping of nanodiamond. Nanoscale **4**, 6792–6799 (2012). doi:10.1039/c2nr31655b
82. K.I. Sotowa, T. Amamoto, A. Sobana, K. Kusakabe, T. Imato, Effect of treatment temperature on the amination of chlorinated diamond. Diam. Relat. Mater. **13**, 145–150 (2004). doi:10.1016/j.diamond.2003.10.029
83. C.L. Huang, H.C. Chang, Adsorption and immobilization of cytochrome c on nanodiamonds. Langmuir **20**, 5879–5884 (2004). doi:10.1021/la0495736
84. W.S. Yeap, S. Chen, K.P. Loh, Detonation nanodiamond: an organic platform for the suzuki coupling of organic molecules. Langmuir **25**, 185–191 (2009). doi:10.1021/la8029787
85. J.R. Bertrand, C. Pioche-Durieu, J. Ayala, T. Petit, H.A. Girard, C. Malvy, E. Le Cam, F. Treussart, J.C. Arnault, Plasma hydrogenated cationic detonation nanodiamonds efficiently deliver to human cells in culture functional siRNA targeting the Ewing sarcoma junction oncogene. Biomaterials **45**, 93–98 (2015). doi:10.1016/j.biomaterials.2014.12.007
86. A. Bolker, C. Saguy, R. Kalish, Transfer doping of single isolated nanodiamonds, studied by scanning probe microscopy techniques. Nanotechnology **25**, 385702 (2014). doi:10.1088/0957-4484/25/38/385702

87. T. Kondo, I. Neitzel, V.N. Mochalin, J. Urai, M. Yuasa, Y. Gogotsi, Electrical conductivity of thermally hydrogenated nanodiamond powders. J. Appl. Phys. **113**, 214307 (2013). doi:10.1063/1.4809549
88. T. Petit, H.A. Girard, M. Combis-Schlumberger, R. Grall, J. Delic, S. Morel-Altmeyer, P. Bergonzo, S. Chevillard, J.C. Arnault, Nanodiamond as a multimodal platform for drug delivery and radiosensitization of tumor cells. in *Proceedings of the 13th IEEE International Conference on Nanotechnology*, Beijing, China, 5–8 Aug 2013
89. R. Grall, H. Girard, L. Saad, T. Petit, C. Gesset, M. Combis-Schlumberger, V. Paget, J. Delic, J.C. Arnault, S. Chevillard, Impairing the radioresistance of cancer cells by hydrogenated nanodiamonds. Biomaterials **61**, 290–298 (2015).
90. J.C. Arnault, T. Petit, H.A. Girard, C. Gesset, M. Combis-Schlumberger, M. Sennour, A. Koscheev, A.A. Khomich, I. Vlasov, O. Shenderova, Surface graphitization of ozone treated detonation nanodiamonds. Phys. Status Solidi A **211**, 2739–2743 (2014). doi:10.1002/pssa.201431397
91. T. Petit, M. Pflüger, D. Tolksdorf, J. Xiao, E.F. Aziz, Valence holes observed in nanodiamonds dispersed in water. Nanoscale **7**, 2987–2991 (2015). doi:10.1039/C4NR06639A
92. Y. Diao, L. Shaw, Z. Bao, S.C.B. Mannsfeld, Morphology control strategies for solution processed organic semiconductor thin films. Energy Environ. Sci. **7**, 2145–2159 (2014). doi:10.1039/c4ee00688g
93. O.A. Williams, Nanocrystalline diamond. Diam. Relat. Mater. **20**, 621–640 (2011). doi:10.1016/j.diamond.2011.02.015
94. J.C. Arnault, S. Saada, O.A. Williams, K. Haenen, P. Bergonzo, M. Nesladek, R. Polini, E. Osawa, Diamond nanoseeding on silicon: stability under H_2 MPCVD exposures and early stages of growth. Diam. Relat. Mater. **17**, 1143–1149 (2008). doi:10.1016/j.diamond.2008.01.008
95. J.C. Arnault, S. Saada, O.A. Williams, K. Haenen, P. Bergonzo, M. Nesladek, R. Polini, E. Osawa, Surface characterisation of silicon substrates seeded with diamond nanoparticles under UHV annealing. Phys. Stat. Sol. (A) **205**, 2108–2113 (2008). doi:10.1002/pssa.200879728
96. S. Zeppilli, J.C. Arnault, C. Gesset, P. Bergonzo, R. Polini, Thermal stability and surface modifications of detonation diamond nanoparticles studied with X-ray photoelectron spectroscopy. Diam. Relat. Mater. **19**, 846–853 (2010). doi:10.1016/j.diamond.2010.02.005
97. M. Daenen, O.A. Williams, J. D'Haen, K. Haenen, M. Nesladek, Seeding, growth and characterization of nanocrystalline diamond films on various substrates. Phys. Sta. Sol A **203**, 3005–3010 (2006). doi:10.1002/pssa.200671122
98. X. Liu, T. Yu, Q. Wei, Z. Yu, X. Xu, Enhanced diamond nucleation on copper substrates by employing an electrostatic self-assembly seeding process with modified nanodiamond particles. Colloids Surf. A Physicochem. Eng. Aspects **412**, 82–89 (2012). doi:10.1007/s00339-014-8355-x
99. J. Hees, N. Heidrich, W. Pletschen, R.E. Sah, M. Wolfer, O.A. Williams, V. Lebedev, C.E. Nebel, O. Ambacher, Piezoelectric actuated micro-resonators based on the growth of diamond on aluminum nitride thin films. Nanotechnology 24, 025601 (7 pp) (2013). doi:10.1088/0957-4484/24/2/025601
100. V.V. Chernov, A.L. Vikharev, A.M. Gorbachev, A.V. Kozlov, A.Y. Vul', A.E. Aleksenskii, The nucleation and growth of nanocrystalline diamond films in millimeter-wave CVD reactor. Fullerenes Nanotubes Carbon Nanostruct. **20**, 600–605 (2012). doi:10.1080/1536383X.2012.656550
101. M. Tsigourakos, T. Hantschel, S.D. Janssens, K. Haenen, W. Vandervorst, Spin-seeding approach for diamond growth on large area silicon-wafer substratesphys. Stat. Sol. A **209**, 1659–1663 (2012). doi:10.1002/pssa.201200137
102. M. Bonnauron, S. Saada, C. Mer, C. Gesset, O.A. Williams, L. Rousseau, E. Scorsone, P. Mailley, M. Nesladek, J.-C. Arnault, P. Bergonzo, Transparent diamond-on-glass

micro-electrode arrays for ex-vivo neuronal study. Phys. Stat. Sol. A **205**, 2126–2129 (2008). doi:10.1002/pssa.200879733
103. H.A. Girard, E. Scorsone, S. Saada, C. Gesset, J.C. Arnault, S. Perruchas, L. Rousseau, S. David, V. Pichot, D. Spitzer, P. Bergonzo, Electrostatic grafting of diamond nanoparticles towards 3D diamond nanostructures. Diam. Relat. Mater. **23**, 83–87 (2012). doi:10.1016/j.diamond.2012.01.021
104. R. Bogdanowicz, M. Śmietana, M. Gnyba, Ł. Gołunski, J. Ryl, M. Gardas, Optical and structural properties of polycrystalline CVD diamond films grown on fused silica optical fibres pre-treated by high-power sonication seeding. Appl. Phys. A **116**, 1927–1937 (2014). doi:10.1007/s00339-014-8355-x
105. S. Ruffinatto, H.A. Girard, F. Becher, J.C. Arnault, D. Tromson, P. Bergonzo, Diamond porous conductive membranes: a new material toward analytical chemistry. Diam. Relat. Mater. (2015). doi:10.1016/diamond.2015.03.008
106. M. Tsigkourakos, T. Hantschel, S.D. Janssens, K. Haenen, W. Vandervorst, Spin-seeding approach for diamond growth on large area silicon-wafer substrates. Phys. Stat. sol a **209**, 1659–1663 (2012). doi:10.1002/pssa.201200137
107. N.A. Feoktistov, V.I. Sakharov, I.T. Serenkov, V.A. Tolmachev, I.V. Korkin, A.E. Aleksenskii, A.Y. Vul', V.G. Golubev, Aerosol Deposition of Detonation Nanodiamonds Used as Nucleation Centers for the Growth of Nanocrystalline Diamond Films and Isolated Particles. Tech. Phys. **56**, 718–724 (2011). doi:10.1134/S1063784211050112
108. O.A. Williams, O. Douheret, M. Daenen, K. Haenen, E. Osawa, M. Takahashi, Enhanced diamond nucleation on monodispersed nanocrystalline diamond. Chem. Phys. Lett. **445**, 255–258 (2007). doi:10.1016/j.cplett.2007.07.091
109. O. Shenderova, S. Hens, G. McGuire, Seeding slurries based on detonation nanodiamond in DMSO. Diam. Relat. Mater. **19**, 260–267 (2010). doi:10.1016/j.diamond.2009.10.008
110. S.C. Hens, G. Cunningham, T. Tyler, S. Moseenkov, V. Kuznetsov, O. Shenderova, Nanodiamond bioconjugate probes and their collection by electrophoresis. Diam. Relat. Mater. **17**, 1858–1866 (2008). doi:10.1016/j.diamond.2008.03.020
111. A. Kromka, O. Babchenko, H. Kozak, K. Hruska, B. Rezek, M. Ledinsky, J. Potmesil, M. Michalka, M. Vanecek, Seeding of polymer substrates for nanocrystalline diamond film growth. Diam. Relat. Mater. **18**, 734–739 (2009). doi:10.1016/j.diamond.2009.01.023
112. W. Zhang, K. Patel, A. Schexnider, S. Banu, A.D. Radadia, Nanostructuring of biosensing electrodes with nanodiamonds for antibody immobilization. ACS Nano **8**, 1419–1428 (2014). doi:10.1021/nn405240g
113. H. Schwertfeger, A. Fokin, P.R. Schreiner, Diamonds are a chemist's best friend: diamondoid chemistry beyond adamantane. Angew. Chem. Int. Ed. **47**, 1022–1036 (2008). doi:10.1002/anie.200701684
114. Y.C. Chen, L. Chang, Chemical vapor deposition of diamond on an adamantane-coated sapphire substrate. RSC Adv. **4**, 18945–18950 (2014). doi:10.1039/c4ra01042f
115. J. Hees, A. Kriele, O.A. Williams, Electrostatic self-assembly of diamond nanoparticles. Chem. Phys. Lett. **509**, 12–15 (2011). doi:10.1016/j.cplett.2011.04.083
116. X.Z. Liu, T. Yu, Q.P. Wei, Z.M. Yu, X.Y. Xu, Enhanced diamond nucleation on copper substrates by employing an electrostatic self-assembly seeding process with modified nanodiamond particles. Colloids Surf. A **412**, 82–89 (2012). doi:10.1016/j.colsurfa.2012.07.020
117. I. Zhitomirsky, Cathodic electrophoretic deposition of diamond particles. Mater. Lett. **37**, 72–78 (1998). doi:10.1016/S0167-577X(98)00074-3
118. A.N. Alimova, N.N. Chubun, P.I. Belobrov, P.Y. Detkov, V.V. Zhirnov, Electrophoresis of nanodiamond powder for cold cathode fabrication. J. Vac. Sci. Technol., B **17**, 715–718 (1999). doi:10.1116/1.590625
119. Y.H. Wang, Q.Z. Chen, J. Cho, A.R. Boccaccini, Electrophoretic co-deposition of diamond/borosilicate glass composite coatings. Surf. Coat. Technology **201**, 7645–7651 (2007). doi:10.1016/j.surfcoat.2007.02.037

120. L. Schmidlin, V. Pichot, S. Josset, R. Pawlak, T. Glatzel, S. Kawai, E. Meyer, D. Spitzer, Two-dimensional nanodiamond monolayers deposited by combined ultracentrifugation and electrophoresis techniques. Appl. Phys. Lett. **101**, 253111 (2012). doi:10.1063/1.4772983]
121. P. Pobedinskas, G. Degutis, W. Dexters, W. Janssen, S.D. Janssens, B. Conings, B. Ruttens, J. D'Haen, H.-G. Boyen, A. Hardy, M.K. Van Bael, K. Haenen, Appl. Phys. Lett. **102**, 201609 (2013). doi:10.1063/1.4807591]
122. V. Pichot, K. Bonnot, N. Piazzon, M. Schaefer, M. Comet, D. Spitzer, Deposition of detonation nanodiamonds by Langmuir-Blodgett technique. Diam. Relat. Mater. **19**, 479–483 (2010). doi:10.1016/j.diamond.2009.10.031
123. E. Scorsone, S. Saada, J.C. Arnault, P. Bergonzo, Enhanced control of diamond nanoparticle seeding using a polymer matrix. J. Appl. Phys. **106**, 14908 (2009). doi:10.1063/1.3153118
124. A. Mamedov, J. Ostrander, F. Aliev, N.A. Kotov, Stratified assemblies of magnetite nanoparticles and montmorillonite prepared by the layer-by-layer assembly. Langmuir **16**, 3941–3949 (2000). doi:10.1021/la990957j
125. W. Xue, T. Cui, Characterization of layer-by-layer self-assembled carbon nanotube multilayer thin films. Nanotechnology **18**, 145709 (2007). doi:10.1088/0957-4484/18/14/145709
126. H.A. Girard, S. Perruchas, C. Gesset, M. Chaigneau, L. Vieille, J.C. Arnault, P. Bergonzo, J. P. Boilot, T. Gacoin, Electrostatic grafting of diamond nanoparticles: a versatile route to nanocrystalline diamond thin films. ACS Appl. Mater. Interfaces **1**, 2738–2746 (2009). doi:10.1021/am900458g
127. J.H. Kim, S.K. Lee, O.M. Kwon, S.I. Hong, D.S. Lim, Thickness controlled and smooth polycrystalline CVD diamond film deposition on SiO2 with electrostatic self assembly seeding process. Diam. Relat. Mater. **18**, 1218–1222 (2009). doi:10.1016/j.diamond.2009.04.012
128. S.K. Lee, J.H. Kim, M.G. Jeong, M.J. Song, D.S. Lim, Direct deposition of patterned nanocrystalline CVD diamond using an electrostatic self-assembly method with nanodiamond particles. Nanotechnology **21**, 505302 (2010). doi:10.1088/0957-4484/21/50/505302
129. E. Chevallier, E. Scorsone, H.A. Girard, V. Pichot, D. Spitzer, P. Bergonzo, Metalloporphyrin-functionalised diamond nano-particles as sensitive layer for nitroaromatic vapours detection at room-temperature. Sens. Actuators B **151**, 191–197 (2010). doi:10.1016/j.snb.2010.09.022
130. H.A. Girard, E. Scorsone, S. Saada, C. Gesset, J.C. Arnault, S. Perruchas, L. Rousseau, S. David, V. Pichot, D. Spitzer, P. Bergonzo, Electrostatic grafting of diamond nanoparticles towards 3D diamond nanostructures. Diam. Relat. Mater. **23**, 83–87 (2012). doi:10.1016/j.diamond.2012.01.021
131. G. Saini, D.S. Jensen, L.A. Wiest, M.A. Vail, A. Dadson, M.L. Lee, V. Shutthanandan, M.R. Linford, Core-shell diamond as a support for solid-phase extraction and high-performance liquid chromatography. Anal. Chem. **82**, 4448–4456 (2010). doi:10.1021/ac1002068
132. A.V. Sumant, O. Auciello, M. Liao, O.A. Williams, MEMS/NEMS based on mono-, nano-, and ultrananocrystalline diamond films. MRS Bull. **39**, 511–516 (2014). doi:10.1557/mrs.2014.98
133. T.M. Babinec, B.J.M. Hausmann, M. Khan, Y. Zhang, J.R. Maze, P.R. Hemmer, M. Loncar, A diamond nanowire single-photon source. Nat. Nanotechnol. **5**, 195–199 (2010). doi:10.1038/NNANO.2010.6
134. X. Checoury, D. Néel, P. Boucaud, C. Gesset, H. Girard, S. Saada, P. Bergonzo, Nanocrystalline diamond photonics platform with high quality factor photonic crystal cavities. Appl. Phys. Lett. **101**, 171115 (2012). doi:10.1063/1.4764548
135. A. Bongrain, E. Scorsone, L. Rousseau, G. Lissorgues, P. Bergonzo, Realisation and characterisation of mass-based diamond micro-transducers working in dynamic mode. Sens. Actuators B **154**, 142–149 (2011). doi:10.1016/j.snb.2009.12.067

136. O. Babchenko, E. Verveniotis, K. Hruska, M. Ledinsky, A. Kromka, B. Rezek, Direct growth of sub-micron diamond structures. Vacuum **86**, 693–695 (2012). doi:10.1016/j.vacuum.2011.08.011
137. O. Shimoni, J. Cervenka, T.J. Karle, K. Fox, B.C. Gibson, S. Tomljenovic-Hanic, A.D. Greentree, S. Prawer, Development of a templated approach to fabricate diamond patterns on various substrates. ACS Appl. Mater. Interfaces **6**, 8894–8902 (2014). doi:10.1021/am5016556
138. S.G. Rao, A. Karim, J. Schartz, N. Antler, T. Schenkel, I. Siddiqi, Directed assembly of nanodiamond nitrogen-vacancy centers on a chemically modified patterned surface. ACS Appl. Mater. Interfaces **6**, 12893–12900 (2014). doi:10.1021/am5027665
139. H. Zhuang, B. Song, T. Staedler, X. Jiang, Microcontact printing of monodiamond nanoparticles: an effective route to patterned diamond structure fabrication. Langmuir **27**, 11981–11989 (2011)
140. T. Vandenryt, L. Grieten, S.D. Janssens, B. Van Grinsven, K. Haenen, B. Ruttens, J. D'Haens, P. Wagner, R. Thoelen, W. De Ceuninck, Rapid fabrication of micron-sized CVD-diamond structures by microfluidic contact printing. Phys. Stat. Sol. A **211**, 1448–1454 (2014). doi:10.1002/pssa.201330665
141. Y.C. Chen, Y. Tzeng, A.J. Cheng, R. Dean, M. Park, B.M. Wilamowski, Inkjet printing of nanodiamond suspensions in ethylene glycol for CVD growth of patterned diamond structures and practical applications. Diam. Relat. Mater. **18**, 146–150 (2009). doi:10.1016/j.diamond.2008.10.004
142. S. Singh, V. Thomas, D. Martyshkin, V. Kozlovskaya, E. Kharlampieva, S.A. Catledge, Spatially controlled fabrication of a bright fluorescent nanodiamond-array with enhanced far-red Si-V luminescence. Nanotechnology **25**, 045302 (2014). doi:10.1088/0957-4484/25/4/045302
143. O. Loh, R. Lam, M. Chen, N. Moldovan, H. Huang, D. Ho, H.D. Espinosa, Nanofountain-Probe-Based High-Resolution Patterning and Single-Cell Injection of Functionalized Nanodiamonds. Small **5**, 1667–1674 (2009). doi:10.1002/smll.200900361
144. A. Albrecht, G. Koplovitz, A. Retzker, F. Jelezko, S. Yochelis, D. Porath, Y. Nevo, O. Shoseyov, Y. Paltiel, M.B. Plenio, Self-assembling hybrid diamond–biological quantum devices. New J. Phys. **16**, 093002 (2014). doi:10.1088/1367-2630/16/9/093002
145. W.X. Wang, D. Pelah, T. Alergand, O. Shoseyov, A. Altmann, Characterization of SP1, a stress-responsive, boiling-soluble, homo-oligomeric protein from aspen. Plant Physiol. **130**, 865–875 (2002). doi:10.1104/pp.002436
146. V. Paget, J.A. Sergent, R. Grall, S. Altmeyer-Morel, H.A. Girard, T. Petit, G. Gesset, M. Mermoux, P. Bergonzo, J.C. Arnault, S. Chevillard, Nanodiamonds are neither cytotoxic nor genotoxic on kidney, intestine, lung and liver human cell lines. Nanotoxicology **8**, 46–56 (2014). doi:10.3109/17435390.2013.855828
147. J.I. Chao, E. Perevedentseva, P.H. Chung, K.K. Liu, C.Y. Cheng, C.C. Chang, C.L. Cheng, Nanometer-sized diamond particle as a probe for biolabeling. Biophys. J. **93**, 2199–2208 (2007). doi:10.1529/biophysj.107.108134
148. J.M. Rosenholm, I.I. Vlasov, S.A. Burinov, T.A. Dolenko, O.A. Shenderova, Nanodiamond-Based composite structures for biomedical imaging and drug delivery. J. Nanosci. Nanotechnol. **15**, 959–971 (2015). doi:10.1166/jnn.2015.9742
149. R.G. Chaudhuri, S. Paria, Core/shell nanoparticles: classes, properties, synthesis mechanisms, characterization, and applications. Chem. Rev. **112**, 2373–2433 (2012). doi:10.1021/cr100449n
150. P. Mélinon, S. Begin-Colin, J.L. Duvail, F. Gauffre, N. Herlin Boime, G. Ledoux, J. Plain, P. Reiss, F. Silly, B. Warot-Fonrose, Engineered inorganic core/shell nanoparticles. Phys. Rep. **543**, 163–197 (2014). doi:10.1016/j.physrep.2014.05.003
151. W. Schärtl, Current directions in core-shell nanoparticle design. Nanoscale **2**, 829–843 (2010). doi:10.1039/c0nr00028k
152. E. von Haartman, H. Jiang, A.A. Khomich, J. Zhang, S.A. Burikov, T.A. Dolenko, J. Ruokolainen, H. Gu, O.A. Shenderova, I.I. Vlasov, J.M. Rosenholm, Core–shell designs of

photoluminescent nanodiamonds with porous silica coatings for bioimaging and drug delivery I: fabrication. J. Mater. Chem. B **1**, 2358–2366 (2013). doi:10.1039/c3tb20308e

153. N. Prabhakar, T. Nareoja, E. von Haartman, D.S. Karaman, H. Jiang, S. Koho, T.A. Dolenko, P.E. Hanninen, D.I. Vlasov, V.G. Ralchenko, S. Hosomi, I.I. Vlasov, C. Sahlgrenbci, J.M. Rosenholm, Core–shell designs of photoluminescent nanodiamonds with porous silica coatings for bioimaging and drug delivery II: application. Nanoscale **5**, 3713–3722 (2013). doi:10.1039/c3nr33926b
154. I. Rehor, J. Slegerova, J. Kucka, V. Proks, V. Petrakova, M.P. Adam, F. Treussart, S. Turner, S. Bals, P. Sacha, M. Ledvina, A.M. Wen, N.F. Steinmetz, P. Cigler, Fluorescent nanodiamonds embedded in biocompatible translucent shells. Small **10**, 1106–1115 (2014). doi:10.1002/smll.201302336
155. S. Oldenburg, R. Averitt, S. Westcott, N. Halas, Nanoengineering of optical resonances. Chem. Phys. Lett. **288**, 243–247 (1998). doi:10.1016/S0009-2614(98)00277-2
156. L. Minati, C.L. Cheng, Y.C. Lin, J. Hees, G. Lewes-Malandrakis, C.E. Nebel, F. Benetti, C. Migliaresi, G. Speranza, Synthesis of novel nanodiamonds-gold core shell nanoparticles. Diam. Relat. Mater. **53**, 23–28 (2015). doi:10.1016/j.diamond.2015.01.004
157. T. Pham, J.B. Jackson, N.J. Halas, T.R. Lee, Preparation and characterization of gold nanoshells coated with self-assembled monolayers. Langmuir **18**, 4915–4920 (2002). doi:10.1021/la015561y
158. W.L. Shi, Y. Sahoo, M.T. Swihart, P.N. Prasad, Gold nanoshells on polystyrene cores for control of surface plasmon resonance. Langmuir **21**, 1610–1617 (2005). doi:10.1021/la047628y
159. I. Rehor, K.L. Lee, K. Chen, M. Hajek, J. Havlik, J. Lokajova, M. Masat, J. Slegerova, S. Shukla, H. Heidari, S. Bals, N.F. Steinmetz, P. Cigler, Plasmonic nanodiamonds: targeted core-shell type nanoparticles for cancer cell thermoablation. Adv. Healthc. Mater. **4**, 460–468 (2015). doi:10.1002/adhm.201400421
160. B.E. Brinson, J.B. Lassiter, C.S. Lewin, R. Bardhan, N. Mirin, N.J. Halas, Nanoshells made easy: improving Au layer growth on nanoparticle surfaces. Langmuir **24**, 14166–14171 (2008). doi:10.1021/la802049p
161. S. Pankasem, J.K. Thomas, M.J. Snowden, B. Vincent, Photophysical studies of poly (N-isopropylacrylamide) microgel structures. Langmuir **10**, 3023–3026 (1994). doi:10.1021/la00021a027
162. T. Hoare, R. Pelton, Titrametric characterization of pH-induced phase transitions in functionalized microgels. Langmuir **22**, 7342–7350 (2006). doi:10.1021/la0608718
163. M. Shibayama, F. Ikkai, S. Inamoto, S. Nomura, C.C. Han, pH and salt concentration dependence of the microstructure of poly (N-isopropylacrylamide-co-acrylic acid) gels. J. Chem. Phys. **105**, 4358–4366 (1996). doi:10.1063/1.472252
164. H.A. Girard, P. Benayoun, C. Blin, A. Trouvé, C. Gesset, J.C. Arnault, P. Bergonzo, Encapsulated nanodiamonds in smart microgels toward self-assembled diamond nanoarrays. Diam. Relat. Mater. **33**, 32–37 (2013). doi:10.1016/j.diamond.2012.12.007
165. Y.P. Sun, B. Zhou, Y. Lin, W. Wang, K.A.S. Fernando, P. Pathak, M.J. Meziani, B.A. Harruff, X. Wang, H. Wang, Quantum-sized carbon dots for bright and colorful photoluminescence. J. Am. Chem. Soc. **128**, 7756–7757 (2006). doi:10.1021/ja062677d
166. X. Zhang, S. Wang, C. Zhu, M. Liu, Y. Ji, L. Feng, L. Tao, Y. Wei, Carbon-dots derived from nanodiamond: Photoluminescence tunable nanoparticles for cell imaging. J. Colloid Interface Sci. **397**, 39–44 (2013). doi:10.1016/j.jcis.2013.01.063
167. O. Shenderova, S. Hens, I. Vlasov, S. Turner, Y.G. Lu, G. Van Tendeloo, A. Schrand, S.A. Burinov, T.A. Dolenko, Carbon dot decorated nanodiamonds. Part. Part. Syst. Charact. **31**, 580–590 (2014). doi:10.1002/ppsc.201300251
168. V.V. Avdeev, N.E. Sorokina, N.V. Maksimova, I.Y. Martnynov, A.V. Sezemin, Synthesis of ternary intercalation compounds in the graphite-HNO_3-R (R = H_2SO_4, H_3PO_4, CH_3COOH) systems. Inorg. Mater. **37**, 366 (2001). doi:10.1023/A:1017527827724
169. M.S. Dresselhaus, G. Dresselhaus, Intercalation compounds of graphite. Adv. Phys. **51**, 1–186 (2002). doi:10.1080/00018730110113644

170. A.E. Aleksensky, M.V. Baidakova, M.A. Yagovkina, V.I. Siklitsky, A.Y. Vul', H. Naramoto, V.I. Lavrentiev, Nanodiamonds intercalated with metals: structure and diamond-graphite phase transitions. Diam. Relat. Mater. **13**, 2076–2080 (2004). doi:10.1016/j.diamond.2004.05.008
171. A.I. Shames, A.M. Panich, VYu. Osipov, A.E. Aleksenskiy, A.Y. Vul', T. Enoki, K. Takai, Structure and magnetic properties of detonation nanodiamond chemically modified by copper. J. Appl. Phys. **107**, 014318 (2010). doi:10.1063/1.3273486
172. A.M. Panich, A. Altman, A.I. Shames, VYu. Osipov, A.E. Aleksenskiy, A.Y. Vul', Proton magnetic resonance study of diamond nanoparticles decorated by transition metal ions. J. Phys. D Appl. Phys. **44**, 125303 (2011). doi:10.1088/0022-3727/44/12/125303
173. A.I. Shames, VYu. Osipov, A.E. Aleksenskiy, E. Ōsawa, A.Y. Vul', Locating inherent unpaired orbital spins in detonation nanodiamonds through the targeted surface decoration by paramagnetic probes. Diamond Relat. Mater. **20**, 318–321 (2011). doi:10.1016/j.diamond.2011.01.007
174. I.D. Gridnev, V.Y. Osipov, A.E. Aleksenskii, A.Y. Vul', T. Enoki, Combined experimental and DFT study of the chemical binding of copper ions on the surface of nanodiamonds. Bull. Chem. Soc. Jpn. **87**, 693–704 (2014). doi:10.1246/bcsj.20130345
175. A. Shakun, J. Vuorinen, M. Hoikkanen, M. Poikelispaa, A. Das, Hard nanodiamonds in soft rubbers: past, present and future—a review. Compos. Part A **64**, 49–69 (2014). doi:10.1016/j.compositesa.2014.04.014
176. Q. Zhang, K. Naito, Y. Tanaka, Y. Kagawa, Grafting polyimides from nanodiamonds. Macromolecules **41**, 536–538 (2008). doi:10.1021/ma702268x
177. I. Cha, K. Shirai, K. Fujiki, T. Yamauchi, N. Tsubokawa, Surface grafting of polymers onto nanodiamond by ligand-exchange reaction of ferrocene moieties of polymers with polycondensed aromatic rings of the surface. Diam. Relat. Mater. **20**, 439–444 (2011). doi:10.1016/j.diamond.2011.01.014
178. S. Morimune, M. Kotera, T. Nishino, K. Goto, K. Hata, Poly (vinyl alcohol) nanocomposites with nanodiamond. Macromolecules **44**, 4415–4421 (2011). doi:10.1021/ma200176r
179. I. Neitzel, V. Mochalin, Y. Gogotsi, Advances in surface chemistry of nanodiamond and nanodiamond–polymer composites, in *Ultrananocrystalline Diamond: Synthesis, Properties and Applications,* 2nd edn, ed. by O.A. Shenderova, D.M. Gruen (William Andrew, 2012), pp. 421–457
180. M.R. Ayatollahi, E. Alishahi, R.S. Doagou, S. Shadlou, Tribological and mechanical properties of low content nanodiamond/epoxy nanocomposites. Compos Part B Eng. **43**, 3425–3430 (2012). doi:10.1016/j.compositesb.2012.01.022
181. K.D. Behler, A. Stravato, V. Mochalin, G. Korneva, G. Yushin, Y. Gogotsi, Nanodiamond-polymer composite fibers and coatings. ACS Nano **3**, 363–369 (2009). doi:10.1021/nn800445z CCC: $40.75
182. I. Neitzel, V.N. Mochalin, J. Niu, J. Cuadra, A. Kontsos, G.R. Palmese, Y. Gogotsi, Maximizing Young's modulus of aminated nanodiamond–epoxy composites measured in compression. Polymer **53**, 5965–5971 (2012). doi:10.1016/j.polymer.2012.10.037
183. V.Y. Dolmatov, Applications of detonation nanodiamond, in *Ultrananocrystalline Diamond*, ed. by A. Shenderova Olga, M. Gruen Dieter (William Andrew Publishing, Norwich, 2006), pp. 477–527
184. A.P. Voznyakovskii, B.M. Ginzburg, D. Rashidov, D.G. Tochil'nikov, S. Tuichiev. Structure, mechanical, and tribological characteristics of polyurethane modified with nanodiamonds. Polym. Sci. Ser. A **52**, 1044–1050 (2010). doi:10.1134/S0965545X10100068
185. E. Roumeli, E. Pavlidou, A. Avgeropoulos, G. Vourlias, D.N. Bikiaris, K. Chrissafis, factors controlling the enhanced mechanical and thermal properties of nanodiamond-reinforced cross-linked high density polyethylene. J. Phys. Chem. B **118**, 11341–11352 (2014). doi:10.1021/jp504531f

186. H.B. Cho, S.T. Nguyen, T. Nakayama, H. Suematsu, T. Suzuki, W. Jiang, S. Tanaka, B.S. Kim, K. Niihara, Polyepoxide-based nanohybrid films with self-assembled linear assemblies of nanodiamonds. Acta Mater. **60**, 7249–7257 (2012). doi:10.1016/j.actamat.2012.09.039
187. Q. Zhang, V.N. Mochalin, I. Neitzel, K. Hazeli, J. Niu, A. Kontsos, J.G. Zhou, P.I. Lelkes, Y. Gogotsi, Mechanical properties and biomineralization of multifunctional nanodiamond-PLLA composites for bone tissue engineering. Biomaterials **33**, 5067–5075 (2012). doi:10.1016/j.biomaterials.2012.03.063
188. R. Liu, F. Zhao, X. Yu, K. Naito, H. Ding, X. Qu, Q. Zhang, Synthesis of biopolymer-grafted nanodiamond by ring-opening polymerization. Diam. Relat. Mater. **50**, 26–32 (2014). doi:10.1016/j.diamond.2014.08.011
189. C.N. Almeida B.C. Ramos, N.S. Da-Silva C. Pacheco-Soares, V.J. Trava-Airoldi, A.O. Lobo, F.R. Marciano, Morphological analysis and cell viability on diamond-like carbon films containing nanocrystalline diamond particles. Appl. Surf. Sci. **275**, 258–263 (2013). doi:10.1016/j.apsusc.2012.12.122
190. J.J. Taha-Tijerina, T.N. Narayanan, C. Sekhar Tiwary, K. Lozano, M. Chipara, P.M. Ajayan, Nanodiamond-based thermal fluids. ACS Appl. Mater. Interfaces **6**, 4778–4785 (2014). doi:10.1021/am405575t
191. M.G. Ivanov, V.V. Kharlamov, V.V. Buznik, D.M. Ivanov, S.V. Pavlyshko, A.K. Tsvetnikov, Tribological properties of the grease containing polytetrafluoroethylene and ultrafine diamond. Friction Wear **25**, 99–103 (2004). INSPEC:8537707
192. V.Y. Dolmatov, Detonation nanodiamonds in oils and lubricants. J. Superhard Mater. **32**, 14–20 (2010). doi:10.3103/S1063457610010028
193. V.I. Zhornik, V.A. Kukareko, M.A. Belotserkovsky, *Tribomechanical Modification of Friction Surface by Running-In in Lubricants with Nano-Sized Diamonds* (Nova Science Publishers, 2011)
194. M. Ivanov, D. Ivanov, Nanodiamond nanoparticles as additives to lubricants (Chap. 8), in *Ultrananocrystalline Diamond*, 2nd edn, ed. by O. Shenderova, D. Gruen (Elsevier, 2012)
195. C.C. Chou, S.H. Lee, Tribological behavior of nanodiamond-dispersed lubricants on carbon steels and aluminum alloy. Wear **269**, 757–762 (2010). doi:10.1016/j.wear.2010.08.001
196. M. Ivanov, Z. Mahbooba, D. Ivanov, S. Smirnov, S. Pavlyshko, E. Osawa, D. Brenner, O. Shenderova, Nanodiamond-based oil lubricants on steel-steel and stainless steel-hard alloy high load contact: investigation of friction surfaces. Nanosystems Phys. Chem. Math. **5**, 160–166 (2014)
197. O. Elomaa, T.J. Hakala, V. Myllymäki, J. Oksanen, H. Ronkainen, V.K. Singh, J. Koskinen, Diam. Relat. Mater. **34**, 89–94 (2013). doi:10.1016/j.diamond.2013.02.008
198. E. Perevedentseva, Y.C. Lin, M. Jani, C.L. Cheng, Biomedical applications of nanodiamonds in imaging and therapy. Nanomedicine **8**, 2041–2060 (2013). doi:10.2217/NNM.13.183
199. V.N. Mochalin, A. Pentecost, X.M. Li, I. Neitzel, M. Nelson, C. Wei, T. He, F. Guo, Y. Gogotsi, Adsorption of drugs on nanodiamond: toward development of a drug delivery platform mol. Pharmaceutics **10**, 3728–3735 (2013). doi:10.1021/mp400213z
200. D. Passeri, F. Rinaldi, C. Ingallina, M. Carafa, M. Rossi, M.L. Terranova, C. Marianecci, Biomedical applications of nanodiamonds: an overview. J. Nanosci. Nanotechnol. **15**, 972–988 (2015). doi:10.1166/jnn.2015.9734
201. P. Metzler, C. von Wilmowsky, B. Stadlinger, W. Zemann, K.A. Schlegel, S. Rosiwal, S. Rupprecht, J. Cranio-Maxillofac. Surg. **41**, 532–538 (2013). doi:10.1016/j.jcms.2012.11.020
202. A. Krueger, D. Lang, Functionality is key: recent progress in the surface modification of nanodiamond. Adv. Funct. Mater. **22**, 890–906 (2012). doi:10.1002/adfm.201102670
203. E.K. Chow, X.Q. Zhang, M. Chen, R. Lam, E. Robinson, H. Huang, D. Schaffer, E. Osawa, A. Goga, D. Ho, Nanodiamond therapeutic delivery agents mediate enhanced chemoresistant tumor treatment. Sci. Transl. Med. **3**, 73ra21 (2011). doi:10.1126/scitranslmed.3001713
204. A. Alhaddad, M.P. Adam, J. Botsoa, G. Dantelle, S. Perruchas, T. Gacoin, C. Mansuy, S. Lavielle, C. Malvy, F. Treussart, J.R. Bertrand, Nanodiamond as a vector for siRNA delivery to Ewing sarcoma cells. Small **7**, 3087–3095 (2011). doi:10.1002/smll.201101193

205. R.A. Shimkunas, E. Robinson, R. Lam, S. Lu, X.Y. Xu, X.Q. Zhang, H.J. Huang, E. Osawa, D. Ho, Nanodiamond-insulin complexes as pH-dependent protein delivery vehicles. Biomaterials **30**, 5720–5728 (2009). doi:10.1016/j.biomaterials.2009.07.004
206. C. Gaillard, H.A. Girard, C. Falck, V. Paget, V. Simic, N. Hugolin, P. Bergonzo, S. Chevillard, J.C. Arnault, Peptide nucleic acid–nanodiamonds: covalent and stable conjugates for DNA targeting. RSC Advances **4**, 3566–3572 (2014). doi:10.1039/c3ra45158e
207. G. Reina, S. Orlanducci, C. Cairone, E. Tamburini, S. Lenti, I. Cianchetta, M. Rossi, M.L. Terranova, Rhodamine/nanodiamond as a system model for drug carrier. J. Nanosci. Nanotechnol. **15**, 1022–1029 (2015). doi:10.1166/jnn.2015.9736
208. B. Guan, F. Zou, J.F. Zhi, Nanodiamond as the pH responsive vehicle for an anticancer drug. Small **6**, 1514–1519 (2010). doi:10.1002/smll.200902305
209. A. Adnan, R. Lam, H. Chen, J. Lee, D. Schaffer, A. Barnard, G.C. Schatz, D. Ho, W.K. Liu, Atomistic simulation and measurement of pH dependent cancer therapeutic interactions with nanodiamond carrier. Mol. Pharm. **8**, 368–374 (2011). doi:10.1021/mp1002398
210. J. Yan, Y. Guo, A. Altawashi, B. Moosa, S. Lecommandoux, N.M. Khashab, Experimental and theoretical evaluation of nanodiamonds as pH triggered drug carriers. New J. Chem. **36**, 1479–1484 (2012). doi:10.1039/c2nj40226b
211. V.N. Mochalin, A. Pentecost, X.M. Li, I. Neitzel, M. Nelson, C. Wei, T. He, F. Guo, Y. Gogotsi, Adsorption of drugs on nanodiamond: toward development of a drug delivery platform. Mol. Pharm. **10**, 3728–3735 (2013). doi:10.1021/mp400213z
212. T.B. Toh, D.K. Lee, W. Hou, L.N. Abdullah, J. Nguyen, D. Ho, E. Kai-Hua, Chow, nanodiamond—mitoxantrone complexes enhance drug retention in chemoresistant breast cancer cells. Mol. Pharm. **11**, 2683–2691 (2014). doi:10.1021/mp5001108
213. A.D. Salaam, P.T.J. Hwang, A. Poonawalla, H.N. Green, H. Jun, D. Dean, Nanodiamonds enhance therapeutic efficacy of doxorubicin in treating metastatic hormone-refractory prostate cancer. Nanotechnology **25**, 425103 (2014). doi:10.1088/0957-4484/25/42/425103
214. L. Moore, E.K.H. Chow, E. Osawa, J.M. Bishop, D. Ho, Diamond-lipid hybrids enhance chemotherapeutic tolerance and mediate tumor regression. Adv. Mater. **25**, 3532–3541 (2013). doi:10.1002/adma.201300343
215. I. Aharonovich, Diamond nanocrystals for photonics and sensing. Japan. J. Appl. Phys. **53**, 05FA01 (2014). doi:10.7567/JJAP.53.05FA01
216. I.I. Vlasov, A.A. Shiryaev, T. Rendler, S. Steinert, S.Y. Lee, D. Antonov, M. Vörös, F. Jelezko, A.V. Fisenko, L.F. Semjonova, J. Biskupek, U. Kaiser, O.I. Lebedev, I. Sildos, P.R. Hemmer, V.I. Konov, A. Gali, J. Wrachtrup, Molecular-sized fluorescent nanodiamonds. Nat. Nanotech. **9**, 54–58 (2014). doi:10.1038/NNANO.2013.255
217. V.N. Mochalin, Y. Gogotsi, Wet chemistry route to hydrophobic blue fluorescent nanodiamond. J. Am. Chem. Soc. **131**, 4594–4595 (2009). doi:10.1021/ja9004514
218. S. Vial, C. Mansuy, S. Sagan, T. Irinopoulou, F. Burlina, J.P. Boudou, G. Chassaing, S. Lavielle, Peptide-grafted nanodiamonds: preparation, cytotoxicity and uptake in cells. Chem-BioChem. **9**, 2113–2119 (2008). doi:10.1002/cbic.200800247, 10.1039/c2nj40226b
219. Y.R. Chang, H.Y. Lee, K. Chen, C.C. Chang, D.S. Tsai, C.C. Fu, T.S. Lim, Y.K. Tzeng, C. Y. Fang, C.C. Han, H.C. Chang, W. Fann, Mass production and dynamic imaging of fluorescent nanodiamonds. Nat. Nanotech. **3**, 284–288 (2008). doi:10.1038/nnano.2008.99
220. N. Mohan, C.S. Chen, H.H. Hsieh, Y.C. Wu, H.C. Chang, In vivo imaging and toxicity assessments of fluorescent nanodiamonds in Caenorhabditis elegans. Nano Lett. **10**, 3692–3699 (2010). doi:10.1021/nl1021909
221. L.P. McGuinness, Y. Yan, A. Stacey, D.A. Simpson, L.T. Hall, D. Maclaurin, S. Prawer, P. Milvaney, J. Wrachtrup, F. Caruso, R.E. Scholten, L.C.L. Hollenberg, Quantum measurement and orientation tracking of fluorescent nanodiamonds inside living cells. Nat. Nanotech. **6**, 358–363 (2011). doi:10.1038/nnano.2011.64
222. D.A. Simpson, A.J. Thompson, M. Kowarsky, N.F. Zeeshan, M.S.J. Barson, L.T. Hall, Y. Yan, S. Kaufmann, B.C. Johnson, T. Ohshima, F. Caruso, R.E. Scholten, R.B. Saint, M. J. Murray, L.C.L. Hollenberg, In vivo imaging and tracking of individual nanodiamonds in

drosophila melanogaster embryos. Biomedical Optics Express **5**, 1250–1261 (2014). doi:10.1364/BOE.5.001250
223. Y.Y. Hui, L.J. Su, O.Y. Chen, Y.T. Chen, T.M. Liu, H.C. Chang, Wide-field imaging and flow cytometric analysis of cancer cells in blood by fluorescent nanodiamond labeling and time gating. Sci. Rep. **4**(5574), 1–7 (2014). doi:10.1038/srep05574
224. S.J. Hollister, W.L. Murphy, Scaffold Translation: Barriers Between Concept and Clinic. Tissue Eng. Part B Rev. **17**, 459–474 (2011). doi:10.1089/ten.teb.2011.0251
225. L. Moore, M. Gatica, H. Kim, E. Osawa, D. Ho, Multi-protein delivery by nanodiamonds promotes bone formation. J. Dent. Res. **92**, 976–981 (2013). doi:10.1177/0022034513504952
226. M. Monaco, M. Giugliano, Carbon-based smart nanomaterials in biomedicine and neuroengineering. Beilstein J. Nanotechnol. **5**, 1849–1863 (2014). doi:10.3762/bjnano.5.196
227. S. Suliman, Z. Xing, X. Wu, Y. Xue, T.O. Pedersen, Y. Sun, A.P. Døskeland, J. Nickel, T. Waag, H. Lygre, A. Finne-Wistrand, D. Steinmüller-Nethl, A. Krueger, K. Mustafa, Release and bioactivity of bone morphogenetic protein-2 are affected by scaffold binding techniques in vitro and in vivo. J. Controlled Release **197**, 148–157 (2015). doi:10.1016/j.jconrel.2014.11.003
228. H. Kato, J. Hees, R. Hoffmann, M. Wolfer, N. Yang, S. Yamasaki et al., Diamond foam electrodes for electrochemical applications. Electrochem. Commun. **33**, 88–91 (2013). doi:10.1016/j.elecom.2013.04.028
229. F. Gao, M.T. Wolfer, C.E. Nebel, Highly porous diamond foam as a thin-film micro-supercapacitor material. Carbon N. Y. **80**, 833–840 (2014). doi:10.1016/j.carbon.2014.09.007
230. K. Purtov, A. Petunin, E. Inzhevatkin, A. Burov, N. Ronzhin, A. Puzyr, V. Bondar, Biodistribution of different sized nanodiamonds in mice. J. Nanosci. Nanotechnol. **15**, 1070–1075 (2015). doi:10.1166/jnn.2015.9746
231. L. Moore, V. Grobarova, H. Shen, H.B. Man, J. Mıcova, M. Ledvina, J. Stursa, M. Nesladek, A. Fiserova, D. Ho, Comprehensive interrogation of the cellular response to fluorescent, detonation and functionalized nanodiamonds. Nanoscale **6**, 11712–11721 (2014). doi:10.1039/c4nr02570a
232. D. Zhu, L.H. Zhang, R.E. Ruther, R.J. Hamers, Photo-illuminated diamond as a solid-state source of solvated electrons in water for nitrogen reduction. Nat. Mater. **12**, 836–841 (2013). doi:10.1038/NMAT3696
233. L.H. Zhang, D. Zhu, G.M. Nathansson, R.J. Hamers, Selective photoelectrochemical reduction of aqueous CO_2 to CO by solvated electrons. Angew. Chem-Int Ed. **53**, 9746 (2014). doi:10.1002/anie.201404328

One-Dimensional Carbon Nanostructures: Low-Temperature Chemical Vapor Synthesis and Applications

Yao Ma, Nianjun Yang and Xin Jiang

Abstract Chemical vapor deposition (CVD) is a powerful method to synthesize various carbon nanostructures (e.g., carbon nanotubes). A conventional CVD process has to be carried out at the temperatures over 600 °C. To extend the applications of carbon nanostructures, for example in the semiconductor industry, low-temperature synthesis processes are thus always pursued. In this chapter we review the CVD growth of carbon nanostructures at low temperatures (<450 °C). These growth processes are discussed in detail with respect to the applied catalyst system, carbon source, reaction atmosphere, catalyst faces, morphology control as well as unique structural characteristics of grown products. For the low-temperature CVD growth, catalytic reaction occurring on the low index faces of a metal catalyst is a crucial issue, and the growth is rate-limited by surface diffusion. Instead of the classical Vapor-Liquid-Solid (VLS) growth mechanism, the growth mechanism at low temperatures is interpreted with a novel Vapor-Facet-Solid (VFS) mechanism. Due to their unique features, the synthesized carbon nanostructures are promising to be applied for interconnects in large-scale integrated circuits, field emission, microwave adsorption, and as the anode material of lithium ion secondary battery, etc.

Keywords Carbon nanostructure · CVD · Low temperature · Catalyst · Growth mechanism

1 Introduction

One dimensional (1D) carbon nanostructures (CNs) usually refer to the structures, which feature large aspect ratios and small diameters (<100 nm). The typical 1D CNs, carbon nanofibers (CNFs) entered people's consciousness initially as harmful byproducts during metal-catalytic synthesis processes, and thus in the early studies researches aimed to inhibit the formation of CNFs. An interesting turning occurred

Y. Ma · N. Yang · X. Jiang (✉)
Institute of Materials Engineering, University of Siegen,
Paul-Bonatz Str. 9-11, 57076 Siegen, Germany
e-mail: xin.jiang@uni-siegen.de

N. Yang et al. (eds.), *Carbon Nanoparticles and Nanostructures*,
Carbon Nanostructures, DOI 10.1007/978-3-319-28782-9_2

in 1991 due to the discovery of a special CNF by Iijima [1], which is constructed by cylindrical graphitic sheets. It was named as carbon nanotube (CNT). Researchers are then attracted by distinct properties of the CNTs such as extremely high strength, aspect ratio, specific surface area, as well as unique optical and electrical features [2–4]. Since then many scientists worldwide have focused on exploring new techniques to synthesize CNTs, because of their extensive application potentials.

Three main techniques have been developed to grow CNTs, including catalytic CVD, arc-discharge [5, 6], and laser-ablation technique [7, 8]. Among them the catalytic CVD strategy has the distinct advantages due to its large-scale production ability of CNTs at relative low temperatures and low cost. Hence it is the most popular technique. In a catalytic CVD growth process, the catalysts (e.g., small-sized metal particles) are firstly heated in a reaction chamber. Carbon containing source gas is then introduced, a series of chemical and physical processes start and lead to the deposition of carbon and finally the formation of CNTs on the catalysts.

In the research fields of catalytic CVD growth of CNTs, lots of other types of carbon nanostructures besides cylindrical CNTs have been synthesized. In term of morphology, they can be sorted to straight nanofiber [9], nanohelix [10], nanocoil [11], nanocone [12], and nanobelt [13], etc. According to structural characteristics (the size and orientation of graphitic sheets constructing the CNs), they feature tubular [14, 15], herringbone-like [16–18], bamboo-like [19–22], and amorphous nature [23].

The morphologies, microstructures and properties of CNs prepared in CVD process are determined by both the catalysts preparation and subsequent growth conditions. In this context, it has been found that transition metals, such as Fe, Co, Ni and their alloys are the effective catalysts for the growth of CNs, and the transition metals with different compositions and morphologies can lead to CNs with different morphologies. As a result, the preparation of effective catalysts has become a significant step for the synthesis of CNs. Numerous methods have been thus employed to prepare these catalysts, including sol-gel method [24, 25], co-reduction of precursors [26, 27], impregnation and incubation [28, 29], ion-exchange-precipitation [30], reverse micelle method [31], thermal decomposition of carbonyl complexes [32, 33], and metal organic chemical vapor deposition (MOCVD) [29].

Different gases, including CH_4, CO, C_2H_2, C_2H_4, benzene were used as carbon source for the growth of CNs. The effects of flow rate, temperature, and additional ingredient such as oxygen and sulfur in reaction atmosphere have been investigated.

Special configuration of the synthesis equipments has been also explored. By using microwave assisted chemical vapor deposition (MWCVD) technique, Fe catalysts were used to catalytically synthesize carbon nanocone [12], carbon nanohelix [34], carbon nanotip [35, 36], and carbon nanobelt [37]. Aligned CNs arrays were often synthesized in electric field induced by bias in the system [38, 39]. Nowadays, the research of CNs is keeping more on the development of their precise morphology control and their extensive applications.

To date, the upsurge of the research about CNFs, CNTs and other CNs has lasted for two decades. During this period the investigations were not only focused on their synthetic methods and applications, the attention was also devoted to the essence of the CVD growth processes. In the commonly accepted growth

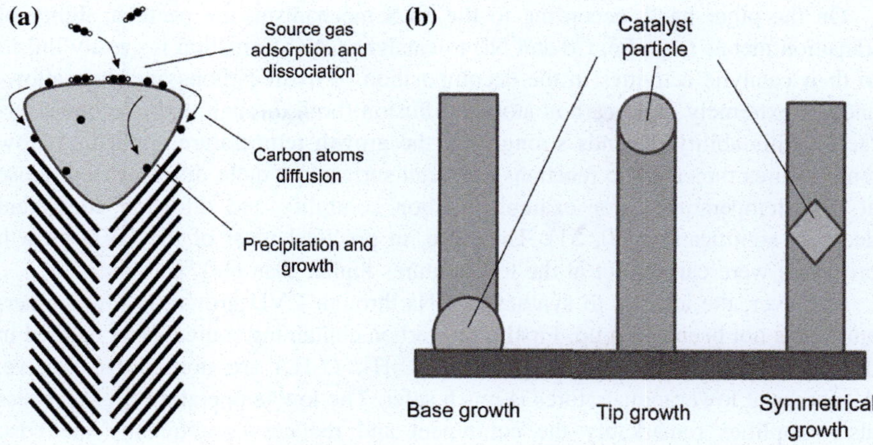

Fig. 1 Sketch of the nanofiber growth: **a** VLS mechanism, **b** three different growth modes

mechanism, carbon dissolves into the catalyst and CNs grow due to the precipitation of excess carbon on the surface of the catalyst (in most cases, it is a metal particle). This idea originates from the vapor-liquid-solid (VLS) mechanism suggested by Wagner et al. to explain the growth of single crystal Si wire [40, 41]. In particular, it is now frequently used to interpret the catalytic CVD growth of 1D CNs. As shown in Fig. 1a, the basic concept is as following: (i) adsorption and dissociation of carbon containing gas on the surface of the catalyst, (ii) diffusion of carbon atoms through the catalyst, and (iii) precipitation of carbon from saturated catalyst particles to form 1D CNs. In this model, the catalyst forms a liquid droplet and preferentially adsorbs the growth species from the surrounding vapor, and the solid carbon wire grows from this supersaturated eutectic liquid.

In the VLS model, the catalyst is the core of the whole reaction system because all important processes progress on or within the catalyst. Any change of the catalyst will be finally reflected on the grown nanostructure. The size of the catalyst (e.g., metal particle) is treated as the determining factor for the diameter of the grown CNs; whilst the crystalline orientation plays a critical role in the determination of CNTs chirality [42, 43]. As for the preparation methods of the catalyst, they influence as well the catalysts' size and their dispersion on the surface of the support and thereby their catalytic properties [44, 45]. For example, the interactions between the metal catalyst and the surface of the support were found to determine the growth mode [14, 46, 47]. Weak interactions yield a tip-growth whereas strong interactions lead to a base-growth. Moreover, distinct activity for hydrocarbon molecule dissociation or carbon atom extrusion on different surfaces of the catalyst [48, 49] is believed to be responsible for a symmetrical growth, as schematically shown in Fig. 1b. These experimental facts suggest that the geometric features of the catalyst play important roles in the growth of CNs. However, in previous studies the effect of the geometric shape of the catalyst on the growth of CNs did not attract enough attention because this concept is beyond the basic framework of the VLS mechanism and certain shaped catalyst is not necessary for a VLS process.

On the other hand, according to the VLS mechanism, the peculiar ability of transition metals (e.g., Fe, Co and Ni) to catalyze CNs' formation is greatly linked to their catalytic activities in the decomposition of hydrocarbons and their allowance of extremely rapid carbon atomic diffusion (both through and over their surfaces). This ability depends strongly on the growth temperature applied. At low temperatures the catalytic reactions on metal surface are quite different from those at high temperatures. For example, carbon solubility and diffusion coefficient decrease significantly [50, 51]. Therefore, in practical most of the CVD growth processes were carried out at the temperatures higher than 600 °C.

However, the attempt to synthesize CNs through CVD growth at low temperatures has not been given up. Firstly, the carbon containing source gases applied in the CVD growth processes (e.g., CH_4, C_2H_2, C_2H_4) are normally explosive. A process at lower temperatures is much safer. The low-temperature CVD process also simplifies remarkably the equipment and reduces consequently the cost. Secondly, the CVD processes are often accompanied by the decomposition of hydrocarbons and coking. The process at lower temperatures mitigates the harmful deposition from the wall of a reaction chamber as well as gas channels. Finally and the most importantly, the growth at low temperatures is of great benefit to a lot of applications when the substrate material cannot withstand a high-temperature process. For example, to apply CNFs in semiconductor industry, the CVD process has to be compatible with the CMOS technology. It means the CVD temperatures must remain below 400–450 °C to avoid mechanical deterioration [52].

Therefore, in this chapter we focus only on the CVD processes which employ temperatures below 450 °C. We summarize recent progress and achievements about the low-temperature CVD growth of 1D CNs, and then discuss the unique growth mechanisms for the growth of these 1D CNs. Some example applications of these 1D CNs are shown as well.

2 Synthesis of 1D Carbon Nanostructure at Low Temperatures

So far, the low-temperature synthesis of CNs has been achieved successfully by applying normal catalytic thermal CVD methods. In a typical thermal CVD process, the growth progresses in a reaction chamber (e.g., quartz tube) which is connected to a gas supplying system. Metal catalyst is pre-placed in the chamber or brought in by a carrier gas during the process. Once the chamber is heated up to reach desired temperature, carbon containing source gas (e.g., C_2H_2, CH_4) is introduced. Consequently CNs grow with the aid of the catalyst. Different temperatures, from 195 to 450 °C, have been used for the growth of various CNs, including CNTs, amorphous CNFs, carbon nanocoils (CNCs), carbon nanohelixs (CNHs), carbon nanosheets (CNSs), and branched CNs. Table 1 lists some representative growth conditions and related CNs achieved.

Table 1 The reported low-temperature thermal CVD processes

Catalyst	Carbon source	Temperature (°C)	Flow	Product	Reference
Ni, Cu–Ni, Co–Ni, Fe–Ni	CH_4	450		CNF	[53]
Ferrocene	C_2H_2 C_8H_{10}	450	Ar, H_2 as carrier	CNT	[54]
Fe_2Co	C_2H_2/CO_2	400		CNT	[55]
Fe–Co–Ni	C_2H_2	400	C_2H_2 10 sccm N_2 100 sccm	MWCNT	[56]
Fe–Co	C_2H_2	400	C_2H_2 50 sccm N_2 250 sccm	CNT	[57]
Ni–Fe	C_2H_2	400		CNT CNF	[58, 59]
Cu–Ni	C_2H_2	400	C_2H_2 3 sccm He 12 sccm	CNF	[60]
Ni	C_2H_2	400		CNT	[61]
Co	Alcohol	400		SWCNT	[52, 62]
Cu	C_2H_2	350	Ar 20 sccm H_2 10 sccm C_2H_2 5/10 sccm NH_3 0/10 sccm	CNF	[63]
Fe	C_2H_2	350		CNF	[64]
Ni_3C	Nickel(II) acetylacetonate	300	H_2 200 sccm	CNF	[65]
$Ni_{0.67}Fe_{0.33}$	C_2H_2	300	C_2H_2 0.5 sccm H_2 50 sccm	CNT	[66]
Cu	C_2H_2	250–350		2-, 3-, 6-, 8-branched CNF	[67]
Cu	C_2H_2	250		CNH CNF	[68, 69]
Cu	C_2H_2	250		CNS	[70]
Cu	C_2H_2	195		CNH	[71]

To get a deep insight into the characteristics as well as the mechanisms of low-temperature growth, we discuss in the following sessions these processes in detail from the aspects of catalyst system, catalyst evolution, the effect of gas composition on the growth as well as the morphology control on the growth products.

2.1 *Types of Catalysts*

It is similar to those conventional high-temperature CVD growths, transition metals such as Fe, Co, Ni and their alloys are often used as the catalysts for the low-temperature growth of CNs. Comparing with pure metals, their alloys show higher catalytic activity. When other experimental parameters remain unchanged, the catalyst from their alloy triggers a growth at lower temperatures, or causes higher growth rates of CNs [56, 66]. However, tri-metallic catalysts do not show any advantage over bimetallic ones [56, 57]. For example, Chiang and Sankaran [66] tested different Fe–Ni alloyed catalysts and gave an optimal composition of approximately 67 % Ni and 33 % Fe.

Cu has been used as the catalyst for the growth of CNs. Notably, in a conventional high-temperature CVD process Cu is not an effective catalyst, because carbon hardly dissolves (diffuses) in Cu [51]. In contrast, in the low-temperature cases Cu is a quite attractive to aid the synthesis of various CNs. For example, by applying a Cu catalyst the growth of CNs was achieved at a temperature as low as 195 °C [71]. It is 100 °C lower than the temperatures for those processes based on Fe, Co, Ni catalysts. Therefore the Cu catalyzed growth of CNs is believed to be a typical case of the low-temperature CVD growth. It might correspond to a unique growth mechanism, different from the VLS mechanism.

Besides the chemical composition of the catalyst, its size is critical for the growth of CNs as well. To effectively catalyze the growth of a nanostructure, the catalysts (e.g., in a form of nanoparticle or nanonetwork) must be well dispersed on a substrate. The size and the dispersion of the catalyst directly decide the dimension and the amount of as-grown CNs. A simple and often used method for the catalyst preparation is a coating-anneal route [57, 61]. Briefly, a metal film with a few to tens of nanometers in thickness is firstly coated on a substrate. In the following annealing the metal film at high temperatures splits and converges the metal film into dispersive nanoparticles. As an example, SEM images of one 20 nm thick Cu film before and after annealing are shown in Fig. 2a and b, respectively. In this way one can use a target with a certain composition during sputter coating process to get the catalyst with a required composition [57, 58]. The size of the catalyst is possible to be adjusted by changing the thicknesses of the metal film and annealing times/temperatures. Other methods are also available for the preparation of these catalysts, for example, sol-gel [67], co-precipitation [55], thermal decomposition [56], arc plasma jet evaporation methods [64], and thermolysis of various complex precursors in hot surfactant solutions [72]. The as-prepared metal catalysts usually have however random near-spherical or polyhedron shape.

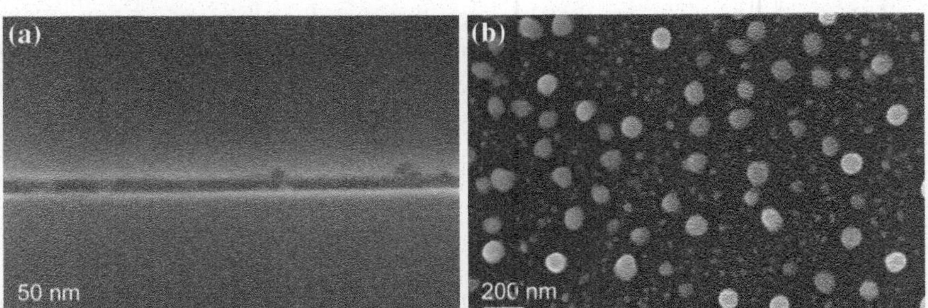

Fig. 2 SEM images of **a** cross section of a 20 nm Cu film coated on Si substrate and **b** dispersive Cu nanoparticles after annealing

The pre-treatment of the catalysts influences their catalytic activities. For example, Shyu et al. reported that by pre-treating the Ni or Fe–Ni catalyst with low-concentrated carbon containing source gas, the growth of CNFs was promoted in the following CVD processes. In contrast, if the catalyst was pre-treated with reductive H_2, the CVD growth was depressed [59]. The reason of such a promotion was believed to be the formation of Ni–C or Fe–Ni–C alloy.

2.2 *Catalyst Faceting*

The growth of CNFs is always accompanied by tremendous shape change of the catalyst. For instance, the Cu catalyst are always faceted with regular shapes after the CNs growth, the majority of which have a rhombic projection. Detailed investigations indicate that the Cu catalyst undergoes a shape change from an irregular to a regular faceted form during the CNs growth. Moreover, their shapes have considerable effects on the final cross-sections of CNs. Based on systemic TEM observations as well as accompanying selected area electron diffraction (SAED) indexes performed by Xia et al., the Cu catalyst (e.g., particles) in as-grown nanostructures can be sorted to three kinds of basic highly symmetric polyhedrons, namely octahedron, triangular prism and tetrahedron, as shown in Fig. 3 [73]. Octahedron catalyst particle is in the left column (Fig. 3a, d, f, i) and triangular prism catalyst in the middle column (Fig. 3b, e, g, j). From upper to bottom row they are the TEM images of a straight nanofiber, the TEM image of a helical nanofiber, the sketch of a catalyst particle, and the projection of a catalyst particle along Cu $\langle 110 \rangle$ zone axis, respectively. In the right column (Fig. 3c, h, k), it shows Cu catalyst with the shape of tetrahedron. From upper to bottom row they are the TEM images of straight nanofiber, the sketch of catalyst particle, and the projection of catalyst particle along Cu $\langle 110 \rangle$ zone axis, respectively. One can see that for these grown nanofibers, regardless of the morphology of the nanofiber and the shape of the catalyst, Cu particles are all encompassed by low-index faces and CNFs always attaches on Cu $\{111\}$ faces. In the TEM image as well as HRTEM images of Fig. 4a, b, one can see the grown carbon product is connected to Cu catalyst with a quite flat interface. When the focus is on the grown carbon product, the HRTEM image shows discontinuous near-parallel fragments, indicating amorphous nature of grown CNfs, as seen in Fig. 4c [73].

The catalyst faceting appears on Fe and Ni based catalyst along with the CNFs growth as well [64, 65]. Figure 5a shows a TEM image of a CNF grown at 350 °C where Fe catalyst is encapsulated in the fiber. The Fe particle has a polygonal projection and features the $\{110\}$ face as the growth front [64]. Figure 5b shows a typical TEM image of a symmetrical CNF grown at 350 °C when a Ni catalyst particle is applied. One can see an approximate rhombus projection in the TEM

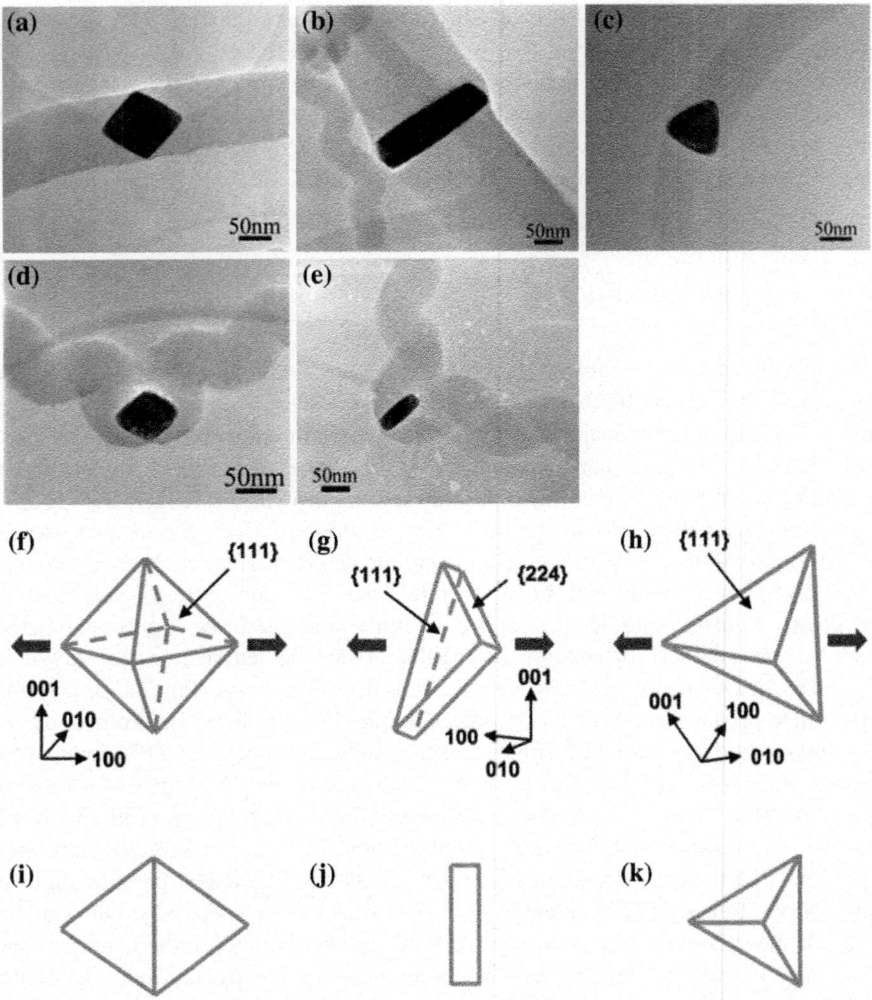

Fig. 3 **a–e** TEM images of different morphologies of nanostructures obtained along Cu ⟨110⟩ zone axis. **f–h** Three kinds of catalyst polyhedrons: octahedron, triangular prism and tetrahedron, respectively. **i–k** The respective projections of the polyhedrons (**f–h**) obtained along Cu ⟨110⟩ zone axis [73]

image. For the Ni_3C assisted growth at a quite low temperature of 300 °C, Yu et al. performed HRTEM observation and stated that the growth of CNFs is initiated from the {001} planes of the Ni_3C particles which exhibit a regular shape [57].

As a result of faceting, the shape of the catalyst evolves until specific stable crystalline faces expose. In particular, the low index faces tend to appear. It indicates that these faces play important roles in the growth of CNFs. Some critical

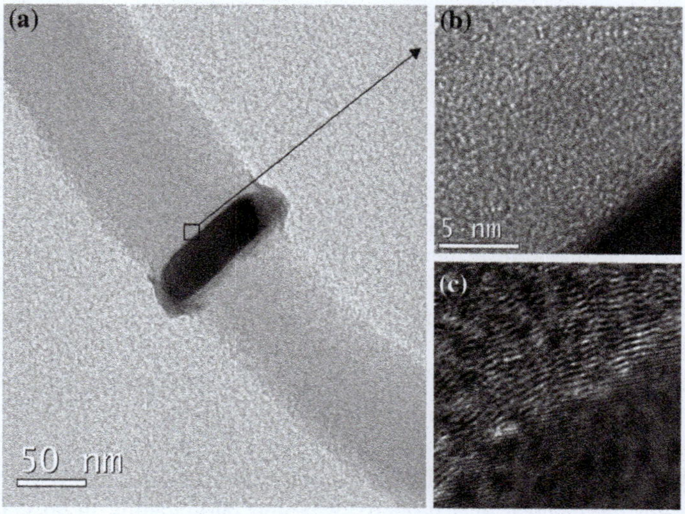

Fig. 4 The enlarged TEM images of a CNF. **a** The TEM image on Cu [112] zone axis. The fiber is grown on the catalyst along Cu ⟨111⟩ crystal direction. **b** HRTEM images of the areas indicated with *red square* in (**a**). It shows a flat interface between grown carbon and the catalyst. **c** HRTEM image focusing on the grown carbon product [73]

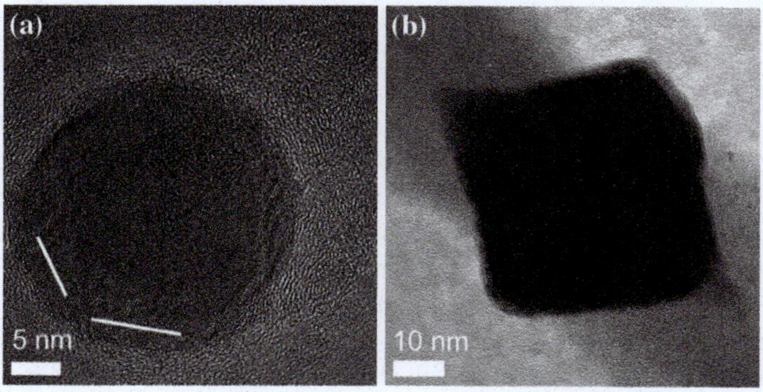

Fig. 5 **a** TEM image of the tip of a CNF. The *dark particle* is Fe catalyst. The *bright lines* mark the (110) plane of the Fe catalyst. **b** TEM image of a symmetrical grown CNF. The *black spot* is Ni catalyst particle

processes such as surface reactions may progress on certain indexed metal planes, e.g., the {111} faces of a Cu particle. Unfortunately, so far relatively limited attention has been paid to the effect of the geometric features of the catalyst on the low-temperature CVD growth of CNs.

2.3 Gas Composition

In the CVD growth of CNs, the carbon feedstock relies on carbon containing source gases. For the conventional high-temperature growth, many carbon containing gases are applicable, such as CH_4, C_2H_2, C_2H_4, CO, benzene etc. [74–78]. On the contrary, only a few carbon sources have been successfully applied in the low-temperature CVD growth up to now. As summarized in Table 1, most of the successful processes were achieved with C_2H_2. Some researchers replaced C_2H_2 with other carbon sources (e.g., C_2H_4, CH_4, CO) as controlled experiments, but only negative results were obtained [58, 71], revealing a high selectivity of carbon sources for the low-temperature CVD growth of CNs. Two hypotheses might be workable for the high activity of C_2H_2. Firstly, at low temperatures, the cleavage of C_2H_2 to generate atomic carbon is easier, in comparison with other carbon containing molecules [79]. Therefore the growth with C_2H_2 can happen at lower temperatures. Another possibility is the unique structure of C_2H_2 allows it adsorbing on the catalyst in a property way and participating in special catalytic surface reactions.

Besides carbon sources, unreactive gases have been also introduced on purpose during the growth process, such as N_2, Ar, He and sometimes H_2. Although they do not contribute to the construction of CNs directly, their roles are somehow critical. For example, in some experimental settings they work as the carrier gas to bring the catalyst into the reaction chamber [54]. Moreover, except few special processes which require ultra-high vacuum (UHV) condition [52, 62], most of the low-temperature CVD growth can progress in a broad pressure range (even in atmospheric pressure [64, 67, 71]). Therefore, the growth is often accompanied with cocking and the generation of a large amount of organic byproducts. The unreactive gas dilutes the reactive atmosphere [56, 57, 60] and then prevent or alleviate the harmful deposition on the wall of the reaction chamber and gas channels. Meanwhile, the reaction byproducts can be taken away if the unreactive gas passes the reaction chamber as a constant flow.

It has been reported that a little amount of water ($\sim$100 ppm) in the reaction atmosphere promoted the CNT growth in a high-temperature CVD process [80, 81]. Based on this idea, a method so-called water assisted CVD was developed [82, 83]. The truth of the promotion is that the mild oxidizer such as water helps to prevent amorphous carbon deposition on the surface of the catalyst [80, 81], thus the poison of the catalyst is efficiently avoided. This method was further extended to applying other oxygen containing gas, for example O_2 [84] and CO_2 [85, 86]. Researches confirmed that NH_3 has the ability to etch amorphous carbon [78, 87, 88]. However, for the low-temperature CVD growth using additional etching gases seems to be unnecessary. For most of reported low-temperature processes, only carbon containing source gases and unreactive gases were applied. It implies that amorphous carbon does not become an obstacle to the growth of CNs at low temperatures. A possible explanation is that at low temperatures the diffusion of carbon atoms and related molecules is mainly through the surface of catalyst, therefore the surface of the catalyst is always scoured by moving reaction products. In other words, the surface of the catalyst is always fresh and active. Recently we found that some gases (e.g., HCl) depressed the growth of CNs at low temperatures, as they eliminate specific active faces on the catalyst. For example, when a little amount of HCl is introduced into the reaction atmosphere of a Cu assisted CVD process, the Cu {110} face enlarges while the Cu{111} face shrinks. It is due to the strong interaction between HCl and the Cu{110} face [89, 90]. As the CNF only grows on the Cu{111} face, consequently the as-grown nanofiber becomes thinner and the total amount of growth product reduces [67].

2.4 *Structural Characteristic of the Carbon Nanostructure Synthesized at Low-Temperature*

Most of CNs synthesized using the low-temperature CVD processes have amorphous natures. From their HTREM images one can see that these CNs feature usually discontinuous fragments rather than integral graphitic sheets [64, 65, 67, 73]. The example TEM images are shown in Fig. 4. Consequently, their Raman spectra often show a D peak with a high intensity, indicating their poor crystallinity, as seen in Fig. 6. It should be pointed out that although a lot of literatures involving the low-temperature CVD growth used ‘CNT’ to define their fiber-like products, they did not give clear evidence (e.g., TEM) to prove these claimed structures. Therefore it is doubtable to classify those products into CNT in standard sense, which normally indicates a highly ordered graphic tubular nanostructure.

In a systematic investigation, we performed CVD growths under atmospheric pressure of C_2H_2 with the assistance of Fe, Co, Ni, and Cu nanopowder catalysts, respectively. The process temperatures were set between 250 and 350 °C. Regardless of the catalyst, all the synthesized CNs contain considerable hydrogen. Normally,

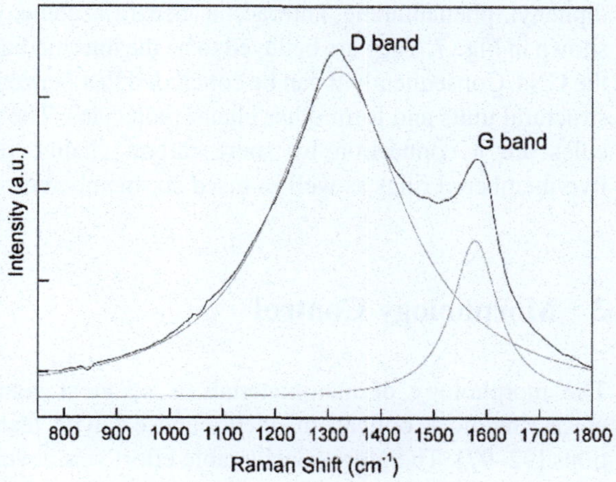

Fig. 6 Raman spectrum of a CNF which was synthesized via a CVD process at 350 °C. Ni and C_2H_2 were used as the catalyst and carbon source, respectively

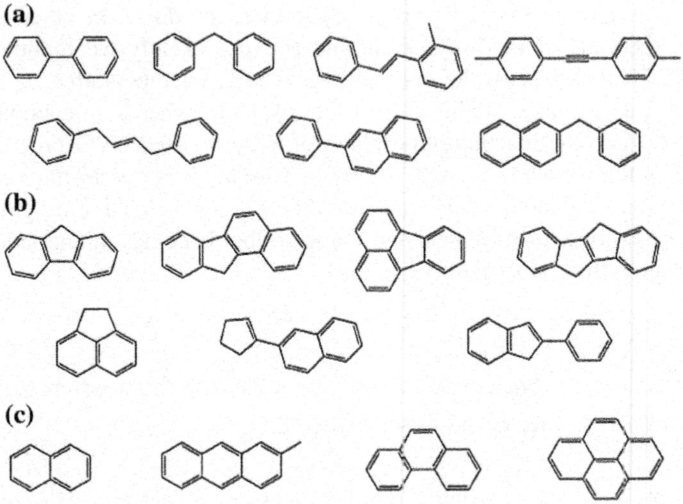

Fig. 7 Molecular fragments presented in the as-grown CNFs, which were identified from the TD-GC/MS method [64, 91]. **a** connected by short carbon chain, **b** connected by five-membered ring, **c** form fused six-membered ring

the lower the growth temperature is, the higher percentage of hydrogen is contained in as-grown CNs [71]. In a typical case, the Cu assisted growth progressing at 250 °C, the product contains about haft hydrogen in atomic percentage. Therefore the CNs synthesized via low-temperature CVD processes shows more or less polymer natures. Furthermore, by means of thermal desorption-gas chromatography/mass spectrometry (TD-GC/MS) technique, more detailed structural characteristics of synthesized CNs were revealed. The synthesized CNs were firstly heated up to release fragments (thermal desorption), and then these evaporated species were injected into GC for separation and final identification by MS. No matter the catalyst is Fe, Co, Ni or Cu, the measurements always reveal a series of planar benzene derivatives, such as diphenyl, phenanthrene, anthracene, benz(a)azulene, 9-Methylene-fluorene, etc., as shown in Fig. 7. They are believed to be the intermediate products or the fragments of the CNs. Consequently it can be concluded that benzene rings act as the fundamental structural units and form other planar molecules. The paths to construct other molecules are a connection by short carbon chains, resulting in the formation of five-membered rings as well as fused six-membered rings [64].

3 Morphology Control

The morphology of nanomaterials is of great significance for their electronic, electrochemical, optical, magnetic, and catalytic features as well as their applications [92–97]. Therefore considerable effort was devoted to the morphology control during synthesizing nanostructures [98–100]. In terms of CNs, by applying high-temperature CVD processes, various CNs have been synthesized, including CNTs [101], nanocones [12], Y-junctions [102, 103], nanocoils [104], fullerenes [105], etc. They are actually quite useful since specific CNs are preferred for some special applications [106–108]. However, comparing with the results achieved on metal nanocrystals [93, 95, 97], metal oxides, and semiconductors [92, 98], the morphology control of CNs is far less precise. Actually if a fact that highly ordered crystalline lattice does not exist inside CNs exists, it is easy to conclude that most shape-control methods developed for nanocrystals are not applicable over CNs. Fortunately, the low-temperature CVD growth grants more chances to deal with the precise shape-control of CNs.

Xia et al. [69] investigated Cu assisted low-temperature CVD growth and reported a relationship between the size of the catalyst and the morphology of as-grown CNFs, namely a size effect. They suggested there is a critical size of the catalyst to divide the growth of two kinds of CNFs, as shown in Fig. 8. Straight CNFs are formed with larger catalysts while helical CNFs are formed with smaller ones. The critical value is ~70 nm for the octahedron shaped catalysts and ~110 nm for the triangular prism ones [69]. Therefore it is highly possible to synthesize CNFs with desired morphology by use of certain sized and shaped catalysts.

We further developed a method to control the growth and get CNs with more complex shapes, so-named multi-branched CNs. The core idea of this method is to tailor the shape of the catalyst which decides the morphology of grown CNs [13, 109, 110]. It is notable that metallic catalysts are intrinsically symmetric and always tend to expose favorable low index facets in thermodynamic equilibrium status [111]. In this context a precise shape control of CNs can be achieved via intentionally tailoring the geometric features of the catalyst and exposing its specific catalytic-active facets.

Figure 9 shows 4 different branched CNs [67]. By means of carrying out the CVD growth with assistance of Cu catalyst at 250, 275 and 350 °C, we obtained 2-, 3-, and 6-branched CNs respectively. In these nanostructures, Cu nanocrystal works as a node and carbon branches grow on {111} or {001} faces of Cu nanoparticle, depending on the process temperatures. In general, the outer surface of Cu

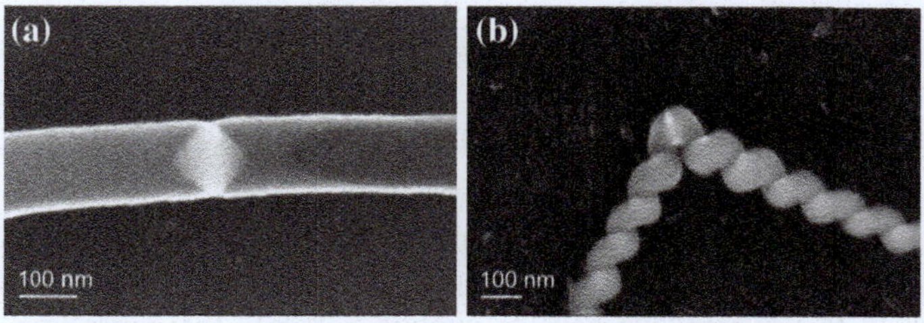

Fig. 8 SEM image of two kinds of CNFs synthesized at 250 °C: **a** a straight nanofiber, **b** a helix nanofiber

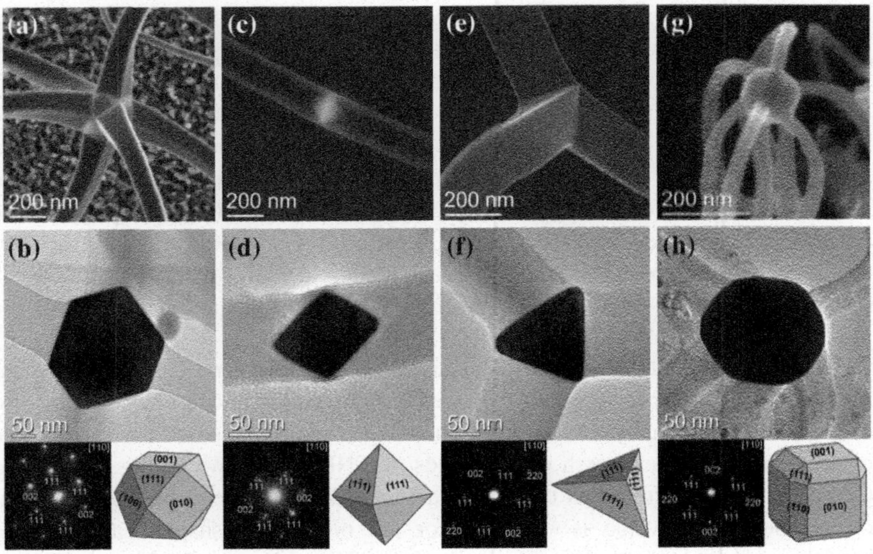

Fig. 9 **a** SEM image and **b** TEM image, electron diffraction pattern, catalyst particle sketch of a 6-branched CN prepared at 350 °C. **c** SEM image and **d** TEM image, electron diffraction pattern, catalyst particle sketch of a 2-branched CN prepared at 250 °C. **e** SEM image and **f** TEM image, electron diffraction pattern, catalyst particle sketch of a 3-branched CN prepared at 275 °C. **g** SEM image and **h** TEM image, electron diffraction pattern, catalyst particle sketch of a 8-branched CN prepared at 250 °C with additional HCl in the reaction atmosphere [67]

nanoparticle is a combination of {001}, {110} and {111} facets [111]. Corresponding to different process temperatures, the Cu particles show unique equilibrium shapes and faceted surface with specific index distribution. The facet distribution was further tailored via adding a little amount of HCl in the CVD chamber. It changed the ratios of {001}, {110} and {111} faces, resulting in the growth of 8-branched CNs. In terms of the carbon branches in these CNs, their numbers and stretching directions follow the intrinsic symmetry of Cu nanocrystals. Their lengths depend on the growth duration. Their diameters correlate to both the size of Cu particles and the distribution of catalysis-active facets on Cu particles. All these features are tailorable via changing experimental parameters.

4 Mechanism of Low-Temperature Thermal CVD Growth

With respect to the mechanism of the low-temperature CVD growth of CNs, the core issue is the approach to convert carbon containing gas into solid carbon structures. In this context we have to explain not only the generation of carbon feedstock but also the route for carbon transportation, for example carbon diffusion through/on the catalyst. For high-temperature CVD growth, the classical VLS model has given appropriate explanation on above issues, see Sect. 1. Unfortunately, the VLS mechanism cannot be used to interpret the low-temperature cases, if the fact that the carbon solubility and diffusion coefficient decrease significantly with the reduction of the growth temperature is noticed [50, 112, 113]. In particular, for the typical Cu assisted growth which produces polymer-like CNs, it completely excludes the possibility of VLS approach, because neither carbon nor hydrocarbon molecules dissolves (diffuses) in the Cu catalyst [51]. Therefore a new model is necessary to interpret the mechanism of the low-temperature thermal CVD growth of CNs on these metal catalysts.

4.1 Fundamental Reactions

Since at lower temperatures a direct dissociation of hydrocarbons is more difficult, various reaction paths have been proposed to interpret carbon feedstock during the CVD growth. For example, Magrez et al. [55] proposed an oxidative dehydrogenation reaction to explain the growth of CNTs below 500 °C

$$C_2H_2 + CO_2 \rightarrow 2C + H_2O + CO$$

This approach is however only applicable for those processes with additional CO_2 introduction into the reaction atmosphere. For most CVD processes which progress in pure C_2H_2, the reactions should follow fully different routes.

Fortunately, the interaction of hydrocarbons such as acetylene with the surface of low-index metal single-crystals under ultra-high vacuum (UHV) conditions has been investigated for a long time. These efforts contribute well to a fundamental understanding of important catalytic surface processes, unravel fundamental mechanisms in heterogeneous catalysis, and identify important surface intermediates in practical reactions. Taking the Cu/C_2H_2 system as an example, the adsorption and surface reactions of acetylene on Cu low-index surfaces ({110}, {001}, and {111}) under UHV conditions have been widely investigated both theoretically and experimentally [114–118]. The coupling of acetylene on Cu surface has been demonstrated, which forms benzene (C_6H_6) on Cu{001} [119], Cu {110} [120], Cu{111} facets [121] and cyclooctatetraene (COT, C_8H_8) on Cu {111} facet [121]. In these investigations, no coverage threshold for the onset of the cyclization reactions was observed on any Cu facets, implying high adsorbate mobility [119–121]. It was also confirmed that the reaction pathways of Cu catalytic reactions in atmospheric pressure of acetylene were in good agreement with the model under UHV conditions [71, 121]. Another aspect that should be taken into consideration is the fact that a graphene sheet matches a Cu{111} facet very well (both are in D_{6h} symmetry and the mismatch of lattice parameter is about 2 %), which can dramatically increase the interactions between a Cu{111} facet and a graphene sheet as well as its derivatives. Therefore, the fundamental reaction in the

Cu assisted CVD growth is proposed to be the trimerization of C_2H_2 on the low-index Cu surface. It was achieved in following path [71, 119–121]

$$C_2H_2 + C_2H_2 \rightarrow C_4H_4$$
$$C_4H_4 + C_2H_2 \rightarrow C_6H_6$$

Based on these reaction pathways various planar benzene derivatives (oligomers) are formed. The oligomers are further polymerized to construct larger planar fragments as well as CNs eventually.

In terms of other catalyst system, above approach is also workable. Previous research carried out under UHV conditions showed that the adsorption of C_2H_2 on a Fe surface induces the formation of various hydrocarbons such as methane, ethane, butane and benzene, but no C3 and C5 products. It was always assumed that the former were created via hydrogen abstraction, C–C bond breaking, dimerization, and trimerization of C2 entities [122]. The literatures have shown that that adsorbed C_2H_2 can dehydrogenize or couple on Ni surface to form a series of hydrocarbon molecules [123–126]. Among these surface reactions, oligomerization [127, 128], e.g., the cyclic trimerization of C_2H_2 on Ni(111) face to from benzene [124], and following polymerization which generates polymer-like products [129] are notable. They can provide fundamental structural units for the construction of CNs.

4.2 Feedstock Transportation/Diffusion Route

For the growth a nanostructure such as a nanofiber, a stable transportation path between reaction products' generation and precipitation/extrusion has to be established on the catalyst. Otherwise the reaction products will simply accumulate on the surface of the catalyst and prevent any further surface reactions as well as the further growth. Therefore the feedstock transportation on the catalyst is a critical issue.

Based on the experimental data of bi- and tri-metallic catalysts assisted low-temperature thermal CVD growth, Pitkänen et al. performed simulation and got very low activation energies for the CVD growth of CNs. It was suggested that the carbon diffusion is the rate-limiting step rather than the hydrocarbon cracking [57]. Yazyev et al. calculated a barrier of 0.39 eV for the diffusion of adsorbed carbon atoms on the Ni(111) surface, which agreed with the activation energy obtained from the experimental CNT growth process. Therefore surface diffusion is the rate-limiting step for low-temperature CVD growth of CNs [130–132].

Certain diffusion mechanisms on base of carbon diffusion on the surface of the catalyst have been thus proposed for CVD growth of 1D CNs [133–139], especially for the low-temperature CVD growth process, which is a low energy path for the growth of CNTs [130]. The surface diffusion channel is also believed to be the reason of increased catalytic activity of nanoparticle alloys at low temperatures [66]. Meanwhile, the calculation of adsorbed atomic carbon diffusion on Cu surfaces featured surprisingly low barriers of 0.07 eV. The energy barrier of atomic carbon diffusion on Cu surfaces is also lower than that on Ni [131]. Thus the relative low-temperature Cu assisted CVD growth of CNs is reasonable, as listed in Table 1.

4.3 VFS Mechanism

Based on the characteristics of low-temperature thermal CVD growth of CNs, we describe a 3-step mechanism to interpret the polymers/CNFs growth on the surface of the metal catalyst, as schematically shown in Fig. 10.

Here we still take the Cu/C_2H_2 system as an example. In first step, at a typical growth temperature of 250 °C which is much lower than the melting point, Cu/CuO particles remain in a solid phase. They initiate from irregular shapes. Along with the reduction process on the Cu surface, the acetylene molecules adsorb on the Cu surface and start to couple and polymerize with each other as well as induce Cu surface to reconstruct. The interaction and reaction route of acetylene molecular with Cu facets differ according to the index of facets. On Cu{001} and Cu{110} facets, the coupling reaction products, oligomers such as benzene and its derivatives, have high mobility; while on Cu{111} facet, in addition to benzene formation, small amount of COT forms and it is immobilized at the surface of the Cu catalyst when the temperature is lower than approximately 300 °C [121]. Therefore, at a relatively low reaction temperature like 250 °C, eventually the Cu{111} facet is blocked by COT and its derivatives because of the strong interaction between the Cu surface and the compound. It leads to two consequences, a terminated coupling reaction and stabilized Cu{111} facets. The reaction activity of the blocked facets dramatically drops, so that the coupling reaction is terminated. As a result, a concentration gradient of movable oligomers and the further derivatives between Cu{001}, Cu{110} facets and Cu{111} facets establishes since reaction products are accumulated unceasingly on Cu{001} and Cu{110} facets whereas the surface reaction terminates on Cu{111} facets. In the second step, the concentration gradient drives the movable oligomers and derivatives to diffuse from Cu{001} and Cu {110} facets to Cu{111} facets through the Cu surface, thus excess oligomers and derivatives precipitate on Cu{111} facets to build up CNs in the third step.

This model could be described as an adsorption (Vapor)—diffusion (Facet)—precipitation (Solid) process and then we name it as a VFS mechanism to distinguish from the classical VLS mechanism and vapor-solid-solid (VSS) mechanism [140], since a surface diffusion process of hydrocarbon dominates the whole process instead of bulk/surface diffusion of carbon atoms. This mechanism features: (i) Competition among different catalyst facets is the core factor (see Fig. 10). The facets on which C_2H_2 molecular is absorbed and coupled are defined as adsorbent facets, while the facets receiving coupling products and finally building up the CNs are defined as the growth facets. The adsorbent facets and the growth facets may reverse, depending on surface reactions and surface energies of the catalysts, which

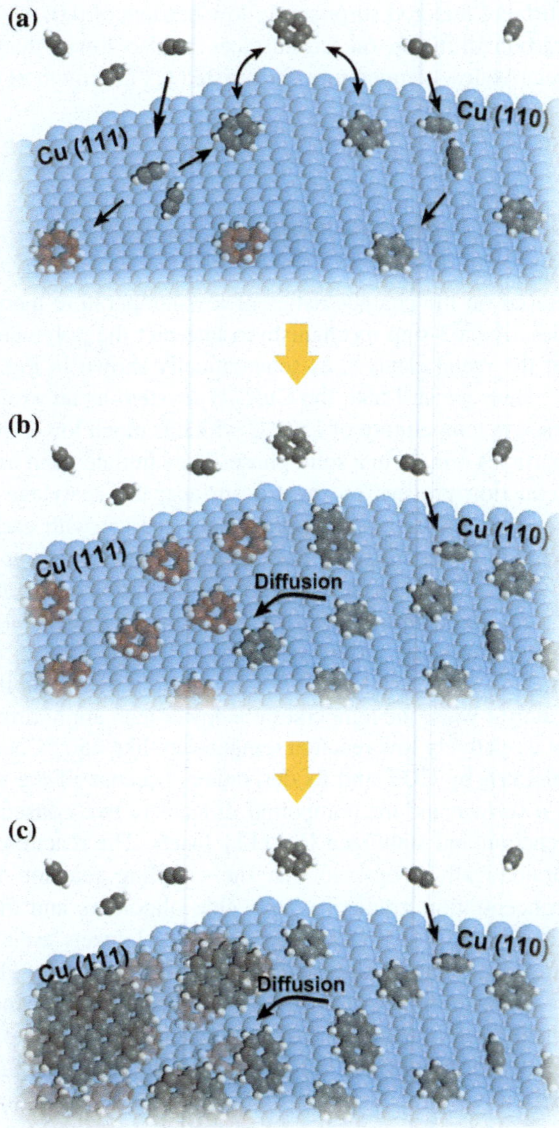

Fig. 10 Sketch of the VFS growth mechanism: **a** C_2H_2 adsorbs on Cu surface, coupling to form different oligomers, movable ones on Cu{110} facets and immobilized ones on Cu {111} facets. **b** The immobilized oligomers block Cu{111} facets, and the movable oligomers diffuse from Cu{110} to Cu{111} facets. **c** Movable oligomers and further derivatives creep to overlay those immobilized oligomers and form nano-sheets during the diffusion [67]

are actually influenced by reaction temperatures and gaseous environment; (ii) The loose stacking nanostructure of polymer sheets allows the new build-up units to insert themselves into the gaps between existing nanofibers and the copper facets, and consequently to lift the old nanofibers up a bit to become a new part of the existing fiber. It presents that the existing nanofibers do not act a barrier blocking further the growth of themselves. (iii) Catalytic nanocrystals have undergone a surface reconstruction process, and turned into a regular appearance with certain low indexes and smooth facets, which correspond to its intrinsic symmetry.

4.4 The Applicability of VFS Mechanism

As the typical case of the low-temperature CVD growth of CNs, Cu catalyzed reaction is the most suitable one for the fundamental modeling research. Based on the model extracted from Cu/C_2H_2 reaction system, the growth with other transition metal catalysts can be interpreted. For example, in a growth which is based on Fe/C_2H_2 system and progressed at 350 °C, the Fe nanoparticle catalyst always exposes {110} face during growth. C_2H_2 is adsorbed on the active {110} face to form oligomers, and then the oligomers diffuse towards {001} face which is blocked by carbon and thus loses catalytic activity [64]. This process finally initiates the formation of CNs. Figure 11 shows related TEM images of the grown CNFs and schematically a possible growth route. Regardless of the detailed surface reactions and diffusion routes, the transportation and precipitation of aromatic species on the surface of the catalyst induce the growth of CNs. The surface processes can be generalized as a competition and cooperation between different catalytic crystalline faces. All these essential commons between the processes based on different catalysts illustrate an extensive adoptability of the proposed VFS mechanism for the explanation of low-temperature CVD growth of CNs on the metal catalyst.

With respect to the explanation of various morphologies of CNs, the VFS mechanism can be employed as well, since they only depend on the size and geometric features of the catalyst. For the highly symmetrical catalyst such as Cu nanocrystal, it causes symmetrically straight or helical nanofiber. It is also responsible for branched nanostructures when the Cu nanocrystal shows a more complex faceted shape. In cases that catalyst features a low symmetry, it results in a tip-grown nanofiber. The nanofiber can be solid or tubular. However, such a tubular product prepared at low temperatures is quite different form the CNT synthesized at high temperatures. The former has actually amorphous natures instead of graphitic walls.

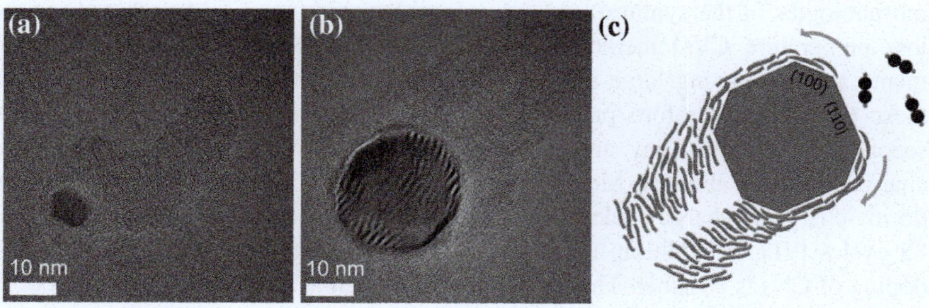

Fig. 11 TEM images of CNFs synthesized at 350 °C using Fe catalyst and C_2H_2 carbon source: **a** a solid nanofiber; **b** a tubular nanofiber; **c** sketch of the growth mechanism [64]

5 Enhanced Low-Temperature CVD Growth

Referred to the low-temperature CVD growth, it's necessary to mention a group of processes which are progressing with some enhancement factors, for example, plasma enhanced chemical vapor deposition (PECVD). By introducing plasma into the CVD growth, the process temperature was reduced remarkably [141–143], and then CNFs were even grown at room temperature [144]. However, the reported growth temperatures have been questioned due to the argument on temperature measurement [145]. In addition to the application of plasma, some other experimental setting also achieved enhancement effect on the CVD growth of CNs. For example, with the assistant of hot filament, CNFs were successfully synthesized at low temperatures [146–148]. The enhanced CVD growth concerns a lot of extra issues such as the complex plasma environment, the non-uniform heating, etc.

6 Applications

In terms of the one-dimensional carbon nanostructures synthesized via the low-temperature CVD method, their special morphologies, unusual atomic architectures as well as the properties associated with the unique growth mechanism grant them extensive applications. For example, Na et al. [61] synthesized densely packed CNT arrays on the conductive substrate at 400 °C. They were used in large-scale integrated circuits as interconnects to carry high density electric current [149]. The low process-temperature is compatible with other materials which are applied in semiconductor industry. Similar vertically aligned CNTs were also applied for field emission [58]. Qin et al. prepared pure carbon nanocoils (CNCs), and then coated these CNCs with Fe, Ni and Al_2O_3 with aid of atomic layer deposition (ALD) technique to synthesize Fe_3O_4/Al_2O_3/CNC and Ni/Al_2O_3/CNC coaxial multilayer nanostructures. The prepared magnetic CNCs showed enhanced microwave absorption properties [108]. This is attributed to the highly controllable morphologies of the synthesized CNCs. Furthermore since CNFs prepared via the low-temperature CVD method often feature disordered stacking graphene fragments, namely having large interlayer distances and abundant diffusion channels, these CNFs are therefore promising to be utilized as the anode of lithium-ion secondary batteries. They are supposed to bring high capacities and super-fast charge/discharge abilities. Meanwhile, they overcome the structural fragmentation during charge/discharge cycles. For example, no deterioration was found even after 85 cycles [91]. In addition, using such a low-temperature growth process in-situ doping of CNs is possible. The doped elements, such as nitrogen, silicon and boron will change the morphologies as well as the band structures of CNs. For instance, when the carbon anode in Li-ion secondary battery is doped by Si, the capacity was doubled, or even more. Doping nitrogen into CNTs not only improved the performance in supercapacitors [150, 151], but also resulted in a new application, namely metal-free catalyst, which can be used in fuel cells to replace expensive Pt and Pd catalysts [152].

7 Summary

The growth of one dimensional carbon nanostructure is possible at relative low temperatures (<450 °C). Nanoparticles from transition metal such as Fe, Co, Ni, their alloys, and Cu are the suitable catalysts for these low-temperature growths. The low-temperature thermal CVD process tends to produce amorphous carbon nanostructures rather than graphitic ones. Considerable hydrogen is often contained in the as-grown carbon nanostructures. Therefore these carbon nanostructures are polymer sheets. By means of tailoring the size and geometric features of the catalyst, precise morphology control of grown carbon nanostructure is achievable.

The classical VLS mechanism is not applicable to the low-temperature CVD process. An alternative model namely VFS mechanism has been proposed. It is based on the coupling of hydrocarbons (e.g., acetylene on metal surface) and subsequent surface diffusion of the coupling products. The core process in this mechanism is generalized as a competition and cooperation between different catalyst crystalline faces. This process results in the generation and transportation of carbon feedstock for the construction of carbon nanostructure. The rate-limiting step of the low-temperature CVD growth is the diffusion of carbon species through a pathway with a low activation energy on the surface of the catalyst, instead of the dissociation of carbon source gas. The fundamental structural unit, which diffuses on the surface of the catalyst and finally precipitates to construct carbon nanostructures, is most likely formed through catalytic reactions occurring on the surface of the metal catalyst.

The one-dimensional carbon nanostructures synthesized via the low-temperature CVD method have controllable special morphology, unusual atomic architecture as well as unique properties associated to the distinct VFS growth mechanism. These nanomaterials have the features of carbon materials as well as the unique properties of nanostructures. They are therefore promising as the absorbent for hazard materials in environmental clean-up, as the electrode for supercapacitors and lithium batteries, as well as the metal-free catalyst for different catalytic reactions.

Acknowledgements The authors would like to acknowledge financial support of this work by the German Research Foundation (DFG JI22/16-1, DFG JI22/21-1).

References

1. S. Iijima, Helical microtubules of graphitic carbon. Nature **354**(6348), 56–58 (1991). doi:10.1038/354056a0
2. J. Hone, M.C. Llaguno, N.M. Nemes, A.T. Johnson, J.E. Fischer, D.A. Walters, M. J. Casavant, J. Schmidt, R.E. Smalley, Electrical and thermal transport properties of

magnetically aligned single walt carbon nanotube films. Appl. Phys. Lett. **77**(5), 666–668 (2000). doi:10.1063/1.127079
3. M.F. Yu, B.S. Files, S. Arepalli, R.S. Ruoff, Tensile loading of ropes of single wall carbon nanotubes and their mechanical properties. Phys. Rev. Lett. **84**(24), 5552–5555 (2000). doi:10.1103/PhysRevLett.84.5552
4. H. Kataura, Y. Kumazawa, Y. Maniwa, I. Umezu, S. Suzuki, Y. Ohtsuka, Y. Achiba, Optical properties of single-wall carbon nanotubes. Synth. Met. **103**(1–3), 2555–2558 (1999). doi:10.1016/s0379-6779(98)00278-1
5. J.L. Hutchison, N.A. Kiselev, E.P. Krinichnaya, A.V. Krestinin, R.O. Loutfy, A.P. Morawsky, V.E. Muradyan, E.D. Obraztsova, J. Sloan, S.V. Terekhov, D.N. Zakharov, Double-walled carbon nanotubes fabricated by a hydrogen arc discharge method. Carbon **39**(5), 761–770 (2001). doi:10.1016/s0008-6223(00)00187-1
6. E.G. Gamaly, T.W. Ebbesen, Mechanism of carbon nanotube formation in the arc-discharge. Phys. Rev. B. **52**(3), 2083–2089 (1995). doi:10.1103/PhysRevB.52.2083
7. C.D. Scott, S. Arepalli, P. Nikolaev, R.E. Smalley, Growth mechanisms for single-wall carbon nanotubes in a laser-ablation process. Appl. Phys. A Mater. Sci. Process. **72**(5), 573–580 (2001). doi:10.1007/s003390100761
8. Y. Zhang, S. Iijima, Formation of single-wall carbon nanotubes by laser ablation of fullerenes at low temperature. Appl. Phys. Lett. **75**(20), 3087–3089 (1999). doi:10.1063/1.125239
9. A. Tanaka, S.H. Yoon, I. Mochida, Formation of fine Fe–Ni particles for the non-supported catalytic synthesis of uniform carbon nanofibers. Carbon **42**(7), 1291–1298 (2004). doi:10.1016/j.carbon.2004.01.029
10. J.H. Xia, X. Jiang, C.L. Jia, C. Dong, Hexahedral nanocementites catalyzing the growth of carbon nanohelices. Appl. Phys. Lett. **92**(6), 063121 (2008). doi:10.1063/1.2842410
11. S. Motojima, Q.Q. Chen, Three-dimensional growth mechanism of cosmo-mimetic carbon microcoils obtained by chemical vapor deposition. J. Appl. Phys. **85**(7), 3919–3921 (1999). doi:10.1063/1.369765
12. G.Y. Zhang, X. Jiang, E.G. Wang, Tubular graphite cones. Science **300**(5618), 472–474 (2003). doi:10.1126/science.1082264
13. X.S. Qi, W. Zhong, Y. Deng, C.T. Au, Y.W. Du, Characterization and magnetic properties of helical carbon nanotubes and carbon nanobelts synthesized in acetylene decomposition over Fe–Cu nanoparticles at 450 °C. J. Phys. Chem. C **113**(36), 15934–15940 (2009). doi:10.1021/jp905387v
14. A.M. Cassell, J.A. Raymakers, J. Kong, H.J. Dai, Large scale CVD synthesis of single-walled carbon nanotubes. J. Phys. Chem. B **103**(31), 6484–6492 (1999). doi:10.1021/jp990957s
15. E. Couteau, K. Hernadi, J.W. Seo, L. Thien-Nga, C. Miko, R. Gaal, L. Forro, CVD synthesis of high-purity multiwalled carbon nanotubes using $CaCO_3$ catalyst support for large-scale production. Chem. Phys. Lett. **378**(1–2), 9–17 (2003). doi:10.1016/s0009-2614(03)01218-1
16. Y.A. Kim, T. Hayashi, S. Naokawa, T. Yanaisawa, M. Endo, Comparative study of herringbone and stacked-cup carbon nanofibers. Carbon **43**(14), 3005–3008 (2005). doi:10.1016/j.carbon.2005.06.037
17. Y.A. Zhu, Z.J. Sui, T.J. Zhao, Y.C. Dai, Z.M. Cheng, W.K. Yuan, Modeling of fishbone-type carbon nanofibers: a theoretical study. Carbon **43**(8), 1694–1699 (2005). doi:10.1016/j.carbon.2005.02.011
18. A. de Lucas, P.B. Garcia, A. Garrido, A. Romero, J.L. Valverde, Catalytic synthesis of carbon nanofibers with different graphene plane alignments using Ni deposited on iron pillared clays. Appl. Catal. A Gen. **301**(1), 123–132 (2006). doi:10.1016/j.apcata.2005.11.026
19. H. Cui, O. Zhou, B.R. Stoner, Deposition of aligned bamboo-like carbon nanotubes via microwave plasma enhanced chemical vapor deposition. J. Appl. Phys. **88**(10), 6072–6074 (2000). doi:10.1063/1.1320024
20. C.J. Lee, J. Park, Growth model of bamboo-shaped carbon nanotubes by thermal chemical vapor deposition. Appl. Phys. Lett. **77**(21), 3397–3399 (2000). doi:10.1063/1.1320851

21. C.J. Lee, J.H. Park, J. Park, Synthesis of bamboo-shaped multiwalled carbon nanotubes using thermal chemical vapor deposition. Chem. Phys. Lett. **323**(5–6), 560–565 (2000). doi:10.1016/s0009-2614(00)00548-0
22. M. Lin, J.P.Y. Tan, C. Boothroyd, K.P. Loh, E.S. Tok, Y.L. Foo, Dynamical observation of bamboo-like carbon nanotube growth. Nano Lett. **7**(8), 2234–2238 (2007). doi:10.1021/nl070681x
23. J.P. Tu, L.P. Zhu, K. Hou, S.Y. Guo, Synthesis and frictional properties of array film of amorphous carbon nanofibers on anodic aluminum oxide. Carbon **41**(6), 1257–1263 (2003). doi:10.1016/s0008-6223(03)00047-2
24. Z.W. Pan, S.S. Xie, B.H. Chang, L.F. Sun, W.Y. Zhou, G. Wang, Direct growth of aligned open carbon nanotubes by chemical vapor deposition. Chem. Phys. Lett. **299**(1), 97–102 (1999). doi:10.1016/s0009-2614(98)01240-8
25. M.J. de Andrede, M.D. Lima, C.P. Bergmann, G.D. Ramminger, N.M. Balzaretti, T.M.H. Costa, M.R. Gallas, Carbon nanotube/silica composites obtained by sol-gel and high-pressure techniques. Nanotechnology **19**(26), 265607 (2008). doi:10.1088/0957-4484/19/26/265607
26. R.R. Bacsa, C. Laurent, A. Peigney, W.S. Bacsa, T. Vaugien, A. Rousset, High specific surface area carbon nanotubes from catalytic chemical vapor deposition process. Chem. Phys. Lett. **323**(5–6), 566–571 (2000). doi:10.1016/s0009-2614(00)00558-3
27. J.P. Pinheiro, M.C. Schouler, P. Gadelle, Nanotubes and nanofilaments from carbon monoxide disproportionation over Co/MgO catalysts I. Growth versus catalyst state. Carbon **41**(15), 2949–2959 (2003). doi:10.1016/s0008-6223(03)00410-x
28. Y.M. Li, W. Kim, Y.G. Zhang, M. Rolandi, D.W. Wang, H.J. Dai, Growth of single-walled carbon nanotubes from discrete catalytic nanoparticles of various sizes. J. Phys. Chem. B **105** (46), 11424–11431 (2001). doi:10.1021/jp012085b
29. D. Venegoni, P. Serp, R. Feurer, Y. Kihn, C. Vahlas, P. Kalck, Parametric study for the growth of carbon nanotubes by catalytic chemical vapor deposition in a fluidized bed reactor. Carbon **40**(10), 1799–1807 (2002). doi:10.1016/s0008-6223(02)00057-x
30. K. Hernadi, A. Fonseca, J.B. Nagy, D. Bernaerts, A. Fudala, A.A. Lucas, Catalytic synthesis of carbon nanotubes using zeolite support. Zeolites **17**(5–6), 416–423 (1996). doi:10.1016/s0144-2449(96)00088-7
31. H. Ago, T. Komatsu, S. Ohshima, Y. Kuriki, M. Yumura, Dispersion of metal nanoparticles for aligned carbon nanotube arrays. Appl. Phys. Lett. **77**(1), 79–81 (2000). doi:10.1063/1.126883
32. Y. Li, J. Liu, Y.Q. Wang, Z.L. Wang, Preparation of monodispersed Fe–Mo nanoparticles as the catalyst for CVD synthesis of carbon nanotubes. Chem. Mater. **13**(3), 1008–1014 (2001). doi:10.1021/cm000787s
33. C.L. Cheung, A. Kurtz, H. Park, C.M. Lieber, Diameter-controlled synthesis of carbon nanotubes. J. Phys. Chem. B. **106**(10), 2429–2433 (2002). doi:10.1021/jp0142278
34. Y. Qin, Z.K. Zhang, Z.L. Cui, Helical carbon nanofibers with a symmetric growth mode. Carbon **42**(10), 1917–1922 (2004). doi:10.1016/j.carbon.2004.03.020
35. N.G. Shang, X. Jiang, Large-sized tubular graphite cones with nanotube tips. Appl. Phys. Lett. **87**(16), 163102 (2005). doi:10.1063/1.2093919
36. N.G. Shang, W.I. Milne, X. Jiang, Tubular graphite cones with single-crystal nanotips and their antioxygenic properties. J. Am. Chem. Soc. **129**(28), 8907–8911 (2007). doi:10.1021/ja071830g
37. G.Y. Zhang, X.C. Ma, D.Y. Zhong, E.G. Wang, Polymerized carbon nitride nanobells. J. Appl. Phys. **91**(11), 9324–9332 (2002). doi:10.1063/1.1476070
38. Y.G. Zhang, A.L. Chang, J. Cao, Q. Wang, W. Kim, Y.M. Li, N. Morris, E. Yenilmez, J. Kong, H.J. Dai, Electric-field-directed growth of aligned single-walled carbon nanotubes. Appl. Phys. Lett. **79**(19), 3155–3157 (2001). doi:10.1063/1.1415412
39. A. Ural, Y.M. Li, H.J. Dai, Electric-field-aligned growth of single-walled carbon nanotubes on surfaces. Appl. Phys. Lett. **81**(18), 3464–3466 (2002). doi:10.1063/1.1518773

40. R.S. Wagner, W.C. Ellis, Vapor-liquid-solid mechanism of single crystal growth (New method growth catalysis from impurity whisker epitaxial + large crystals Si E). Appl. Phys. Lett. **4**(5), 89 (1964). doi:10.1063/1.1753975
41. H. Kanzow, A. Ding, Formation mechanism of single-wall carbon nanotubes on liquid-metal particles. Phys. Rev. B **60**(15), 11180–11186 (1999). doi:10.1103/PhysRevB.60.11180
42. M.A. Ermakova, D.Y. Ermakov, A.L. Chuvilin, G.G. Kuvshinov, Decomposition of methane over iron catalysts at the range of moderate temperatures: the influence of structure of the catalytic systems and the reaction conditions on the yield of carbon and morphology of carbon filaments. J. Catal. **201**(2), 183–197 (2001). doi:10.1006/jcat.2001.3243
43. Y. Shibuta, S. Maruyama, Molecular dynamics simulation of formation process of single-walled carbon nanotubes by CCVD method. Chem. Phys. Lett. **382**(3–4), 381–386 (2003). doi:10.1016/j.cplett.2003.10.080
44. S. Tsunekawa, S. Ito, Y. Kawazoe, J.T. Wang, Critical size of the phase transition from cubic to tetragonal in pure zirconia nanoparticles. Nano Lett. **3**(7), 871–875 (2003). doi:10.1021/ni034129t
45. S.S. Fan, W.J. Liang, H.Y. Dang, N. Franklin, T. Tombler, M. Chapline, H.J. Dai, Carbon nanotube arrays on silicon substrates and their possible application. Physica E **8**(2), 179–183 (2000). doi:10.1016/s1386-9477(00)00136-3
46. R.T.K. Baker, Catalytic growth of carbon filaments. Carbon **27**(3), 315–323 (1989). doi:10.1016/0008-6223(89)90062-6
47. J. Gavillet, A. Loiseau, C. Journet, F. Willaime, F. Ducastelle, J.C. Charlier, Root-growth mechanism for single-wall carbon nanotubes. Phys. Rev. Lett. **87**(27), 275504 (2001). doi:10.1103/PhysRevLett.87.275504
48. N.M. Rodriguez, A. Chambers, R.T.K. Baker, Catalytic engineering of carbon nanostructures. Langmuir **11**(10), 3862–3866 (1995). doi:10.1021/la00010a042
49. R.T. Yang, J.P. Chen, Mechanism of carbon-filament growth on metal-catalysts. J. Catal. **115** (1), 52–64 (1989). doi:10.1016/0021-9517(89)90006-7
50. J.A. Lobo, G.H. Geiger, Thermodynamics and solubility of carbon in ferrite and ferritic Fe–Mo alloys. Metall. Trans. A Phys. Metall. Mater. Sci. **7**(9), 1347–1357 (1976). doi:10.1007/bf02658820
51. C.P. Deck, K. Vecchio, Prediction of carbon nanotube growth success by the analysis of carbon-catalyst binary phase diagrams. Carbon **44**(2), 267–275 (2006). doi:10.1016/j.carbon.2005.07.023
52. T. Maruyama, K. Sato, Y. Mizutani, K. Tanioku, T. Shiraiwa, S. Naritsuka, Low-temperature synthesis of single-walled carbon nanotubes by alcohol gas source growth in high vacuum. J. Nanosci. Nanotechnol. **10**(6), 4095–4101 (2010). doi:10.1166/jnn.2010.2000
53. J. Highfield, Y.S. Loo, Z. Zhong, B. Grushko, Thermogravimetric studies of carbon nanofiber formation from methane at low temperature over Ni-based skeletal catalysts and the effect of substrate pre-carburization. Carbon **45**(13), 2597–2607 (2007). doi:10.1016/j.carbon.2007.08.012
54. D. He, J. Bai, Acetylene-enhanced growth of carbon nanotubes on ceramic microparticles for multi-scale hybrid structures. Chem. Vap. Depos. **17**(4–6), 98–106 (2011). doi:10.1002/cvde.201006878
55. A. Magrez, J.W. Seo, R. Smajda, B. Korbely, J.C. Andresen, M. Mionic, S. Casimirius, L. Forro, Low-temperature, highly efficient growth of carbon nanotubes on functional materials by an oxidative dehydrogenation reaction. ACS Nano **4**(7), 3702–3708 (2010). doi:10.1021/nn100279j
56. N. Halonen, A. Sapi, L. Nagy, R. Puskas, A.-R. Leino, J. Maklin, J. Kukkola, G. Toth, M.-C. Wu, H.-C. Liao, W.-F. Su, A. Shchukarev, J.-P. Mikkola, A. Kukovecz, Z. Konya, K. Kordas, Low-temperature growth of multi-walled carbon nanotubes by thermal CVD. Physica Status Solidi B Basic Solid State Phys. **248**(11), 2500–2503 (2011). doi:10.1002/pssb.201100137
57. O. Pitkanen, N. Halonen, A.R. Leino, J. Maklin, A. Dombovari, J.H. Lin, G. Toth, K. Kordas, Low-temperature growth of carbon nanotubes on bi- and tri-metallic catalyst templates. Top. Catal. **56**(9–10), 522–526 (2013). doi:10.1007/s11244-013-0047-9

58. Y.M. Shyu, F.C.N. Hong, Low-temperature growth and field emission of aligned carbon nanotubes by chemical vapor deposition. Mater. Chem. Phys. **72**(2), 223–227 (2001). doi:10.1016/s0254-0584(01)00441-2
59. Y.M. Shyu, F.C.N. Hong, The effects of pre-treatment and catalyst composition on growth of carbon nanofibers at low temperature. Diam. Relat. Mater. **10**(3–7), 1241–1245 (2001). doi:10.1016/s0925-9635(00)00550-1
60. K. Aoki, T. Yamamoto, H. Furuta, T. Ikuno, S. Honda, M. Furuta, K. Oura, T. Hirao, Low-temperature growth of carbon nanofiber by thermal chemical vapor deposition using CuNi catalyst. Jpn. J. Appl. Phys. Part 1 Regular Pap. Brief Commun. Rev. Pap. **45**(6A), 5329–5331 (2006). doi:10.1143/jjap.45.5329
61. N. Na, D.Y. Kim, Y.-G. So, Y. Ikuhara, S. Noda, Simple and engineered process yielding carbon nanotube arrays with 1.2×10^{13} cm^{-2} wall density on conductive underlayer at 400 °C. Carbon **81**, 773–781 (2015). doi:10.1016/j.carbon.2014.10.023
62. K. Tanioku, T. Maruyama, S. Naritsuka, Low temperature growth of carbon nanotubes on Si substrates in high vacuum. Diam. Relat. Mater. **17**(4–5), 589–593 (2008). doi:10.1016/j.diamond.2007.10.028
63. X. Li, Z. Xu, One-step catalytic growth of carbon nanofiber arrays vertically aligned on carbon substrate. Mater. Res. Bull. **47**(6), 1557–1561 (2012). doi:10.1016/j.materresbull.2012.02.027
64. Y. Ma, C. Weimer, N. Yang, L. Zhang, T. Staedler, X. Jiang, Low-temperature growth of carbon nanofiber using a vapor–facet–solid process. Mater. Today Commun. **2**, e55–e61 (2015). doi:10.1016/j.mtcomm.2014.12.003
65. B. Yu, S. Wang, Q. Zhang, Y. He, H. Huang, J. Zou, Ni_3C-assisted growth of carbon nanofibres 300 °C by thermal CVD. Nanotechnology **25**(32), 325602 (2014). doi:10.1088/0957-4484/25/32/325602
66. W.-H. Chiang, R.M. Sankaran, Synergistic effects in bimetallic nanoparticles for low temperature carbon nanotube growth. Adv. Mater. **20**(24), 4857–4861 (2008). doi:10.1002/adma.200801006
67. Y. Ma, X. Sun, N. Yang, J. Xia, L. Zhang, X. Jiang, Shape-controlled growth of carbon nanostructures: yield and mechanism. Chem. Eur. J. **21**, 12370–12375 (2015). doi:10.1002/chem.201500440
68. Y. Qin, Q. Zhang, Z.L. Cui, Effect of synthesis method of nanocopper catalysts on the morphologies of carbon nanofibers prepared by catalytic decomposition of acetylene. J. Catal. **223**(2), 389–394 (2004). doi:10.1016/j.jcat.2004.02.004
69. J.H. Xia, X. Jiang, C.L. Jia, The size effect of catalyst on the growth of helical carbon nanofibers. Appl. Phys. Lett. **95**(22), 223110 (2009). doi:10.1063/1.3271031
70. Y. Qin, M. Eggers, T. Staedler, X. Jiang, Symmetric growth of carbon nanosheets on Cu nanowires by a surface diffusion mechanism. Nanotechnology **18**(34), 345607 (2007). doi:10.1088/0957-4484/18/34/345607
71. Y. Qin, X. Jiang, Z.L. Cui, Low-temperature synthesis of amorphous carbon nanocoils via acetylene coupling on copper nanocrystal surfaces at 468 K: a reaction mechanism analysis. J. Phys. Chem. B **109**(46), 21749–21754 (2005). doi:10.1021/jp054412b
72. X. Sun, Y.W. Zhang, R. Si, C.H. Yan, Metal (mn Co, and Cu) oxide nanocrystals from simple formate precursors. Small **1**(11), 1081–1086 (2005). doi:10.1002/smll.200500119
73. J.H. Xia, Growth of carbon nanofibers studied by using transmission electron microscopy. Shaker Verlag, D-52018 Aachen (2010)
74. A.J. Hart, A.H. Slocum, L. Royer, Growth of conformal single-walled carbon nanotube films from $Mo/Fe/Al_2O_3$ deposited by electron beam evaporation. Carbon **44**(2), 348–359 (2006). doi:10.1016/j.carbon.2005.07.008
75. Y.J. Tian, Z. Hu, Y. Yang, X.Z. Wang, X. Chen, H. Xu, Q. Wu, W.J. Ji, Y. Chen, In situ TA-MS study of the six-membered-ring-based growth of carbon nanotubes with benzene precursor. J. Am. Chem. Soc. **126**(4), 1180–1183 (2004). doi:10.1021/ja037561i

76. B. Zheng, C.G. Lu, G. Gu, A. Makarovski, G. Finkelstein, J. Liu, Efficient CVD growth of single-walled carbon nanotubes on surfaces using carbon monoxide precursor. Nano Lett. **2**(8), 895–898 (2002). doi:10.1021/nl025634d
77. A.J. Hart, A.H. Slocum, Rapid growth and flow-mediated nucleation of millimeter-scale aligned carbon nanotube structures from a thin-film catalyst. J. Phys. Chem. B. **110**(16), 8250–8257 (2006). doi:10.1021/jp055498b
78. M. Jung, K.Y. Eun, J.K. Lee, Y.J. Baik, K.R. Lee, J.W. Park, Growth of carbon nanotubes by chemical vapor deposition. Diam. Relat. Mater. **10**(3–7), 1235–1240 (2001). doi:10.1016/s0925-9635(00)00446-5
79. A.V. Vasenkov, D. Sengupta, M. Frenklach, Multiscale modeling catalytic decomposition of hydrocarbons during carbon nanotube growth. J. Phys. Chem. B. **113**(7), 1877–1882 (2009). doi:10.1021/jp808346h
80. G.D. Nessim, A. Al-Obeidi, H. Grisaru, E.S. Polsen, C.R. Oliver, T. Zimrin, A.J. Hart, D. Aurbach, C.V. Thompson, Synthesis of tall carpets of vertically aligned carbon nanotubes by in situ generation of water vapor through preheating of added oxygen. Carbon **50**(11), 4002–4009 (2012). doi:10.1016/j.carbon.2012.04.043
81. D.N. Futaba, K. Hata, T. Yamada, K. Mizuno, M. Yumura, S. Iijima, Kinetics of water-assisted single-walled carbon nanotube synthesis revealed by a time-evolution analysis. Phys. Rev. Lett. **95**(5), 4 (2005). doi:10.1103/PhysRevLett.95.056104
82. S. Hussain, R. Amade, E. Bertran, Study of CNTs structural evolution during water assisted growth and transfer methodology for electrochemical applications. Mater. Chem. Phys. **148**(3), 914–922 (2014). doi:10.1016/j.matchemphys.2014.08.070
83. M. Bansal, C. Lal, R. Srivastava, M.N. Kamalasanan, L.S. Tanwar, Comparison of structure and yield of multiwall carbon nanotubes produced by the CVD technique and a water assisted method. Physica B Condens. Matter **405**(7), 1745–1749 (2010). doi:10.1016/j.physb.2010.01.031
84. C.-S. Chen, C.-K. Hsieh, Oxygen-assisted low-pressure chemical vapor deposition for the low-temperature direct growth of graphitic nanofibers on fluorine-doped tin oxide glass as a counter electrode for dye-sensitized solar cell. Jpn. J. Appl. Phys. **53**(11), 11RE02 (2014). doi:10.7567/jjap.53.11re02
85. I.H. Son, H.J. Song, S. Kwon, A. Bachmatiuk, S.J. Lee, A. Benayad, J.H. Park, J.-Y. Choi, H. Chang, M.H. Ruemmeli, CO_2 enhanced chemical vapor deposition growth of few-layer graphene over NiO_x. ACS Nano **8**(9), 9224–9232 (2014). doi:10.1021/nn504342e
86. J.Q. Huang, Q. Zhang, M.Q. Zhao, F. Wei, Process intensification by CO_2 for high quality carbon nanotube forest growth: double-walled carbon nanotube convexity or single-walled carbon nanotube bowls? Nano Res. **2**(11), 872–881 (2009). doi:10.1007/s12274-009-9088-6
87. Z. Zhu, H. Jiang, T. Susi, A.G. Nasibulin, E.I. Kauppinen, The use of NH_3 to promote the production of large-diameter single-walled carbon nanotubes with a narrow (n, m) distribution. J. Am. Chem. Soc. **133**(5), 1224–1227 (2011). doi:10.1021/ja1087634
88. T. Susi, A.G. Nasibulin, P. Ayala, Y. Tian, Z. Zhu, H. Jiang, C. Roquelet, D. Garrot, J.-S. Lauret, E.I. Kauppinen, High quality SWCNT synthesis in the presence of NH_3 using a vertical flow aerosol reactor. Physica Status Solidi B Basic Solid State Phys. **246**(11–12), 2507–2510 (2009). doi:10.1002/pssb.200982338
89. A.F. Carley, P.R. Davies, K.R. Harikumar, R.V. Jones, M.W. Roberts, Oxygen states at magnesium and copper surfaces revealed by scanning tunneling microscopy and surface reactivity. Top. Catal. **24**(1–4), 51–59 (2003). doi:10.1023/B:TOCA.0000003076.82649.c4
90. P.R. Davies, D. Edwards, D. Richards, Possible role for Cu(II) compounds in the oxidation of malonyl dichloride and HCl at Cu (110) surfaces. J. Phys. Chem. C **113**(24), 10333–10336 (2009). doi:10.1021/jp903042f
91. Y. Ma, Vapor-facet-solid (VFS) mechanism: a new route for catalytic CVD growth of one-dimensional nanostructures at low temperature. Schriftenreihe der Arbeitsgruppe des Lehrstuhls für Oberfächen- und Werkstofftechnologie im Institut für Werkstofftechnik. **4** (2015)

92. J.T. Hu, L.S. Li, W.D. Yang, L. Manna, L.W. Wang, A.P. Alivisatos, Linearly polarized emission from colloidal semiconductor quantum rods. Science **292**(5524), 2060–2063 (2001). doi:10.1126/science.1060810
93. A.X. Yin, X.Q. Min, Y.W. Zhang, C.H. Yan, Shape-selective synthesis and facet-dependent enhanced electrocatalytic activity and durability of monodisperse sub-10 nm Pt-Pd tetrahedrons and cubes. J. Am. Chem. Soc. **133**(11), 3816–3819 (2011). doi:10.1021/ja200329p
94. S. Mostafa, F. Behafarid, J.R. Croy, L.K. Ono, L. Li, J.C. Yang, A.I. Frenkel, B.R. Cuenya, Shape-dependent catalytic properties of Pt nanoparticles. J. Am. Chem. Soc. **132**(44), 15714–15719 (2010). doi:10.1021/ja106679z
95. H. Zhang, M.S. Jin, Y.J. Xiong, B. Lim, Y.N. Xia, Shape-controlled synthesis of Pd nanocrystals and their catalytic applications. Acc. Chem. Res. **46**(8), 1783–1794 (2013). doi:10.1021/ar300209w
96. R. Narayanan, M.A. El-Sayed, Catalysis with transition metal nanoparticles in colloidal solution: nanoparticle shape dependence and stability. J. Phys. Chem. B. **109**(26), 12663–12676 (2005). doi:10.1021/jp051066p
97. Y.H. Leng, Y.H. Zhang, T. Liu, M. Suzuki, X.G. Li, Synthesis of single crystalline triangular and hexagonal Ni nanosheets with enhanced magnetic properties. Nanotechnology **17**(6), 1797–1800 (2006). doi:10.1088/0957-4484/17/6/042
98. Y.W. Jun, J.S. Choi, J. Cheon, Shape control of semiconductor and metal oxide nanocrystals through nonhydrolytic colloidal routes. Angewandte Chemie International Edition **45**(21), 3414–3439 (2006). doi:10.1002/anie.200503821
99. Y.N. Xia, Y.J. Xiong, B. Lim, S.E. Skrabalak, Shape-controlled synthesis of metal nanocrystals: simple chemistry meets complex physics? Angewandte Chemie International Edition **48**(1), 60–103 (2009). doi:10.1002/anie.200802248
100. A.R. Tao, S. Habas, P.D. Yang, Shape control of colloidal metal nanocrystals. Small **4**(3), 310–325 (2008). doi:10.1002/smll.200701295
101. E.F. Kukovitsky, S.G. L'Vov, N.A. Sainov, VLS-growth of carbon nanotubes from the vapor. Chem. Phys. Lett. **317**(1–2), 65–70 (2000). doi:10.1016/s0009-2614(99)01299-3
102. B.C. Satishkumar, P.J. Thomas, A. Govindaraj, C.N.R. Rao, Y-junction carbon nanotubes. Appl. Phys. Lett. **77**(16), 2530–2532 (2000). doi:10.1063/1.1319185
103. J. Li, C. Papadopoulos, J. Xu, Nanoelectronics—growing Y-junction carbon nanotubes. Nature **402**(6759), 253–254 (1999). doi:10.1038/46214
104. H. Takikawa, M. Yatsuki, R. Miyano, M. Nagayama, T. Sakakibara, S. Itoh, Y. Ando, Amorphous carbon fibrilliform nanomaterials prepared by chemical vapor deposition. Jpn. J. Appl. Phys. Part 1 Regular Pap. Short Notes Rev. Pap. **39**(9A), 5177–5179 (2000). doi:10.1143/jjap.39.5177
105. K. Inomata, N. Aoki, H. Koinuma, Production of fullerenes by low-temperature plasma chemical-vapor-deposition under atmospheric-pressure. Jpn. J. Appl. Phys. Part 2 Lett. **33**(2A), L197–L199 (1994). doi:10.1143/jjap.33.l197
106. Y. Suda, Y. Shimizu, M. Ozaki, H. Tanoue, H. Takikawa, H. Ue, K. Shimizu, Y. Umeda, Electrochemical properties of fuel cell catalysts loaded on carbon nanomaterials with different geometries. Mater. Today Commun. **3**, 96–103 (2015). doi:10.1016/j.mtcomm.2015.02.003
107. G. Wang, G. Ran, G. Wan, P. Yang, Z. Gao, S. Lin, C. Fu, Y. Qin, Size-selective catalytic growth of nearly 100 % pure carbon nanocoils with copper nanoparticles produced by atomic layer deposition. ACS Nano **8**(5), 5330–5338 (2014). doi:10.1021/nn501709h
108. G. Wang, Z. Gao, S. Tang, C. Chen, F. Duan, S. Zhao, S. Lin, Y. Feng, L. Zhou, Y. Qin, Microwave absorption properties of carbon nanocoils coated with highly controlled magnetic materials by atomic layer deposition. ACS Nano **6**(12), 11009–11017 (2012). doi:10.1021/nn304630h

109. K. Hernadi, A. Fonseca, J.B. Nagy, D. Bernaerts, A.A. Lucas, Fe-catalyzed carbon nanotube formation. Carbon **34**(10), 1249–1257 (1996). doi:10.1016/0008-6223(96)00074-7
110. D. Chen, K.O. Christensen, E. Ochoa-Fernandez, Z.X. Yu, B. Totdal, N. Latorre, A. Monzon, A. Holmen, Synthesis of carbon nanofibers: effects of Ni crystal size during methane decomposition. J. Catal. **229**(1), 82–96 (2005). doi:10.1016/j.jcat.2004.10.017
111. P.L. Hansen, J.B. Wagner, S. Helveg, J.R. Rostrup-Nielsen, B.S. Clausen, H. Topsoe, Atom-resolved imaging of dynamic shape changes in supported copper nanocrystals. Science **295**(5562), 2053–2055 (2002). doi:10.1126/science.1069325
112. C.A. Wert, Diffusion coefficient of C in α-iron. Phys. Rev. **79**(4), 601–605 (1950). doi:10.1103/PhysRev.79.601
113. J.J. Lander, H.E. Kern, A.L. Beach, Solubility and diffusion coefficient of carbon in nickel-reaction rates of nickel-carbon alloys with barium oxide. J. Appl. Phys. **23**(12), 1305–1309 (1952). doi:10.1063/1.1702064
114. B.C. Stipe, M.A. Rezaei, W. Ho, Single-molecule vibrational spectroscopy and microscopy. Science **280**(5370), 1732–1735 (1998). doi:10.1126/science.280.5370.1732
115. B.C. Stipe, M.A. Rezaei, W. Ho, Coupling of vibrational excitation to the rotational motion of a single adsorbed molecule. Phys. Rev. Lett. **81**(6), 1263–1266 (1998). doi:10.1103/PhysRevLett.81.1263
116. J. Szanyi, M.T. Paffett, Dimerization and trimerization of acetylene over a model Sn/Pt catalyst. J. Am. Chem. Soc. **117**(3), 1034–1042 (1995). doi:10.1021/ja00108a020
117. S. Helveg, C. Lopez-Cartes, J. Sehested, P.L. Hansen, B.S. Clausen, J.R. Rostrup-Nielsen, F. Abild-Pedersen, J.K. Norskov, Atomic-scale imaging of carbon nanofibre growth. Nature **427**(6973), 426–429 (2004). doi:10.1038/nature02278
118. T.E. Fischer, S.R. Kelemen, Influence of the substrate structure on the bonding of chemisorbed acetylene to transition metal surfaces. Surf. Sci. **74**(47), 47–53 (1978). doi:10.1016/0039-6028(78)90270-4
119. J. Dvorak, J. Hrbek, Adsorbate ordering effects in the trimerization reaction of acetylene on Cu (100). J. Phys. Chem. B. **102**(47), 9443–9450 (1998). doi:10.1021/jp981956n
120. J.R. Lomas, C.J. Baddeley, M.S. Tikhov, R.M. Lambert, Ethyne cyclization to benzene over Cu (110). Langmuir **11**(8), 3048–3053 (1995). doi:10.1021/la00008a033
121. G. Kyriakou, J. Kim, M.S. Tikhov, N. Macleod, R.M. Lambert, Acetylene coupling on Cu (111): formation of butadiene, benzene, and cyclooctatetraene. J. Phys. Chem. B. **109**(21), 10952–10956 (2005). doi:10.1021/jp044213c
122. W. Alter, D. Borgmann, M. Stadelmann, M. Worn, G. Wedler, Interaction of acetylene with films of the transition-metals iron, nickel, and palladium. J. Am. Chem. Soc. **116**(22), 10041–10049 (1994). doi:10.1021/ja00101a024
123. F. Zaera, R.B. Hall, High-resolution electron energy loss spectroscopy and thermal programmed desorption studies of the chemisorption and thermal decomposition of ethylene and acetylene on Ni (100) single-crystal surfaces. J. Phys. Chem. **91**(16), 4318–4323 (1987). doi:10.1021/j100300a023
124. J.C. Bertolini, J. Massardier, G. Dalmaiimelik, Evolution of adsorbed species during C_2H_2 adsorption on Ni (111) in relation to their vibrational spectra. J. Chem. Soc. Faraday Trans. I. **74**, 1720–1725 (1978). doi:10.1039/f19787401720
125. J.A. Stroscio, S.R. Bare, W. Ho, The chemisorption and decomposition of ethylene and acetylene on Ni (110). Surf. Sci. **148**(2–3), 499–525 (1984). doi:10.1016/0039-6028(84)90596-x
126. A. Benninghoven, P. Beckmann, D. Greifendorf, M. Schemmer, Investigation of surface-reactions by SIMS and TDMS—interaction of ethylene and acetylene with hydrogen on polycrystalline nickel. Appl. Surf. Sci. **6**(3–4), 288–296 (1980). doi:10.1016/0378-5963(80)90018-5
127. P.M. Mattlis, The oligomerization of acetylenes induced by metals of the nickel triad. Pure Appl. Chem. **30**(3–4), 427–448 (1972). doi:10.1351/pac197230030427

128. D.L. Trimm, I.O.Y. Liu, N.W. Cant, The oligornerization of acetylene in hydrogen over Ni/SiO_2 catalysts: product distribution and pathways. J. Mol. Catal. A Chem. **288**(1–2), 63–74 (2008). doi:10.1016/j.molcata.2008.03.022
129. B. Lesiak, A. Jablonski, W. Palczewska, I. Kulszewiczbajer, M. Zagorska, Identification of the carbonaceous residues at nickel and platinum surfaces on the basis of the carbon Kll spectra. Surf. Interf. Anal. **18**(6), 430–438 (1992). doi:10.1002/sia.740180610
130. S. Hofmann, G. Csanyi, A.C. Ferrari, M.C. Payne, J. Robertson, Surface diffusion: the low activation energy path for nanotube growth. Phys. Rev. Lett. **95**(3), 036101 (2005). doi:10.1103/PhysRevLett.95.036101
131. O.V. Yazyev, A. Pasquarello, Effect of metal elements in catalytic growth of carbon nanotubes. Phys. Rev. Lett. **100**(15), 4 (2008). doi:10.1103/PhysRevLett.100.156102
132. K. Bartsch, K. Biedermann, T. Gemming, A. Leonhardt, On the diffusion-controlled growth of multiwalled carbon nanotubes. J. Appl. Phys. **97**(11), 7 (2005). doi:10.1063/1.1922067
133. Z.Y. Juang, J.F. Lai, C.H. Weng, J.H. Lee, H.J. Lai, T.S. Lai, C.H. Tsai, On the kinetics of carbon nanotube growth by thermal CVD method. Diam. Relat. Mater. **13**(11–12), 2140–2146 (2004). doi:10.1016/j.diamond.2004.03.007
134. O.A. Louchev, Y. Sato, H. Kanda, Multiwall carbon nanotubes: self-organization and inhibition of step-flow growth kinetics. J. Appl. Phys. **89**(6), 3438–3446 (2001). doi:10.1063/1.1347407
135. S. Hofmann, B. Kleinsorge, C. Ducati, A.C. Ferrari, J. Robertson, Low-temperature plasma enhanced chemical vapour deposition of carbon nanotubes. Diam. Relat. Mater. **13**(4–8), 1171–1176 (2004). doi:10.1016/j.diamond.2003.11.046
136. O.A. Louchev, T. Laude, Y. Sato, H. Kanda, Diffusion-controlled kinetics of carbon nanotube forest growth by chemical vapor deposition. J. Chem. Phys. **118**(16), 7622–7634 (2003). doi:10.1063/1.1562195
137. O.A. Louchev, Formation mechanism of pentagonal defects and bamboo-like structures in carbon nanotube growth mediated by surface diffusion. Physica Status Solidi A Appl. Res. **193**(3), 585–596 (2002). doi:10.1002/1521-396x(200210)193:3<585:aid-pssa585>3.0.co;2-y
138. O.A. Louchev, Y. Sato, H. Kanda, Growth mechanism of carbon nanotube forests by chemical vapor deposition. Appl. Phys. Lett. **80**(15), 2752–2754 (2002). doi:10.1063/1.1468266
139. D.C. Li, L.M. Dai, S.M. Huang, A.W.H. Mau, Z.L. Wang, Structure and growth of aligned carbon nanotube films by pyrolysis. Chem. Phys. Lett. **316**(5–6), 349–355 (2000). doi:10.1016/s0009-2614(99)01334-2
140. S.A. Dayeh, E.T. Yu, D. Wang, III-V nanowire growth mechanism: V/III ratio and temperature effects. Nano Lett. **7**(8), 2486–2490 (2007). doi:10.1021/nl0712668
141. P.B. Amama, O. Ogebule, M.R. Maschmann, T.D. Sands, T.S. Fisher, Dendrimer-assisted low-temperature growth of carbon nanotubes by plasma-enhanced chemical vapor deposition. Chem. Commun. **27**, 2899–2901 (2006). doi:10.1039/b602623k
142. H.Y. Wang, J.J. Moore, Low temperature growth mechanisms of vertically aligned carbon nanofibers and nanotubes by radio frequency-plasma enhanced chemical vapor deposition. Carbon **50**(3), 1235–1242 (2012). doi:10.1016/j.carbon.2011.10.041
143. S. Hofmann, C. Ducati, J. Robertson, B. Kleinsorge, Low-temperature growth of carbon nanotubes by plasma-enhanced chemical vapor deposition. Appl. Phys. Lett. **83**(1), 135–137 (2003). doi:10.1063/1.1589187
144. T.M. Minea, S. Point, A. Granier, M. Touzeau, Room temperature synthesis of carbon nanofibers containing nitrogen by plasma-enhanced chemical vapor deposition. Appl. Phys. Lett. **85**(7), 1244–1246 (2004). doi:10.1063/1.1781352
145. M. Meyyappan, A review of plasma enhanced chemical vapour deposition of carbon nanotubes. J. Phys. D Appl. Phys. **42**(21), 15 (2009). doi:10.1088/0022-3727/42/21/213001
146. Y. Ishikawa, K. Ishizuka, Growth of single-walled carbon nanotubes by hot-filament assisted chemical vapor deposition below 500 °C. Appl. Phys. Express **2**(4), 3 (2009). doi:10.1143/apex.2.045001

147. N.G. Shang, Y.Y. Tan, V. Stolojan, P. Papakonstantinou, S.R.P. Silva, High-rate low-temperature growth of vertically aligned carbon nanotubes. Nanotechnology **21**(50), 6 (2010). doi:10.1088/0957-4484/21/50/505604
148. Y. Ishikawa, H. Jinbo, Synthesis of multiwalled carbon nanotubes at temperatures below 300 °C by hot-filament assisted chemical vapor deposition. Jpn. J. Appl. Phys. Part 2 Lett. Express Lett. **44**(12–15), L394–L397 (2005). doi:10.1143/jjap.44.l394
149. Y. Awano, S. Sato, M. Nihei, T. Sakai, Y. Ohno, T. Mizutani, Carbon nanotubes for VLSI: interconnect and transistor applications. Proc. IEEE **98**(12), 2015–2031 (2010). doi:10.1109/JPROC.2010.2068030
150. C.L. Long, D.P. Qi, T. Wei, J. Yan, L.L. Jiang, Z.J. Fan, Nitrogen-doped carbon networks for high energy density supercapacitors derived from polyaniline coated bacterial cellulose. Adv. Funct. Mater. **24**(25), 3953–3961 (2014). doi:10.1002/adfm.201304269
151. N.P. Wickramaratne, J.T. Xu, M. Wang, L. Zhu, L.M. Dai, M. Jaroniec, Nitrogen enriched porous carbon spheres: attractive materials for supercapacitor electrodes and CO_2 adsorption. Chem. Mater. **26**(9), 2820–2828 (2014). doi:10.1021/cm5001895
152. W. Wei, H.W. Liang, K. Parvez, X.D. Zhuang, X.L. Feng, K. Mullen, Nitrogen-doped carbon nanosheets with size-defined mesopores as highly efficient metal-free catalyst for the oxygen reduction reaction. Angewandte Chemie-International Edition **53**(6), 1570–1574 (2014). doi:10.1002/anie.201307319

Carbon Nanohorns and Their High Potential in Biological Applications

Minfang Zhang and Masako Yudasaka

Abstract Carbon nanohorns, also called single-wall carbon nanohorns (SWNHs), are single-graphene tubules with horn-shaped tips, and were first reported by Iijima and colleagues in 1999 [1]. The tubule lengths and diameters range from 30 to 50 nm and 2 to 5 nm, respectively, and therefore, SWNHs are not uniform in size. Thousands of SWNHs assemble to form an aggregate, which in turn has an average diameter of ~80–100 nm. SWNHs are produced in large quantities (1 kg/day) by laser ablation of graphite. This process does not require a metal catalyst, and thus it is possible to prepare SWNHs with high purity (>95 %). Owing to their large surface area, molecular sieving effects and photo-thermal conversion characteristics, SWNHs show promise for applications in gas adsorption and storage, biosensor and nanomedicine such as drug delivery and photo-hyperthermia cancer therapy. In this chapter, we briefly introduce nanohorn production methods, biomaterial properties, and functionalization, and then highlight the potential use of SWNHs in various biological research fields. Issues concerning toxicity and biodegradation are also discussed.

Keywords Biomaterial · Biosensor · Carbon nanohorn · Drug delivery system · Nanomedicine · Photo-hyperthermia

M. Zhang (✉)
Nanotube Application Research Center, National Institute of Advanced Industrial Science and Technology (AIST), Tsukuba, Japan
e-mail: m-zhang@aist.go.jp

M. Yudasaka
Nanomaterials Research Institute, National Institute of Advanced Industrial Science and Technology (AIST), Tsukuba, Japan
e-mail: m-yudasaka@aist.go.jp

M. Yudasaka
Meijo University, Nagoya, Japan

N. Yang et al. (eds.), *Carbon Nanoparticles and Nanostructures*,
Carbon Nanostructures, DOI 10.1007/978-3-319-28782-9_3

1 Introduction

Single-wall carbon nanohorns (SWNHs) were first reported in 1999 [1]. SWNHs have a single-graphene tubule structure, conferring unique characteristics to the biomaterial. The nanohorns are produced in large quantities at high purity without using metal catalysts. The superior pore structure of SWNHs (i.e., large surface area and an ample interior nanospace) suggests their prospective use in gas storage, catalyst support, and drug delivery applications. Moreover, like other carbon nanomaterials (e.g., single-walled carbon nanotubes (SWNTs) [2] and multiwall carbon nanotubes (MWNTs)) [3], SWNHs absorb light in a wide spectrum of wavelengths, ranging from infrared to ultraviolet. They also generate heat, which could be beneficial for their utilization as photo-hyperthermia agents in cancer therapy and as photo-thermal energy conversion devices for the regeneration of lost heat. In this chapter, we briefly introduce the production, biomaterial properties, and functionalization of SWNHs, and then focus on current and proposed applications of SWNHs in biological research fields.

1.1 SWNH Production

Two representative methods are currently employed for the production of SWNHs with high purity: carbon dioxide (CO_2) laser ablation and arc discharge. The size distribution and purity of the SWNHs can be changed by varying production parameters, such as temperature, pressure, and laser or arc power.

Laser ablation The earliest report [1] of SWNH fabrication by Iijima and colleagues [1] utilized a CO_2 laser ablation technique for the generation of carbon nanohorns at room temperature without a metal catalyst (Fig. 1). The SWNH generator employed in this study consisted of two parts: a high-power CO_2 laser source (wavelength = ~0.6 μm; maximum power = 5 kW) and a reaction chamber. Argon (Ar) gas was introduced into and flowed throughout the inner chamber

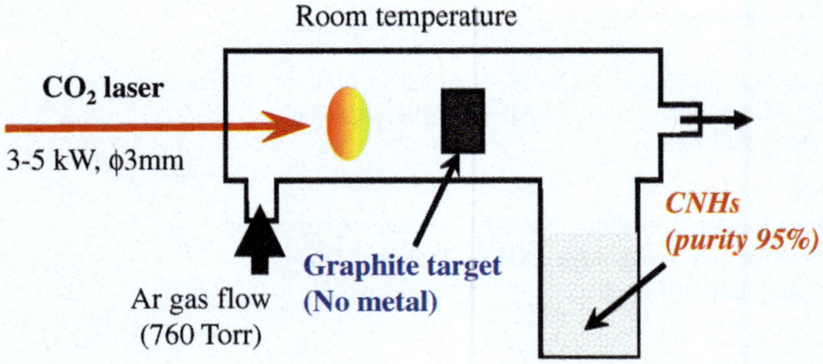

Fig. 1 Schematic diagram of SWNH production by laser ablation

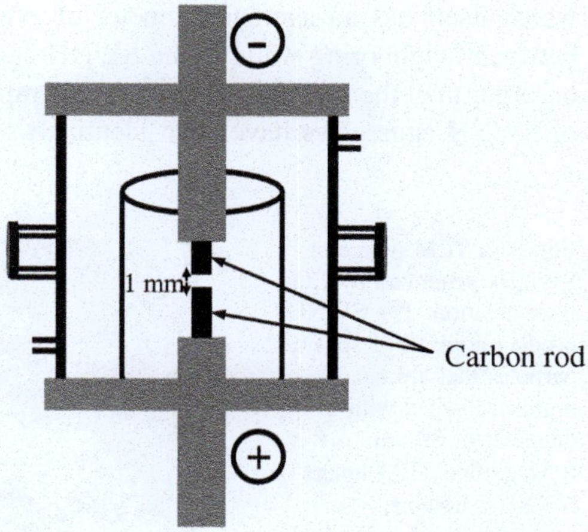

Fig. 2 Arc discharge device for SWNH production [6]

to carry the nanohorn products to the collection chamber under a pressure of 760 Torr at room temperature. The graphite target was placed in the middle of the reaction chamber and exposed to the laser beam [1, 4]. For large-scale production of SWNHs, the graphite target was replaced by a graphite rod, which was continuously rotated and moved along its axis so that a new surface was constantly exposed to the laser beam [5]. This setup resulted in a production rate of ~1 kg/day and a purity of the resultant as-grown SWNHs of ~95 % [5].

Arc discharge SWNHs are also prepared by direct current (DC) arc discharge [6–9]. DC arc discharge is carried out in a water-cooled stainless steel chamber [6] (Fig. 2). Two electrodes corresponding to pure carbon rods are separated from each other by a constant distance of ~1 mm. The arc discharge between the carbon rods is conducted under an atmospheric pressure of air [6, 7], nitrogen (N_2) [8, 9], water [10], CO_2, or carbon monoxide (CO) [7]. In the arc discharge system described in [6], the electric power density on the carbon rod surface was 9 kW/cm^2, and the SWNHs were deposited on the surface of the chamber. The mean size of the prepared SWNH particles was approximately 50 nm, which is smaller than that of the as-grown SWNHs prepared by CO_2 laser ablation. Pre-heating of the carbon rod to a temperature of up to 1000 °C before ignition of the arc improved the quality of manufactured SWNHs [6].

1.2 SWNH Properties

Structure The SWNH [1] is a single-graphene tubule of an irregular shape with non-uniform diameters of 2–5 nm, a length of 30–50 nm, and a horn-shaped tip (Fig. 3a). The tips often have cone angles of ~19° indicating the existence of five pentagonal rings. Thousands of SWNHs assemble to form a nanohorn aggregate, which itself has an average diameter of ~80–100 nm (Fig. 3b, c). The distance between neighboring walls of the SWNH aggregate is approximately 0.4 nm [11], or larger than the basal plane distance of graphite (0.335 nm). Three different types of SWNH aggregates have been identified so far: the "dahlia-like" aggregate, the

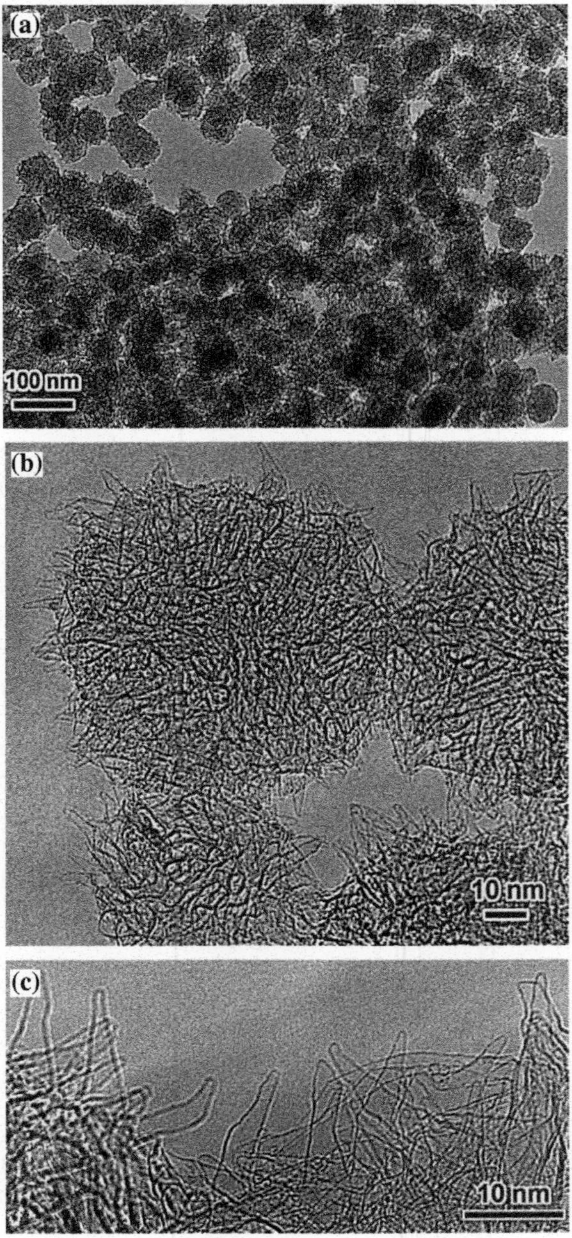

Fig. 3 **a** TEM image of SWNHs generated by CO_2 laser ablation. The SWNH product consists of spherical particles that are nearly uniform in size, with a diameter of 80 nm. **b** Magnified TEM image of SWNHs showing aggregations of individual tubule-like structures with protruding tips. **c** Highly magnified TEM image of the edge regions of the graphitic aggregate showing conical horn-like protrusions of ≤20 nm in length along the aggregate surface, with some modified shapes [1]

"bud-like" aggregate, and the "seed-like" aggregate [1]. In the first type of aggregate, SWNH tips protrude from the aggregate surface (Fig. 3b), while in the second and third types, the SWNHs appear to develop inside the aggregate itself. Different aggregate types are selectively produced by laser ablation with different gases. For example, "dahlia-like" SWNH aggregates are produced with a yield of 95 % when the buffer gas is Ar (760 Torr), while "bud-like" SWNH aggregates are formed with a yield of 70–80 % when either helium (He) or N_2 gas is employed at 760 Torr [12].

Porosity and hole-opening The pore structure of SWNHs has been extensively studied through simulation and adsorption experiments [13]. Isotherm measurements of N_2 adsorption at 77 K and high-pressure He buoyancy at 303 K reveal that as-grown SWNHs with closed tubules have a specific surface area of $\sim$300 m^2/g [14] and a total pore volume of $\sim$0.40 ml/g [15].

Nanoscale holes or windows can be generated on the tubule walls or tips by oxidation of SWNHs with oxygen (O_2) [16–19], CO_2 [20], or oxidative acids [15, 21, 22]. The number and size of the nanowindows so-generated are controlled by oxidation conditions [17]. Oxidation with a slow temperature increase of 1 °C/min in a low O_2 concentration (21 % in air) yields hole openings with little generation of carbonaceous dust [18]. After such oxidation-induced holes are created on the SWNH wall, the resultant nanohorn surface areas, total pore volume, and particle density are reportedly 1450–1460 m^2/g [15, 16], 1.05 ml/g [15], and 2.05 g/ml [17], respectively.

The SWNH possesses three adsorption sites: the inter-SWNH micropore, the interior wall surface, and the interior space; the volume ratio of the three is about 1:2:2 [17]. Approximately 11 and 36 % of the intraparticle pore spaces are opened by oxidation at 573 and 623 K, respectively. Treatment of SWNHs with nitric acid (HNO_3) induces the intercalation of HNO_3 into narrow interstitial spaces, further increasing pore volume through the development of microporosity [15]. The obtained ultra-microporous SWNH aggregates show a high storage capacity (100 mg/g) for methane [15].

Holes opened on the SWNH tubule walls are visible by high resolution transmission electron microscopy (TEM). Figure 4a shows holes generated by oxidation in O_2, and Fig. 4b shows the size distributions of the holes (0.5–1.0 nm at the tip, and 0.5–1.5 nm on the sidewalls) measured on the TEM images [19]. Therefore, materials of a size smaller than 1.5 nm, such as fullerenes, can be readily incorporated inside the SWNHs.

Holes or nanowindows introduced in the SWNH wall are differentially closed by high temperatures. Constant-temperature tight-binding molecular dynamic (TBMD) simulations indicate that holes in the sidewalls are not easily closed by thermal annealing, unlike those at the tips [23]. This finding provides an explanation for why some SWNH holes remain open following exposure to high temperatures [24]. A proposed hole-closing mechanism suggests that the holes at the tips undergo a closed-open-closed evolution when the initial hole size is >0.7–0.9 nm, while the same holes close rapidly and remain in that state when the initial hole sizes are ≤0.7–0.9 nm [24].

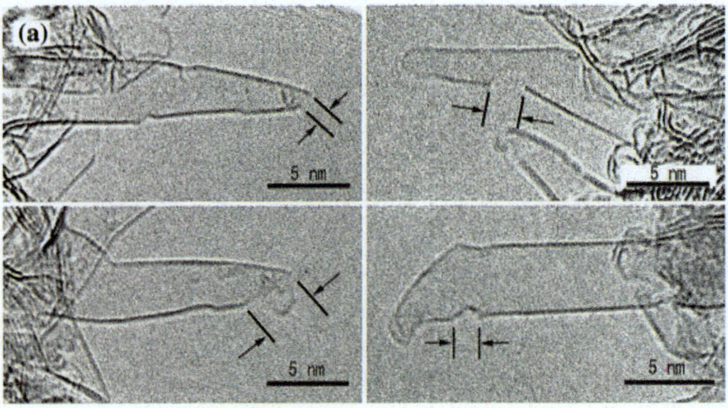

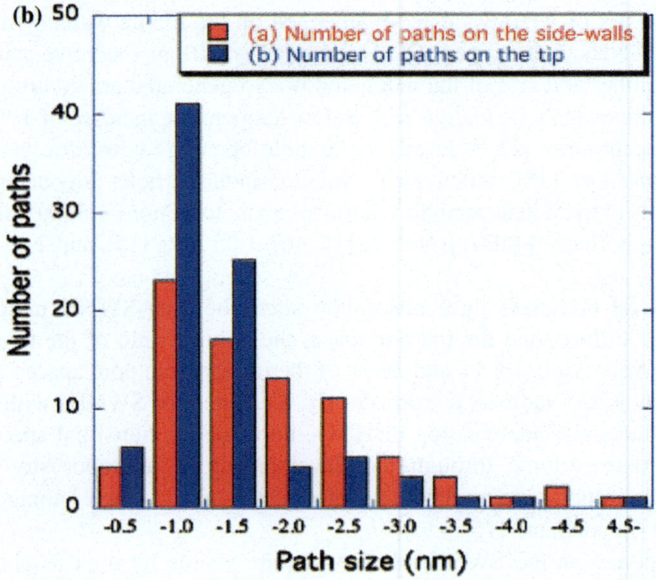

Fig. 4 **a** TEM images of SWNHs after heat treatment in O_2. The pathway holes are clearly visible at the nanohorn tips and sidewalls. **b** Size distributions of the holes in the SWNH sidewalls (*red*) and tips (*blue*) [19]

1.3 Isolation of SWNHs from Aggregates

As described above, as-grown SWNH aggregates are assembled from thousands of individual SWNHs. The individual nanoparticles in these aggregates are held together by strong forces, and cannot be separated without difficulty. However, the individual SWNHs or small-sized carbon nanohorns can be readily isolated by ultrasonication, followed by density gradient centrifugation [25, 26]. To separate individual SWNHs, oxidized SWNH aggregates are dispersed in a surfactant of

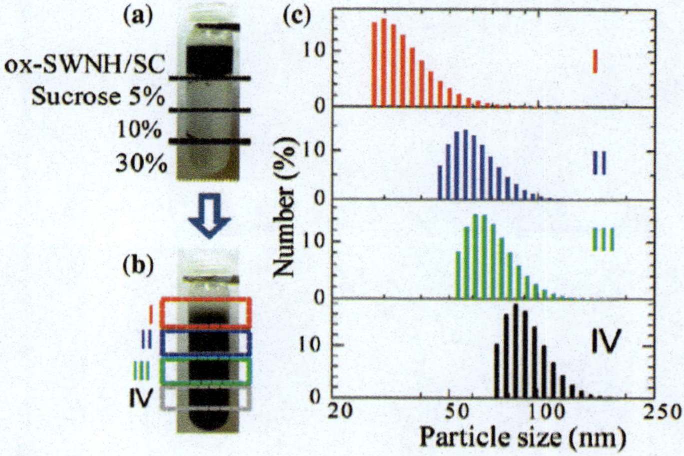

Fig. 5 **a** Centrifuge tube containing a sodium cholate dispersion of an oxidized SWNH aggregate (ox-SWNH) placed on top of a sucrose gradient made up of three layers with varying sucrose concentrations (5, 10, and 30 %). **b** The same tube after centrifugation for 4 h at 4600 × g. Four zones are illustrated (I–IV). **c** Size distributions of the particles in layers I–IV. Particle diameters were measured by dynamic light scattering [26]. Ox-SWNH/SC, oxidized SWNH aggregate/sucrose

sodium dodecylbenzenesulfonate by ultrasonication, and then subjected to sucrose density gradient centrifugation (Fig. 5a) [26]. This treatment yields three populations of SWNHs: individual SWNHs and SWNH aggregates with a diameter of ~30–40 nm at the top of the centrifuge tube (layers I Fig. 5b), SWNH aggregates with a diameter of ~70 nm in the middle of the centrifuge tube (layers II and III), and SWNH aggregates with a diameter of ~100 nm at the bottom of the centrifuge tube (layer IV). The shapes of the individual SWNHs correspond to straight and two- and three-way branched forms. The lengths of the straight SWNHs and each arm of the branched forms are about 10–70 nm, and their diameters are about 2–10 nm (Fig. 6).

Similar structures of small-sized SWNHs in larger quantities (yield >20 %) can be successfully separated by a surfactant-free oxidation protocol [27], which is similar to Hummer's oxidative exfoliation technique for graphene oxide separation from graphite [28]. The particle sizes of the obtained small-sized SWNH aggregates (S-SWNHs) are ~20–50 nm, as estimated by dynamic light scattering (Fig. 7a) [27]. The typical morphologies of the S-SWNHs and accompanying large-sized SWNH aggregates (L-SWNHs, ~80–120 nm) are shown in the TEM images in Fig. 7b. S-SWNHs obtained by this oxidation method are hydrophilic, and show stable dispersion in aqueous solutions (Fig. 7a, inset).

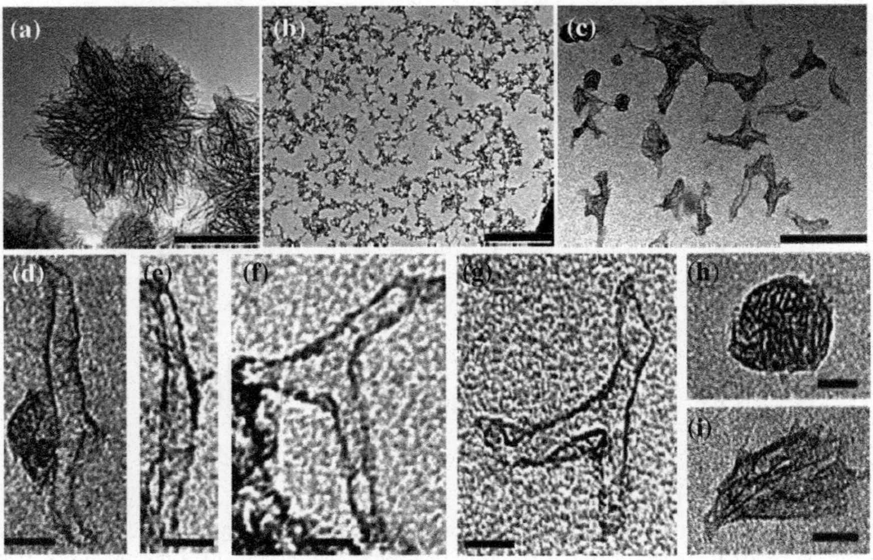

Fig. 6 **a** Typical TEM image of an oxidized SWNH aggregate. **b**, **c** TEM images of the particles in layer I (Fig. 5b). **d**, **e** TEM images of straight particles. **f**, **g** TEM images of branched forms. **h**, **i** TEM images of small aggregates. Scale bars, 50 nm (**a**, **b**), 100 nm (**c**), and 10 nm (**d**–**i**) [26]

1.4 Surface Functionalization

Covalent modification SWNHs have a hydrophobic graphite surface, which limits their applications in biological research fields. Many efforts have been made to change the nanohorn surface from hydrophobic to hydrophilic by functionalization with hydrophilic groups. The treatment of SWNHs with strong acids (e.g., HNO_3 or sulfuric acid (H_2SO_4), either singly or in combination) [15, 21, 22, 29] induces the formation of defects or holes, and also generates numerous carboxyl, carbonyl, and other oxygenated groups at the site of the defects and the edges of the holes. Hydrogen peroxide (H_2O_2)-mediated oxidation assisted by light irradiation also effectively generates carboxylic groups on the SWNH surface [21]. The carboxylic groups are then reacted via amidation with amino-containing compounds, including proteins (e.g., bovine serum albumin (BSA)), organic amines, alcohols, and thiols [22, 30, 31]. The obtained BSA-SWNHs exhibit high biocompatibility, stable dispersion in aqueous solutions, and good cell entry by endocytosis without cytotoxicity [22].

Another kind of chemical functionalization is initiated through the modification of as-grown SWNH sidewalls by direct attack of an amine (nucleophilic addition) [31, 32] or by 1,3-dipolar cycloaddition [33, 34]. The reacted SWNH derivatives show high solubility in common organic solvents, such as chloroform, dimethylformamide [33], and toluene [34].

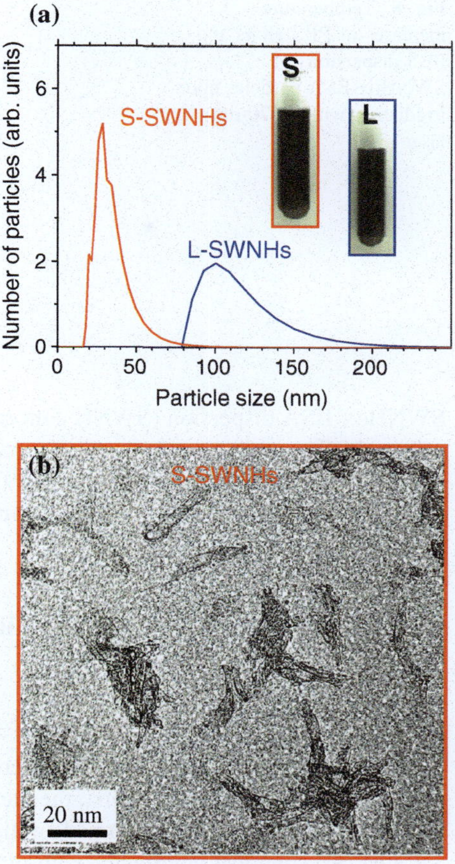

Fig. 7 **a** Dynamic light scattering measurements of aqueous dispersions of S-SWNHs and L-SWNHs. **b** TEM images of S-SWNHs [27]

Noncovalent coating. Noncovalent coating of SWNHs with polyethylene glycol (PEG)-based amphiphilic molecules can disperse nanohorn aggregates in aqueous solutions due to the hydrophilic properties of the PEG moieties. Physical modification of SWNHs with PEG successfully reduces non-specific adsorption of proteins onto SWNHs [35–38]. Different types of PEG-based macromolecules show different dispersion capabilities toward SWNH aggregates [37, 38].

1.5 Photo-Thermal Conversion Efficiency of SWNHs

SWNHs exhibit high photo-thermal energy conversion efficiency and high absorption cross-sections in a 650–1100-nm wavelength region of the so-called diagnostic window [39–46]. Oxidized SWNH particles complexed with BSA (SWNHox-BSA) and dispersed in aqueous solutions can effectively convert light energy into thermal energy [39, 40]. The temperature changes accompanying

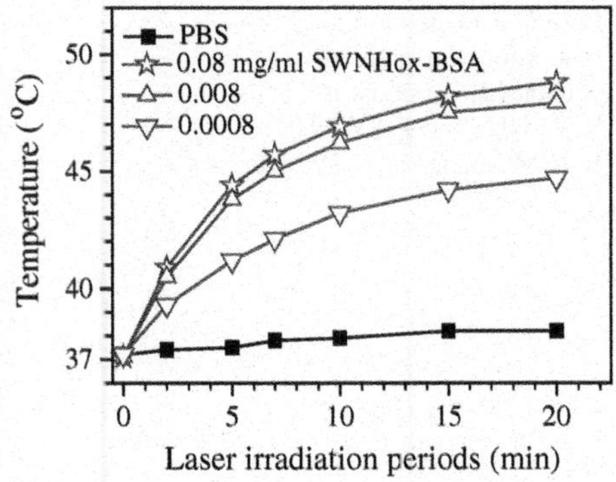

Fig. 8 Temperature increases of PBS alone and PBS dispersions of SWNHox-BSA particles upon irradiation with a 670-nm laser [39]

SWNHox-BSA dispersion (SWNH concentration = 8 μg/mL in phosphate buffered saline (PBS)) exhibit a temperature increase of ~10 °C upon irradiation with a 670-nm laser irradiation for ~10 min [39], whereas a control PBS solution without the SWNHox-BSA complex shows no such increase in temperature (Fig. 8).

2 Application of SWNHs in Biological Research Fields

2.1 *Drug Delivery*

SWNHs are regarded as promising carriers in drug delivery systems [47–54] due to the following characteristics. First, the nanohorns have large surface areas and total pore volumes, enabling abundant adsorption or storage of guest molecules. Second, SWNHs are spherical aggregates with diameters of ~100 nm, making them ideal for passive tumor-targeting conditions. Third, SWNHs can be produced in large quantities with high purity in the absence of metal catalysts, and in this way differ from carbon nanotubes. Drug incorporation inside an oxidized SWNHs (SWNHox) were initially demonstrated for dexamethasone (DEX) [47], an anti-inflammatory drug. Numerous DEX molecules were adsorbed onto the interior surface of the SWNHox, and the resultant DEX-SWNHox particle showed slow drug release kinetics, both in buffer solution and in cell culture medium.

The well-known anticancer agent, cisplatin (CDDP), can also be incorporated with up to 50 wt% of SWNHs efficiency inside the SWNHox particle to yield a CDDP@SWNHox delivery system (Fig. 9) [50]. The high anticancer efficacy of the derivatized particle has been tested in vitro and in vivo [49, 50], and stems from slow CDDP release from the SWNHox particle and a tendency of CDDP@SWNHox to attach to the cell surface. When CDDP@SWNH was locally injected into tumors, its anticancer actions were more pronounced than those of parental CDDP (Fig. 10), both due to slow drug release kinetics and the long particle retention time at the tumor site. These characteristics were attributed to the

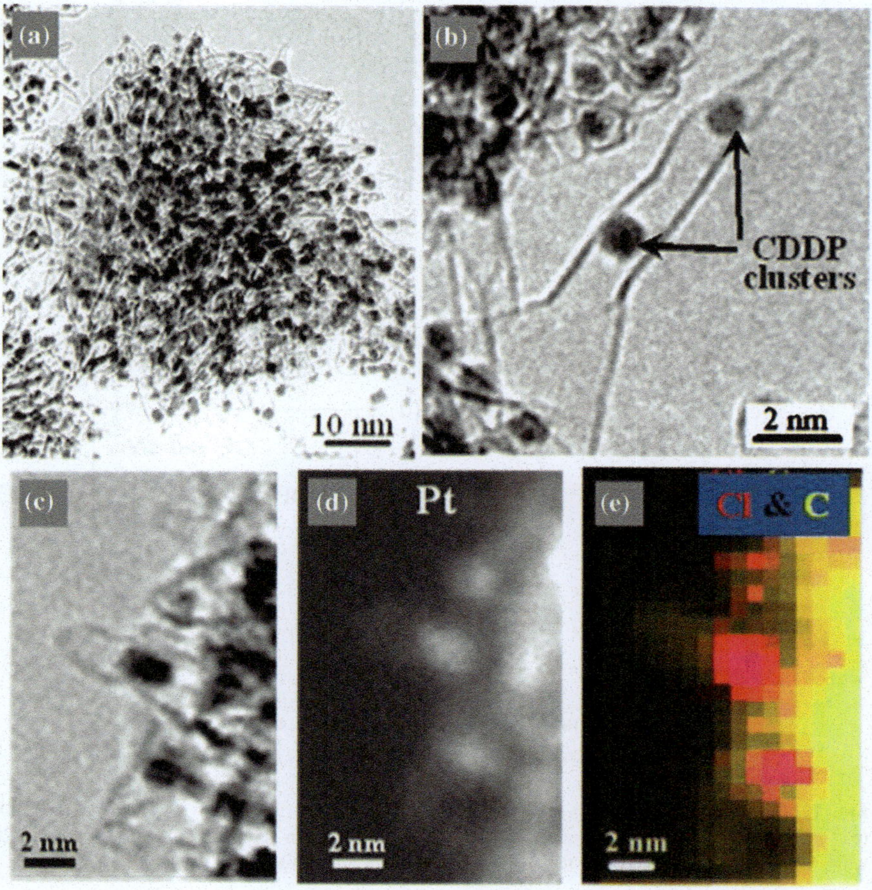

Fig. 9 TEM (**a**, **b**), scanning TEM (**c**), and Z-contrast (**d**) images of CDDP@SWNHox particles. C (carbon) and Cl (chloride) mapping was clarified by electron energy loss spectrum measurements (**e**). Observations or measurements of the images (**c**, **e**) were fixed at the same area. *Black* particles (**a**, **c**) indicate CDDP clusters. Two CDDP clusters within one CNHox sheath are indicated by *arrows* in (**b**). *Bright spots* in (**d**) correspond to platinum (Pt) atoms in the CDDP clusters. *Yellow* and *magenta* areas in (**e**) indicate the presence of C and Cl, respectively [50]

size of the CDDP@SWNHox particle, which precluded its ready drainage through the lymphatic system [50].

2.2 *Use of SWNHs as Photo-Hyperthermia Agents*

Besides their use as drug carriers, SWNHs are expected to be advantageous as photo-hyperthermia (PHT) agents for cancer therapy [39, 41, 46, and 55] and the

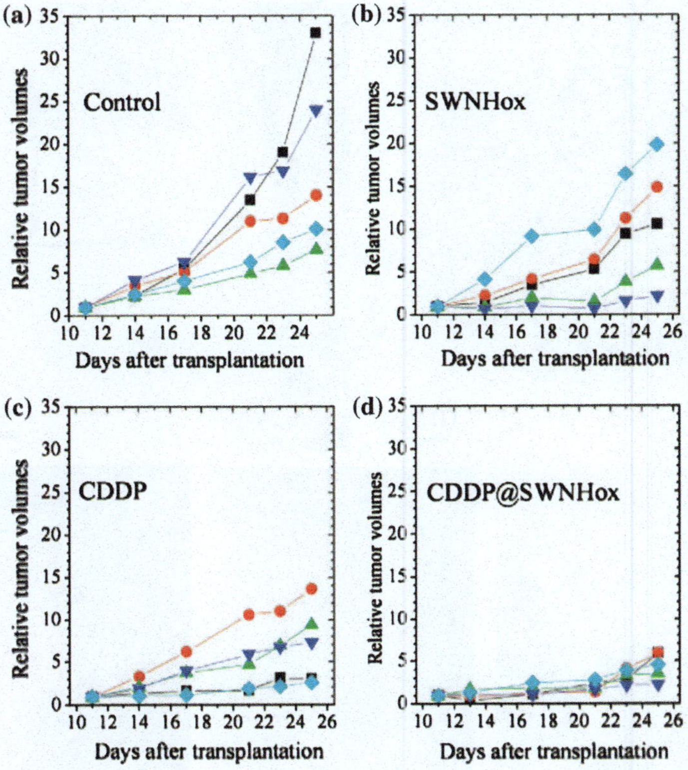

Fig. 10 Relative tumor volumes normalized at unity on day 11. Saline (**a**), SWNHox (**b**), CDDP (**c**), and CDDP@SWNHox (**d**) were intratumorally injected on day 11. The dosages of CDDP were 0.5 mg/kg. In each graph, the results of five mice are shown [50]

elimination of microbes [42] and viruses [43]. This is because the nanohorns absorb light in the phototherapy window (650–1100 nm), potentially transforming light energy into thermal energy and triggering cell death by a localized photo-thermal or PHT effect.

To evaluate the utility of SWNHs as PHT agents, a zinc phthalocyanine (ZnPc)-SWNHox-BSA conjugate was fabricated for the realization of such functions (Fig. 11). Here, ZnPc, a forthcoming agent for photodynamic therapy (PDT), was loaded inside and on the outer surface of a SWNHox with BSA attached to increase its biocompatibility [39]. Double PHT/PDT phototherapy with ZnPc-SWNHox-BSA particles and single-wavelength laser irradiation previously revealed the highly therapeutic impact of the derivatized biomaterial [39]. In in vivo tests, ZnPc-SWNHox-BSA was locally injected into tumors subcutaneously transplanted into mice. The nanohorn-loaded tumors were then subjected to laser irradiation (15 min/day for 10 days), prompting tumor disappearance (Fig. 12). These findings

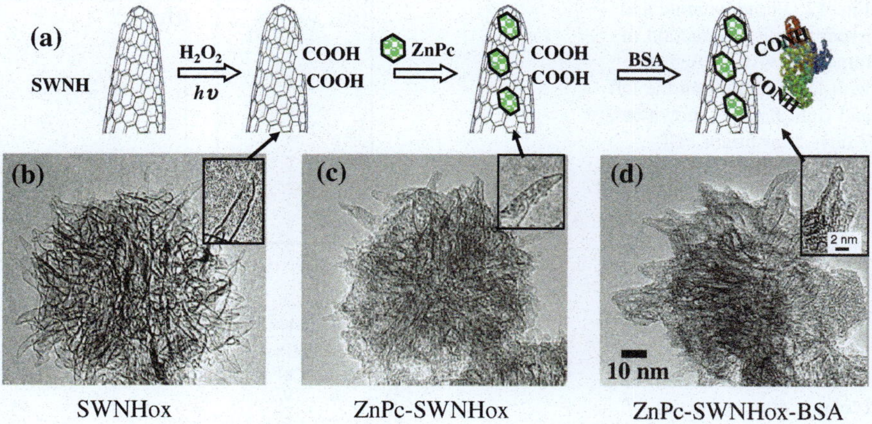

Fig. 11 Preparation of ZnPc-SWNHox-BSA particles. **a** Multistep production procedure. The nanohorns were visualized by TEM at each stage of production. **b** SWNHox, **c** ZnPc-SWNHox, and **d** ZnPc-SWNHox-BSA particles. *Insets* show magnified images [39]

were not replicated by ZnPc or SWNHox-BSA alone, suggesting that the enhanced antitumor efficiency was due to double phototherapy.

The photodynamic mechanism of ZnPc-SWNHox-BSA was investigated in detail via photophysical studies [39, 55]. Excitation of ZnPc by light-stimulated electron transfer to the SWNHox particle engenders a charge-separation state in the ZnPc-SWNHox-BSA system. In the presence of oxygen, electrons from the charge-separated ZnPc-SWNHox-BSA particle are transferred to O_2 to generate $O_2^{\bullet-}$ with the subsequent production of additional reactive species, such as hydroxyl radicals. The presence of these reactive species can then instigate the death of nearby cancer cells [39]. The photochemical processes involving ZnPc-SWNHox-BSA under aerobic conditions are illustrated in Fig. 13 [39, 55].

2.3 *Photo-Thermal Conversion Engineering*

The characteristics of photo-thermal energy conversion might also be useful in gene expression engineering [40], or the production of photoinduced nanomodulators for selective eradication of microbes [42]. In this regard, BSA-SWNHs can supply the thermal energy necessary for heat shock promoter-mediated gene expression when introduced into various cells and transgenic medaka (*Oryzias latipes*) [40].

Recently, functionalization of SWNHs with an infrared dye, IRDye800CW, permitted the fabrication of a light-driven nanomodulator for controlling calcium ion flux and membrane currents at the single-cell level [56]. IRDye800CW was chosen for labeling the nanohorns as it can generate reactivate oxygen species (ROS) through near infrared (NIR) light absorption, and promote successive

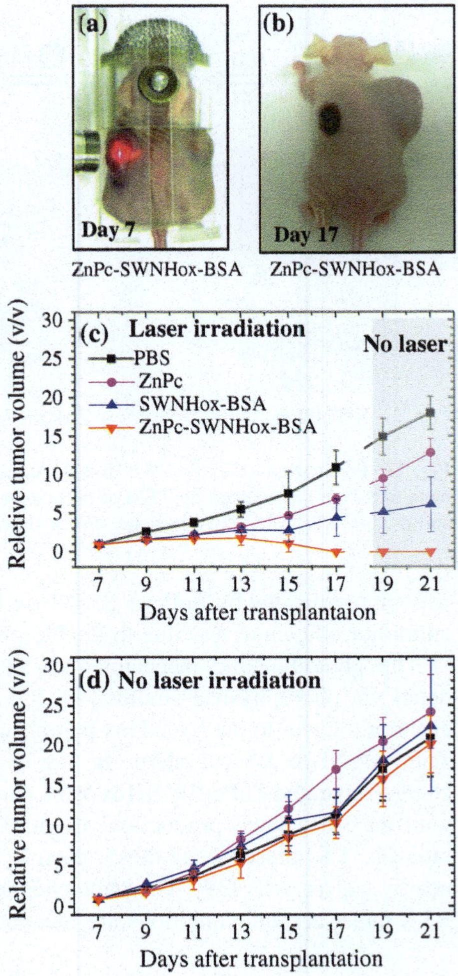

Fig. 12 Photodynamic and hyperthermic destruction of tumors in vivo. **a** A mouse with large tumors on the left and right flank is shown at 7 days after tumor cell transplantation (day 7). **b** A mouse after 10 days of treatment (day 17) with intratumorally-injected ZnPc-SWNHox-BSA particles and 670-nm laser irradiation of the tumor on the left flank. **c** Relative volume of the tumor on the left flank (volume after irradiation/volume before irradiation) following injection with PBS (*black line*) or PBS dispersions of ZnPc (*magenta line*), SWNHox-BSA particles (*blue line*), or ZnPc-SWNHox-BSA particles (*red line*) and irradiation with a 670-nm laser. **d** Relative volume of the tumor on the right flank following injection with PBS (*black line*) or PBS dispersions of ZnPc (*magenta line*), SWNHox-BSA particles (*blue line*), or ZnPc-SWNHox-BSA particles (*red line*), but not subjected to laser irradiation [39]

electron/energy transfer to the generated ROS [39, 55, 57, 58]. ROS-mediated regulation of calcium and other ion channel activities plays a central role in many free radical-driven processes, including stress, hormone signaling, and immunological responses.

After labeling with IRDye800CW, the obtained dye-labeled SWNHs (dye-SWNHs) generated both heat and ROS. Mouse neuroblastoma-derived ND7/23 hybrid cells were used to study the actions of dye-SWNH, because these cells express temperature-activated transient receptor potential ion channels (thermo-TRPs) and calcium ion channels (e.g., the inositol triphosphate (IP3) receptor). Incubation of ND7/23 cells with dye-SWNHs and irradiation with a NIR laser caused a local temperature increase, generation of high ROS levels, and the opening of thermo-TRPs and calcium ion channels. Channel opening then caused an influx and/or release of Ca^{2+} from intracellular compartments (Fig. 14a) [56].

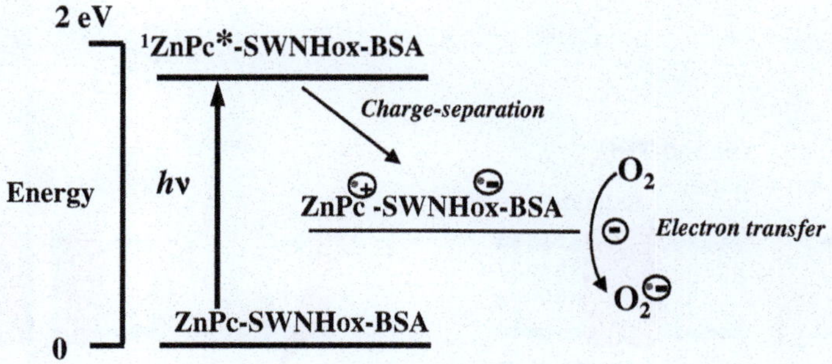

Fig. 13 Photochemical processes involving ZnPc-SWNHox-BSA in the presence of O_2. Under aerobic conditions both in vivo and in vitro, O_2 accepts an electron from the charge-separated ZnPc-SWNHox-BSA particle, generating $O_2^{\bullet-}$ [39, 55]

Next, the authors confirmed whether the nanomodulator-generated local temperature and ROS content were sufficiently increased to regulate channel activity. This was done by observing intracellular calcium flux through the use of a fluorescent calcium probe, Fluo-8. Following internalization of dye-SWNH nanomodulators into ND7/23 cells and cellular irradiation at 808 nm (laser power = 204 μW, or ~104 μW/μm^2), a bright green fluorescence due to calcium influx was observed. Fluo-8 fluorescence intensity was diminished in low TRP-expressing RAW264.7 and HeLa cells relative to ND7/23 cells, and fluorescence intensity was also reduced for dye-only and SWNH-only controls (Fig. 14b) [56].

Laser-triggered remote bioexcitation was further investigated in a model of a *Xenopus laevis* paw (Fig. 15) [56]. *X. laevis* have nerves that express thermo-TRPs and a variety of other channel proteins [59]. NIR light can penetrate into the tissue for a distance of ≤10 cm [60], allowing noninvasive excitation of dye-SWNH nanomodulators within a significant area of the thigh. Dye-SWNHs were injected under the thigh of a euthanized frog. Immediately after the initiation of irradiation (wavelength = 800 nm; laser power = 2.1 W, or ~292 mW/mm^2), paw twitching was observed. The same movement was absent when underivatized NH2-SWNHs, IRDyeCW800 alone, or Ringer solution alone without SWNHs was injected. These results indicate that dye-SWNHs can mediate cell stimulation in *X. laevis* tissues, facilitating significant bioexcitation [56].

2.4 *Biosensor Applications*

SWNHs are useful in electrochemical and biosensor applications [61–66], because they are metal-free, biocompatible, easily functionalized, and possess very large surface areas. Xu and colleagues first reported the use of SWNHs for construction of a glucose biosensor in 2008 [61]. The authors fashioned a biosensor by encapsulating glucose oxidase in a Nafion-SWNHs composite film. Ferrocene monocarboxylic acid was used as a redox mediator to decrease detection potential.

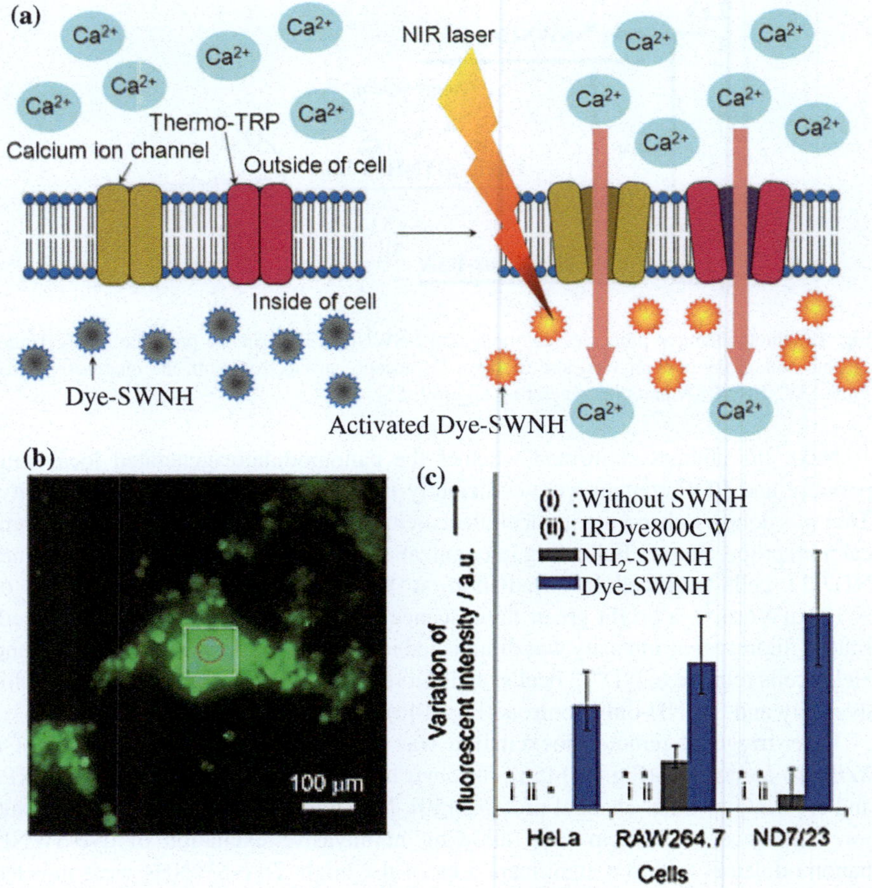

Fig. 14 Cellular stimulation by nanomodulators. **a** Illustration of the control of cellular activity by photoinduced nanomodulators. **b** Calcium imaging in ND7/23 cells incubated with dye-SWNHs, as assessed by fluorescence emission of the calcium probe, Fluo-8, after laser irradiation. The *white square* shows the site of fluorescence analysis, and the *red circle* indicates the irradiated site. **c** Analysis of fluorescence emission by irradiated cells with internalized dye-SWNHs. *No significant difference in fluorescence intensity [56]

The mediated biosensor demonstrated good electro-catalytic activity toward oxidation of glucose, and showed a linear range from 0 to 6.0 mM. The biosensor also showed high sensitivity (1.06 μA/mM) and stability, and avoided the commonly coexisting interference. Later, the same group reported development of a H_2O_2 biosensor based on the direct electrochemistry of SWNHs noncovalently modified with soybean peroxidase and deposited on a modified electrode surface [62]. This biosensor exhibited high sensitivity (16.625 μAL/mmol) with a detection limit of 5.0×10^{-7} mmol/L and excellent stability, indicating that noncovalent

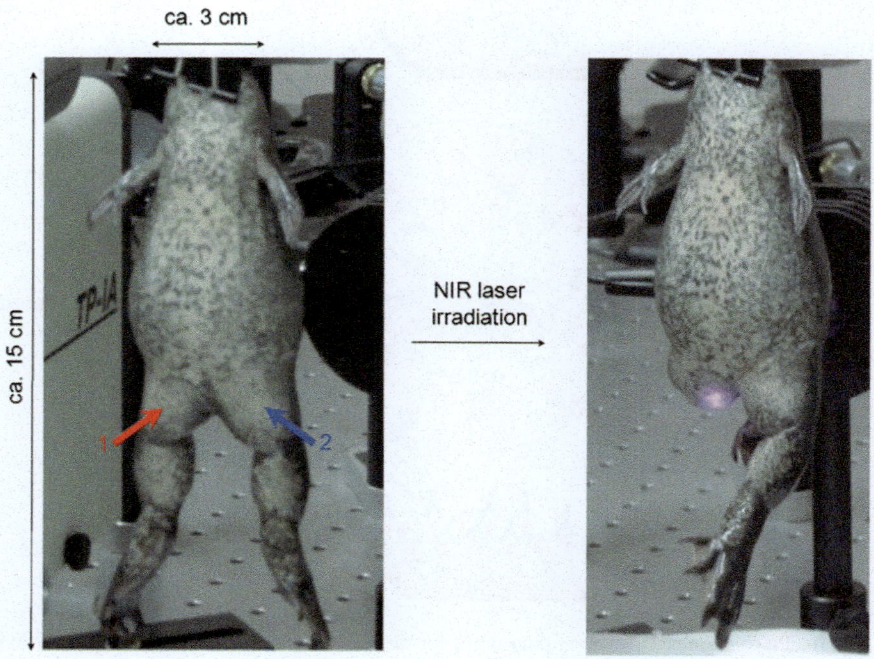

Fig. 15 Laser-driven remote stimulation of a frog (*X. laevis*) before (*left*) and after (*right*) laser irradiation. *Red* and *blue arrows* show the location of dye-SWNH (1) and Ringer solution (2) injection. Sample concentrations of dye-SWNH, NH_2CNH, and IRDye800CW were 300, 225, and 75 mg/mL, respectively [56]

functionalization of SWNHs can facilitate their applications in biosensor and electrochemistry research fields.

Recently, sensitive electrochemical immunosensors incorporating SWNHs were developed for detection of toxins, glycoproteins, and cancer biomarkers [64–66]. For example, SWNHs were functionalized by covalently binding microcystin-LR (MC-LR) to the carboxylic groups on the nanohorn tips for MC-LR toxin-detection (Fig. 16) [64]. An antibody against horseradish peroxidase (HRP)-labeled MC-LR was then prepared and utilized in a competitive immunoassay. Subsequently, the immunosensor exhibited a wide linear response to MC-LR under optimal conditions, ranging from 0.05 to 20 μg/L with a detection limit of 0.03 μg/L at a signal-to-noise ratio of 3, which was much lower than the World Health Organization/WHO provisional guideline limit of 1 μg/L for MC-LR in drinking water [67]. These results suggest that SWNHs can provide a useful platform for preparation of immunosensors for small-toxin molecules, which could readily be extended toward the on-site monitoring of hazardous components in food and environmental matrices.

Toward the detection of cancer biomarkers, a novel sandwich-type electrochemical immunosensor based on functionalized nanomaterial labels and bi-enzyme

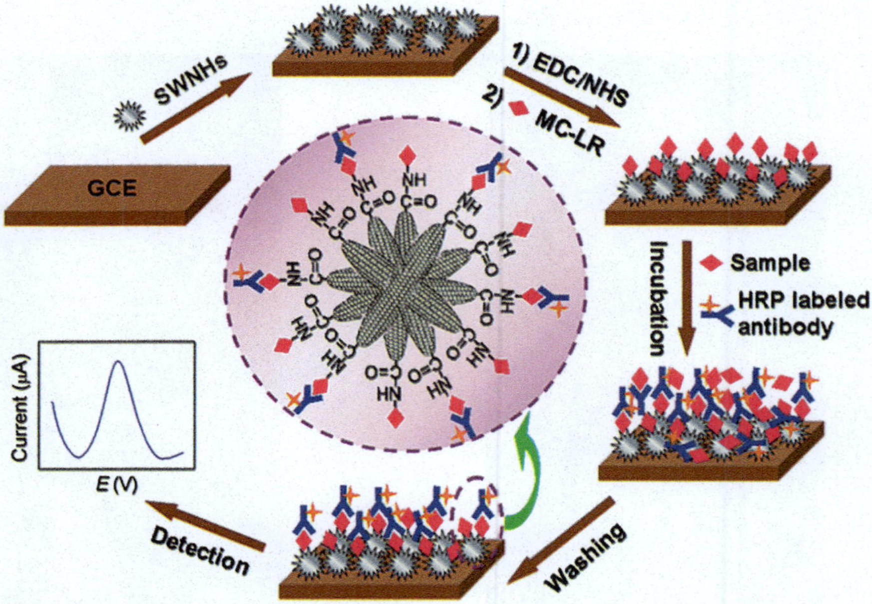

Fig. 16 Preparation and detection procedures for the MC-LR immunosensor by using functionalized SWNHs. EDC, 1-ethyl-3-(3-dimethylaminopropyl)carbodiimide; NHS, N-hydroxysuccimide [64]

(horseradish peroxidase and glucose oxidase)-catalyzed precipitation was developed for the detection of α-fetoprotein (AFP) [66]. The enzymes were linked to functionalized SWNHs and used as biocatalysts to accelerate 4-chloro-1-naphthol oxidation by H_2O_2 to yield a insoluble precipitate on the electrode surface. Mass loading of the precipitate led to a significantly enhanced signal relative to conventional loading techniques. Under optimal conditions, the immunosensor showed high sensitivity with a low detection limit of 0.33 pg/mL, and a wide linear range from 0.001 to 60 ng m/L. The immunosensor also exhibited good selectivity, acceptable stability, and good reproducibility. Therefore, the proposed amplification strategy described in Ref. [66] is promising for SWNH-based clinical screening of tumor biomarkers.

3 SWNH Toxicity and Biodegradation

3.1 Toxicity

The toxicity of SWNHs, SWNHoxs, and their functionalized forms has been investigated in vitro and in vivo [22, 32, 46–54]. Importantly, no serious cell death was found in in vitro cellular viability assessments. Tailored toxicity tests [58]

Table 1 SWNH toxicological tests [68]

Test	Test organism/animal	Dosage	Findings
Reverse mutation (Ames) test	S. typhimurium and *E. coli* strains	78–1250 g/plate	No positive increase in revertants; no growth inhibitory effect
Chromosomal aberration test	Chinese-hamster lung fibroblast cell line	0.010–0.078 or 0.313–2.5 mg/mL	Negligible positive incidences of structural chromosomal aberrations or polyploidy
Skin primary irritation test	Rabbits	0.015 g/site	Primary irritation index (P.I.I.) = 0; no clinical signs of abnormalities; normal body weight gain
Eye irritation test	Rabbits	0.02 g/eye	Draize irritation score = 0; no clinical signs of abnormalities; normal body weight gain
Skin sensitization (adjuvant and patch) test	Guinea pigs	0.02 g/site (induction); 0.01 g/site (challenge)	Mean response score 0; no clinical signs of abnormalities; normal body weight gain
Peroral administration test	Rats	2000 mg/kg	No mortality; no clinical signs of abnormalities; normal body weight gain
Intratracheal instillation test	Rats	2.25 mg/animal (17.3 mg/kg)	No mortality; rale for all animals including control group; normal body weight gain; black lung spots and anthracosis; foamy macrophage in intra-alveolar spaces

likewise showed no abnormal signs in animals (Table 1); and histological studies, blood tests, and cytokine measurements showed no appreciable abnormalities in most organs [69, 70], hematological outcomes, or immune responses [70] after intravenous injection of functionalized SWNHs and SWNHoxs.

The availability of large quantities of SWNHs was exploited by assessing cytotoxicity and immune responses following the abundant uptake of these structures by RAW264.7 murine macrophages [71]. High cellular uptake of the nanohorns was accompanied by localization of SWNHs at lysosomes, destabilization pf lysosomal membranes, and ROS generation with ensuing apoptotic and necrotic cell death. Despite these findings, only low levels of cytokines were released by the SWNH-loaded macrophages [71]. Next, the nanohorn-triggered cell death process was investigated in more detail to identify an underlying mechanism of ROS-provoked apoptosis [72]. The results showed that SWNHs accumulated in the lysosomes of RAW264.7 macrophages, where they induced lysosomal membrane permeabilization and the subsequent release of cathepsins and other lysosomal proteases, in turn triggering mitochondrial dysfunction and production of ROS in the mitochondria. Nicotinamide adenine dinucleotide phosphate oxidase was not directly involved in SWNH-directed ROS production, and ROS generation was not regulated by the mitochondrial electron transport chain. An ROS feed-back loop further amplified the mitochondrial dysfunction, leading to the activation of caspases and cell apoptosis [72].

3.2 Biodistribution

Labeling of SWNHs with gadolinium(III) oxide (Gd_2O_3) Although SWNHs show high potential for use in drug delivery systems and other biomedical applications [46–54], several issues must be addressed prior to their practical employment. These issues include SWNH biodistribution, excretion, and degradation, and bring into question not only the safety of SWNHs, but also the efficiency of in vivo nanohorn delivery. SWNHs have no unique characteristics that are useful for detection, and thus, they cannot be easily quantified in cells or whole-body systems based on intrinsic properties. To overcome this problem, a Gd_2O_3 label was incorporated into SWNHs to yield Gd_2O_3@SWNHs through holes opened by oxidation [73–77]. The holes were closed by heat treatment after Gd_2O_3 incorporation, permitting stable labeling of the biomaterial without leaching of Gd_2O_3 into the surrounding biological environment.

Measurement of Gd content in Gd_2O_3@SWNHs by inductively coupled plasma atomic emission spectroscopy (ICP-AES) permitted an estimation of nanohorn accumulation in every mouse organ and tissue [76]. In addition, the embedded Gd_2O_3 in the SWNH aggregates also facilitated ultrastructural localization of SWNHs by TEM, because Gd is not found naturally within the body. The high electron-scattering ability of Gd provided high contrast for clear detection by TEM (Fig. 17), and energy dispersive X-ray spectroscopy (EDX) or electron energy loss spectroscopy (EELS) further assisted in the identification of the SWNH aggregates. TEM coupled with elemental analysis by EDX and EELS is a conventional technique for localizing fine particles derived from various materials in living tissues [78–81].

Biodistribution of SWNHs with glucose-coating Gd_2O_3@SWNHs were dispersed in glucose and intravenously injected into mice. The Gd_2O_3@SWNHs were rapidly entrapped by the liver and the spleen, and these organs retained 70–80 and 10 % of the initially injected dose, respectively, in <30 min. Ultrastructural localization of Gd_2O_3@SWNHs in the murine liver revealed their presence in cellular phagosomes and phagolysosomes [76].

Biodistribution of SWNHs with PEG coating In another experiment, our research group dispersed Gd_2O_3@SWNHs into a solution of DSPE-PEG to enhance the stealth properties of SWNHs. As expected, the PEG-coated Gd_2O_3@SWNHs had a relatively long circulatory half-life (5–6 h) (Fig. 18a) after intravenous injection into mice [77] compared with SWNHs dispersed in glucose [76].

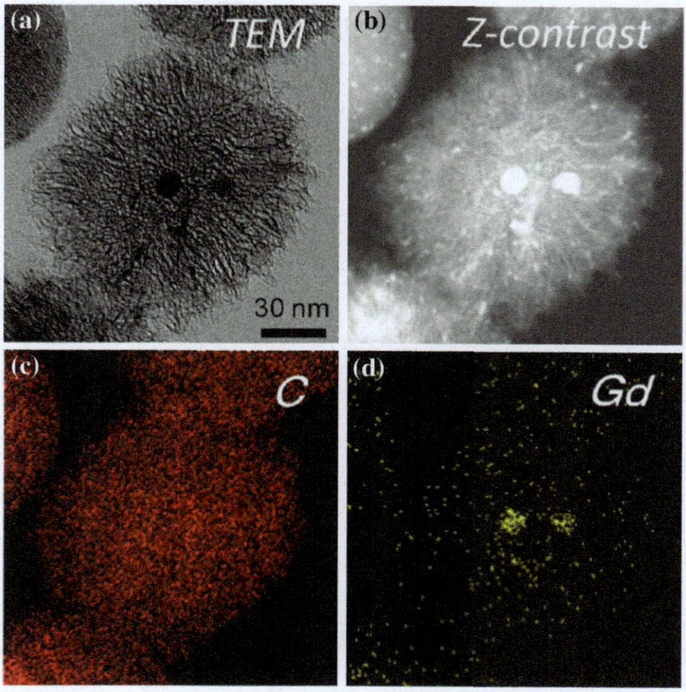

Fig. 17 Microscopic structure of the Gd_2O_3@SWNH powder. **a** TEM images and **b** Z-contrast images. **c** Carbon (C) and **d** Gd maps measured in the same area shown in (**a**) and (**b**) [76]

To understand the fate of SWNHs within the body, we analyzed SWNH content in the whole body and every organ of the mouse at seven time points over a period of 4 months. SWNH contents in various organs as a percentage of the initial dose are shown in Fig. 18b. SWNH content in the liver increased and reached a maximum of ~70 % at a post-injection time (PIT) of 24 h, before decreasing to ~30 % at a PIT of 30 days. Thereafter, no further appreciable changes in SWNH content were observed up until the end of the study period (120 days) (Fig. 18b). The changes in nanohorn content were paralleled by color changes in the liver; namely, the liver became gradually darker in color from PIT 1–24 h due to the presence of labeled SWNHs, but gradually returned to baseline color at PIT 30 d, with little change occurring thereafter. SWNH content in the spleen increased from 6 % at 1 h to 10 % at 7 days, with little change discerned after the first week of the experiment. SWNH content in the kidney was ~3–5 % throughout the study period. No darkening of the kidney tissue was observed, but histological observations revealed black particles in the renal corpuscles [77].

SWNH content and associated color deposition showed no significant changes in other parts of the body during the observation period, including the heart, lung, and stomach (Fig. 18b). Interestingly, SWNH content in the whole body of the mouse decreased to 60 % at a PIT of 30 days, with no further changes at later times. These

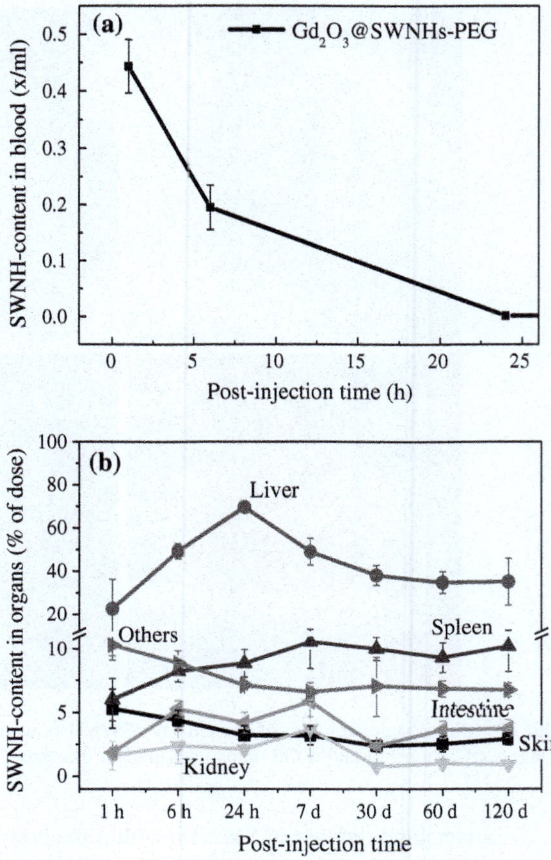

Fig. 18 **a** Blood-circulation behavior and **(b)** biodistribution of Gd_2O_3@SWNHs-PEG in the whole body of the mouse from 1 h to 4 months after a single intravenous injection of the biomaterial (dose = 5 mg/kg). **b** SWNH content in tissues and organs was determined by measuring the Gd content via ICP-AES. All data are given as a percentage of the original total injected dose (means ± the SD, n = 5) [77]

results signify that ~40 % of the injected SWNH dose was lost from the mouse body within 30 days (Fig. 19a). SWNH content in the feces was also measured, and was found to be ~15 % of the original dose at 60 days post-injection (Fig. 19b). We presumed that the other 25 % of the original dose lost from the body was degraded.

Size-dependent biodistribution The biodistribution of SWNHs administered to mice seems to depend on particle or aggregate size [70]. Two different sizes of SWNH aggregates (30–50-nm aggregates (S-SWNHs) and 80–120-nm aggregates (large-sized SWNH aggregates, or L-SWNHs)) were used to investigate nanohorn biodistribution and toxicity by histological analysis and blood testing for 7 days after intravenous injection into mice. Consequently, S-SWNHs accumulated more slowly in the liver and the spleen than L-SWNHs (Fig. 20), suggesting a longer time spent circulating in the blood.

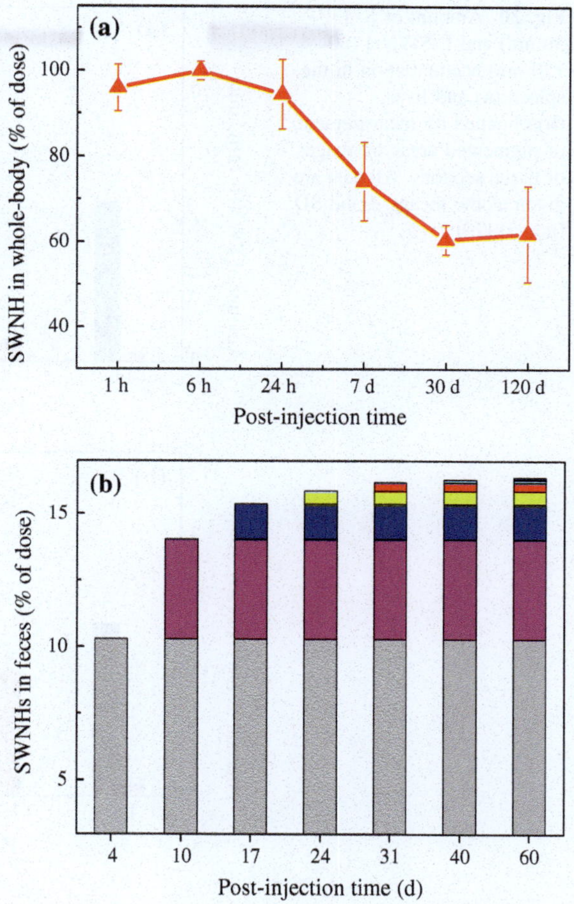

Fig. 19 **a** SWNH content in the whole body of the mouse after a single intravenous injection of Gd2O3@SWNHs-PEG from PIT 1 h to 4 months. Data are given as the means ± the SD (n = 5). **b** Average SWNH content in the feces collected from PIT 4 days to 2 months after a single intravenous injection of Gd2O3@SWNHs-PEG (n = 5) [77]

3.3 *Biodegradation*

Historically, carbon nanomaterials have been generally regarded as non-biodegradable in tissues due to their stable and inert graphitic structures; however, recent studies indicate that peroxidase-based enzymatic processes facilitate the oxidation and biodegradation of carbon nanotubes (CNTs) and graphene by cells and cell-derived catalysts [82–91]. For instance, oxidized SWNTs were enzymatically degraded by stimulated neutrophils without causing obvious pulmonary inflammation when the products of degradation were instilled in the lungs [84, 91].

Another study investigated SWNH degradation via enzymatic oxidation by isolated macrophage-derived myeloperoxidase (MPO) [92]. Here, SWNH dispersions were treated with a solution of human MPO supplemented with a low concentration of H_2O_2 (800 μM). The dark dispersion of MPO-/H_2O_2-treated SWNHs gradually lightened over 24 h, while the dark color of a control H_2O_2-treated

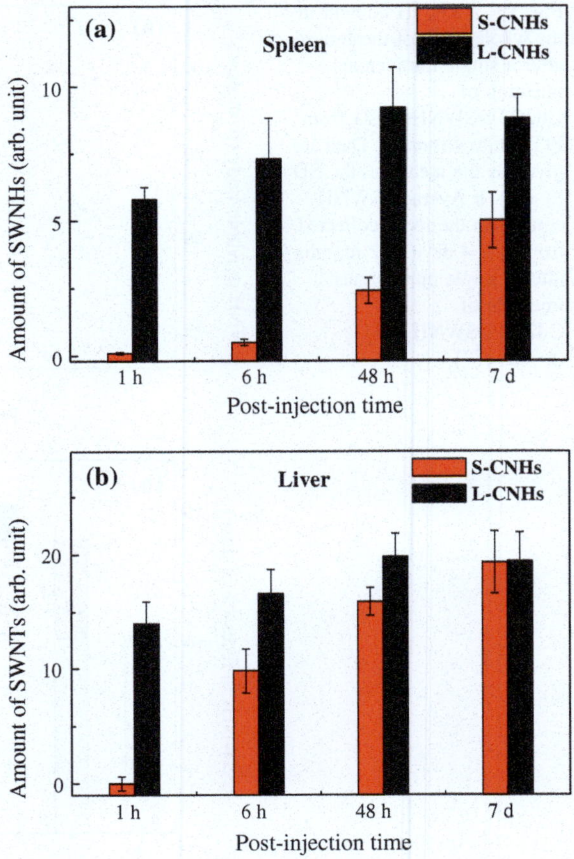

Fig. 20 Amount of S- (30–50 nm) and L-SWNH (80–120 nm) accumulation in the spleen (**a**) and liver (**b**) obtained by measurement of pigmented areas in images of tissue sections. All data are given as the means ± the SD (n = 5) [70]

SWNH dispersion showed no such change during the same time period (Fig. 21a). Optical absorption measurements [92] revealed that SWNH content decreased by approximately 40 wt% and 60 wt% after treatment with a combination of MPO and H_2O_2 for 5 h and 24 h, respectively (Fig. 21b). Additionally, TEM images showed that any SWNHs remaining after enzymatic oxidation for ~24 h were severely damaged, with collapse of the horn-shaped tips and spherical forms. The structures of the incompletely degraded SWNHs resembled amorphous carbon or graphite-like carbon nanoparticles [90], implying that MPO-mediated biodegradation occurs from the aggregate periphery to the center.

The cellular degradation of SWNHs by RAW 264.7 macrophages and THP-1 monocyte-derived macrophages was also investigated in vitro. The cells were incubated for 24 h with SWNHox (10 μg/mL), washed with PBS, and then re-seeded with SWNH-free fresh culture medium for an additional 1–9 days (recovery period). Images of the THP-1 cells after the recovery period showed a reduction in intracellular accumulation of SWNHs (Fig. 22a). Optical absorption

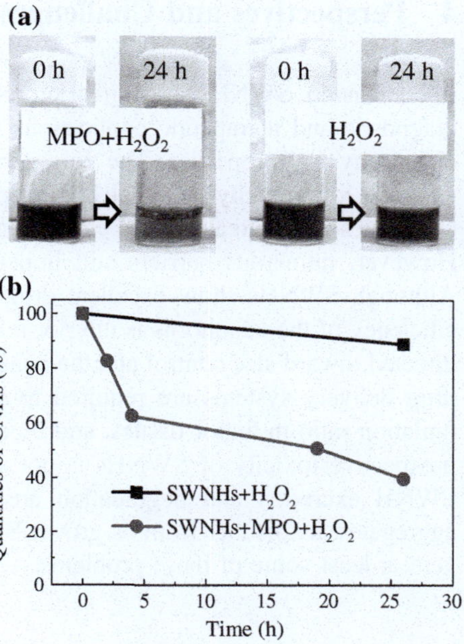

Fig. 21 Human MPO-mediated degradation of SWNHs. **a** Photographs of SWNH dispersions before and after treatment with 800 μM H_2O_2 in the presence or absence of MPO. **b** SWNH content as a percentage of the original dose after treatment with H_2O_2 in the presence or absence of MPO for the indicated time. SWNH content was determined based on optical absorbance at 700 nm [92]

measurements (shown for THP-1 cells in Fig. 22b) revealed that ~30 % of the internalized SWNHs disappeared within the 1–9-day time period for both macrophage cell lines [92].

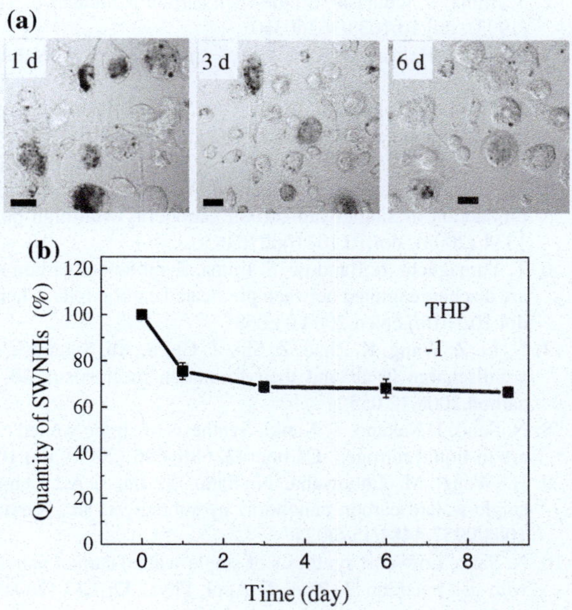

Fig. 22 Degradation of SWNHs in THP-1 monocyte-derived macrophages. Cells were differentiated by adding phorbol myristate acetate, incubated with SWNHs for 24 h, and then re-seeded in fresh medium without SWNHs and incubated for a further 1–9 days. **a** Confocal microscopy differential contrast images of THP-1 cells at days 1, 3, and 6. The black spots represent SWNHs. **b** SWNH accumulatation within the cells on days 1–9. Data represent SWNH content as a percentage of the original dose, and are given as the means ± the SD (n = 3 independent replicates) [92]

4 Perspectives and Challenges

In conclusion, SWNHs show great promise for use in biological research fields and diagnostic and therapeutic applications owing to their unique structure and associated physical properties. The pore geometry, photo-thermal transfer characteristics, and low toxicity of SWNHs offer potential solutions for many of the current challenges in the treatment of cancer, cardiovascular diseases, and other illnesses. However, numerous barriers still impede the practical employment of SWNHs. Although SWNHs show excellent drug-loading capabilities, the disease-targeting efficiency of the nanohorns is inferior relative to that of SWNTs. Additional efforts directed toward size control and the construction of multi-functional, SWNH-based drug delivery systems are required to increase the blood circulation time, accumulation ratio in tumor tissues, and biocompatibility of SWNHs. Furthermore, the prospective toxicity of SWNHs is not yet clear, especially in the long term, and SWNH excretion and degradation are difficult to study. Smaller-sized SWNH aggregates, as distinct from as-grown SWNHs, could possibly be used to circumvent at least some of these problems.

References

1. S. Iijima, M. Yudasaka, R. Yamada, S. Bandow, K. Suenaga, F. Kokai, K. Takahashi, Nano-aggregates of single-walled graphitic carbon nano-horns. Chem. Phys. Lett. **309**, 165–170 (1999). doi:10.1016/S0009-2614(99)00642-9
2. S. Iijima, T. Ichihashi, Single-shell carbon nanotubes of 1-nm diameter. Nature **363**, 603–605 (1993). doi:10.1038/363603a0
3. S. Iijima, Helical microtubules of graphitic carbon. Nature **354**, 56–58 (1991). doi:10.1038/354056a0
4. F. Kokai, K. Takahashi, D. Kasuya, M. Yudasaka, S. Iijima, Growth dynamics of single-wall carbon nanotubes and nanohorn aggregates by CO_2 laser vaporization at room temperature. Appl. Surf. Sci. **197–198**, 650–655 (2002). doi:10.1016/S0169-4332(02)00434-8
5. T. Azami, D. Kasuya, R. Yuge, M. Yudasaka, S. Iijima, T. Yoshitake, Y. Kubo, Large-scale production of single-wall carbon nanohorns with high purity. J. Phys. Chem. C **112**, 1330–1334 (2008). doi:10.1021/jp076365o
6. T. Yamaguchi, S. Bandow, S. Iijima, Synthesis of carbon nanohorn particles by simple pulsed arc discharge ignited between pre-heated carbon rods. Chem. Phys. Lett. **389**, 181–185 (2004). doi:10.1016/j.cplett.2004.03.068
7. N. Li, Z. Wang, K. Zhao, Z. Shi, Z. Gu, S. Xu, Synthesis of single-wall carbon nanohorns by arc-discharge in air and their formation mechanism. **48**, 1580–1585 (2010). doi:10.1016/j.carbon.2009.12.055
8. N. Sano, J. Nakano, T. Kanki, Synthesis of single-walled carbon nanotubes with nanohorns by arc in liquid nitrogen. Carbon **42**, 688–696 (2003). doi:10.1016/j.carbon.2003.12.078
9. H. Wang, M. Chhowalla, N. Sano, S. Jia, G.A.J. Amaratunga, Large-scale synthesis of single-walled carbon nanohorns by submerged arc. Nanotechnology **15**, 546 (2004). doi:10.1088/0957-4484/15/5/024
10. N. Sano, Low-cost synthesis of single-walled carbon nanohorns using the arc in water method with gas injection. J. Phys. D Appl. Phys. **37**, L17 (2004). doi:10.1088/0022-3727/37/8/L01

11. S. Bandow, F. Kokai, K. Takahashi, M. Yudasaka, L. Qin, S. Iijima, Interlayer spacing anomaly of single-wall carbon nanohorn aggregate. **321**, 514–519 (2000). doi:10.1016/S0009-2614(00)00353-5
12. D. Kasuya, M. Yudasaka, K. Takahashi, F. Kokai, S. Iijima, Selective production of single-wall carbon nanohorn aggregates and their formation mechanism. J. Phy. Chem. B **106**, 4947–4951 (2002). doi:10.1021/jp020387n
13. M. Inagaki, K. Kaneko, T. Nishizawa, Nanocarbons-recent research in Japan. Carbon **42**, 8–9 (2004). doi:10.1016/j.carbon.2004.02.032
14. K. Murata, K. Kaneko, F. Kokai, K. Takahashi, M. Yudasaka, S. Iijima, Pore structure of single-wall carbon nanohorn aggregates. Chem. Phys. Lett. **331**, 14–20 (2000). doi:10.1016/S0009-2614(00)01152-0
15. C. Yang, H. Noguchi, K. Murata, M. Yudasaka, A. Hashimoto, S. Iijima, K. Kaneko, Highly ultramicroporous single-walled carbon nanohorn assemblies. Adv. Mater. **17**, 866–870 (2005). doi:10.1002/adma.200400712
16. S. Utsumi, J. Miyawaki, H. Tanaka, Y. Hattori, T. Ito, N. Ichikuni, H. Kanoh, M. Yudasaka, S. Iijima, K. Kaneko, Opening mechanism of internal nanoporosity of single-wall carbon nanohorn. J. Phys. Chem. B **109**, 14319–14324 (2005). doi:10.1021/jp0512661
17. K. Murata, K. Kaneko, W. Steele, F. Kokai, K. Takahashi, D. Kasuya, K. Hirahara, M. Yudasaka, S. Iijima, Molecular potential structures of heat-treated Single-Wall Carbon Nanohorn Assemblies. J. Phys. Chem. B **105**, 10210–10216 (2001). doi:10.1021/jp010754f
18. J. Fan, M. Yudasaka, J. Miyawaki, K. Ajima, K. Murata, S. Iijima, Control of hole opening in single-wall carbon nanotubes and single-wall carbon nanohorns using oxygen. J. Phys. Chem. B **110**, 1587–1591 (2006). doi:10.1021/jp0538870
19. K. Ajima, M. Yudasaka, K. Suenaga, D. Kasuya, T. Azami, S. Iijima, Material storage mechanism in porous nanocarbon. Adv. Mater. **16**, 397–401 (2004). doi:10.1002/adma.200306142
20. E. Bekyarova, K. Kaneko, M. Yudasaka, D. Kasuya, S. Iijima, A. Huidobro, F. Rodriguez-Reinoso, Controlled opening of single-wall carbon nanohorns by heat treatment in carbon dioxide. J. Phys. Chem. B **107**, 4479–4484 (2003). doi:10.1021/jp026737n
21. C. Yang, D. Kasuya, M. Yudasaka, S. Iijima, K. Kaneko, Microporosity development of single-wall carbon nanohorn with chemically induced coalescence of the assembly structure. J. Phys. Chem. B **106**, 17775–17782 (2004). doi:10.1021/jp048391h
22. M. Zhang, M. Yudasaka, K. Ajima, J. Miyawaki, Sumio Iijima, Light-assisted oxidation of single-wall carbon nanohorns for abundant creation of oxygenated groups that enable chemical modifications with proteins to enhance biocompatibility. ACS Nano **1**, 265–272 (2007). doi:10.1021/nn700130f
23. T. Kawai, Y. Miyamoto, O. Sugino, Y. Koga, Nanotube and nanohorn nucleation from graphitic patches: Tight-binding molecular-dynamics simulations. Phys. Rev. B **66**, 033404 (2002). doi:10.1103/PhysRevB.66.033404
24. J. Miyawaki, R. Yuge, T. Kawai, M. Yudasaka, S. Iijima, Evidence of thermal closing of atomic-vacancy holes in single-wall carbon nanohorns. J. Phys. Chem. C **111**, 1553–1555 (2007). doi:10.1021/jp067283n
25. M. Zhang, M. Yudasaka, J. Miyawaki, J. Fan, S. Iijima, Isolating single-wall carbon nanohorns as small aggregates through a dispersion method. J. Phys. Chem. B **109**, 22201–22204 (2005). doi:10.1021/jp054793t
26. M. Zhang, T. Yamaguchi, S. Iijima, M. Yudasaka, Individual single-wall carbon nanohorns separated from aggregates. J. Phys. Chem. C **113**, 11184–11186 (2009). doi:10.1021/jp9037705
27. M. Zhang, X. Zhou, S. Iijima, M. Yudasaka, Small-sized carbon nanohorns enabling cellular uptake control. Sm Iijima all **8**, 2524–2531 (2012). doi:10.1002/smll.201102595
28. W. Hummer, R. Offeman, Preparation of graphitic oxide. J. Am. Chem. Soc. **80**, 1339 (1958). doi:10.1021/ja01539a017

29. S. Zhu, Z. Liu, L. Hu, Y. Yuan, G. Xu, Turn-on fluorescence sensor based on single-walled-carbon-nanohorn–peptide complex for the detection of thrombin. Chem. Eur. J. **18**, 16556–16561 (2012). doi:10.1002/chem.201201468
30. G. Pagona, N. Tagmatarchis, J. Fan, M. Yudasaka, S. Iijima, Cone-end functionalization of carbon nanohorns. Chem. Mater. **18**, 3918–3920 (2006). doi:10.1021/cm0604864
31. C. Cioffi, S. Campidelli, C. Sooambar, M. Marcaccio, G. Marcolongo, M. Meneghetti, D. Paolucci, F. Paolucci, C. Ehli, G.M. Rahman, V. Sgobba, D.M. Guldi, M. Prato, Synthesis, characterization, and photoinduced electron transfer in functionalized single wall carbon nanohorns. J. Am. Chem. Soc. **129**, 3938–3945 (2007). doi:10.1021/ja068007
32. H. Isobe, T. Tanaka, R. Maeda, E. Noiri, N. Solin, M. Yudasaka, S. Iijima, E. Nakamura, Preparation, purification, characterization, and cytotoxicity assessment of water-soluble, transition-metal-free carbon nanotube aggregates. Angew. Chem. Int. Ed. **45**, 6676–6680 (2006). doi:10.1002/ange.200601718
33. C. Cioffi, S. Campidelli, F.G. Brunetti, M. Meneghetti, M. Prato, Functionalisation of carbon nanohorns, ChemComm 2129–2131 (2006). doi:10.1039/B601176D
34. N. Tagmatarchis, A. Maigne, M. Yudasaka, S. Iijima, Functionalization of carbon nanohorns with azomethine ylides: towards solubility enhancement and electron-transfer processes. Small **2**, 490–494 (2006). doi:10.1002/smll.200500393
35. T. Murakami, J. Fan, M. Yudasaka, S. Iijima, K. Shiba, Solubilization of single-wall carbon nanohorns using a PEG—doxorubicin conjugate. Mol. Pharm. **3**, 407–414 (2006). doi:10.1021/mp060027a
36. S. Matsumura, S. Sato, M. Yudasaka, A. Tomida, T. Tsuruo, S. Iijima, K. Shiba, Prevention of carbon nanohorn agglomeration using a conjugate composed of comb-shaped polyethylene glycol and a peptide aptamer. Mol. Pharm. **6**, 441–447 (2009). doi:10.1021/mp800141v
37. J. Xu, S. Iijima, M. Yudasaka, Appropriate PEG compounds for dispersion of single wall carbon nanohorns in salted aqueous solution. Appl. Phys. A **99**, 15–21 (2010). doi:10.1007/s00339-010-5582-7
38. M. Yang, M. Wada, M. Zhang, K. Kostarelos, R. Yuge, S. Iijima, M. Masuda, M. Yudasaka, A high poly(ethylene glycol) density on graphene nanomaterials reduces the detachment of lipid–poly(ethylene glycol) and macrophage uptake. Acta Biomater. **9**, 4744–4753 (2013). doi:10.1016/j.actbio.2012.09.012
39. M. Zhang, M.T. Murakami, K. Ajima, K. Tsuchida, O. Ito, S. Iijima, M. Yudasaka, Fabrication of ZnPc/protein nanohorns for double photodynamic and hyperthermic cancer phototherapy. Proc. Natl. Acad. Sci. U.S.A. **105**, 14773–14778 (2008). doi:10.1073/pnas.0801349105
40. E. Miyakoa, T. Deguchi, Y. Nakajima, M. Yudasaka, Y. Hagihara, M. Horie, M. Shichiri, Y. Higuchi, F. Yamashita, M. Hashida, Y. Shigeri, Y. Yoshida, S. Iijima, Photothermic regulation of gene expression triggered by laser-induced carbon nanohorns. Proc. Natl. Acad. Sci. USA, **109**, 7523–7528 (2012). doi:10.1073/pnas.1204391109
41. J.R. Whitney, S. Sarkar, J. Zhang, T. Do, T. Young, M.K. Manson, T. Campbell, A. Puretzky, C. Rouleau, K. More, D. Geohegan, C. Rylander, H. Dorn, M.N. Rylander, Single-walled carbon nanohorns as photothermal cancer agents. Lasers Surg. Med. **43**, 43–51 (2011). doi:10.1002/lsm.21025
42. E. Miyako, H. Nagata, K. Hirano, Y. Makita, K. Nakayama, T. Hirotsu, Near-infrared laser-triggered carbon nanohorns for selective elimination of microbes. Nanotechnology **18**, 475103 (2007). doi:10.1088/0957-4484/18/47/475103
43. E. Miyako, H. Nagata, K. Hirano, K. Sakamoto, Y. Makita, K. Nakayama, T. Hirotsu, Photoinduced antiviral carbon nanohorns. Nanotechnology **19**, 075106 (2008). doi:10.1088/0957-4484/19/7/075106
44. E. Miyako, C. Hosokawa, M. Kojima, M. Yudasaka, R. Funahashi, I. Oishi, Y. Hagihara, M. Shichiri, M. Takashima, K. Nishio, Y. Yoshida, A photo-thermal-electrical convertor based on carbon nanotubes for bioelectronic applications. Angew. Chem. Int. Ed. **51**, 2266–12270 (2011). doi:10.1002/ange.201106136

45. E. Miyako, H. Nagata, H. Hirano, T. Hirotsu, Carbon nanotube–polymer composite for light-driven microthermal control. Angew. Chem. Int. Ed. Engl. **47**, 3610–3613 (2008). doi:10.1002/anie.200800296
46. S. Chechetka, B. Pichon, M. Zhang, M. Yudasaka, S. Begin-Colin, A. Bianco, E. Miyako, Multifunctional carbon nanohorn complexes for cancer treatment. Chem. Asian J. **10**, 160–165 (2015). doi:10.1002/asia.201403059
47. T. Murakami, K. Ajima, J. Miyawaki, M. Yudasaka, S. Iijima, K. Shiba, Drug-loaded carbon nanohorns: adsorption and release of dexamethasone in vitro. Mol. Pharm. **1**, 399–405 (2004). doi:10.1021/mp049928e
48. K. Ajima, T. Murakami A. Maigne, K. Shiba, S. Iijima, Carbon nanohorns as anticancer drug carriers. Mol. Pharm. 2:475–80 (2005). doi:10.1021/mp0500566
49. K. Ajima, A. Maigné, M. Yudasaka, Sumio Iijima, Optimum Hole-opening condition for cisplatin incorporation in single-wall carbon nanohorns and its release. J. Phys. Chem. B **110**, 19097–19099 (2006). doi:10.1021/jp064915x
50. K. Ajima, T. Murakami, Y. Mizoguchi, K. Tsuchida, T. Ichihashi, S. Iijima, M. Yudasaka, Enhancement of in vivo anticancer effects of cisplatin by incorporation inside single-wall carbon nanohorns. ACS Nano **2**, 2057–2064 (2008). doi:10.1021/nn800395t
51. T. Murakami, H. Sawada, G. Tamura, M. Yudasaka, S. Iijima, K. Tsuchida, Water-dispersed single-wall carbon nanohorns as drug carriers for local cancer chemotherapy. Nanomed. Lond. **3**, 453–463 (2008). doi:10.2217/17435889.3.4.453
52. J. Xu, M. Yudasaka, S. Kouraba, M. Sekido, Y. Yamamoto, S. Iijima, Single wall carbon nanohorn as a drug carrier for controlled release. Chem. Phys. Lett. **461**, 189–192 (2008). doi:10.1016/j.cplett.2008.06.077
53. K. Tsuchida, T. Murakami, Recent advances in inorganic nanoparticle-based drug delivery systems. Mini Rev. Med. Chem. **8**, 175–183 (2008). doi:10.2174/138955708783498078
54. M. Nakamura, Y. Tahara, Y. Ikehara, T. Murakami, K. Tsuchida, S. Iijima, I. Waga, M. Yudasaka, Single-walled carbon nanohorns as drug carriers: adsorption of prednisolone and anti-inflammatory effects on arthritis. Nanotechnology **22**, 465102 (2011). doi:10.1088/0957-4484/22/46/465102
55. A.S.D. Sandanayaka, O. Ito, M. Zhang, K. Ajima, S. Iijima, M. Yudasaka, T. Murakami, K. Tsuchida, Photoinduced electron transfer in zinc phthalocyanine loaded on single-walled carbon nanohorns in aqueous solution. Adv. Mater. **21**, 4366–4371 (2009). doi:10.1002/adma.200901256
56. E. Miyako, J. Russier, M. Mauro, C. Cebrian, H. Yawo, C. Ménard-Moyon, J. Hutchison, M. Yudasaka, S. Iijima, L. De Cola, A. Bianco, Photofunctional nanomodulators for bioexcitation. Angew. Chem. Int. Ed. **53**, 13121–13125 (2014). doi:10.1002/annie.201407169
57. M. Mitsunaga, M. Ogawa, N. Kosaka, L. Rosenblum, P. Choyke, H. Kobayashi, Cancer cell–selective in vivo near infrared photoimmunotherapy targeting specific membrane molecules. Nat. Med. **17**, 1685–1691 (2011). doi:10.1038/nm.2554
58. X. Peng, D. Draney, W. Volcheck, G. Bashford, D. Lamb, D. Grone, Y. Zhang, C. Johnson, Phthalocyanine dye as an extremely photostable and highly fluorescent near-infrared labeling reagent. Proc. SPIE Int. Soc. Opt. Eng. **6097**, 113–124 (2006). doi:10.1117/12.669173
59. M. Ohkita, S. Saito, T. Imagawa, K. Takahashi, M. Tominaga, T. Ohta, Molecular cloning and functional characterization of xenopus tropicalis frog transient receptor potential vanilloid 1 reveal its functional evolution for heat, acid, and capsaicin sensitivities in terrestrial vertebrates. J. Biol. Chem. **287**, 2388–2397 (2012). doi:10.1074/jbc.M111.305698
60. R. Weissleder, A clearer vision for in vivo imaging. Nat. Biotechnol. **19**, 316–317 (2001). doi:10.1038/86684
61. X. Liu, L. Shi, W. Niu, H. Li, S. Han, J. Chen, G. Xu, Amperometric glucose biosensor based on single-walled carbon nanohorns. Biosens. Bioelectron. **23**, 1887–1890 (2008). doi:10.1016/j.bios.2008.07.001

62. L. Shi, X. Liu, W. Niu, H. Li, S. Han, J. Chen, G. Xu, Hydrogen peroxide biosensor based on direct electrochemistry of soybean peroxidase immobilized on single-walled carbon nanohorn modified electrode. Biosens. Bioelectron. **24**(2009), 1159–1163 (2009)
63. X. Liu, H. Li, F. Wanga, S. Zhu, Y. Wang, G. Xu, Functionalized single-walled carbon nanohorns for electrochemical biosensing. Biosens. Bioelectron. **25**, 2194–2199 (2010). doi:10.1016/j.bios.2010.02.027
64. J. Zhang, J. Lei, C. Xu, L. Ding, H. Ju, Carbon nanohorn sensitized electrochemical immunosensor for rapid detection of microcystin-LR. Anal. Chem. **82**, 1117–1122 (2010). doi:10.1021/ac902914r
65. I. Ojeda, B. Garcinuñ, M. Moreno-Guzman, A. Gonzalez-Cortes, M. Yudasaka, S. Iijima, F. Langa, P. Yanez-Sedeno, J. Pingarron, Carbon nanohorns as a scaffold for the construction of disposable electrochemical immunosensing platforms. Application to the determination of fibrinogen in human plasma and urine. Anal. Chem. **86**, 7749–7756 (2014). doi:10.1021/ac501681n
66. F. Yang, J. Han, Y. Zhuo, Z. Yang, Y. Chai, R. Yuan, Highly sensitive impedimetric immunosensor based on single-walled carbon nanohorns as labels and bienzyme biocatalyzed precipitation as enhancer for cancer biomarker detection. Biosens. Bioelectron. **55**, 360–365 (2014). doi:10.1016/j.bios.2013.12.040
67. WHO, *Guidelines for drinking-water quality, Addendum to Volume 2, Health Criteria and Other Supporting Information* (World Health Organization, Geneva, Switzerland, 1998)
68. J. Miyawaki, M. Yudasaka, T. Azami, Y Kubo, S. Iijima, Toxicity of single-walled carbon nanohorns, ACS Nano, **2**(2), 213–226 (2008). doi:10.1021/nn700185t
69. Y. Taharal, J. Miyawaki,, M. Zhang, M. Yang, I. Waga, S. Iijima, H. Irie, M. Yudasaka, Histological assessments for toxicity and functionalization-dependent biodistribution of carbon nanohorns. Nanotechnology **22**, 265106, 8 (2011). doi:10.1088/0957-4484/22/26/265106
70. M. Zhang, T. Yamaguchi, S. Iijima, M. Yudasaka, Size-dependent biodistribution of carbon nanohorns in vivo, Nanomedicine. NBM **9**, 657–664 (2013). doi:10.1016/j.nano.2012.11.011
71. Y. Tahara, M. Nakamura, M. Yang, M. Zhang, S. Iijima, M. Yudasaka, Lysosomal membrane destabilization induced by high accumulation of single-walled carbon nanohorns in murine macrophage RAW 264.7. Biomaterials **33**, 2762–2769 (2012). doi:10.1016/j.biomaterials.2011.12.023
72. M. Yang, M. Zhang, Y. Tahara, S. Chechetka, E. Miyako, S. Iijima, M. Yudasaka, Lysosomal membrane permeabilization: Carbon nanohorn-induced reactive oxygen species generation and toxicity by this neglected mechanism. Toxicol. Appl. Pharmacol. **280**, 117–126 (2014). doi:10.1016/j.taap.2014.07.022
73. A. Hashimoto, H. Yorimitsu, K. Ajima, K. Suenaga, H. Isobe, J. Miyawaki, M. Yudasaka, S. Iijima, E. Nakamura, Selective deposition of a gadolinium(III) cluster in a hole opening of single-wall carbon nanohorn Proc. Natl. Acad. Sci. U.S.A. **101**, 8527–8530 (2004). doi:10.1073/pnas.0400596101
74. J. Miyawaki, M. Yudasaka, H. Imai, H. Yorimitsu, H. Isobe, E. Nakamura, S. Iijima, Synthesis of ultrafine Gd2O3 nanoparticles inside single-wall carbon nanohorns. J. Phys. Chem. B **110**, 5179–5181 (2006). doi:10.1021/jp0607622
75. R. Yuge, R. Ichihashi, J. Miyawaki, T. Yoshitake, S. Iijima, M. Yudasaka, Hidden caves in an aggregate of single-wall carbon nanohorns found by using Gd2O3 Probes. J. Phys. Chem. C **113**, 2741–2744 (2009)
76. J. Miyawaki, S. Matsumura, R. Yuge, T. Murakami, S. Sato, A. Tomida, T. Tsuruo, T. Ichihashi, T. Fujinami, H. Irie, K. Tsuchida, S. Iijima, K. Shiba, M. Yudasaka, Biodistribution and ultrastructural localization of single-walled carbon nanohorns determined in vivo with embedded Gd2O3 labels. ACS Nano **3**, 1399–1406 (2009). doi:10.1021/nn9004846
77. M. Zhang, Y. Tahara, M. Yang, X. Zhou, S. Iijima, M. Yudasaka, Quantification of whole body and excreted carbon nanohorns intravenously injected into mice. Advanced Healthcare Materials **3**, 239–244 (2013). doi:10.1002/adhm.201300192

78. H. Irie, W. Mori, Long term effects of thorium dioxide (Thorotrast) administration on heman liver. Acta Pathol. Jpn. **34**, 221–228 (1984). doi:10.1111/j.1440-1827.1984.tb07551.x
79. M. Yamamoto, K. Kato, Y. Ikada, Ultrastructure of the interface between cultured osteoblasts and surface modified polymer substrates. J. Biomed. Mater. Res. **37**, 29–36 (1997). doi:10.1002/(SICI)1097-4636(199710)37:1<29:AID-JBM4>3.0.CO;2-L
80. A.E. Porter, M. Gass, K. Muller, J.N. Skepper, P.A. Midgley, M. Welland, Direct imaging of single-walled carbon nanotubes in cells. Nat. Nanotechnol. **2**, 713–717 (2007). doi:10.1038/nnano.2007.347
81. C. Thakaral, J.L. Abraham, Automated scanning electron microscopy and X-ray microanalysis for in-situ quantification of gadolinium deposits in skin. J. Electron Microsc. **56**, 181–187 (2007). doi:10.1093/jmicro/dfm020
82. B.L. Allen, P.D. Kichambare, P. Gou, I. Vlasova, A. Kapralov, N. Konduru, V.E. Kagan, A. Star, Biodegradation of single-walled carbon nanotubes through enzymatic catalysis. Nano Lett. **8**, 3899–3903 (2008). doi:10.1021/nl802315h
83. B.L. Allen, G.P. Kotchey, Y. Chen, N. Yanamala, J. Klein-Seetharaman, V.E. Kagan, A. Star, Mechanistic investigations of horseradish peroxidase-catalyzed degradation of single-walled carbon nanotubes. J. Am. Chem. Soc. **131**, 17194–17205 (2009). doi:10.1021/ja9083623
84. V. Kagan, N. Konduru, W. Feng, B. Allen, J. Conroy, Y. Volkov, I. Vlasova, N. Belikova, N. Yanamala, A. Kapralov, Y. Tyurina, J. Shi, E. Kisin, A. Murray, J. Franks, D. Stolz, P. Gou, J. Klein-Seetharaman, B. Fadeel, A. Star, A. Shvedova, Carbon nanotubes degraded by neutrophil myeloperoxidase induce less pulmonary inflammation. Nat. Nanotechnol. **5**, 354–359 (2010). doi:10.1038/nnano.2010.44
85. Y. Zhao, B. Allen, A. Star, Enzymatic degradation of multiwalled carbon nanotubes. J. Phys. Chem. A **115**, 9536–9544 (2011). doi:10.1021/jp112324d
86. G.P. Kotchey, Y. Zhao, V.E. Kagan, A. Star, Peroxidase-mediated biodegradation of carbon nanotubes in vitro and in vivo. Adv. Drug Delivery Rev. **65**, 1921–1932 (2013). doi:10.1016/j.addr.2013.07.007
87. G.P. Kotchey, S.A. Hasan, A. Kapralov, S. Ha, K. Kim, A. Shvedova, V.E. Kagan, A. Star, A natural vanishing act: the enzyme-catalyzed degradation of carbon nanomaterials. Acc. Chem. Res. **45**, 1770–1781 (2012). doi:10.1021/ar300106h
88. A. Shvedova, A. Kapralov, W. Feng, E. Kisin, A. Murray, R. Mercer, C. St Croix, M. Lang, S. Watkins, N. Konduru, B. Allen, J. Conroy, G. Kotchey, B. Mohamed, A. Meade, Y. Volkov, A. Star, B. Fadeel, V. Kagan, Impaired clearance and enhanced pulmonary inflammatory/fibrotic response to carbon nanotubes in myeloperoxidase-deficient mice. PLoS ONE **7**, e30923 (2012). doi:10.1371/journal.pone.0030923
89. A. Nunes, C. Bussy, L. Gherardini, M. Meneghetti, M.A. Herrero, A. Bianco, M. Prato, T. Pizzorusso, K. Al-Jamal, K. Kostarelos, In vivo degradation of functionalized carbon nanotubes after stereotactic administration in the brain cortex. Nanomed. Lond. **7**, 1485–1494 (2012). doi:10.2217/nnm.12.33
90. Y. Sato, A. Yokoyama, Y. Nodasaka, T. Kohgo, K. Motomiya, H. Matsumoto, E. Nakazawa, T. Numata, M. Zhang, M. Yudasaka, H. Hara, R. Araki, O. Tsukamoto, H. Saito, T. Kamino, F. Watari, K. Tohji, Long-term biopersistence of tangled oxidized carbon nanotubes inside and outside macrophages in rat subcutaneous tissue. Sci. Rep. **3**, 2516 (2013). doi:10.1038/srep0216
91. V.E. Kagan, A. Kapralov, C. St, S.C. Croix, E.R. Watkins, G.P. Kisin, K.Balasubramanian Kotchey, I. Vlasova, J. Yu, K. Kim, W. Seo, R.K. Mallampalli, A. Star, A. Shvedova, Lung macrophages "digest" carbon nanotubes using a superoxide/peroxynitrite oxidative pathway. ACS Nano **8**, 5610–5621 (2014). doi:10.1021/nn406484b
92. M. Zhang, M. Yang, C. Bussay, S. Iijima, K. Kostarelos, M. Yudasaka, Biodegradation of carbon nanohorns in macrophage cells. Nanoscale **7**, 2834–3840 (2015). doi:10.1039/c4nr06175f

Bioimaging and Quantum Sensing Using NV Centers in Diamond Nanoparticles

Yuen Yung Hui, Chi-An Cheng, Oliver Y. Chen and Huan-Cheng Chang

Abstract Diamond nanoparticle hosting negatively-charged nitrogen vacancy (NV^-) center has unique chemical, optical and spin properties in a wide range of nanotechnology applications. For instance, diamond nanoparticles containing NV centers have been well-known as Fluorescent NanoDiamond (FND) for fluorescence imaging. Recently the NV^- center has been applied for nanothermometry. In this chapter we are going to discuss the recent advances of the NV^- center for bioimaging and quantum sensing.

Keywords Bioimaging · Fluorescent nanodiamond · Nitrogen vacancy center · Optically detected magnetic resonance · Quantum sensing

1 Introduction

Fluorescent NanoDiamond (FND) containing negatively-charged nitrogen vacancy (NV^-) center has recently attracted much attention in a wide range of nanotechnology applications [1]. FND owes its quantum properties to NV^- center's unique triplet ground state, whose spin not only can be optically detected but also optically manipulated [2]. Most remarkably, a NV^- center, locked within the diamond matrix, retains these quantum properties and has a spin coherence time as long as 10 ms even at ambient environments [3]. Combined with nanodiamond's chemical inertness and biocompatibility, FND makes the perfect candidate to import quantum physics into biology and opens up new territories and possibilities for bioimaging [4]. We are going to focus on the recent advance of the NV centers hosting in diamond nanoparticle [5] in this chapter, since the NV center in bulk diamond crystal has been recently comprehensively reviewed by Schirhagl et al. [1]. The

Y.Y. Hui (✉) · C.-A. Cheng · O.Y. Chen · H.-C. Chang (✉)
Institute of Atomic and Molecular Sciences, Academia Sinica, Taipe 106, Taiwan
e-mail: yyhui@pub.iams.sinica.edu.tw

H.-C. Chang
e-mail: hchang@gate.sinica.edu.tw

N. Yang et al. (eds.), *Carbon Nanoparticles and Nanostructures*,
Carbon Nanostructures, DOI 10.1007/978-3-319-28782-9_4

basic properties of a NV center are elaborated in Sect. 2. Then we will discuss the emerging technologies of the NV^- center for bioimaging and quantum sensing in Sects. 3 and 4 respectively. We also provide the perspective of the center in the final section.

2 Unique Properties of Negative Nitrogen-Vacancy (NV^-) Center

2.1 Electronic Structure of NV Center

A NV center is composed of a carbon vacancy site adjacent to one substitutional nitrogen atom. An imaginary line that passes through the vacancy site and the nitrogen atom is a symmetry axis, assuming along $\langle 111 \rangle$ direction (Fig. 1). The symmetry of the NV center belongs to the C_{3v} point group. In diamond, carbon's valence electrons occupy the sp^3 hybridized orbitals. In defectless diamond, all the orbitals are completely filled with 2 electrons. If one carbon atom were to be removed from the diamond matrix, it effectively removes 4 valence electrons along with +4-charged ion core. This defect, known as GR1 center, remains charge neutral and creates 4 holes in the previously completely filled orbitals. Moreover, if one of the carbon atoms near the vacancy is replaced by a nitrogen atom, then the one extra valence electron from a nitrogen atom will fill out one hole. This neutral NV defect center, known as NV^0, has 3 holes and zero net charge [6]. Furthermore, if this defect center traps one extra electron from an electron-rich source, possibly nearby defects or crystal surface [1], then another hole is annihilated. This NV^- center now has 2 holes and net charge of −1. Both neutral and negative NV center in diamond share similar optical properties for bioimaging (like red fluorescence and photostability), but only the negative NV center has demonstrated unique spin properties for quantum sensing.

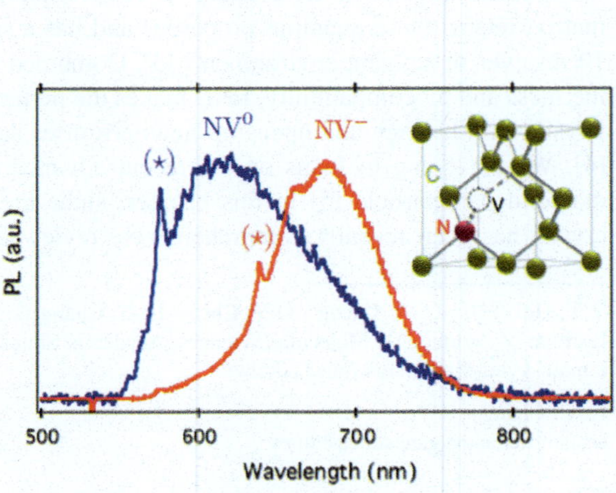

Fig. 1 PL spectra of NV^0 center and NV^- center. The zero-phonon line of NV^0 center and NV^- center are 575 nm and 637 nm respectively. *Inset*: Structure of NV center in diamond lattice. (Reprinted from Rondin et al. [7])

2.2 *Common Optical Properties of NV Center*

The NV^0 center and the NV^- center produce red fluorescence with zero phonon line (ZPL) at 576 nm and 637 nm respectively [7], shown in Fig. 1. Both ZPLs are accompanied with broad phonon sidebands red-shifted by about 50 nm. (It has further found that the NV^0 center can be converted to NV^- center in diamond by annealing in oxygen environment [7, 8].) Since both NV centers are hosted in the diamond lattice, they are photostable, compared with other organic fluorophores [9]. Apart from the excellent photostability, nanodiamond is biocompatible with animal [10–13]. Hence, the NV centers in nanodiamond enable long-term fluorescence tracking in biological systems.

2.3 *Unique Spin Properties of NV^- Center*

The NV^- center in diamond has unique spin properties, because of the triplet ground state and excited state. The experimental and theoretical constructions for the energy state diagram have been recently reviewed [14, 15]. The simplified energy level diagram is displayed in Fig. 2, which can describe the photo-physics of the NV^- center. The NV^- center is treated as a three-level system with a triplet ground state $^3A_2$, a triplet excited state ^3E and intermediate singlet states. The energy separation between the triplet ground state and the triplet excited states is 1.945 eV, which corresponds to the zero phonon line of NV^- center at 637 nm. Both triplet states of the NV^- center show a zero-field splitting among the spin sublevels with $m_s = 0$ and $m_s = \pm 1$ respectively. The energy splitting of the ground

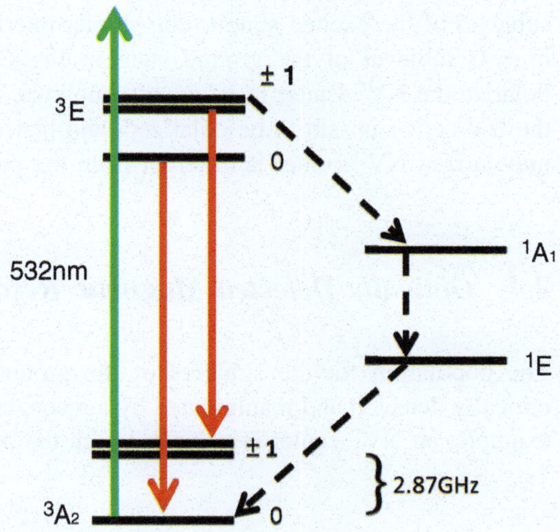

Fig. 2 Energy level diagram for a NV^- center

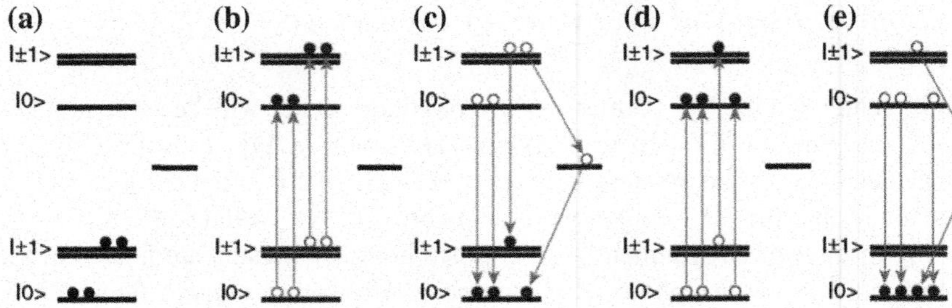

Fig. 3 Schematic representation for the polarization mechanism of a NV^- center. (Reprinted from Bradac et al. [2])

state (without an external field) has been measured as 2.87 GHz which corresponds to the microwave excitation.

Moreover, the spin-state of a NV^- center can be optically polarized [16]. In other words, optical excitation can induce a non-Boltzmann steady-state spin alignment of the NV^- center in the ground state, and hence the $m_s = 0$ sublevel of the ground state can be preferentially populated. We will illustrate the polarization mechanism in Fig. 3 [2]. Initially the NV^- center is unpolarized, wherein the $m_s = 0$ and $m_s = \pm 1$ sublevels of the ground state are equally populated in Fig. 3a. When the NV^- center is excited by a green laser, the population in the $m_s = 0$ and $m_s = \pm 1$ sublevels of the ground state shifts to the corresponding sublevels of the excited state. The optical transition is spin-conserving, and hence the spin number conserves in Fig. 3b. Furthermore, the population in the excited state can return to the ground state either by radiative transition or by non-radiative transition. The radiative transition produces the emission of red photons with ZPL at 637 nm. However, the non-radiative transition involving the singlet state is not necessarily spin-conserving. The dominant non-radiative decay path travels from the $m_s = \pm 1$ sublevel of the excited state to the singlet intermediate state and eventually to the $m_s = 0$ sublevel of the ground state in Fig. 3c [17]. A few optical cycles can polarize the NV^- center to its $m_s = 0$ sublevel, shown in Fig. 3e. The spin-state of the NV center is said to be polarized, and hence the population distribution of the unpolarized NV^- center is different from the polarized NV^- center.

2.4 Optically Detected Magnetic Resonance (ODMR)

The population for the sublevel of the ground state in the NV^- center can be optically detected and manipulated by a resonant microwave of 2.87 GHz [18]. For example, an NV^- center is polarized with the $m_s = 0$ sublevel of the ground state.

The application of a resonant microwave increases the population for the $m_s = \pm 1$ sublevel of the ground state, similar to magnetic resonance. If the NV^- center is further probed with a green laser, the fluorescence intensity will decrease because of the non-radiative decay via the intermediate singlet state. Hence, the fluorescence intensity from the NV^- center is correlated with the population for the sublevel of the ground state. The fluorescence intensity from carries the information of spin state of the NV^- center. We can slowly sweep an auxiliary microwave over the diamond sample to detect the spin state of the NV^- center optically, and obtain an optically detected magnetic resonance (ODMR) spectrum.

When the NV^- center is polarized and the $m_s = 0$ sublevel of the ground state is preferentially occupied, the center produce maximum fluorescence intensity. On the other hand, when the center is unpolarized, the $m_s = 0$ and $m_s = \pm 1$ sublevels of the ground state are both equally populated. Under optical excitation, the center will undergo the non-radiative decay via the transition between $m_s = \pm 1$ to the singlet state and so its fluorescence intensity is 20–30 % lower than the maximum intensity. This ODMR effect is a characteristic of the NV^- center, which has only been observed for a handful of molecules and defects.

2.5 *Production of NV Centers in Diamond Nanoparticles*

In 2001, single NV^- center in diamond nanoparticles was produced for quantum communication [19]. Later Yu et al. demonstrated that diamond nanoparticles hosting multiple NV centers are applicable for cellular fluorescence imaging [20]. Since then, there have been a large number of researches aiming to produce small and bright FND [21–24]. (For sake of bioimaging application, the FND particle should be small, and contain as many NV centers as possible.) NV centers can be created by irradiating diamond nanoparticles with high energy particles (electrons, neutrons, protons, helium ions) and followed by vacuum annealing at 600–800 °C. High-energy particle irradiation forms vacancies in the diamond structure, then vacuum annealing allows these vacancies to migrate and to be trapped by nitrogen atoms to form NV centers. Mass production of FND has also been developed for biomedical applications since 2008 [25]. More specifically, synthetic type Ib diamond powders (with mean sizes of 35 and 140 nm respectively) typically contain 100 ppm of atomically dispersed nitrogen atoms as the major internal impurity. Chang et al. applied 40 keV He^+ ion bombardment at a dose of about 1×10^{13} ions cm^{-2} on synthetic type Ib diamond powders for the creation of radiation damage. It is noted that He^+ ion owns some merits as the irradiation source. Firstly, helium atoms are chemically inert, and embedding these atoms in a diamond lattice does not significantly change the photophysical properties of the FNDs. Secondly, helium atoms possess remarkably high damage efficiency which reduces the dosage for ion irradiation. (For instance, a single 40 keV He^+ ion can create 40 vacancies in diamond, in contrast to the 0.1 and 13 vacancies generated by 2 meV e^- and 3 meV H^+ ion respectively.)

Based on density functional theory, the NV center concentration was predicted to increase nonlinearly with the crystal size [26]. The probability of producing NV centers was less than 5 % in High-Pressure-High-Temperature nanodiamond with 30 nm in size, due to the vacancy annihilation at the surface. In addition, quantitative analysis using Monte Carlo simulations for the size effect was performed [27]. The predicted probability of forming NV centers in nanodiamond with 5 nm in size was about 4.5 and 25 times lower than 20 nm and 55 nm respectively. Although the feasibility of producing NV centers in smaller ND particles is predicted to be low, many efforts have been contributed to produce ultra-small nanodiamond for bio-imaging. For example, Boudou et al. reduced the size of FND by ball milling [28, 29]. The procedure was shown in Fig. 4. The starting material was microdiamond instead of nanodiamond, because it offered a long vacancy migration path, which significantly increases the probability of vacancy trapping by nitrogen atom. Briefly, microdiamond powders were irradiated in a linear particle accelerator consisting of an electron beam with energy of 10 meV. Then, the irradiated particles were annealed at 750 °C to form fluorescent microdiamonds. The authors further applied nitrogen jet milling and ball milling in order to further convert fluorescent microdiamonds into smaller particles. The final size of the milled particles was smaller than 10 nm [29]. On the other hand, the size of the nanodiamond can also be reduced by air oxidation [30, 31]. In addition, Rabeau et al. produced NV^- centers in diamond nanocrystals by chemical vapour deposition (CVD). The NDs are grown directly on quartz cover slips in a microwave plasma CVD reactor, during when the substrate temperature was maintained at 800 °C. Single NV^- centers are observed in the resultant nanocrystals [32].

Furthermore, Mich et al. reported a microwave plasma-assisted chemical vapor deposition diamond growth technique on (111)-oriented bulk diamond substrates, and successfully produced perfect alignment of as-grown NV centers along a single crystallographic direction [33]. Now the aligned NV centers were formed by the incorporation of nitrogen atoms in the (111) growth surface and then followed by the neighboring vacancy on top. Meanwhile, Edmonds et al. produced NV center with two orientations, instead of four, by chemical vapour deposition [34]. (There are four possible orientations of the NV centers in the diamond crystal.) The achieved homogeneity of the grown NV centers in bulk diamond is anticipated to benefit quantum information and metrology applications. Multiple NV centers with a preferential direction in nanodiamond can impose significant impact to bioimaging and magnetometry [3, 35].

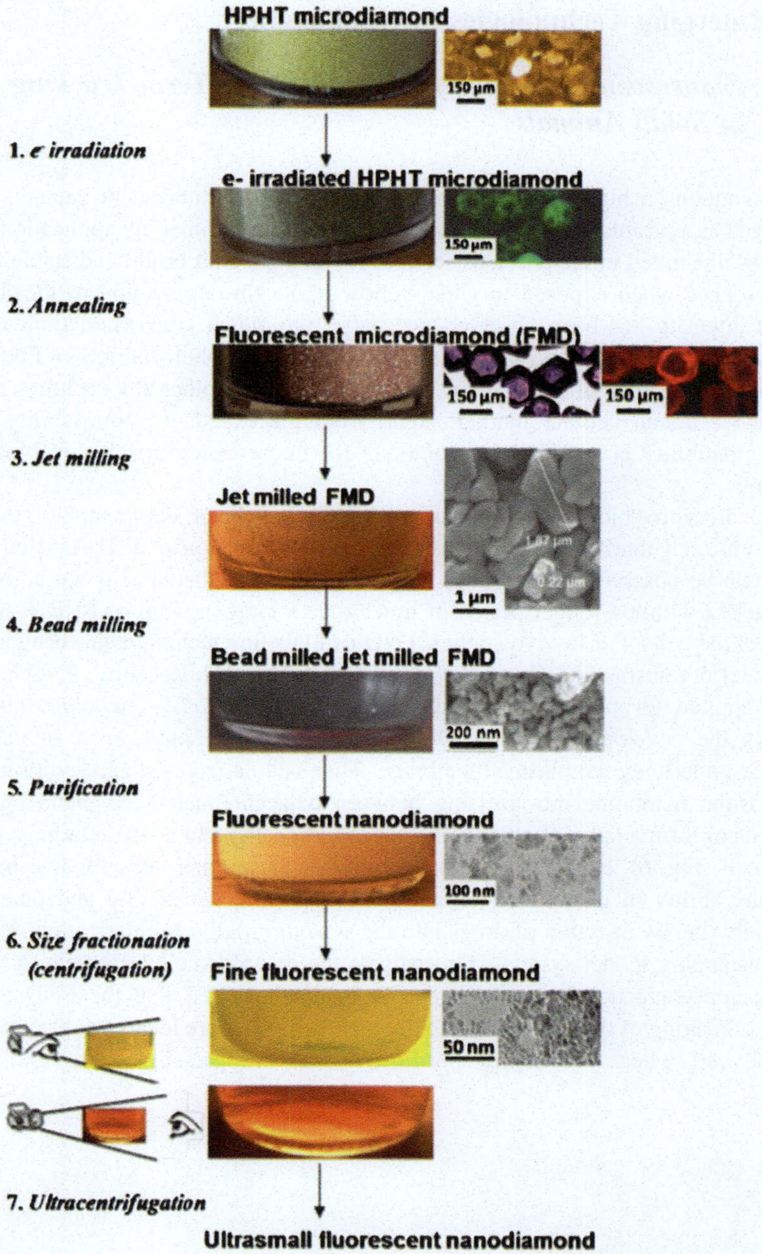

Fig. 4 Flow chart illustrating for the FND preparation. Sample images obtained at each step are provided in the center column. In contrast to previously synthesized nanodiamonds, a dispersion of fine fluorescent nanodiamonds in water is translucent and appears *orange* with a *white lamp* in the *inset* on the *left-hand side*. Microscope images on the *right-hand column*. (Reprinted from Boudou et al. [29])

3 Emerging Technologies for Bioimaging

3.1 Fluorescence Imaging for In Vivo Long-Term Tracking in Small Animals

FND contains a high concentration of NV defects as fluorescent centers. It has emerged as a promising new nanoparticle platform for bioimaging applications [36–38]. As illustrated in the previous section, NV centers emit bright and stable far red fluorescence when exposed to green-yellow light. No degradation of the fluorescence intensity has been observed even after prolonged laser irradiation of individual FND particles at room temperature [39]. Extreme photostability of FND is an outstanding feature of FND to be distinguished from other fluorophores and to become a potent cellular marker. Apart from the excellent photostability, good biocompatibility is another merit of FND for fluorescence imaging in biological systems.

For in vitro biological applications, FND has been demonstrated to be a long-term cell marker [40]. For instance, the cell division of FND-labeled HeLa cells can be observed up to 6 days (Fig. 5). The FND-labeled cells are allowed to be tracked within a longer period of time, once a large amount of FND is applied to label the cells. For in vivo studies, FND-cell labeling technique has been applied to model organisms like *Caenorhabditis elegans* [11, 41]. Recently, FND has also been applied to study another model organism, *Drosophila melanogaster* [42]. During the development of the *Drosophila* embryo, a single layer of nuclei is formed underlying the plasma membrane. Then cellularization begins with furrows of plasma membrane introgressing between adjacent nuclei and enclosing each nucleus to form a layer of outer cells (blastoderm cells) and a single large syncytial yolk cell (Fig. 6). Based on the accessibility of these outer cells for live imaging and the ability to manipulate cellular processes both genetically and pharmacologically (i.e. by injection of drugs into the syncytium), the cellularization serves as an outstanding model system to investigate the complex cellular processes underlying embryonic development. The authors demonstrated that the FND enables in vivo imaging of the embryos [42]. Most of the FND were located at the periphery of the cell where the membrane furrow was introgressing (Fig. 6c). Taking

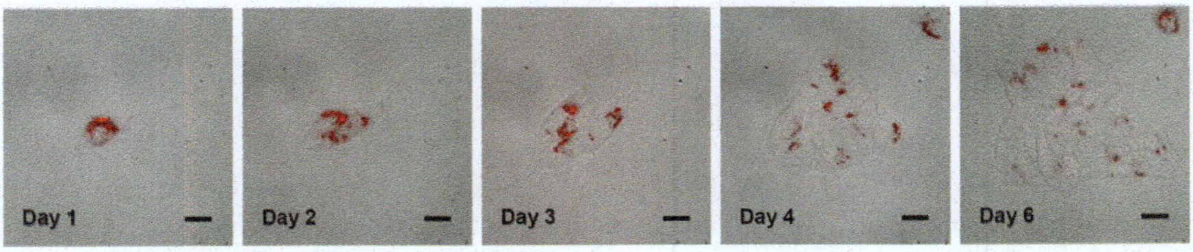

Fig. 5 Tracking for cell division of FND-labeled HeLa cells by DIC/epi-fluorescence microscopy. The tracking was conducted for up to 6 days of post-labeling incubation. Scale bars are 20 μm. (Reprinted from Fang et al. [40])

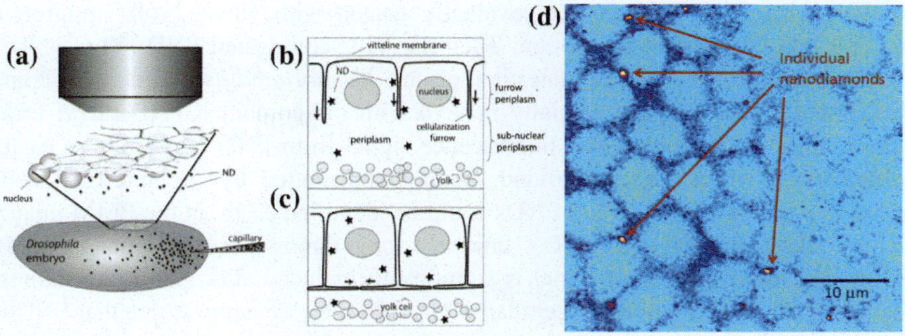

Fig. 6 **a** Schematic of the micro-injection of nanodiamonds into the Drosophila embryo. Early **b** and late **c** Stage 5 embryos showing the cellularization furrows introgressing between nuclei, which invade the yolk-free periplasm during the later syncytial divisions indicated as *arrows* in **b**. Nanodiamonds that have diffused into the yolk-free periplasm can become internalized in the blastoderm cells at the completion of Stage 5. **d** Scanning confocal fluorescence image of individual nanodiamonds in the blastoderm cells during Stage 5 of development. The image shows the auto-fluorescence from the introgressing cellularization furrows defining each blastoderm cell as well as the strong fluorescent signal from individual nanodiamonds which in the majority of cases is localized to the cell periphery. (Reprinted from Simpson et al. [42])

advantage of the photostability of the NV^- center in FND, the authors did the single particle tracking of individual FND particles in the blastoderm cells of developing *Drosophila* embryos. Since the FND fluorescence could be clearly distinguished from the embryo autofluorescence, the dynamics of individual FND particles were easily probed. The authors first took wide-field fluorescence images of the embryo which were divided into two sections. The first section of images was measured at the level of the nuclei and ingressing furrow, and the second section of images was measured in the periplasm underlying the nuclei. The results showed that the transport mechanisms of FND in the furrow were similar to the sub-nuclear periplasm.

3.2 *Time-Gated Fluorescence Imaging and Fluorescence Lifetime Imaging*

During fluorescence imaging, the biological systems always produce an autofluorescence background which interferes the FND signal and deteriorates the signal-to-background ratio. Another parameter that can be manipulated to enhance the image contrast of single FND in biological environments is the relatively long-lived fluorescence lifetime of the NV center, i.e. 11.6 ns in bulk diamond [43]. The lifetime of NV centers is substantially longer than that of cell and tissue autofluorescence (about 1–4 ns). Time-gating fluorescence imaging and Fluorescence Lifetime imaging (FLIM) can eliminate the autofluorescence background signal during fluorescence detection.

For instance, FND was non-covalently coated with YLCs (yolk lipoprotein complexes) by physical adsorption. Then the YLC-conjugated FND (YLC-FNDs) was micro-injected into the intestine of the *Caenorhabditis elegans* [41] and monitored temporally and spatially in vivo with the combined FND-FLIM technique. Before time-gating, the fluorescence signal from FND could not be easily differentiated from the background, mainly contributed by the intestinal cells (Fig. 7c). In contrast, the YLC-FNDs could be identified easily in the FLIM image, which takes advantage of FND's uniquely long fluorescence lifetime (Fig. 7d). Moreover, the background signal is significantly reduced after gating the fluorescence signal at the lifetime longer than 10 ns (Fig. 7e). To demonstrate that FND is biocompatible even when performing long-term in vivo tracking, the authors microinjected the GFP::YLC (a GFP molecule fused to YLC) -FND particles into the distal arm of the adult gonad. After tracing the particles for about 12 h, the GFP::YLC-FND-targeted oocytes were observed to grow into mature oocytes, suggesting that the oocytes with FND nanoparticles could develop normally without any influence.

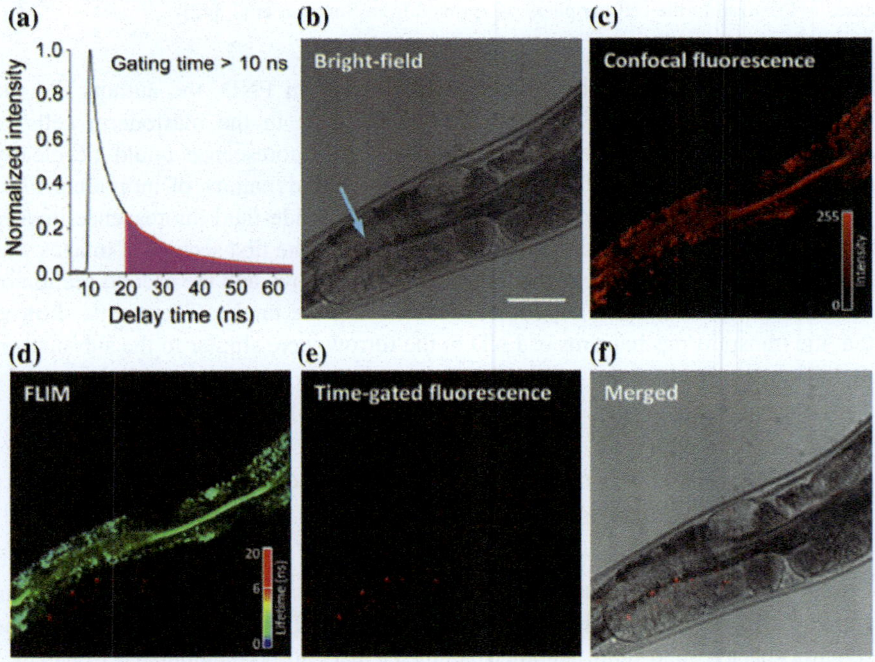

Fig. 7 Observation of GFP::YLC-FNDs in *C. elegans* by FLIM. **a** A fluorescence decay time trace of 100-nm FNDs suspended in water. The area *shaded in magenta* represents the fluorescence signal collected at the gating time (s) longer than 10 ns. **b** Bright-field, **c** confocal fluorescence, **d** FLIM, **e** time-gated fluorescence image at gating time > 10 ns, and **f** merged bright-field and time-gated fluorescence images of a worm microinjected with GFP::YLC-FNDs at the distal gonad. A *blue arrow* indicates the site of injection. Anterior is *left* and dorsal is up in all figures. Scale bar is 50 mm. (Reprinted from Kuo et al. [41])

The lung is another complex organ comprising more than 40 various cell lineages, which produce strong background during fluorescence imaging. Several stem cell niches have been identified in terminal bronchioles or bronchoalveolar junctions [44]. Lung stem cells (LSCs) are responsible for epithelial repair as well as tissue homeostasis, and are able to renew and proliferate to form epithelial cells in vivo. To enable the therapeutic applications of LSCs, it is crucial to determine their tissue-specific engraftment [45] and regenerative capacity [46] in animals. Recently, Chang's group has successfully demonstrated the homing property of LSCs by using the FND labeling technique [47]. Compared to organic dye and fluorescent proteins (with low transfection efficiency) which are susceptible to photobleaching and interfere with the tissue autofluorescence, FND serves as a highly photostable imaging probe that enable long-term tracking of the FND labeled cells in vivo. The authors also showed that LSCs can be spontaneously labeled with FNDs by endocytosis, and the labeling did not affect cellular functions. By combining FND labeling with FLIM technique, they were able to identify transplanted cells in histological lung sections after intravenous (i.v.) injection of the FND-labeled LSCs into mice for more than a week, with single-cell resolution. More importantly, the authors applied lung injury mice to investigate the tissue-specific engraftment capacity of LSCs.

In vivo stem cell tracking was tested in healthy normal mice and injured mice respectively. Firstly, about 5×10^5 FND-labeled LSCs were injected into the tail veins of the healthy mice. Cells in lungs, kidneys, liver and spleen were collected on days 1, 4 and 7 after injection [47]. By gating the fluorescence signal at 9 ns, the authors could readily distinguish the background noises from the FND fluorescence signals. Hence, the location of FND-labeled LSCs could be clearly revealed (Fig. 8). Due to the photostability, of the NV fluorophores, the authors further confirmed that FNDs with prolonged excitation did not result in any significant decrease in fluorescence intensity. It is demonstrated that the FND-labeled LSCs in the mice could be tracked for 7 days after i.v. injection.

It is known that the regenerative capacity of LSCs is determined both by their intrinsic developmental potential and their interaction with other cell elements in their niches [46]. This capacity could be significantly activated after tissue injury. To illustrate this effect, the authors tracked LSCs in lung-injured mice. In this experiment, 5×10^5 FND-labeled LSCs were injected into the mice after lung injury for 2 days. The extent of the injury and the repair of the bronchiolar epithelium could be examined by immunostaining against club cell secretory protein (CCSP) [48], which was expressed by LSCs. Further analysis for the H&E staining bright image and the corresponding time-gated fluorescence (τ = 9–18 ns) image revealed that the FND-labeled LSCs were preferentially localized in the terminal bronchioles of the lung-injured mice (Fig. 9), in contrast to the localization of the cells in the subepithelium of bronchiolar airways in uninjured mice (Fig. 8). Our results showed that the lung epithelia of the injured mice were restored more rapidly after transplantation of the FND-labeled LSCs than with saline control (Fig. 9). The authors also found that the percentage of the transplanted LSCs engrafted to the lung was estimated to be $\sim$23 % on Day 1, and the percentage

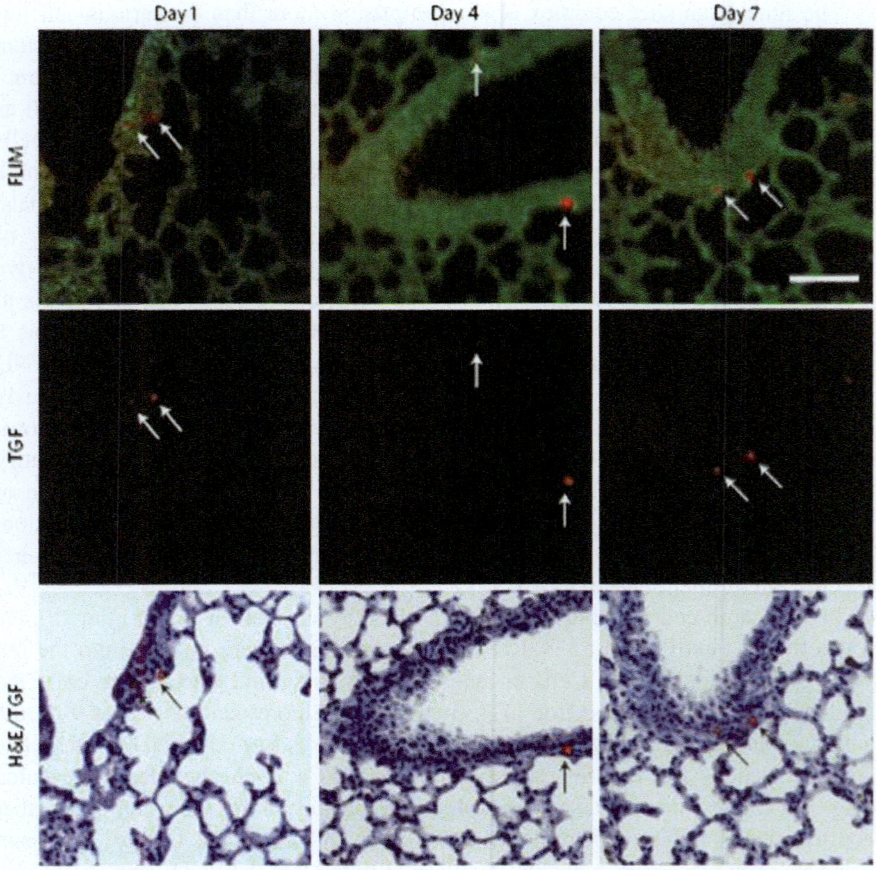

Fig. 8 FND-labeled LSCs in uninjured mice on different days. Representative FLIM, TGF and bright-field H&E staining images of the same lung tissue sections from uninjured mice. The merged H&E and TGF images in the *last row* show that the FND-labelled cells (denoted by *black arrows*) are primarily located in the subepithelium of bronchiolar airways. Scale bar = 50 μm. (Reprinted from Wu et al. [47])

markedly declined to 1.7 % on day 7 (Fig. 8). In contrast, the percentage of the transplanted LSCs engrafted to the lung for the lung injury model exhibited a much smaller decrease, with the percentage of 13 % on day 1 and 11 % on Day 7 (Fig. 9). The distinct contrast between these two results confirms that the homing phenomenon of the transplanted FND-labeled cells in the injury models is a 'pro-active' tissue-specific engraftment, instead of non-specific or passive entrapment [47].

In addition, to examine whether the FND-labeled LSCs engrafted in lung were viewed as foreign substances in the living system and engulfed by macrophages

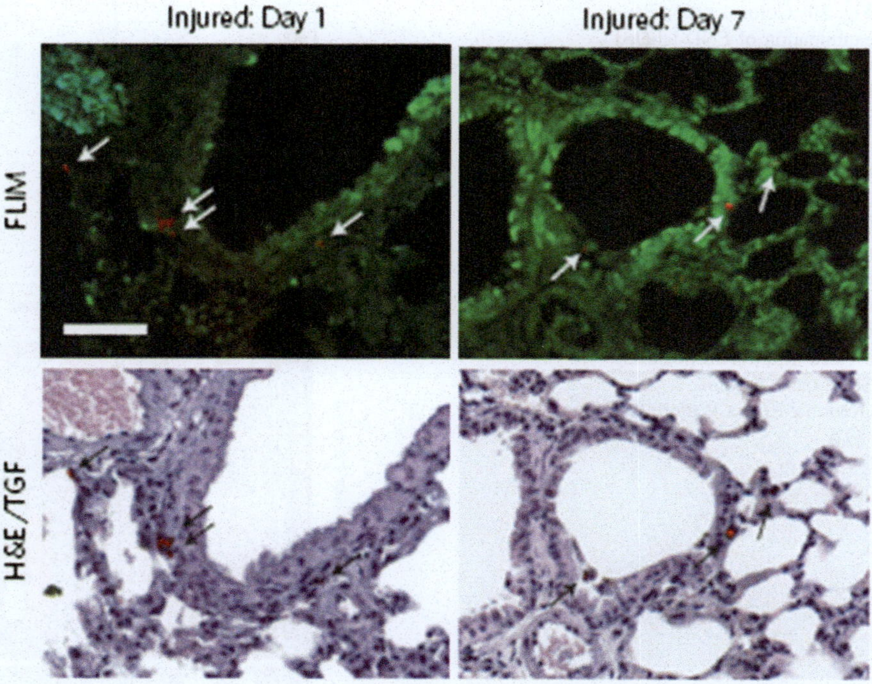

Fig. 9 FND-labeled LSCs in lung-injured mice on different days. Representative FLIM, TGF and bright-field H&E staining images of the same lung tissue sections, showing the location of FND-labeled LSCs (denoted by *white and black arrows*) in terminal bronchioles of the lungs. Scale bar = 50 μm. (Reprinted from Wu et al. [47])

through a process called "phagocytosis" [49], Chang's group stained the lung tissue sections with the macrophage-specific antibody, F4/80. They further did haematoxylin and eosin (H&E) staining and the fluorescence imaging. Overlapping of the bright-field and time-gated fluorescence images (Fig. 10) showed no sign of FND co-localization with the F4/80-stained macrophages, which excluded the possibility that the observed FND-labeled LSCs were phagocytosed after i.v. injection. It should be noted that such identification could not been made with organic dyes such as carboxyfluorescein succinimidyl ester (CFSE), because the lifetimes for the CFSE and the background fluorescence were similar.

In this work, Chang's group demonstrated that the FND labeling technique enables quantitative assessment of the distribution of transplanted LSCs in tissue, due to the excellent chemical stability and photostability of the nanomaterial. Additionally, the FND-labeled cells could be counted visually from the images. Although the mechanism how the injected LSCs reach lung tissues and further engraft to the terminal bronchioles are not yet fully understood, the results support

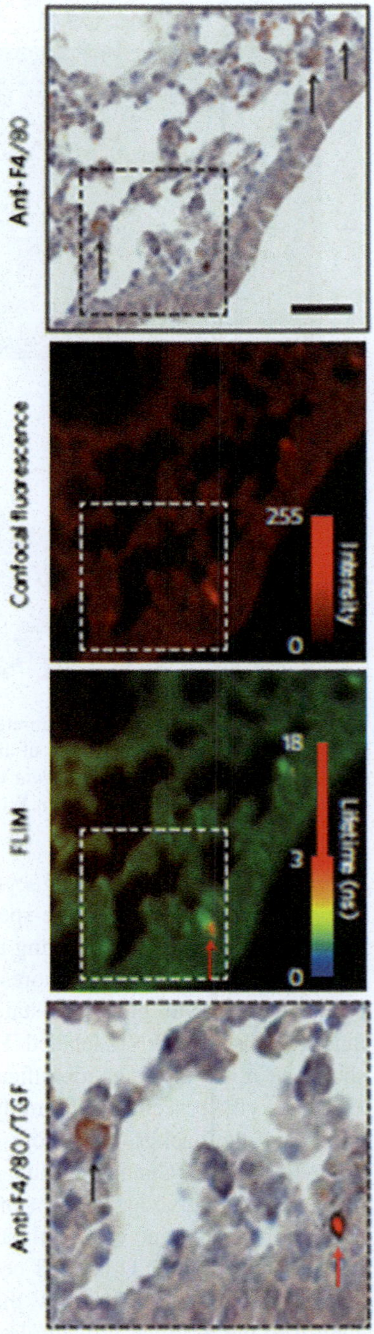

Fig. 10 Colocalization examination of FND-labeled LSCs and macrophages in a lung tissue section immunostained with macrophage-specific antigen F4/80 and H&E staining. The enlarged view shows that the transplanted FND-labeled cells (*red* and indicated by *red arrows*) and alveolar macrophages (*brown* and indicated by *black arrows*) are located at different positions. Scale bar = 10 μm. (Reprinted from Wu et al. [47])

that specific microenvironments (or niches) in lungs plays an important role in the regulation of tissue-specific engraftment, structural support and the signals for self-renewal and differentiation of the stem cells [47].

Most recently, Hui et al. presented another approach to achieve in vivo background-free fluorescence imaging of FNDs by an intensified charge-coupled device (ICCD) as the detector. ICCD is useful to suppress short-lifetime autofluorescence backgrounds and provide time-gated fluorescence images with high contrast [50]. With ICCD and a Raman shifter, the authors demonstrated the first application for wide-field fluorescence imaging of the FND-labeled cells in living animals. The authors labeled mouse lung cancer cells with FNDs and then introduced 1×10^5 labeled cells into a mouse via tail vein injection. Then, fluorescence imaging was carried out in a main blood vessel of the mouse ear. Figure 11a, b are bright-field and time-gated fluorescence images of the ear tissue respectively. About 10 min after intravenous injection of the FND-labeled lung cancer cells into the mouse, bright moving objects with an average speed of 0.4 μm/s could be clearly identified in the blood vessel at a frame rate of 2 Hz (Fig. 11c). It is noted that this measured speed was much lower than the blood flow velocity by about 3 orders of magnitude, and thus most likely associated with rolling, instead of flowing, of the FND-labeled cells in the blood vessel. The technique provides real-time imaging and tracking of transplanted stem cells in tissue repair and regeneration in vivo. Also, the real-time tracking of FND-labeled cancer cells during metastasis in vivo is plausible. The next interesting issue is the application of FND for in vivo study of circulating tumor cells in blood [51]. It is deemed feasible in the near future with FNDs conjugated with bioactive ligands or grafted with high-specificity antibodies against tumor-specific biomarkers through biotin-avidin interactions [52].

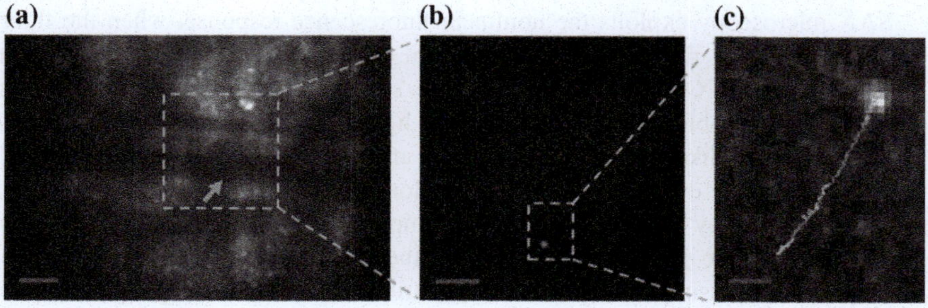

Fig. 11 **a** Bright-field image of a mouse ear tissue. The *green arrow* indicates the position of an FND-labeled lung cancer cell in the blood vessel of ~50 μm in diameter. **b** Enlarged view of the fluorescence image of the *square green* region in **a**. The *bright spot* corresponds to the FND-labeled lung cancer cell. **c** Enlarged view of the fluorescence image of the *rectangular green* region in **b**, showing the trajectory of the FND-labeled lung cancer cell moving in the vessel. The average speed of the cell movement is 0.4 μm/s. The frame rate is 2 Hz and the objective lens is 10×. The red scale bar corresponds to 100, 50, and 10 μm in **a**, **b** and **c**, respectively. (Reprinted from Hui et al. [50])

3.3 Superresolution Imaging

One of the unique feature for FND is its perfect photostability, which provides an excellent fluorophore for superresolution imaging like stimulated emission depletion (STED) microscopy [53–58], saturated excitation (SAX) microscopy [59], localization microscopy [60], Deterministic emitter switch microscopy (DESM) [61] and scanning near field microscopy [62–64] and Dual-point illumination AND-gate microscopy [65]. We focus on STED and SAX, which have already been reported for bio-imaging.

The STED microscopy utilizes two laser beams at different wavelengths [66]. The main laser beam brings the fluorophore of interest to its excited state. The second laser beam, also called the STED beam, de-excites the fluorophore from the excited state to the ground state through stimulated emission processes. The STED beam has a doughnut shape which excites the outer region around the focus. During imaging, the fluorescence of the emitter excited by the main laser stays unaffected in the center of the doughnut spot but diminishes at the outer ring. Compared to the other superresolution techniques, STED microscopy does not require any mathematical post-processing. Accordingly, it is most suitable for high-resolution imaging of biological cells in real time and three dimensions. Tzeng et al. has taken advantage of the FND photostability, and applied STED to study FND labeling for HeLa cells [55]. To prevent agglomeration in cell medium, FND particles of 30 nm in size were first coated non-covalently with bovine serum albumin (BSA) and then delivered to the cell cytoplasm by endocytosis. With the green light for the excitation and a doughnut-shaped 740-nm laser beam for the depletion, the authors were able to achieve a resolution close to 40 nm. They identified the individual FND particles in cells and distinguished them from FND aggregates trapped in endosomes. Prabhakar et al. applied STED to study the endosome merging into late endosomes and estimate the drug load taken up by an individual cell [56].

SAX microscopy exploits the nonlinear fluorescence response, when the fluorophore is intensely illuminated. The excellent photostability of the NV^- center in FNDs enables SAX microscopy to achieve superresolution bioimaging without photobleaching problems, even under intense light illumination [59]. To investigate cellular processes like intracellular transport and cellular uptake, Yamanaka et al. demonstrated three-dimensional high-resolution imaging of FNDs distributed in a macrophage cell by SAX microscopy. Compared with the confocal image for FNDs, SAX image significantly improves the spatial resolution, which enables more accurate measurement of the FND distribution in the macrophage cell after intracellular transport.

3.4 Multiphoton Excitation Microscopy

FND tracking in biological tissue still presents challenging, because the tissue components are extremely heterogeneous and scatter light severely [67]. The strong scattering of the various tissue components has restricted high-resolution optical imaging to the skin surface. Multiphoton excitation microscopy excites the sample with infrared light and offers an attractive alternative to reduce the tissue scattering. As the infrared light scatters in tissue much weaker than the visible light, the multiphoton microscopy has a longer penetration depth to provide better optical sectioning for thick tissue than the one-photon counterpart. Moreover, the multiphoton absorption cross section is typically small and the excitation power density has to be high. Hence, the MPE is spatially confined to the vicinity of the focal spot which reduces the out-of-focus excitation and also the photo-damage to cells. Two-photon microscopy has already been demonstrated to improve the image contrast of FND in cellular level [25, 68]. We anticipate that multiphoton excitation microscopy can be applicable for FND tracking (about few mm) below the skin surface.

3.5 *Tissue Imaging with Microwave Modulation Only*

Utilizing the spin properties of the NV^- center, Igarashi et al. have established a selective imaging protocol to improve the image contrast of FNDs in vitro and in vivo, because the fluorescence of FND can also be darkened by the microwave at 2.87 GHz [69]. The NV^- center of in a FND particle has a triplet ground state with the $m_s = 0$ sublevel and the $m_s = \pm 1$ sublevel. When there is no external magnetic field, the zero-field energy splitting between the $m_s = 0$ and $m_s = \pm 1$ sublevel is 2.87 GHz, which corresponds to the microwave frequency. In this protocol, the microwave turns on and is resonant with the transition, the electron population in the triplet ground state will shift from the $m_s = 0$ to the $m_s = \pm 1$. The population in the $m_s = \pm 1$ sublevel has higher probability to go to the intermediate state than the $m_s = 0$ sublevel, and then return to the $m_s = 0$ sublevel via non-radiative decay. Hence, when the microwave turns from off to on, the FND signal will decrease, from bright to dark. In this protocol, the wide-field fluorescence images were first recorded with or without the microwave irradiation at 2.87 GHz (Fig. 12). Then the

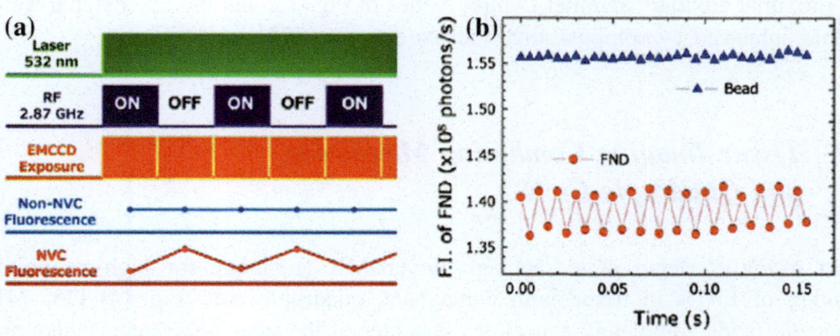

Fig. 12 **a** Time chart of the laser excitation, microwave irradiation and image acquisition, along with the expected intensity profiles of non-NV and NV fluorescence used for selective imaging of FNDs. **b** Time trace of the observed fluorescence intensities of FNDs and fluorescent beads. (Reprinted from Igarashi et al. [69])

subtraction between these two images at every pixel was carried out to yield selective images of the FND particles. Since the alternative microwave irradiation modulated only the fluorescence intensity of the NV^- centers, the subtraction between these two images effectively removed autofluorescence background signals of the specimen and thus significantly improved the image contrast. The protocol was shown applicable to a wide variety of living systems from single cells to whole animals like *Caenorhabditis elegans* [70] and mice.

3.6 Tissue Imaging with Magnetic Field Modulation Only

Apart from a microwave, a magnetic field can also decrease the fluorescence intensity of FND, because a magnetic field mixes NV center's spin sublevels and eliminate the selectivity of intersystem crossing [71]. (The mechanism for FND darkening by magnetic field is different from the microwave in the previous subsection.) When the NV center undergoes continuous excitation without magnetic field, the intersystem crossing from the singlet state to the $m_s = 0$ sublevel of the ground state is dominant than the $m_s = \pm1$ sublevels. Hence the NV center is polarized under continuous excitation and the bright state is predominately occupied. However, if the NV center undergoes continuous excitation in the presence of magnetic field, the intersystem crossing rate becomes approximately equal for all three sublevels. The NV center loses its spin-polarizability and has higher probability for non-radiative transition. Thus, FNDs in the presence of magnetic field appear 10–20 % darker in the absence of magnetic field, and the fluorescence intensity of FND is dependent on the magnetic field. Using image subtraction method [71], the magnetic-independent background can be eliminated, leaving the background-free FND signals. Chapman et al. observed FND in vitro with strong autofluorescence background and improved the image contrast by this image subtraction method. Sarkar et al. injected silica-coated FNDs in the front foot pad of intact mice and imaged sentinel lymph nodes. The FND particles were drained to the proximal auxiliary sentinel Lymph Nodes in Fig. 13, and the FND signal were further enhanced by computational lock-in detection [72].

3.7 Tissue Imaging Combining Microwave and Quadruple Coils

Most recently, Hegyi et al. applied the ODMR technique for high-resolution imaging of FNDs in tissue with a magnetic quadruple coil (Fig. 14) [73, 74]. Magnetic fields completely cancel at the center of the imaging system, where are

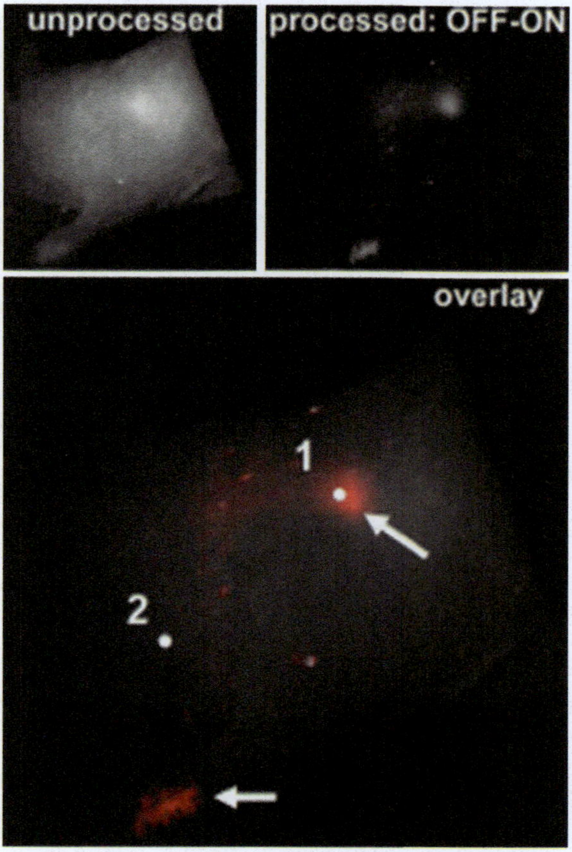

Fig. 13 In vivo imaging of sentinel lymph nodes of a mouse with FND. The mouse is obtained by pairwise subtracting with and without the magnetic field, and averaged with 475 images. The processed image (*top right inset*, *red* in *overlay*) was overlaid on an unprocessed image obtained with the magnetic field off (*top left inset*, *white* in *overlay*). The *white arrows* point to the injection site in the footpad and the location of the auxiliary lymph node. Signal from the FNDs in the lymph node is clearly detected. (Reprinted from Sarkar et al. [72])

called field-free line [73]. Only at this field-free line, FND will resonate with the microwave at 2.87 GHz. In other region with non-vanishing magnetic field, the FND will not resonate with 2.87 GHz because the resonance frequency of the transition is split by the magnetic field. Thus, in the presence of this microwave, which is turned on and off periodically, the variation in the fluorescence intensity, which is further measured with a lock-in amplifier, is proportional to the FND concentration at the field-free region. So the authors swept this field-free point across a region and record the intensities with the microwave on and off respectively. The intensity difference would then correspond to FND concentration at the field-free point. A quantitative map of FND concentration was then generated. This technique vastly improves the resolution in deep tissue optical imaging. They imaged large letter written by FND coated tape under 5 mm chicken breast [73].

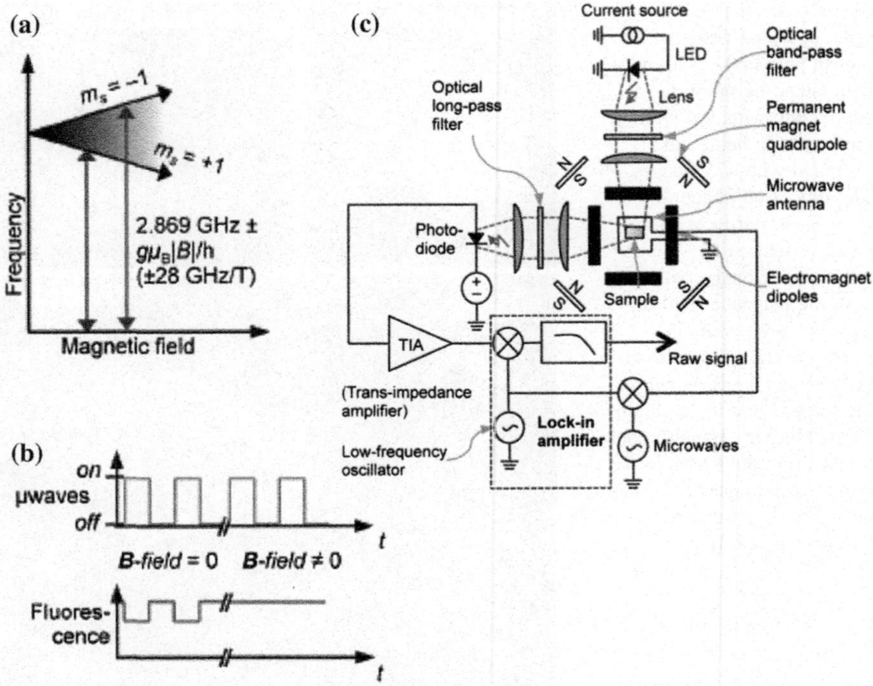

Fig. 14 Schematic diagram of an imaging system for NV⁻ centers in FNDs based on ODMR detection. **a** At zero magnetic field, spin transition frequency is 2.869 GHz. The upper spin sublevels split with increasing magnetic field. **b** With |B| = 0, turning on microwave at 2.869 GHz decreases the observed fluorescence from FND. In a strong B-field (|B| > 0), the microwave does not modulate the fluorescence. **c** "Magnetic well" created by the quadruple coils, with non-vanishing magnetic field everywhere except at the center of imaging system. Adapted with the permission of [73]. (Reprinted from Hegyi et al. [73])

4 Emerging Technologies for Quantum Sensing

4.1 Magnetic Field Sensor

The NV^- center has a magnetically sensitive triplet ground state. The unperturbed electronic energy structure of the center consists of a triplet ground state with the $m_s = 0$ sublevel and $m_s = \pm 1$ sublevel. The latter two sublevels are degenerate at zero external magnetic field, and are energetically higher than the $m_s = 0$ sublevel by a crystal field splitting of 2.87 GHz. When an external magnetic field is applied to the center, the field will split the $m_s = \pm 1$ sublevels and the ODMR spectrum of the NV^- center will change accordingly. Hence, an NV^- center is magnetically sensitive and applicable as a magnetic probe. Gruber, et al. demonstrated that the ODMR spectrum of a single NV^- center can be optically read out at the single molecule level. The sublevel population of the triplet ground state can be inferred from the change in the fluorescence signal between the $^3A_2$ and ^3E transition. This unique spin system, combined with the ODMR technique, suggests ultrasensitive and rapid detection of single electronic spin states under ambient conditions, for nanoscale magnetic resonance imaging. Balasubramanian et al. incorporated a nanodiamond hosting a single nitrogen-vacancy center into a cantilever of an atomic force microscope, and further applied the cantilever with the ODMR setup as a scanning probe magnetometer to achieve subwavelength imaging resolution [75].

Rondin et al. reported nanoscale magnetic imaging with a similar nanodiamond probe which is all-optical and sensitive to large off-axis magnetic fields. Most importantly, it does not require microwave control therefore extends the operation range of diamond-based magnetometry [76]. A few research groups have also applied nanodiamond to detect single spin as well as weak magnetic field, and their outstanding works for high resolution magnetometry have been elaborated in a recent review [3, 77]. Most recently Horowitz et al. have applied an infrared laser as an optical tweezer for the three dimensional control of the FND particle in liquid solution [78]. Then the crystal field splitting of the NV^- centers in the FND particle was read out by simultaneous application of a green laser and microwave. As the applied frequency was tuned across 2.87 GHz, the fluorescence intensity showed a significant decrease ($\sim$10 %). Hence, the optically trapped FNDs enable three-dimensional mapping of the magnetic fields in solution in Fig. 15, and probing of the magnetic fields in complex environments like the interiors of microfluidic

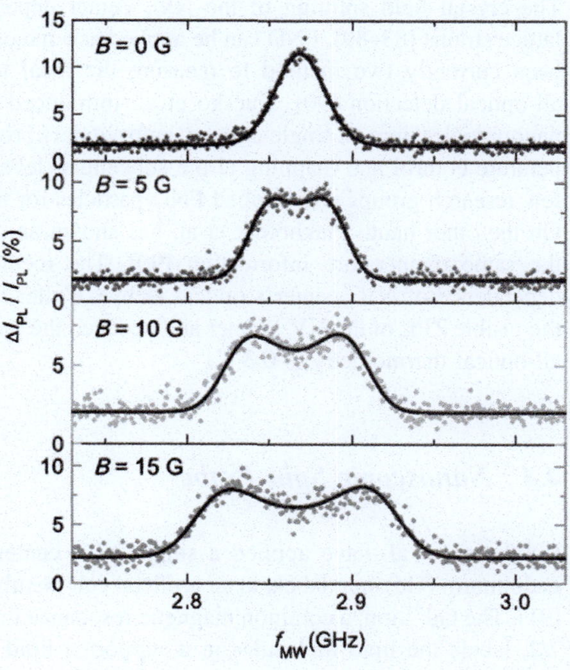

Fig. 15 ODMR spectra of trapped nanodiamond ensembles at calibrated low field strengths. The ODMR measurements are taken at the calibrated applied magnetic fields of 0 G, 5 G, 10 G and 15 G respectively. (Reprinted from Horowitz et al. [78])

channels. Geiselmann et al. have also shown the deterministic trapping and three-dimensional spatial manipulation of individual NDs hosting NV centers for local magnetic field measurement [79].

4.2 Orientation Tracker for NV Axis

The diamond crystal field separates the $m_s = 0$ sublevel of the ground state from the degenerate $m_s = \pm 1$ sublevel by 2.87 GHz. Furthermore, the energy splitting depends on the angle between the NV axis and an external magnetic field. Hence, a single NV^- center can act as an intrinsic compass under the magnetic field. McGuinness et al. can identify the orientation of the NV axis by applying the microwave field to examine the energy splitting of the FND hosting a single NV^- center [80, 81]. They further applied a uniform magnetic field to live HeLa cells to measure the peak positions of the ODMR spectra and successfully resolved the rotational motion of a FND particle in the cell at the millisecond timescale. The orientation of the NV^- center was continuously monitored for more than 16 h in a HeLa cell.

4.3 Nano-Thermometry [82]

The crystal field splitting of the NV^- center depends on the thermally induced lattice strains [83–89]. FND can be applied as a nanoscale thermometer. There have been currently two method to measure the local temperature, ODMR [89] and all-optical detection [90]. Kucsko et al. introduced both nanodiamonds and gold nanoparticles into a single human embryonic fibroblast, and demonstrated temperature control and mapping at the subcellular level. Combining ODMR setup, a few research groups also applied FND particles for nanoscale temperature sensing. On the other hand, Plakhotnik et al. has shown an all-optical technique to extract the same temperature information [90]. The method exploited the temperature dependence of NV^- center's optical Debye-Waller factor. In short, they measured the visible ZPL of the NV^- center and deduced the temperature. The accuracy of the all-optical thermometry is 0.3 K.

4.4 Nanoscopic Spin Probe

Kaufmann et al. have applied a single NV^- center in a nanodiamond to detect gadolinium (Gd) spin labels in an artificial cell membrane under ambient conditions [91]. The Gd^{3+} ion, a common magnetic resonance imaging contrast agent with spin 7/2, labels the lipid molecules in a supported lipid bilayer. The Gd-labeled lipid

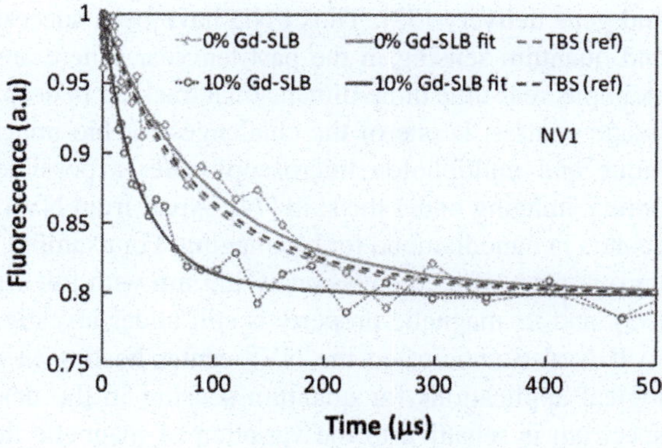

Fig. 16 Detection of spin-labeled lipids in a supported lipid bilayer (SLB) using the relaxation time T_1 of single NV^- spins in nanodiamonds. Relaxation of the spin of a single NV^- center in a nanodiamond in a SLB without spin labels (*green*) and in a SLB with 10 % Gd spin labels (*blue*). (Reprinted from Kaufmann et al. [91])

molecules produces characteristic magnetic fluctuations in the artificial membrane, by cross-lipid Gd–Gd spin dipole interactions, motional diffusion of individual Gd-labeled lipids and intrinsic Gd spin relaxation effects. Kaufmann et al. monitored the relaxation time T_1 of the single NV^- center in the lipid bilayer by optically polarizing the $m_s = 0$ state and then measuring the probability P(t) of finding the NV^- in the $m_s = 0$ state at a later time. The relaxation time is shortened when the artificial membrane is labeled with 10 % Gd spin-labeled lipids in Fig. 16. The average reduction in the T_1 relaxation time from the reference value is 74 ± 6 % and hence the nanodiamond probe is sensitive to near individual (4 ± 2) proximal Gd labels.

Meanwhile, Ermakova et al. demonstrated that single nanodiamond hosting multiple NV^- centers can also detect individual ferritin molecules [92]. (Ferritin has a small fraction of uncompensated Fe^{3+} spin with a magnetic moment of 300 μ_B. The ferritin concentration in blood serum is correlated with the total amount of iron stored in the body.) They first attached the ferritin molecules to the surface of the FND by non-covalent binding to the amino groups of the nanodiamond, and hence produced diamond-ferritin complexes. Then they compared the relaxation time T_1 between the free nanodiamond and the diamond-ferritin complex, and observed a significant decline of the relaxation time for the free nanodiamond. (Compared with current magnetic force microscopy techniques, these nanodiamond probes do not require cryogenic temperature and vacuum.) The successful detection for such small numbers of spins opens a pathway for in situ detection nanoscale detection of dynamical processes in biology.

5 Conclusions and Perspectives

NV^- center in nanodiamond has unique optical properties (like photostability and biocompatibility) as well as spin properties. It is also shown that FND particles could be conjugated with biomolecules which enable FND to serve as a vehicle for drug and gene delivery [52]. Thus FND have been successfully applied for bio-imaging and quantum sensing in the past ten year. There are a few challenges ahead. For example, real-time three-dimensional tracking in deep tissue with better fluorescence image contrast is one of the challenges for bio-imaging. Combining ODMR technique and multiphoton microscopy offer a possible solution for high-resolution optical imaging under the skin [74]. Apart from NV center, there are potential color centers in nanodiamond for bioimaging. For example, the silicon-vacancy center has narrow near infrared emission at 739 nm with full width at half maximum of 4 nm [35], and its magnetic property is still under investigation [93].

It is also anticipated the NV^- center hosting in nanodiamond finds more biological applications for quantum sensing in the near future, because the ODMR spectrum is sensitive to the variation of magnetic field, electric field, temperature and pressure. The NV^- center in nanodiamond has been already operated with the microwave excitation and developed as a nanoprobe for local temperature and magnetic field measurement respectively. To extend the operation range in the biomedical field, it is necessary to develop the FND as an all-optical nanoprobe without the microwave excitation [90]. In summary, we foresee that the applications of FND are going to create more breakthroughs in bioimaging and quantum sensing.

References

1. R. Schirhagl, K. Chang, M. Loretz, C.L. Degen, Nitrogen-vacancy centers in diamond: nanoscale sensors for physics and biology. Annu. Rev. Phys. Chem. **65**, 83–105 (2014). doi:10.1146/annurev-physchem-040513-103659
2. C. Bradac, T. Gaebel, J.R. Rabeau, Nitrogen-vacancy color centers in diamond: properties, synthesis, and applications. in *Optical Engineering of Diamond*, First edn. ed by R.P. Mildren, J.R. Rabeau. (Boschstr. 12, 69496, Wiley-VCH Verlag GmbH & Co. KGaA., Weinheim, Germany, 2013). doi: 10.1002/9783527648603.ch5
3. L. Rondin, J.P. Tetienne, T. Hingant, J.F. Roch, P. Maletinsky, V. Jacques, Magnetometry with nitrogen-vacancy defects in diamond. Rep. Prog. Phys. **77**, 056503 (2014). doi:10.1088/0034-4885/77/5/056503
4. Y.Y. Hui, H.C. Chang, Recent developments and applications of nanodiamonds as versatile bioimaging agents. J. Chin. Chem. Soc. **161**, 67–76 (2014). doi:10.1002/jccs.201300346
5. V.N. Mochalin, O. Shenderova, D. Ho, Y. Gogotsi, The properties and applications of nanodiamonds. Nat. Nanotechnol. **7**, 11–23 (2012). doi:10.1038/nnano.2011.209
6. J.R. Maze, A. Gali, E. Togan, Y. Chu, A. Trifonov, E. Kaxiras, M.D. Lukin, *Properties of nitrogen vacancy centers in diamond: the group theoretic approach* (Phys, New J, 2011). doi:10.1088/1367-2630/13/2/025025
7. L. Rondin, G. Dantelle, A. Slablab, F. Grosshans, F. Treussart, P. Bergonzo, S. Perruchas, T. Gacoin, M. Chaigneau, H.C. Chang, V. Jacques, J.F. Roch, Surface-induced charge state conversion of nitrogen-vacancy defects in nanodiamonds. Phys. Rev. B **82**, 115449 (2010). doi:10.1103/PhysRevB.82.115449
8. K.-M.C. Fu, C. Santori, P.E. Barclay, R.G. Beausoleil, Conversion of neutral nitrogen-vacancy centers to negatively charged nitrogen-vacancy centers through selective oxidation. Appl. Phys. Lett. **96**, 121907 (2010). doi:10.1063/1.3364135

9. C.C. Fu, H.Y. Lee, K. Chen, T.S. Lim, H.Y. Wu, P.K. Lin, P.K. Wei, P.H. Tsao, H.C. Chang, W. Fann, Characterization and application of single fluorescent nanodiamonds as cellular biomarkers. Proc. Natl. Acad. Sci. U.S.A. **104**, 727–732 (2007). doi:10.1073/pnas.0605409104
10. V. Vaijayanthimala, Y.K. Tzeng, H.C. Chang, C.L. Li, The biocompatibility of fluorescent nanodiamonds and their mechanism of cellular uptake. Nanotech. **20**, 425103 (2009). doi:10.1088/0957-4484/20/42/425103
11. N. Mohan, C.S. Chen, H.H. Hsieh, Y.C. Wu, H.C. Chang, In vivo imaging and toxicity assessments of fluorescent nanodiamonds in *Caenorhabditis elegans*. Nano Lett. **10**, 3692–3699 (2010). doi:10.1021/nl1021909
12. V. Vaijayanthimala, P.Y. Cheng, S.H. Yeh, K.K. Liu, C.H. Hsiao, J.I. Chao, H.C. Chang, The long-term stability and biocompatibility of fluorescent nanodiamond as an in vivo contrast agent. Biomater. **33**, 7794 (2012). doi:10.1016/j.biomaterials.2012.06.084
13. D. Passeri, F. Rinaldi, C. Ingallina, M. Carafa, M. Rossi, M.L. Terranova, C. Marianecci, biomedical applications of nanodiamonds: an overview. J Nanosci Nanotech. **15**(2), 972–988 (2015). doi:10.1166/jnn.2015.9734
14. V.M. Acosta, A. Jarmola, E. Bauch, D. Budker, Optical properties of the ntrogen-vacancy singlet levels in diamond. Phys. Rev. B **82**, 201202 (2010). doi:10.1103/PhysRevB.82.201202
15. M.W. Doherty, N.B. Manson, P. Delaney, F. Jelezko, J. Wrachtrup, L.C. Hollenberg, The nitrogen vacancy colour centre in diamond. Phys. Rep. **528**, 1–45 (2013). doi:10.1016/j.physrep.2013.02.001
16. A. Gruber, A. Drabenstedt, C. Tietz, L. Fleury, J. Wrachtrup, C. von Borczyskowski, Scanning Confocal Optical Microscopy and Magnetic Resonance on Single Defect Centers. Sci. **276**, 2012–2014 (1997). doi:10.1126/science.276.5321.2012
17. N.B. Manson, J.P. Harrison, M.J. Sellars, Nitrogen-vacancy center in diamond: model of the electronic structure and associated dynamics. Phy. Rev B **74**, 104303 (2006). doi:10.1103/PhysRevB.74.104303
18. F. Jelezko, J. Wrachtrup, Single defect centres in diamond: a review. Phys. Stat. Solidus A **203**, 3207–3225 (2006). doi:10.1002/pssa.200671403
19. A. Beveratos, R. Brouri, T. Gacoin, J.P. Poizat, P. Grangier, Nonclassical radiation from diamond nanocrystals. Phys. Rev. A **64**, 061802 (2002). doi:10.1103/PhysRevA.64.061802
20. S.J. Yu, M.W. Kang, H.C. Chang, K.M. Chen, Y.C. Yu, Bright fluorescent nanodiamonds: no photobleaching and low cytotoxicity. J. Am. Chem. Soc. **127**, 17604–17605 (2005). doi:10.1021/ja0567081
21. O. Faklaris, D. Garrot, V. Joshi, F. Druon, J.P. Boudou et al., Detection of single photoluminescent diamond nanoparticles in cells and study of the internalization pathway. Small **4**, 2236–2239 (2008). doi:10.1002/smll.200800655
22. J. Tisler, G. Balasubramanian, B. Naydenov, R. Kolesov, B. Grotz et al., Fluorescence and spin properties of defects in single digit nanodiamonds. ACS Nano **3**, 1959–1965 (2009). doi:10.1021/nn9003617
23. N. Mohan, Y.K. Tzeng, L. Yang, Y.Y. Chen, Y.Y. Hui, C.Y. Fang, H.C. Chang, Sub-20 nm fluorescent nanodiamonds as photostable biolabels and fluorescence resonance energy transfer donors Adv. Mater. **21**, 1–5 (2010). doi:10.1002/adma.200901596
24. J. Havlik, V. Petrakova, I. Rehor, V. Petrak, M. Gulka et al., Boosting nanodiamond fluorescence: towards development of brighter probes. Nanoscale **5**, 3208–3211 (2013). doi:10.1039/C2NR32778C
25. Y.R. Chang, H.Y. Lee, K. Chen, C.C. Chang, D.S. Tsai, C.C. Fu, T.S. Lim, Y.K. Tzeng, C.Y. Fang, C.C. Han, H.C. Chang, W. Fann, Mass production and dynamic imaging of fluorescent nanodiamonds *Nat*. Nanotech. **3**, 284–288 (2008). doi:10.1038/nnano.2008.99
26. C. Bradac, G. Torsten, N. Naidoo, J.R. Rabeau, A.S. Barnard, Prediction and measurement of the size-dependent stability of fluorescence in diamond over the entire nanoscale. Nano Lett. **9**, 3555–3564 (2009). doi:10.1021/nl9017379

27. B.R. Smith, D.W. Inglis, B. Sandnes, J.R. Rabeau, A.V. Zvyagin, D. Gruber, C.J. Noble, R. Vogel, E. Osawa, T. Plakhotnik, Five-nanometer diamond with luminescent nitrogen-vacancy defect centers. Small **5**(14), 1649–1653 (2009). doi:10.1002/smll.200801802
28. J.P. Boudou, P.A. Curmi, F. Jelezko, J. Wrachtrup, P. Aubert, M. Sennour, G. Balasubramanian, R. Reuter, A. Thore, E. Gaffet, High yield fabrication of fluorescent nanodiamonds. Nanotechnology **20**, 235602 (2009). doi:10.1088/0957-4484/20/23/235602
29. J.P. Boudou, J.J. Tisler, R. Reuter, A. Thorel, P.A. Curmi, F. Jelezko, J. Wrachtrup, Fluorescent nanodiamonds derived from HPHT with a size of less than 10 nm. Diam. Relat. Mater. **37**, 80–86 (2013). doi:10.1016/j.diamond.2013.05.006
30. B.R. Smith, D. Gruber, T. Plakhotnik, The effects of surface oxidation on luminescence of nanodiamonds. Diam. Relat. Mater. **19**, 314–318 (2010). doi:10.1016/j.diamond.2009.12.009
31. T. Gaebel, C. Bradac, J. Chen, J.M. Say, L. Brown, P. Hemmer, J.R. Rabeau, Size-reduction of nanodiamonds via air oxidation. Diam. Relat. Mater. **21**, 28–32 (2012). doi:10.1016/j.diamond.2011.09.002
32. J.R. Rabeau, A. Stacey, A. Rabeau, S. Prawer, F. Jelezko, I. Mirza, J. Wrachtrup, Single nitrogen vacancy centers in chemical vapor deposited diamond nanocrystals. Nano Lett. **7**, 3433–3437 (2007). doi:10.1021/nl0719271
33. J. Michl, T. Teraji, S. Zaiser, I. Jakobi, G. Waldherr, F. Dolde, P. Neumann, M.W. Doherty, N. B. Manson, J. Isoya, J. Wrachtrup, Perfect alignment and preferential orientation of nitrogen-vacancy centers during chemical vapor deposition diamond growth on (111) surfaces. Appl. Phys. Lett. **104**, 102407 (2014). doi:10.1063/1.4868128
34. A.M. Edmonds, U.F.S. D'Haenens-Johansson, R.J. Cruddace, M.E. Newton, K.-M.C. Fu, C. Santori, R.G. Beausoleil, D.J. Twitchen, M.L. Markham, Production of oriented nitrogen-vacancy color centers in synthetic diamond. Phy. Rev. B **86**, 035201 (2012). doi:10.1103/PhysRevB.86.035201
35. I. Aharonovich, Diamond nanocrystals for photonics and sensing. J.J. Appl. Phys. **53**(5), 05FA01 (2014). doi.:10.7567/JJAP.53.05FA01
36. Y.C. Lin, E. Perevedentseva, L.W. Tsai, K.T. Wu, C.L. Cheng, Nanodiamond for intracellular imaging in the microorganisms in vivo. J. Biophotonics **5**, 838–847 (2012). doi:10.1002/jbio.201200088
37. E. Perevedentseva, Y.C. Lin, M. Jani, C.L. Cheng, Biomedical applications of nanodiamonds in imaging and therapy. Nanomed. **8**(12), 2041–2060 (2013). doi:10.2217/nnm.13.183
38. T. C. Hsu, K. K. Liu, H. C. Chang, E. Hwang, J. I. Chao, Labeling of neuronal differentiation and neuron cells with biocompatible fluorescent nanodiamonds. Sci. Rep. 4 (2014). doi: 10.1038/srep05004
39. O. Faklaris, J. Botsoa, T. Sauvage, J.F. Roch, F. Treussart, Photoluminescent nanodiamonds: comparison of the photoluminescence saturation properties of the NV color center and a cyanine dye at the single emitter level, and study of the color center concentration under different preparation conditions. Diam. Relat. Mater. **19**(7–9), 988–995 (2010). doi:10.1016/j.diamond.2010.03.002
40. C.Y. Fang, V. Vaijayanthimala, C.A. Cheng, S.H. Yeh, C.F. Chang, C.L. Li, H.C. Chang, The exocytosis of fluorescent nanodiamond and its use as a long-term cell tracker. Small **7**(23), 3363–3370 (2011). doi:10.1002/smll.201101233
41. Y. Kuo, T.Y. Hsu, Y.C. Wu, H.C. Chang, Fluorescent nanodiamond as a probe for the intercellular transport of proteins in vivo. Biomat. **34**, 8352–8360 (2013). doi:10.1016/j.biomaterials.2013.07.043
42. D.A. Simpson, A.J. Thompson, M. Kowarsky, N.F. Zeeshan, M.S.J. Barson, L.T. Hall, Y. Yan, S. Kaufmann, B.C. Johnson, T. Ohshima, F. Caruso, R.E. Scholten, R.B. Saint, M. J. Murray, L.C.L. Hollenberg, In vivo imaging and tracking of individual nanodiamonds in drosophila melanogaster embryos. Biomed. Opt. Exp. **5**(4), 1250–1261 (2014). doi:10.1364/BOE.5.001250
43. C. Kurtsiefer, S. Mayer, P. Zarda, H. Weinfurter, Stable solid-state source of single photons. Phys. Rev. Lett. **85**(2), 290–293 (2000). doi:10.1103/PhysRevLett.85.290

44. A. Giangreco, S.D. Reynolds, B.R. Stripp, Terminal bronchioles harbor a unique airway stem cell population that localizes to the bronchoalveolar duct junction. Am. J. Pathol. **161**, 173–182 (2002). doi:10.1016/S0002-9440(10)64169-7
45. D.A. Chistiakov, Endogenous and exogenous stem cells: a role in lung repair and use in airway tissue engineering and transplantation. J. Biomed. Sci. **17**, 92 (2010). doi:10.1186/1423-0127-17-92
46. A.N. Lau, M. Goodwin, C.F. Kim, D.J. Weiss, Stem cells and regenerative medicine in lung biology and diseases. Mol. Ther. **20**, 1116–1130 (2012). doi:10.1038/mt.2012.37
47. T.J. Wu, Y.K. Tzeng, W.W. Chang, C.A. Cheng, Y. Kuo, C.H. Chien, H.C. Chang, J. Yu, Tracking the engraftment and regenerative capabilities of transplanted lung stem cells using fluorescent nanodiamonds. Nature Nanotech. **8**, 682–689 (2013). doi:10.1038/nnano.2013.147
48. B.R. Stripp, K. Maxson, R. Mera, G. Singh, Plasticity of airway cell proliferation and gene expression after acute naphthalene injury. Am. J. Physiol. Lung Cell Mol. Physiol. **269**, L791–L799 (1995)
49. M. Jonathan, Austyn and Siamon Gordon, F4/80, a monoclonal antibody directed specifically against the mouse macrophage. Eur. J. Immunol **11**, 805–815 (1981). doi:10.1002/eji.1830111013
50. Y.Y. Hui, L.J. Su, O.Y. Chen, Y.T. Chen, T.M. Liu, H.C. Chang, Wide-field imaging and flow cytometric analysis of cancer cells in blood by fluorescent nanodiamond labeling and time gating. Sci. Rep. **4**, 5574 (2014). doi:10.1038/srep05574
51. E. Galanzha, E.V. Shashkov, T. Kelly, J.W. Kim, L. Yang, V.P. Zharov, In vivo magnetic enrichment and multiplex photoacoustic detection of circulating tumour cells. Nat. Nanotech. **4**, 855–860 (2009). doi:10.1038/NNANO.2009.333
52. B.M. Chang, H.H. Lin, L.J. Su, W.D. Lin, R.J. Lin, Y.K. Tzeng, R.T. Lee, Y.C. Lee, A.L. Yu, H.C. Chang, Highly fluorescent nanodiamonds protein-functionalized for cell labeling and targeting. Adv. Funct. Mater. **23**, 5737–5745 (2013). doi:10.1002/adfm.201301075
53. K.Y. Han, K.I. Willig, E. Rittweger, F. Jelezko, C. Eggeling, S.W. Hell, Three-Dimensional stimulated emission depletion microscopy of nitrogen-vacancy centers in diamond using continuous-wave light. Nano Lett. **9**, 3323–3329 (2009). doi:10.1021/nl901597v
54. Y.K. Tzeng, O. Faklaris, B.M. Chang, Y. Kuo, J.H. Hsu, H.C. Chang, Superresolution imaging of albumin-conjugated fluorescent nanodiamonds in cells by stimulated emission depletion. Angew. Chem. Int. Ed. **50**, 2262 (2011). doi:10.1002/anie.201007215
55. N. Prabhakar, T. Näreoja, E. von Haartman, D.Ş. Karaman, H. Jiang, S. Koho, T.A. Dolenko, P.E. Hänninen, D.I. Vlasov, V.G. Ralchenko, S. Hosomi, I.I. Vlasov, C. Sahlgrenbci, J.M. Rosenholm, Core–shell designs of photoluminescent nanodiamonds with porous silica coatings for bioimaging and drug delivery II: application. Nanoscale **5**, 3713–3722 (2013). doi:10.1039/c3nr33926b
56. N.D. Lai, O. Faklaris, D. Zheng, V. Jacques, H.C. Chang, J.F. Roch, F. Treussart, Quenching nitrogen–vacancy center photoluminescence with an infrared pulsed laser. New J. Phys. **15**, 033030 (2013). doi:10.1088/1367-2630/15/3/033030
57. S. Arroyo-Camejo, M.P. Adam, M. Besbes, J.P. Hugonin, V. Jacques, J.J. Greffet, J.F. Roch, S.W. Hell, F. Treussart, Stimulated emission depletion microscopy resolves individual nitrogen vacancy centers in diamond nanocrystals. ACS Nano **7**, 10912–10919 (2013). doi:10.1021/nn404421b
58. X. Yang, Y.K. Tzeng, Z. Zhu, Z. Huang, X. Chen, Y. Liu, H.C. Chang, L. Huang, W.D. Li, P. Xi, Sub-diffraction imaging of nitrogen-vacancy centers in diamond by stimulated emission depletion and structured illumination. RSC Adv. **4**(11305–11310), 2014 (2014). doi:10.1039/c3ra47240j
59. M. Yamanaka, Y.K. Tzeng, S. Kawano, N.I. Smith, S. Kawata, H.C. Chang, K. Fujita, SAX microscopy with fluorescent nanodiamond probes for high-resolution fluorescence imaging. Biomed. Optics Exp. **2**, 1946–1954 (2011). doi:10.1364/BOE.2.001946
60. M. Gu, Y. Cao, S. Castelletto, B. Kouskousis, X. Li, Super-resolving single nitrogen vacancy centers within single nanodiamonds using a localization microscope. Opt. Exp. **21**(15), 17639–17646 (2013). doi:10.1364/OE.21.017639

61. E.H. Chen, O. Gaathon, M.E. Trusheim, D. Englund, Wide-field multispectral super-resolution imaging using spin-dependent fluorescence in nanodiamonds. Nano Lett. **13**, 2073–2077 (2013). doi:10.1021/nl400346k
62. Y.Y. Hui, Y.C. Lu, L.J. Su, C.Y. Fang, J.H. Hsu, H.C. Chang, Tip-enhanced sub-diffraction fluorescence imaging of nitrogen-vacancy centers in nanodiamonds. Appl. Phys. Lett. **102**, 013102 (2013). doi:10.1063/1.4773364
63. R. Beams, D. Smith, T.W. Johnson, S.H. Oh, L. Novotny, A.N. Vamivakas, Nanoscale fluorescence lifetime imaging of an optical antenna with a single diamond NV center. Nano Lett. **13**(8), 3807–3811 (2013). doi:10.1021/nl401791v
64. A.W. Schell, P. Engel, J.F.M. Werra, C. Wolff, K. Busch, O. Benson, Scanning single quantum emitter fluorescence lifetime imaging: quantitative analysis of the local density of photonic states. Nano lett. **14**(5), 2623–2627 (2014). doi:10.1021/nl500460c
65. J. Kwon, Y. Lim, J. Jung, S.K. Kim, New sub-diffraction-limit microscopy technique: dual-point illumination AND-gate microscopy on nanodiamonds. Opt Exp **20**, 13347–13356 (2014). doi:10.1364/OE.20.013347
66. G. Vicidomini, G. Moneron, K.Y. Han, V. Westphal, H. Ta, M. Reuss, J. Engelhardt, C. Eggeling, S.W. Hell, Sharper low-power, STED nanoscopy by time gating. Nat. Meth. **8**, 571–575 (2011). doi:10.1038/nmeth.1624
67. F. Helmchen1, W. Denk, Deep tissue two-photon microscopy. Nat. Meth. **2**(12), 932–940 (2005). doi:10.1038/NMETH818 2 photon review
68. Y.Y. Hui, B. Zhang, Y.C. Chang, C.C. Chang, H.C. Chang, J.H. Hsu, K. Chang, F.H. Chang, Two-photon fluorescence correlation spectroscopy of lipid-encapsulated fluorescent nanodiamonds in living cells. Optics Express **18**, 5896–5905 (2010). doi:10.1364/OE.18.005896
69. R. Igarashi, Y. Yoshinari, H. Yokota, T. Sugi, F. Sugihara, K. Ikeda, H. Sumiya, S. Tsuji, I. Mori, H. Tochio, Y. Harada, M. Shirakawa, Real-time background-free selective imaging of fluorescent nanodiamonds in Vivo. Nano Lett. **12**, 5726–5732 (2012). doi:10.1021/nl302979d
70. Y. Yoshinari, S. Mori, R. Igarashi, T. Sugi, H. Yokota, K. Ikeda, H. Sumiya, I. Mori, H. Tochio, Y. Harada, M. Shirakawa, Optically detected magnetic resonance of nanodiamonds in vivo; implementation of selective imaging and fast sampling. J. Nanosci. Nanotechnol **15**, 1014–1021 (2015). doi:10.1166/jnn.2015.9739
71. R. Chapman, T. Plakhoitnik, Background-free imaging of luminescent nanodiamonds using external magnetic field for contrast enhancement. Optics Lett. **38**(11), 1847–1849 (2013). doi:10.1364/OL.38.001847
72. S.K. Sarkar, A. Bumb, X. Wu, K.A. Sochacki, P. Kellman, M.W. Brechbiel, K.C. Neuman, Wide-field in vivo background free imaging by selective magnetic modulation of nanodiamond fluorescence. Biomed. Opt. Exp. **5**, 1190–1202 (2014). doi:10.1364/BOE.5.001190
73. A. Hegyi, E. Yablonovitch, Molecular imaging by optically detected electron spin resonance of nitrogen-vacancies in nanodiamonds. Nano. Lett. **13**, 1173 (2013). doi:10.1021/nl304570b
74. A. Hegyi, E. Yablonovitch, Nanodiamond molecular imaging with enhanced contrast and expanded field of view. J. Biomed. Opt. **19**, 011015 (2014). doi: 10.1117/1.JBO.19.1.011015
75. G. Balasubramanian, I.Y. Chan, R. Kolesov, M. Al-Hmoud, J. Tisler et al., Nanoscale imaging magnetometry with diamond spins under ambient conditions. Nature **455**, 648–651 (2008). doi:10.1038/nature07278
76. L. Rondin, J.P. Tetienne, P. Spinicelli, C. Dal Savio, K. Karrai, G. Dantelle, A. Thiaville, S. Rohart, J.F. Roch, V. Jacques, Nanoscale magnetic field mapping with a single spin scanning probe magnetometer. Appl. Phys. Lett. **100**, 153118 (2012). doi:10.1063/1.3703128
77. J.M. Taylor, P. Cappellaro, L. Childress, L. Jiang, D. Budker, P.R. Hemmer, A. Yacoby, R. Walsworth, M.D. Lukin, High-sensitivity diamond magnetometer with nanoscale resolution. Nat. Phy. **4**, 810–816 (2008). doi:10.1038/nphys1075
78. V. R. Horowitz, B. J. Alemán, D. J. Christle, A. N. Cleland, D. D. Awschalom, Electron spin resonance of nitrogen-vacancy centers in optically trapped nanodiamonds. Proc. Nat. Acad. Sci. USA, **109**, 13493 (2012). doi:10.1073/pnas.1211311109M

79. M.L. Geiselmann, J. Juan, J.M. Renger, L.J. Say, F.J.G. de Brown, F. Abajo, R. Koppens, Quidant, three-dimensional optical manipulation of a single electron spin. Nat. Nantech. **8**, 175–179 (2013). doi:10.1038/nnano.2012.259
80. L.P. McGuinness, Y. Yan, A. Stacey, D.A. Simpson, L.T. Hall et al., Quantum measurement and orientation tracking of fluorescent nanodiamonds. Nat. Nanotechnol. **6**, 358–363 (2011). doi:10.1038/nnano.2011.64
81. D. Maclaurin, L.T. Hall, A.M. Martin, L.C.L. Hollenberg, Nanoscale magnetometry through quantum control of nitrogen–vacancy centres in rotationally diffusing nanodiamonds. New J. Phys. **15**, 013041 (2013). doi:10.1088/1367-2630/15/1/013041
82. G. Baffou, H. Rigneault, D. Marguet, L. Jullien, A critique of methods for temperature imaging in single cells. Nat. Meth. **11**(9), 899–901 (2014). doi:10.1038/nmeth.3073
83. V.M. Acosta, E. Bauch, M.P. Ledbetter, A. Waxman, L.S. Bouchard, D. Budker, Temperature dependence of the nitrogen-vacancy magnetic resonance in diamond. Phys. Rev. Lett. **104**, 070801 (2010). doi:10.1103/PhysRevLett.104.070801
84. T. Plakhotnik, D. Gruber, Luminescence of nitrogen-vacancy centers in nanodiamonds at temperatures between 300 and 700 K: perspectives on nanothermometry. Phys. Chem. Chem. Phys. **12**, 9751–9756 (2010). doi:10.1039/c001132k
85. X.D. Chen, C.H. Dong, F.W. Sun, C.L. Zou, J.M. Cui et al., Temperature dependent energy level shifts of nitrogen-vacancy centers in diamond. Appl. Phys. Lett. **99**, 161903 (2011). doi:10.1063/1.3652910
86. D.M. Toyli, D.J. Christle, A. Alkauskas, B.B. Buckley, C.G. van de Walle, D.D. Awschalom, Measurement and control of single nitrogen-vacancy center spins above 600 K. Phys. Rev. X **2**, 031001 (2012). doi:10.1103/PhysRevX.2.031001
87. P. Neumann, I. Jakobi, F. Dolde, C. Burk, R. Reuter et al., High-precision nanoscale temperature sensing using single defects in diamond. Nano Lett. **13**, 2738–2742 (2013). doi:10.1021/nl401216y
88. D.M. Toyli, C.F. de las Casas, D.J. Christle, V.V. Dobrovitski, D.D. Awschalom, Fluorescence thermometry enhanced by the quantum coherence of single spins in diamond. Proc. Natl. Acad. Sci. USA **110**, 8417–8421 (2013). doi:10.1073/pnas.1306825110
89. G. Kucsko, P.C. Maurer, N.Y. Yao, M. Kubo, H.J. Noh, P.K. Lo, H. Park, M.D. Lukin, Nanometer scale quantum thermometry in a living cell. Nature **500**, 54–58 (2013). doi:10.1038/nature12373
90. T. Plakhotnik, M.W. Doherty, J.H. Cole, R. Chapman, N.B. Manson, All-optical thermometry and thermal properties of the optically detected spin resonances of the NV–center in nanodiamond. Nano Lett. **14**, 4989–4996 (2014). doi:10.1021/nl501841d
91. S. Kaufmann, D.A. Simpson, L.T. Hall, V. Perunicic, P. Senn, S. Steinertf, L.P. McGuinnessa, B.C. Johnsong, T. Ohshimag, F. Carusod, J. Wrachtrup, R.E. Scholtenh, P. Mulvaney, L. Hollenberg, Detection of atomic spin labels in a lipid bilayer using a single-spin nanodiamond probe. Proc. Natl. Acad. Sci. USA **110**, 10894–10898 (2013). doi:10.1073/pnas.1300640110
92. A. Ermakova, G. Pramanik, J. Cai, G. Algara-Siller, U. Kaiser, T. Weil, Y.K. Tzeng, H.C. Chang, L.P. McGuinness, M.B. Plenio, B. Naydenov, F. Jelezko, Detection of a few metallo-protein molecules using color centers in nanodiamonds. Nano Lett. **13**, 3305–3309 (2013). doi:10.1021/nl4015233
93. C. Hepp, T. Müller, V. Waselowski, J.N. Becker, B. Pingault, H. Sternschulte, D. Steinmüller-Nethl, A. Gali, J.R. Maze, M. Atatüre, C. Becher, Electronic structure of the silicon vacancy color center in diamond. Phys. Rev. Lett. **112**, 036405 (2014). doi:10.1103/PhysRevLett.112.036405

Polyglycerol-Functionalized Nanoparticles for Biomedical Imaging

Naoki Komatsu and Li Zhao

Abstract Polyglycerol (PG) functionalization on the surface of nanoparticle is one of the most effective methods to well disperse the particle in a physiological environment. The functionality also provides the nanoparticle with scaffold for further derivatization to add more functions. In this chapter, we will describe PG functionalization of nanoparticles including detonation nanodiamond (dND), superparamagnetic iron oxide nanoparticle (SPION) and fluorescence nanodiamond (fND), and their further derivatization for biomedical imaging agents in magnetic resonance and fluorescence imaging.

Keywords Nanodiamond · Iron oxide nanoparticle · MRI · Fluorescence · Polyglycerol

1 Introduction

Nanomedicine is newly constructed field of science and is rapidly growing [1–3]. In this research field, nanoparticle plays a central role and is mostly applied to biomedical imaging and drug delivery system (DDS) [4–7]. Since pristine nanoparticle cannot work as an imaging probe and a drug carrier, it should be functionalized to add the following requisite functions; (1) enough dispersibility and stability in a physiological environment, (2) enough duration of circulation in a living body, (3) recognition of and bonding to a target, and (4) visualization of the target for diagnostic purpose or release of a drug for therapeutic one. In order to fulfill the above requirements, we developed methodology in chemical modification of nanoparticles such as nanodiamond (ND) [8–12], superparamagnetic iron oxide

N. Komatsu (✉) · L. Zhao
Graduate School of Human and Environmental Studies, Kyoto University,
Sakyo-ku, Kyoto 606-8501, Japan
e-mail: komatsu.naoki.7w@kyoto-u.ac.jp

N. Yang et al. (eds.), *Carbon Nanoparticles and Nanostructures*,
Carbon Nanostructures, DOI 10.1007/978-3-319-28782-9_5

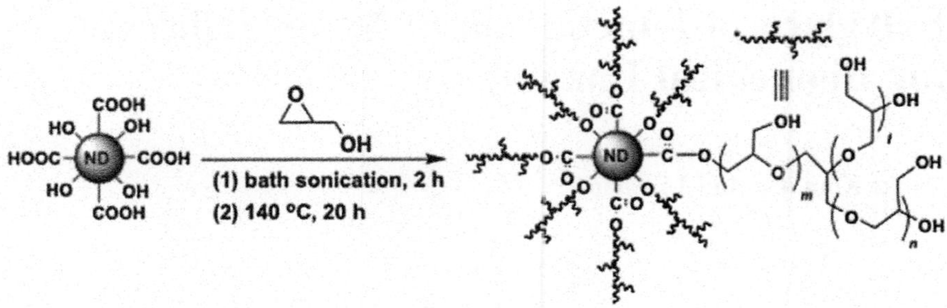

Scheme 1 Synthesis of ND functionalized with hyperbranched PG through the ring-opening polymerization of glycidol

nanoparticle (SPION) [13], and zinc oxide nanoparticle [14]. In order to add enough hydrophilicity in a physiological environment, we first adopted common methodology by use of polyethylene glycol (PEG) [15]. The PEG coating afforded only a little dispersibility to ND with 30 nm size. Then, we found that polyglycerol (PG) improved the dispersibility of ND significantly because of a large number of hydroxyl groups and branched structure in the PG (Scheme 1) [10]. The high aqueous dispersibility of the PG-functionalized ND (ND-PG) allowed chromatography for size separation and solution-phase NMR spectroscopies for structural characterization. We also demonstrated the generality of PG functionalization for various nanoparticles including carbon nanomaterials and metal oxide nanoparticles [13, 14]. Herein, we describe further chemical functionalization of PG-functionalized nanomaterials for biomedical imaging.

2 PG-Functionalized Detonation ND (dND-PG) Conjugated with Gadolinium Complexes for Magnetic Resonance Imaging (MRI)

MRI is a noninvasive diagnostic tool giving anatomical images with high depth penetration and spatial resolution. In order to enhance the contrast, a contrast agent is frequently used in MRI and Gd(III) complexes have been widely employed as well as SPION in clinic [16, 17]. The clinically used Gd(III) complexes such as Gd ($DTPA^{2-}$) or Magnevist® are relatively small molecules and still have room to improve relaxivity, in vivo circulation, and targeting efficacy. Therefore, metal complexes conjugated with various bulky platforms have been proposed, because the platforms are expected to restrict the motion of the complexes, increasing the relaxivities [18], and to prolong the in vivo circulation [4–6]. Various kinds of platforms have been proposed so far, including dendrimers [19], polymers [20], proteins [21], and nanoparticles [22]. Recently, a ND platform has been reported as a Gd(III) complex-conjugated MRI contrast agent to increase the relaxivity significantly [23, 24]. However, the dispersibility of the ND-Gd(III) conjugates in physiological media should be improved for in vivo applications. This motivated us to graft ND with PG (Scheme 1) and immobilize Gd complexes covalently on the surface [10, 12].

2.1 Synthesis and Characterization of dND-PG-Gd(III)

Since PG layer affords a versatile scaffold for surface engineering as well as large hydrophilicity [25], functionalization of dND-PG with Gd(III) complexes was performed through multistep organic transformations [15], starting at hydroxyl groups in the PG layer, followed by Gd(III) complexation (Scheme 2) [10, 12]. After tosylation of the hydroxyl groups in dND-PG, the resulting tosylates in dND-PG-OTs were substituted nucleophilically by azido (dND-PG-N_3). The azido groups were subsequently reduced to amine (dND-PG-NH_2) with triphenylphosphine. The ligand for Gd(III), diethylenetriaminepentaacetic acid (DTPA), was covalently bound through thiourea by reacting the isothiocyanate (*p*-SCN-Bn-DTPA) with dND-PG-NH_2. The resulting dND-PG-DTPA formed complex with Gd(III) ions under mild acidic condition to give dND-PG-Gd(III).

All the dND-PG derivatives synthesized in the transformations from dND-PG to dND-PG-Gd(III) shown in Scheme 2 were characterized by FTIR (Fig. 1). The formation of dND-PG-OTs from dND-PG was confirmed by the asymmetric and

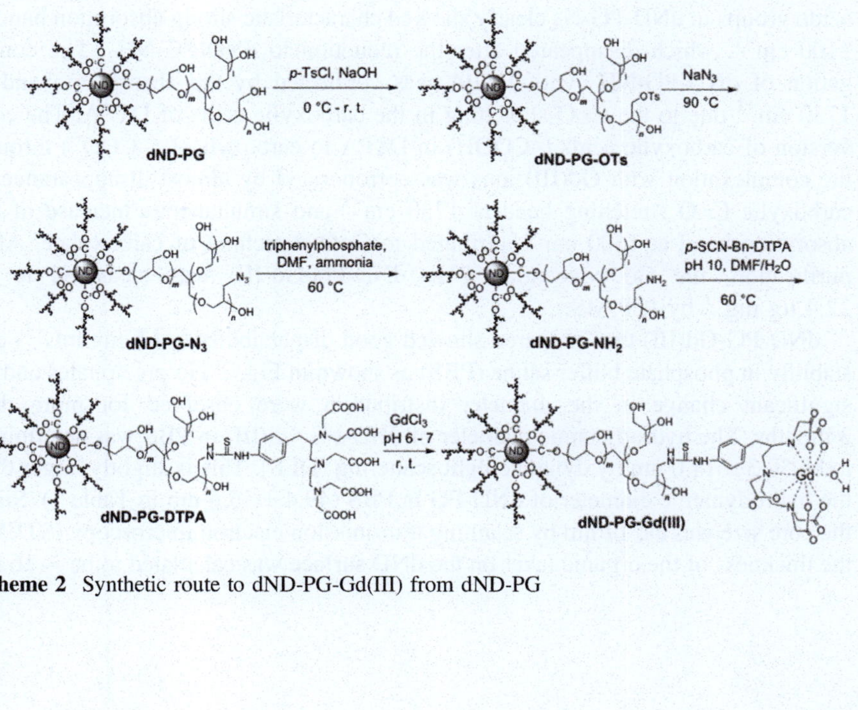

Scheme 2 Synthetic route to dND-PG-Gd(III) from dND-PG

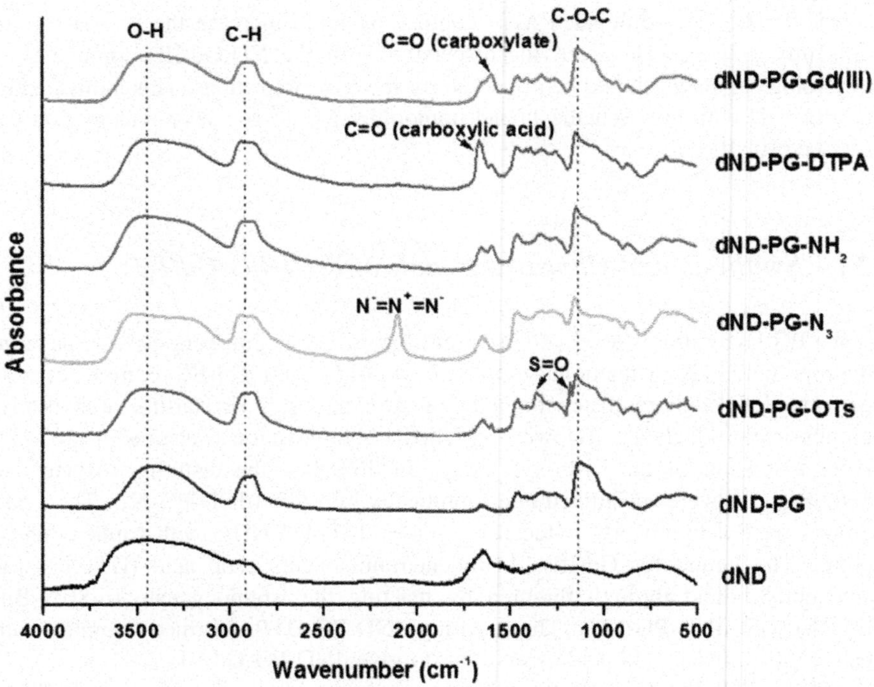

Fig. 1 FTIR spectra of pristine dND, dND-PG, and dND-PG derivatives. *Arrows* indicate new absorption bands in each step

symmetric stretchings of S→O bonds at 1350 and 1176 cm^{-1}, respectively. The azido groups in dND-PG-N_3 clearly showed characteristic strong absorption band at 2100 cm^{-1}, which disappeared after the reduction to dND-PG-NH_2. The conjugation of dND-PG-NH_2 with DTPA was confirmed by the absorption band at 1730 cm^{-1} due to the C=O stretching in the carboxylic acids of DTPA. The conversion of carboxylic acids (–COOH) in DTPA to carboxylates (–COO^-) through the complexation with Gd(III) ions was corroborated by almost disappearance of carboxylic C=O stretching band at 1730 cm^{-1} and simultaneous increase of the absorption band at 1600 cm^{-1} attributed to C=O stretching of carboxylate. After purification, the Gd concentration in dND-PG-Gd(III) was measured to be 22.4 μg mg^{-1} by ICP-MS.

dND-PG-Gd(III) thus prepared showed good dispersibility (>4.5 mg mL^{-1}) and stability in phosphate buffer saline (PBS) as shown in Fig. 2. No precipitates and no significant change in the diameter distribution were observed for more than 3 months. The hydrodynamic diameter of dND-PG-Gd(III) in PBS was determined to be 50.3 ± 14.0 nm by dynamic light scattering (DLS). This is slightly larger than the hydrodynamic diameter of dND-PG in PBS (49.4 ± 15.6 nm in Table 1). Since the core size was ca. 17 nm by scanning transmission electron microscopy (STEM), the thickness of the organic layer on the dND surface was calculated to be ~16 nm

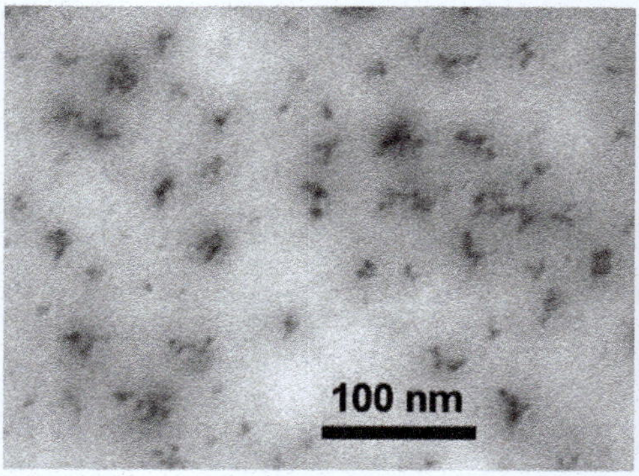

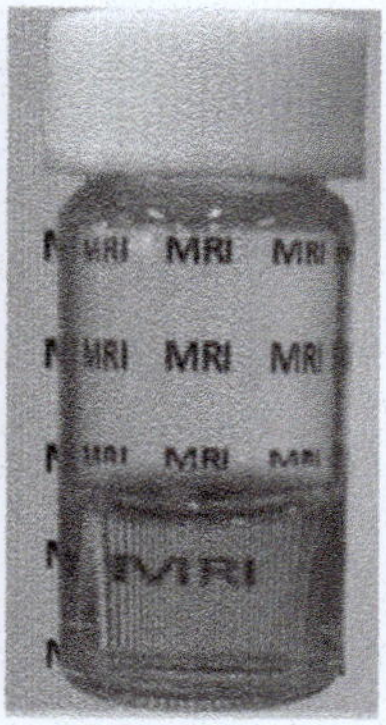

Fig. 2 *Left* STEM image of dND-PG-Gd(III) and *right* photograph of dND-PG-Gd(III) well dispersed in PBS (4.5 mg mL^{-1})

Table 1 Particle sizes of dND-PG and dND-PG-Gd(III) determined by STEM and DLS

Particle	Core size (nm)[a]	Hydrodynamic size (nm) in PBS[b]	Size difference (nm)[c]	Thickness of PG (nm)[d]
dND-PG	16.2 ± 12.0	49.4 ± 15.6	33.2	16.6
dND-PG-Gd(III)	17.8 ± 12.2	50.3 ± 14.0	32.5	16.2

[a]Average core diameter ±SD was determined by more than 100 particles in the STEM images
[b]Mean diameter ±SD was determined by DLS on the basis of number distribution
[c]Difference between core and hydrodynamic sizes
[d]Half of the size difference

in both dND-PG and dND-PG-Gd in PBS (Table 1). In addition, no large aggregates were found in the STEM image of dND-PG-Gd(III) (Fig. 2).

2.2 *MRI Relaxivity of dND-PG-Gd(III)*

The relaxivities in aqueous solutions containing dND-PG-Gd(III) were measured at 1.5, 3.0, and 7.0 T. All the dispersions of dND-PG-Gd(III) shown in Fig. 3 exhibited brighter MR images than pure water at 3.0 T. This is probably because the Gd(III) complexes reduce the proton longitudinal relaxation time (T_1) of adjacent water molecules. In addition, the MRI images of dND-PG-Gd(III) dispersions were brighter than those of Magnevist® solutions at the same Gd concentrations (0.50, 0.25, and 0.12 mM). This indicates superior contrast efficiency of dND-PG-Gd(III) to the clinically used Magnevist®.

The T_1 relaxivity of a Gd(III)-based contrast agent was calculated by the following equation:

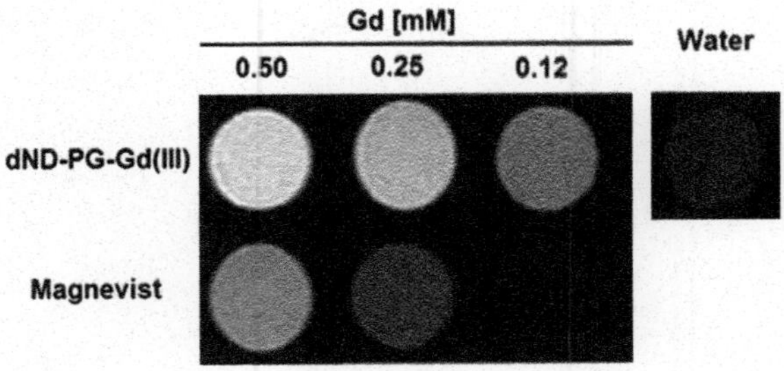

Fig. 3 T_1-weighted MR images of the aqueous dispersions of dND-PG-Gd (III) and solutions of Magnevist® at 3.0 T

Table 2 T_1 relaxivity (r_1) of dND-PG-Gd(III) and Magnevist® at different magnetic field strength

Magnetic filed (T)	r_1 (mM^{-1} s^{-1})[a]	
	dND-PG-Gd(III)	Magnevist®
1.5	19.4	3.7
3.0	16.7	3.5
7.0	8.2	3.4

[a]The r_1 is determined by Eq. 1

$$1/T_1 = 1/T_1^{\circ} + r_1[\mathrm{Gd}] \tag{1}$$

where $1/T_1$ is the observed relaxation rate in the presence of a Gd(III)-based contrast agent, $1/T^{\circ}_1$ is the relaxation rate in pure water, [Gd] is the Gd concentration, and r_1 is the longitudinal relaxation rate representing the efficiency or T_1 relaxivity of a Gd(III)-based contrast agent [16, 17]. The r_1 determined by Eq. 1 is summarized in Table 2. dND-PG-Gd(III) has much larger r_1 than Magnevist® probably because of the restriction in the motion of the Gd(III) complex moiety by the dND-PG platform as mentioned above [17]. The r_1 of dND-PG-Gd(III) at 1.5 T was found to be 19.4 mM^{-1} s^{-1}, which is more than 5 times larger than that of Magnevist®. As the magnetic field strength increased to 7.0 T, the r_1 of dND-PG-Gd(III) decreased to 8.2 mM^{-1} s^{-1}, which is still 2.4 times larger than the r_1 of Magnevist®. These results showed that dND-PG-Gd(III) has superior relaxivity to Magnevist® over a wide range of magnetic fields and that lower magnetic field is more effective to increase the r_1 [26, 27].

3 PG-Functionalized SPION (SPION-PG) for MRI

SPION is one of the most well-known nanoparticles in the field of nanomedicine as an MRI probe for diagnosis, and as a drug carrier and a hyperthermic agent for therapy [28–34]. As mentioned in the Introduction, good dispersibility and stability of the nanoparticle in a physiological environment and strict control of the size are preferable in the theranostic applications of SPION [35, 36]. Actually, most of the commercial SPIONs are wrapped with dextran and its derivatives [32]. Although dextran-coating provides SPION with good aqueous solubility, this may cause increase of the hydrodynamic diameter and broadening of the size distribution compared to those of the pristine SPION. Quite recently, the hydroxyl groups on the hydrophilic SPION prepared via polyol process [37, 38] were used as scaffolds for multi-step covalent transformations on the surface [39]. This implies that we can utilize the hydroxyl groups on the surface to graft hydrophilic polymers directly [40]. In this section, we will describe the PG functionalization on the surface of SPION to impart high dispersibility in a physiological medium and the relationship between the particle size, controlled by size exclusion chromatography [10, 41, 42], and the magnetic property [13, 41, 43–46].

3.1 Synthesis and Characterization of SPION-PG

As in the case of dND mentioned above [12], SPION was also grafted with PG through ring opening polymerization of glycidol as shown in Scheme 3 [13]. The resulting SPION-PG was characterized by FTIR, STEM and DLS. The nanoparticle after the reaction, shown in STEM image (Fig. 4c), had the same shape (sphere), average diameter, and standard deviation (SD) (8.8 ± 2.3 nm) as that (8.8 ± 2.2 nm) before the reaction (Fig. 4a). Therefore, we concluded that the core is individual SPION. The PG grafting for SPION was confirmed by large increase of the absorption bands at 3400, 2900, and 1100 cm^{-1} corresponding to O–H, C–H, and C–O–C stretchings, respectively (Fig. 5), as in the case of dND-PG mentioned above (Fig. 1) [12]. The FTIR spectrum of SPION-PG was also similar to that of free PG (Fig. 5), which is prepared by ring-opening polymerization of glycidol without SPION. The mean hydrodynamic diameter and the SD of the

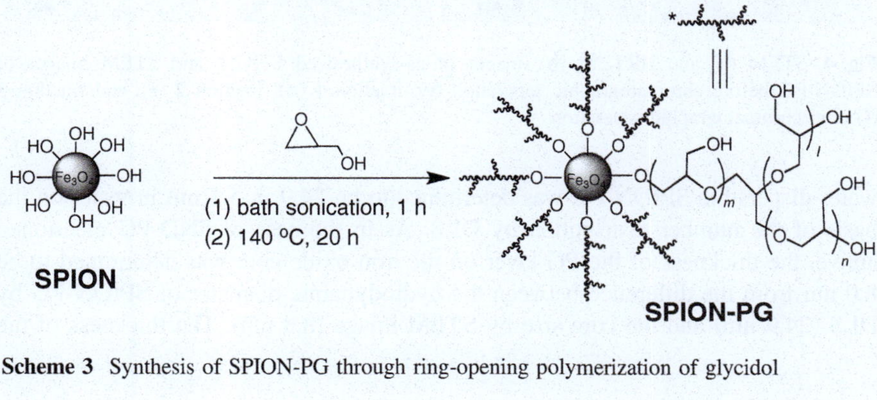

Scheme 3 Synthesis of SPION-PG through ring-opening polymerization of glycidol

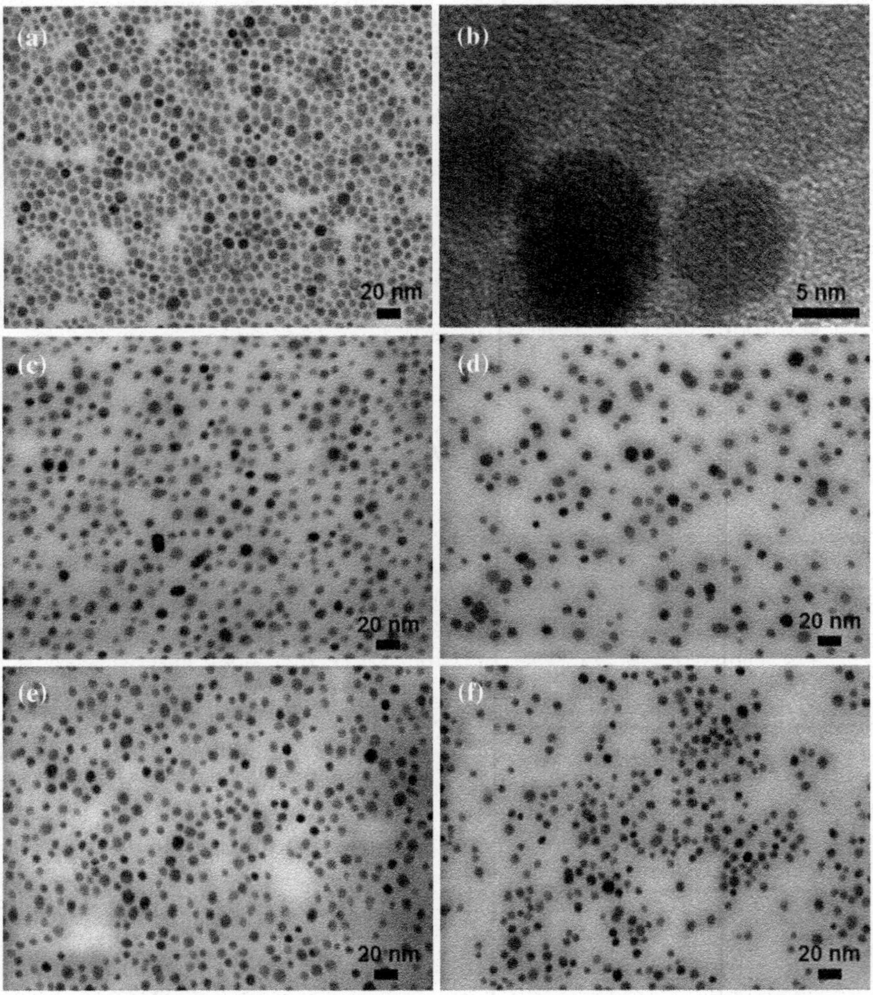

Fig. 4 STEM (**a**) and HRTEM (**b**) images of as-synthesized SPION, and STEM images of SPION-PG before chromatographic separation (**c**), fraction 1 (**d**), fraction 2 (**e**), and fraction 3 (**f**) after chromatographic separation

water-dispersible SPION-PG was determined to be 24.9 ± 5.1 nm in water on the basis of the number distribution by DLS. As in the case of dND-PG mentioned above, the thickness of the PG layer on the iron oxide core was determined to be 8.0 nm from the difference between the hydrodynamic diameter of SPION-PG by DLS (24.9 nm) and the core size by STEM image (8.8 nm). The thickness of the

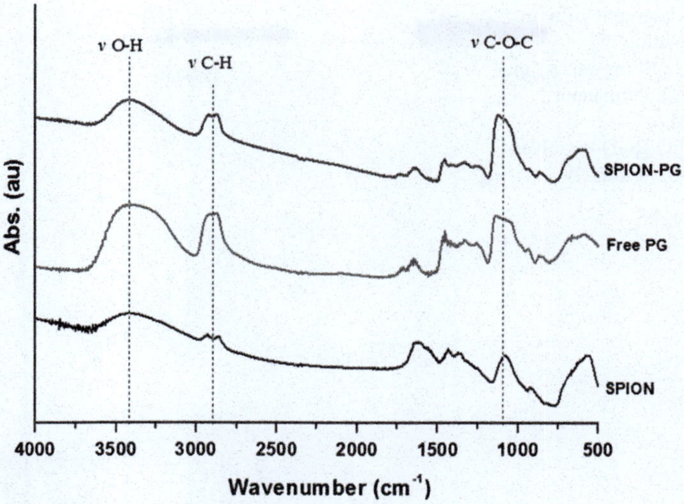

Fig. 5 FTIR spectra of as-synthesized SPION, free PG, and SPION-PG

Table 3 Structural and magnetic properties of SPION-PG before and after chromatographic separation

Sample	Retention time (min)[a]	Core size (nm)[b]	Hydrodynamic size (nm)[c]	Size difference (nm)[d]	Thickness of PG (nm)[e]	r_2 (mM^{-1} s^{-1})
Before separation	–	8.8 ± 2.3	24.1 ± 4.4, 24.9 ± 5.1[f]	16.3 ± 3.8, 17.8 ± 4.6[f]	8.2 ± 1.9, 8.9 ± 2.3[f]	86.30
Fraction 1	21.0–24.0	10.2 ± 2.7	28.9 ± 5.8	18.7 ± 5.1	9.4 ± 2.6	91.97
Fraction 2	24.0–26.0	9.1 ± 1.9	24.5 ± 4.5	15.4 ± 4.1	7.7 ± 2.0	86.91
Fraction 3	26.0–29.0	7.8 ± 1.7	19.4 ± 3.8	11.6 ± 3.4	5.8 ± 1.7	77.91

[a]Upon SEC separation
[b]Average core size of SPION-PG determined by more than 200 particles in the STEM images (Fig. 4c–f)
[c]Mean diameter of the number distribution determined by DLS in buffer, unless otherwise noted
[d]Difference between core and hydrodynamic sizes
[e]Half of size difference
[f]In Milli-Q water

PG layer on SPION is almost half of that on dND, which is probably because of the difference in the size of the core (dND: 16.6 nm in Table 1, SPION: 8.8 nm in Table 3). Accordingly, the dispersibility of SPION-PG in PBS (>25 mg mL^{-1}) is less than that of dND-PG (>80 mg mL^{-1}). Strong hydrophilicity and superparamagnetism of SPION-PG are simultaneously demonstrated in Fig. 6 [13].

Fig. 6 Photographs of **a** an aqueous solution of SPION-PG (40 mg mL^{-1}) in response to a permanent magnet and **b** SPION-PG-RGD dissolved in PBS (1.0 mg mL^{-1})

3.2 MRI Relaxivity in SPION-PG

It has been reported that not only the size of SPION but also the thickness of the surface coating affects the magnetic properties of hydrophilic SPIONs [28, 43, 44, 47, 48]. In order to correlate the size and thickness of the SPION-PG with the magnetic properties, MRI transverse relaxivity (r_2) were determined for SPION-PG with various mean diameters, separated by size exclusion chromatography [13], in a similar manner to r_1 (Eq. 1) by the following equation (Eq. 2);

$$1/T_2 = 1/T_2^{\circ} + r_2[\mathrm{Fe}] \tag{2}$$

where T_2 and T°_2 are the observed transverse relaxation times of SPION-PG and pure water, respectively, and [Fe] is the iron concentration of the solutions. The core size, the thickness of PG layer, and r_2 are summarized in Table 3. The r_2 of as-synthesized SPION-PG was found to be 86.30 Fe mM^{-1} s^{-1}, which is similar to that of the SPION prepared in almost the same process (82.68 Fe mM^{-1} s^{-1}) [38]. Since the core size of our SPION-PG (average diameter: 8.8 ± 2.3 nm in Fig. 4a and Table 3) is similar to that of the SPION reported by Cai et al. (particle size:

8 ± 1.1 nm) [38], the similar r_2 of the SPION and SPION-PG can be attributed to little or no influence of the hydrophilic PG functionality on the r_2. On the basis of the discussion, the r_2 decrease from fraction 1 to 3 shown in Table 3 can be attributed to size decrease of the core in SPION-PG, because the large magnetic particles possess higher magnetic moments and hence distort the magnetic field more efficiently to enhance r_2 [45, 48, 49]. Size effect of SPION on r_2 was concluded to exceed the effect of the PG coating in the case of SPION-PG, at least, in this range of the particle size.

4 PG-Functionalized Fluorescence ND (fND-PG) for Fluorescent Cellular Imaging

fND is one of the ideal nanoparticles as a fluorescence imaging agent [50, 51] thanks to its high biocompatibility, high extensibility of the surface functionality, size tunability, and intrinsic non-bleaching and non-blinking fluorescence [52, 53]. Therefore, fND has been used for cell labeling [54, 55], and even in vivo imaging [56]. Behind these excellent applications, fundamental technique to afford good dispersibility to fND in a physiological environment has been investigated by PG functionalization (Scheme 1) [10, 57], protein coating [58], and silica encapsulation [59]. In this section, we will describe synthesis, derivatization, and characterization of fND-PG and its application to the targeted cell labeling with fND-PG by conjugation with the targeting peptide [9].

4.1 Preparation and Characterization of fND-PG

The intrinsic fluorescence of fND provides an ideal tool for cell labeling and intracellular tracking, because an organic fluorescence dye is not required to be bound on the surface of the nanoparticle [15]. The fND-PG and its derivatives were prepared in the same process as that shown in Schemes 1 and 2 [10, 12]. Before PG functionalization and its derivatization, fND was prepared according to the method reported by Chang et al. [52]. Since annealing process constructing nitrogen-vacancy (N-V) center graphitized the fND surface, air-oxidization and mixed acid treatment were performed to regenerate the oxygen containing functional groups such as carboxyl and hydroxyl groups on the surface [60, 61]. The oxygen containing functional groups can initiate the ring-opening polymerization of glycydol [10].

TEM, photoluminescence (PL) spectroscopy, and fluorescence microscopy were measured to characterize fND thus prepared. The mean diameter of fND was determined to be 48.2 ± 13.4 nm by STEM (Fig. 7a). The intrinsic emission of fND was confirmed to be 550–800 nm by PL spectroscopy at the excitation of 488 nm

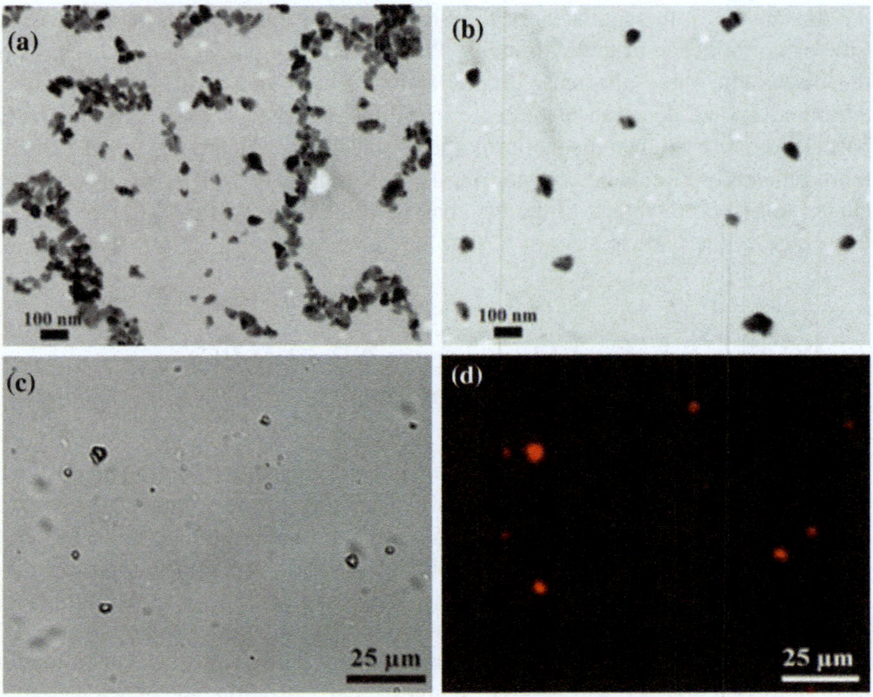

Fig. 7 STEM images of **a** fND and **b** fND-PG, **c** bright-field image of fND powder, and **d** fluorescence image under cy3 mode

Table 4 Size characterization of ND50, fND, and their derivatives by STEM and DLS

Particle	Core size (nm)[a]	Hydrodynamic size (nm) in water[b]	Thickness of PG layer[c]
ND50	52.2 ± 14.4	52.8 ± 20.2	–
ND50-PG	52.7 ± 13.9	66.9 ± 14.8	7.1
fND	48.2 ± 13.4	128.4 ± 49.8	–
fND-PG	48.4 ± 12.8	63.4 ± 14.9	7.5

[a]Average core diameter ±SD is determined by more than 100 particles in the STEM images
[b]Mean diameter ±SD is determined by DLS on the basis of number distribution
[c]Calculated from the difference of core size and hydrodynamic diameter

[62]. Fluorescence microscopy (Fig. 7d) showed bright red fluorescence from the fND clusters which appeared as gloomy dots in bright-field image (Fig. 7c).

The as-prepared fND was grafted with PG in the same method as that of ND (Scheme 1) [10, 57]. However, when fND was sonicated in glycidol before initiating the ring opening polymerization, a large amount of precipitates were observed even after sonication for 2 h. Actually, DLS measurements in water revealed that fND contained large aggregates; the mean size of fND (128 ± 50 nm) was 2.4 times larger than that of ND50 (52.8 ± 20.2 nm) as shown in Table 4. The aggregation reduced specific surface area of fND, resulting in five times less aqueous dispersibility (4.0 mg mL^{-1}) than that of ND50-PG (≥20.0 mg mL^{-1}) because of insufficient PG coverage of fND [9]. In addition, the dispersion was not so stable that precipitation occurred after standing overnight.

In order to increase the aqueous dispersibility and stability, the PG functionalization was employed once again under the same conditions to grow further the PG layer on the fND surface. The synthesized fND-PG was characterized by STEM, DLS, and TGA. STEM clearly showed deaggregation of fND after PG grafting twice (Fig. 7b) with no significant change in the fND core size (Table 4). The deaggregation was also supported by DLS (Table 4), exhibiting decrease in the mean hydrodynamic diameter from fND (128 ± 50 nm) to fND-PG (63.4 ± 14.9 nm). As compared with ND50-PG prepared by PG grafting of ND50 [9], fND-PG had quite similar thickness in the PG layer and weight ratio in PG:ND. That is, the thickness calculated by the hydrodynamic and core sizes is 7.5 nm in fND-PG and 7.1 nm in ND50-PG (Table 4), and the weight ratio determined by TGA is 39:61 in fND-PG and 37:63 in ND50-PG (Fig. 8). As expected from the above results, fND-PG had similar dispersibility to that of ND50-PG. Actual dispersibility of fND-PG in water, PBS and Dulbecco's modified Eagle medium (DMEM) is ≥20, 15 and 12 mg mL^{-1}, respectively. Since only a few precipitates were observed over 3 months, these fND-PG dispersions were concluded to have sufficient dispersibility.

The fND-PG was further functionalized by the RGD peptide (Scheme 4) to confirm the targeting property to the specific tumor cell by use of the intrinsic fluorescent of fND. The synthetic process is similar to that shown in Scheme 2 (–OH → –OTs → –N_3 → –RGD) [10, 12]. The resulting fND-PG-RGD exhibited good dispersibility and stability in water and DMEM (>1.0 mg mL^{-1}).

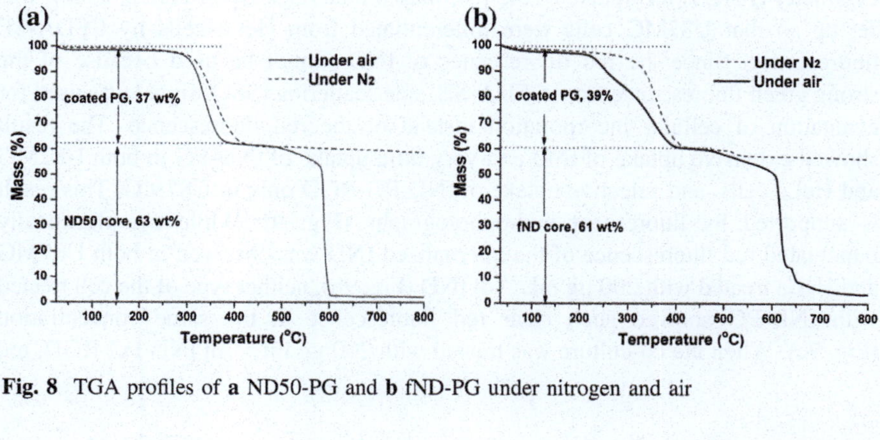

Fig. 8 TGA profiles of **a** ND50-PG and **b** fND-PG under nitrogen and air

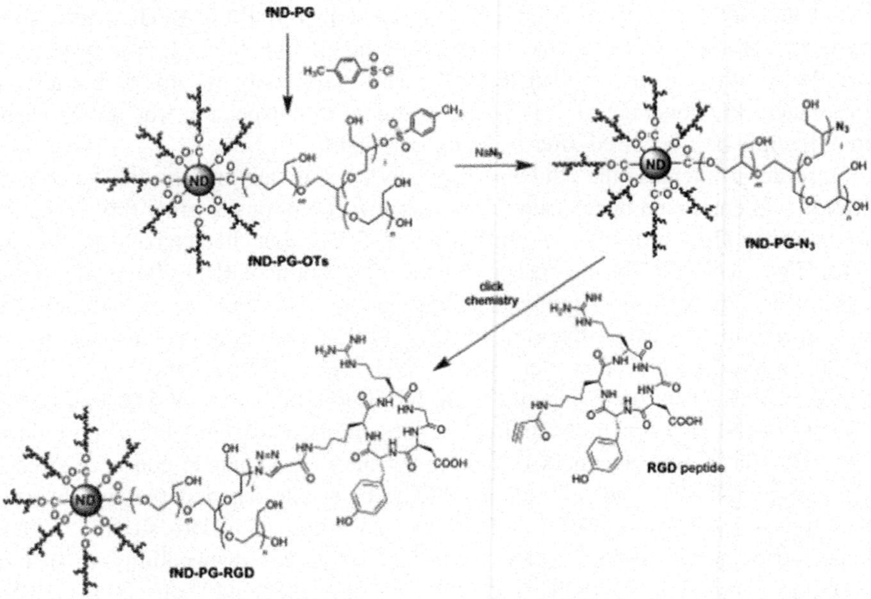

Scheme 4 Surface engineering on fND-PG for conjugation with RGD peptide

4.2 Fluorescence Cellular Imaging with fND-PG-RGD

In order to demonstrate the targeting property of fND-PG-RGD, U87MG cells labeled with 5(6)-carboxyfluorescein diacetate succinimidyl ester (CFDA-SE), a green fluorescent dye, were grown with HeLa cells. After treatment with fND, fND-PG, and fND-PG-RGD, the cells in the mixed culture were analyzed by flow cytometry (FACS) for uptake of the fND-based materials. Upon FACS, gating was set up so that U87MG cells were differentiated from HeLa cells by CFDA-SE fluorescence. Since the red fluorescence of fND cannot be used because of the strong green fluorescence from CFDA-SE, side scattering (SSC) signal was used for evaluation of cellular incorporation instead of the red fluorescence. The result showed extensive uptake of fND and very little uptake of fND-PG in both U87MG and HeLa cells, and selective uptake of fND-PG-RGD only in U87MG. This result is supported by fluorescent microphotographs (Fig. 9). While cytoplasmically punctuated red fluorescence of the internalized fND was observed in both U87MG and HeLa treated with 200 μg mL^{-1} of fND (Fig. 9b), neither type of the cell treated with fND-PG showed such clear red fluorescence at the same concentration (Fig. 9c). When the co-culture was treated with 200 μg mL^{-1} of fND-PG-RGD, red

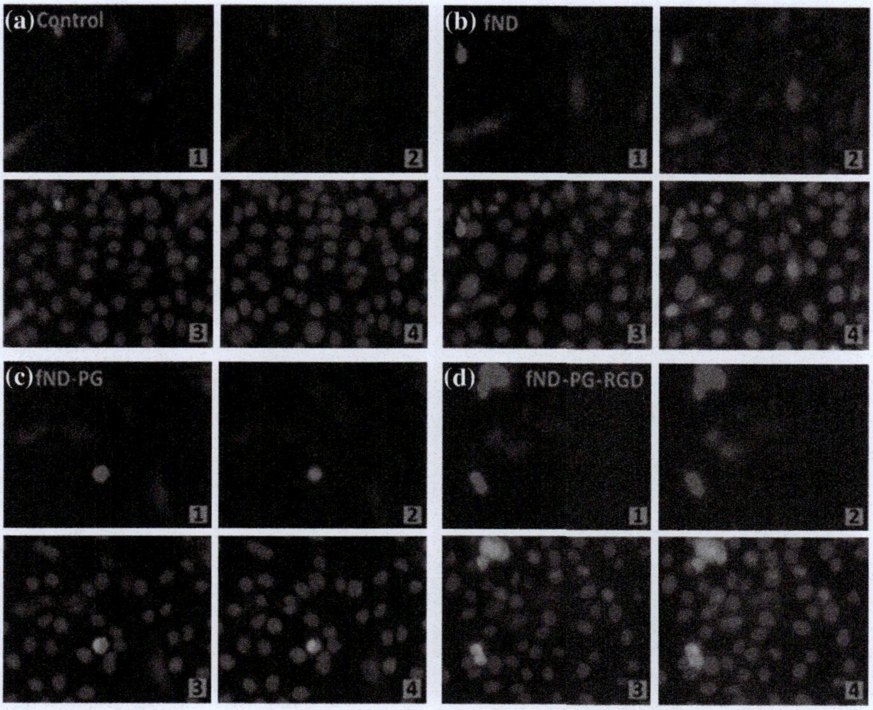

Fig. 9 Fluorescent microphotographs of co-cultured U87MG and HeLa cells after 24-h treatment of fND, fND-PG and fND-PG-RGD (200 μg mL^{-1}). Three images of the same field were taken using different filters. *1* and *3*: *Green* fluorescence was from U87MG stained with CFDA-SE. *2* and *4*: Punctuated *red* fluorescence was from internalized fluorescent ND particles. *3* and *4*: *Blue* fluorescence was from the nuclei of both U87MG and HeLa cells

fluorescence was observed only in the cytoplasm of U87MG cells, but not in HeLa (Fig. 9d).

The lysosomal compartment is commonly reported as a major intracellular depositing site for internalized particles [55, 63]. In our experiment, fluorescence microscopy revealed localization of fND-PG-RGD in the lysosomes. The lysosome of control U87MG cells was stained with LysoTracker® Blue (Invitrogen) against mitochondrial counterstained by rhodamine 123 (Fig. 10a). LysoTracker® Blue staining was then applied to U87MG cells treated with 200 μg mL^{-1} of fND-PG-RGD for 24 h. As shown in Fig. 10b, the red fluorescence originating from the internalized fND-PG-RGD is found to co-localize with the blue staining of lysosomes.

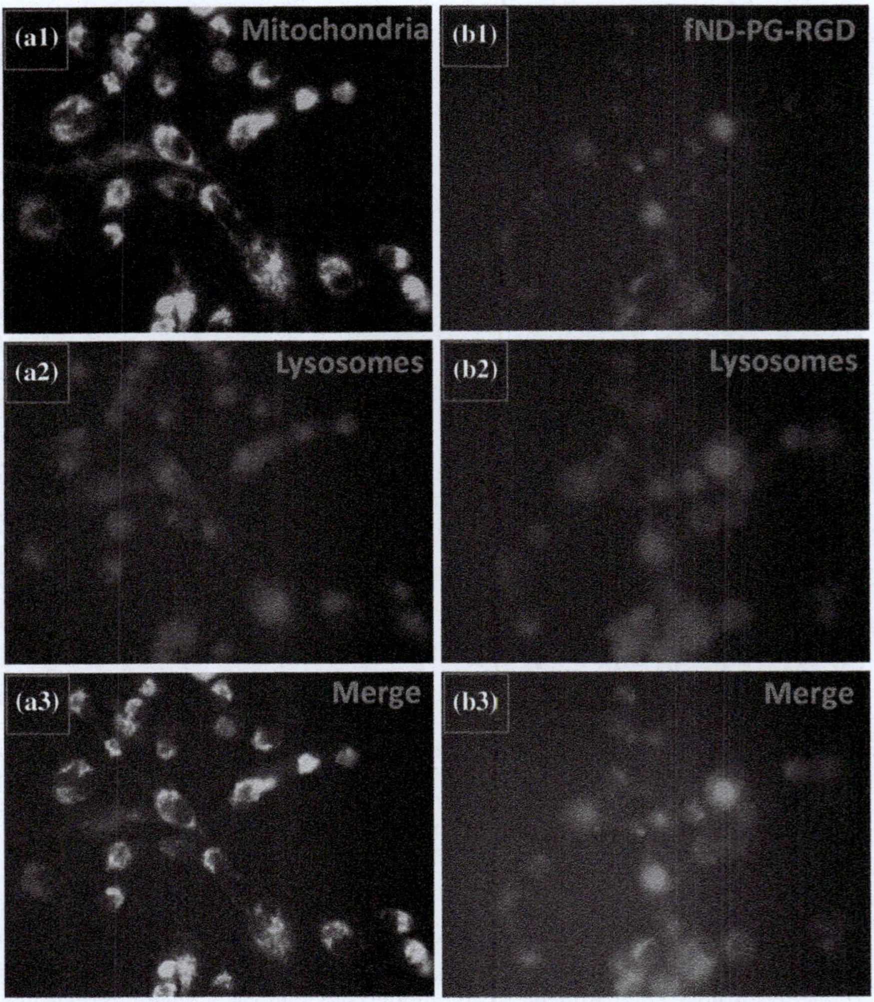

Fig. 10 **a** and **b**: Subcellular localization of internalized fND-PG-RGD in U87MG cells. **a1** and **a2**: Lysosomal staining by LysoTracker® Blue with counter staining of mitochondria by rhodamine 123 in control cells. **a3**: **a1** merged with **a2**. **b1**: Fluorescence of internalized fND-PG-RGD. **b2**: Lysosomal staining by LysoTracker® Blue. **b3**: **b1** merged with **b2**

5 Concluding Remarks

Although PG has been known as a hydrophilic polymer [25, 40, 64, 65], we have demonstrated that PG has advantage over the other hydrophilic functionalities such as PEG, especially in the following aspects; (1) more hydrophilic property due to a number of hydroxyl groups and branched structure to cover the surface of nanoparticle densely, (2) good extensibility by using the hydroxyl groups as scaffolds for further derivatization, and (3) generality for various kinds of nanoparticles such as nanocarbon materials and metal oxide nanoparticles. In terms of (1), we realized 400 times larger dispersibility of ND after PG functionaliztion as compared with PEG-functionalized ND [10, 15]. Immobilization of the functionalities such as a targeting peptide and/or a drug has been demonstrated herein through substitution for some of the hydroxyl groups in the PG. PG functionalization has been applied to not only ND, but also fND, SPION, and other nanomaterials such as ZnO nanoparticle, single-walled carbon nanotubes, and graphene. Since we have also found that the PG-coated ND has non-specific uptake from cells referred to as stealth effect [9, 66], PG coating should be very promising in the application of nanoparticle for an imaging probe and a drug carrier, and make significant progress in the field of nanomedicine.

References

1. V. Wagner, A. Dullaart, A.-K. Bock, A. Zweck, The emerging nanomedicine landscape. Nat. Biotechnol. **24**, 1211–1217 (2006). doi:10.1038/nbt1006-1211
2. K. Riehemann, S.W. Schneider, T.A. Luger, B. Godin, M. Ferrari, H. Fuchs, Nanomedicine—challenge and perspectives. Angew. Chem. Int. Ed. **48**, 872–897 (2009). doi:10.1002/anie.200802585
3. L. Cheng, C. Wang, L. Feng, K. Yang, Z. Liu, Functional nanomaterials for phototherapies of cancer. Chem. Rev. **114**, 10869–10939 (2014). doi:10.1021/cr400532z
4. Y. Liu, H. Miyoshi, M. Nakamura, Nanomedicine for drug delivery and imaging: a promising avenue for cancer therapy and diagnosis using targeted functional nanoparticles. Int. J. Cancer **120**, 2527–2537 (2007). doi:10.1002/ijc.22709
5. O.C. Farokhzad, R. Langer, Nanomedicine: developing smarter therapeutic and diagnostic modalities. Adv. Drug Delivery Rev. **58**, 1456–1459 (2006). doi:10.1016/j.addr.2006.09.011
6. T. Sun, Y.S. Zhang, B. Pang, D.C. Hyun, M. Yang, Y. Xia, Engineered nanoparticles for drug delivery in cancer therapy. Angew. Chem. Int. Ed. **53**, 12320–12364 (2014). doi:10.1002/anie.201403036
7. E.-K. Lim, T. Kim, S. Paik, S. Haam, Y.-M. Huh, K. Lee, Nanomaterials for theranostics: recent advances and future challenges. Chem. Rev. **115**, 327–394 (2015). doi:10.1021/cr300213b
8. L. Zhao, Y.-H. Xu, T. Akasaka, S. Abe, N. Komatsu, F. Watari, X. Chen, Nanodiamond with stealth polyglycerol coating: a macrophage-evading platform for selective drug delivery in cancer cells. Biomaterials **35**, 5393–5406 (2014). doi:10.1016/j.biomaterials.2014.03.041
9. L. Zhao, Y.-H. Xu, H. Qin, S. Abe, T. Akasaka, T. Chano, F. Watari, T. Kimura, N. Komatsu, X. Chen, Platinum on nanodiamond: a promising prodrug conjugated with stealth polyglycerol, targeting peptide and acid-responsive antitumor drug. Adv. Funct. Mater. **24**, 5348–5357 (2014). doi:10.1002/adfm.201304298
10. L. Zhao, T. Takimoto, M. Ito, N. Kitagawa, T. Kimura, N. Komatsu, Chromatographic separation of highly soluble diamond nanoparticles prepared by polyglycerol grafting. Angew. Chem. Int. Ed. **50**, 1388–1392 (2011). doi:10.1002/anie.201006310
11. L. Zhao, Y. Nakae, H. Qin, T. Ito, T. Kimura, H. Kojima, L. Chan, N. Komatsu, Polyglycerol-functionalized nanodiamond as a platform for gene delivery: derivatization, characterization, and hybridization with DNA. Beilstein J. Org. Chem. **10**, 707–713 (2014). doi:10.3762/bjoc.10.64

12. L. Zhao, A. Shiino, H. Qin, T. Kimura, N. Komatsu. Synthesis, characterization, and magnetic resonance evaluation of polyglycerol-functionalized detonation nanodiamond conjugated with gadolinium(III) complex. J. Nanosci. Nanotechnol. **15**, 1076–1082 (2015). doi:10.1166/jnn.2015.9738
13. L. Zhao, T. Chano, S. Morikawa, Y. Saito, A. Shiino, S. Shimizu, T. Maeda, T. Irie, S. Aonuma, H. Okabe, T. Kimura, T. Inubushi, N. Komatsu, Hyperbranched polyglycerol-grafted superparamagnetic iron oxide nanoparticles: synthesis, characterization, functionalization, size separation, magnetic properties, and biological applications. Adv. Funct. Mater. **22**, 5107–5117 (2012). doi:10.1002/adfm.201201060
14. L. Zhao, T. Takimoto, T. Kimura, N. Komatsu, Polyglycerol functionalization of ZnO nanoparticles for stable hydrosol in physiological media. J. Indian Chem. Soc. **88**, 1787–1790 (2011)
15. T. Takimoto, T. Chano, S. Shimizu, H. Okabe, M. Ito, M. Morita, T. Kimura, T. Inubushi, N. Komatsu, Preparation of fluorescent diamond nanoparticles stably dispersed under physiological environment through multi-step organic transformations. Chem. Mater. **22**, 3462–3471 (2010). doi:10.1021/cm100566v
16. P. Caravan, J.J. Ellison, T.J. McMurry, R.B. Lauffer, Gadolinium(III) chelates as MRI contrast agents: structure, dynamics, and applications. Chem. Rev. **99**, 2293–2352 (1999). doi:10.1021/cr980440x
17. Z. Zhou, Z.-R. Lu, Gadolinium-based contrast agents for magnetic resonance cancer imaging. Wiley Interdiscip. Rev. Nanomed. Nanobiotechnol. **5**, 1–18 (2013). doi:10.1002/wnan.1198
18. P. Caravan, Strategies for increasing the sensitivity of gadolinium based MRI contrast agents. Chem. Soc. Rev. **35**, 512–523 (2006). doi:10.1039/B510982P
19. E. Wiener, M.W. Brechbiel, H. Brothers, R.L. Magin, O.A. Gansow, D.A. Tomalia, P.C. Lauterbur, Dendrimer-based metal chelates: a new class of magnetic resonance imaging contrast agents. Magn. Reson. Med. **31**, 1–8 (1994). doi:10.1002/mrm.1910310102
20. R.C. Brasch, Rationale and applications for macromolecular Gd-based contrast agents. Magn. Reson. Med. **22**, 282–287 (1991). doi:10.1002/mrm.1910220225
21. F.M. Cavagna, F. Maggioni, P.M. Castelli, M. Dapra, L.G. Imepratori, V. Lorusso, B.G. Jenkins, Gadolinium chelates with weak binding to serum proteins: a new class of high-efficiency, general purpose contrast agents for magnetic resonance imaging. Invest. Radiol. **32**, 780–796 (1997)
22. P.J. Endres, T. Paunesku, S. Vogt, T.J. Meade, G.E. Woloschak, DNA-TiO_2 nanoconjugates labeled with magnetic resonance contrast agents. J. Am. Chem. Soc. **129**, 15760–15761 (2007). doi:10.1021/ja0772389
23. L.M. Manus, D.J. Mastarone, E.A. Waters, X.-Q. Zhang, E.A. Schultz-Sikma, K.W. MacRenaris, D. Ho, T.J. Meade, Gd(III)-nanodiamond conjugates for MRI contrast enhancement. Nano Lett. **10**, 484–489 (2009). doi:10.1021/nl903264h
24. T. Nakamura, T. Ohana, H. Yabuno, R. Kasai, T. Suzuki, T. Hasebe, Simple fabrication of Gd(III)-DTPA-nanodiamond particles by chemical modification for use as magnetic resonance imaging (MRI) contrast agent. Appl. Phys. Express **6**, 015001 (2013). doi:10.7567/APEX.6.015001
25. M. Calderón, M.A. Quadir, S.K. Sharma, R. Haag, Dendritic polyglycerols for biomedical applications. Adv. Mater. **22**, 190–218 (2010). doi:10.1002/adma.200902144
26. P. Caravan, C.T. Farrar, L. Frullano, R. Uppal, Influence of molecular parameters and increasing magnetic field strength on relaxivity of gadolinium- and manganese-based T_1 contrast agents. Contrast Media Mol. Imaging **4**, 89–100 (2009). doi:10.1002/cmmi.267
27. P.L. de Sousa, J.B. Livramento, L. Helm, A.E. Merbach, W. Même, B.-T. Doan, J.-C. Beloeil, M.I.M. Prata, A.C. Santos, C.F.G.C. Geraldes, É. Tóth, In vivo MRI assessment of a novel GdIII-based contrast agent designed for high magnetic field applications. Contrast Media Mol. Imaging **3**, 78–85 (2008). doi:10.1002/cmmi.233
28. E. Amstad, M. Textor, E. Reimhult, Stabilization and functionalization of iron oxide nanoparticles for biomedical applications. Nanoscale **3**, 2819–2843 (2011). doi:10.1039/C1NR10173K

29. J. Xie, G. Liu, H.S. Eden, H. Ai, X. Chen, Surface-engineered magnetic nanoparticle platform for cancer imaging and therapy. Acc. Chem. Res. **44**, 883–892 (2011). doi:10.1021/ar200044b
30. R. Hao, R. Xing, Z. Xu, Y. Hou, S. Gao, S. Sun, Synthesis, functionalization, and biomedical applications of multifunctional magnetic nanoparticles. Adv. Mater. **22**, 2729–2742 (2010). doi:10.1002/adma.201000260
31. A.H. Latham, M.E. Williems, Controlling transport and chemical functionality of magnetic nanopartiles. Acc. Chem. Res. **41**, 411–420 (2008). doi:10.1021/ar700183b
32. S. Laurent, D. Forge, M. Port, A. Roch, C. Robic, L. Vander Elst, R.N. Muller, Magnetic iron oxide nanoparticles: synthesis, stabilization, vectorization, physicochemical characterizations, and biological applications. Chem. Rev. **108**, 2064–2110 (2008). doi:10.1021/cr068445e
33. Y. Jun, J.-H. Lee, J. Cheon, Chemical design of nanoparticle probes for high-performance magnetic resonance imaging. Angew. Chem. Int. Ed. **47**, 5122–5135 (2008). doi:10.1002/anie.200701674
34. A.K. Gupta, M. Gupta, Synthesis and surface engineering of iron oxide nanoparticles for biomedical applications. Biomaterials **26**, 3995–4021 (2005). doi:10.1016/j.biomaterials.2004.10.012
35. K. Chen, J. Xie, H. Xu, D. Behera, M.H. Michalski, S. Biswal, A. Wang, X. Chen, Triblock copolymer coated iron oxide nanoparticle conjugate for tumor integrin targeting. Biomaterials **30**, 6912–6919 (2009). doi:10.1016/j.biomaterials.2009.08.045
36. F.M. Kievit, Z.R. Stephen, O. Veiseh, H. Arami, T. Wang, V.P. Lai, J.O. Park, R.G. Ellenbogen, M.L. Disis, M. Zhang, Targeting of primary breast cancers and metastases in a transgenic mouse model using rationally designed multifunctional spions. ACS Nano **6**, 2591–2601 (2012). doi:10.1021/nn205070h
37. W. Cai, J. Wan, Facile synthesis of superparamagnetic magnetite nanoparticles in liquid polyols. J. Colloid Interface Sci. **305**, 366–370 (2007). doi:10.1016/j.jcis.2006.10.023
38. J. Wan, W. Cai, X. Meng, E. Liu, Monodisperse water-soluble magnetite nanoparticles prepared by polyol process for high-performance magnetic resonance imaging. Chem. Commun. **47**, 5004–5006 (2007). doi:10.1039/B712795B
39. N. Miguel-Sancho, O. Bomati-Miguel, G. Colom, J.-P. Salvador, M.-P. Marco, J. Santamaria, Development of stable, water-dispersible, and biofunctionalizable superparamagnetic iron oxide nanoparticles. Chem. Mater. **23**, 2795–2802 (2011). doi:10.1021/cm1036452
40. L. Wang, K.G. Neoh, E.T. Kang, B. Shuter, S.-C. Wang, Superparamagnetic hyperbranched polyglycerol-grafted Fe_3O_4 nanoparticles as a novel magnetic resonance imaging contrast agent: an in vitro assessment. Adv. Funct. Mater. **19**, 2615–2622 (2009). doi:10.1002/adfm.200801689
41. U.I. Tromsdorf, O.T. Bruns, S.C. Salmen, U. Beisiegel, H. Weller, A highly effective, nontoxic T_1 MR contrast agent based on ultrasmall PEGylated iron oxide nanoparticles. Nano Lett. **9**, 4434–4440 (2009). doi:10.1021/nl902715v
42. H. Wei, N. Insin, J.Y. Lee, H.-S. Han, J.M. Cordero, W. Liu, M.G. Bawendi, Compact zwitterion-coated iron oxide nanoparticles for biological applications. Nano Lett. **12**, 22–25 (2012). doi:10.1021/nl202721q
43. S. Tong, S. Hou, Z. Zheng, J. Zhou, G. Bao, Coating optimization of superparamagnetic iron oxide nanoparticles for high T_2 relaxivity. Nano Lett. **10**, 4607–4613 (2010). doi:10.1021/nl102623x
44. H. Duan, M. Kuang, X. Wang, Y.A. Wang, H. Mao, S. Nie, Reexamining the effects of particle size and surface chemistry on the magnetic properties of iron oxide nanocrystals: new insight into spin disorder and proton relaxivity. J. Phys. Chem. C **112**, 8127–8131 (2008). doi:10.1021/jp8029083
45. Y. Jun, Y.-M. Huh, J. Choi, J.-H. Lee, H.-T. Song, S. Kim, S. Yoon, K.-S. Kim, J.-S. Shin, J.-S. Suh, J. Cheon, Nanoscale size effect of magnetic nanocrystals and their utilization for cancer diagnosis via magnetic resonance imaging. J. Am. Chem. Soc. **127**, 5732–5733 (2005). doi:10.1021/ja0422155
46. L. Zhao, N. Komatsu, in *Magnetic Nanoparticles Synthesis, Physicochemical Properties and Role in Biomedicine*, ed. by N.P. Sabbas (Nove publisher, New York, 2014), pp. 95–111

47. L.E.W. LaConte, N. Nitin, O. Zurkiya, D. Caruntu, C.J. O'Connor, X. Hu, G. Bao, Coating thickness of magnetic iron oxide nanoparticles affects r2 relaxivity. J. Magn. Reson. Imaging **26**, 1634–1641 (2007). doi:10.1002/jmri.21194
48. J. Huang, L. Bu, J. Xie, K. Chen, Z. Cheng, X. Li, X. Chen, Effects of nanoparticle size on cellular uptake and liver MRI with polyvinylpyrrolidone-coated iron oxide nanoparticles. ACS Nano **4**, 7151–7160 (2010). doi:10.1021/nn101643u
49. H.B. Na, G. Palui, J.T. Rosenberg, X. Ji, S.C. Grant, H. Mattoussi, Multidentate catechol-based polyethylene glycol oligomers provide enhanced stability and biocompatibility to iron oxide nanoparticles. ACS Nano **6**, 389–399 (2012). doi:10.1021/nn203735b
50. V.N. Mochalin, O. Shenderova, D. Ho, Y. Gogotsi, The properties and applications of nanodiamonds. Nat. Nanotechnol. **7**, 11–23 (2012). doi:10.1038/nnano.2011.209
51. Y. Xing, L. Dai, Nanodiamond for nanomedicine. Nanomedicine **4**, 207–218 (2009). doi:10.2217/17435889.4.2.207
52. Y.-R. Chang, H.-Y. Lee, K. Chen, C.-C. Chang, D.-S. Tsai, C.-C. Fu, T.-S. Lim, Y.-K. Tzeng, C.-Y. Fang, C.-C. Han, H.-C. Chang, W. Fann, Mass production and dynamic imaging of fluorescent nanodiamonds. Nat. Nanotechnol. **3**, 284–288 (2008). doi:10.1038/nnano.2008.99
53. R.J. Narayan, R.D. Boehm, A.V. Sumant, Medical applications of diamond particles & surfaces. Mater. Today **14**, 154–163 (2011). doi:10.1016/S1369-7021(11)70087-6
54. C.-C. Fu, H.-Y. Lee, K. Chen, T.-S. Lim, H.-Y. Wu, P.-K. Lin, P.-K. Wei, P.-H. Tsao, H.-C. Chang, W. Fann, Characterization and application of single fluorescent nanodiamonds as cellular biomarkers. Proc. Nat. Acad. Sci. **104**, 727–732 (2007). doi:10.1073/pnas.0605409104
55. O. Faklaris, V. Joshi, T. Irinopoulou, P. Tauc, M. Sennour, H. Girard, C. Gesset, J.-C. Arnault, A. Thorel, J.P. Boudou, P.A. Curmi, F. Treussart, Photoluminescent diamond nanoparticles for cell labeling: study of the uptake mechanism in mammalian cells. ACS Nano **3**, 3955–3962 (2009). doi:10.1021/nn901014j
56. N. Mohan, Y.-K. Tzeng, L. Yang, Y.-Y. Chen, Y.Y. Hui, C.-Y. Fang, H.-C. Chang, Sub-20-nm fluorescent nanodiamonds as photostable biolabels and fluorescence resonance energy transfer donors. Adv. Mater. **22**, 843–847 (2010). doi:10.1002/adma.200901596
57. J.-P. Boudou, M.-O. David, V. Joshi, H. Eidi, P.A. Curmi, Hyperbranched polyglycerol modified fluorescent nanodiamond for biomedical research. Diamond Relat. Mater. **38**, 131–138 (2013). doi:10.1016/j.diamond.2013.06.019
58. Y.-K. Tzeng, O. Faklaris, B.-M. Chang, Y. Kuo, J.-H. Hsu, H.-C. Chang, Superresolution imaging of albumin-conjugated fluorescent nanodiamonds in cells by stimulated emission depletion. Angew. Chem. Int. Ed. **50**, 2262–2265 (2011). doi:10.1002/anie.201007215
59. A. Bumb, S.K. Sarkar, N. Billington, M.W. Brechbiel, K.C. Neuman, Silica encapsulation of fluorescent nanodiamonds for colloidal stability and facile surface functionalization. J. Am. Chem. Soc. **135**, 7815–7818 (2013). doi:10.1021/ja4016815
60. S. Osswald, G. Yushin, V. Mochalin, S.O. Kucheyev, Y. Gototsi, Control of sp^2/sp^3 carbon ratio and surface chemistry of nanodiamond powders by selective oxidation in air. J. Am. Chem. Soc. **128**, 11635–11642 (2006). doi:10.1021/ja063303n
61. J. Havlik, V. Petrakova, I. Rehor, V. Petrak, M. Gulka, J. Stursa, J. Kucka, J. Ralis, T. Rendler, S.-Y. Lee, R. Reuter, J. Wrachtrup, M. Ledvina, M. Nesladek, P. Cigler, Boosting nanodiamond fluorescence: towards development of brighter probes. Nanoscale **5**, 3208–3211 (2013). doi:10.1039/C2NR32778C
62. S.-J. Yu, M.-W. Kang, H.-C. Chang, K.-M. Chen, Y.-C. Yu, Bright fluorescent nanodiamonds: no photobleaching and low cytotoxicity. J. Am. Chem. Soc. **127**, 17604–17605 (2005). doi:10.1021/ja0567081
63. S. Manchun, C.R. Dass, P. Sriamornsak, Targeted therapy for cancer using pH-responsive nanocarrier systems. Life Sci. **90**, 381–387 (2012). doi:10.1016/j.lfs.2012.01.008
64. D. Wilms, S.-E. Stiriba, H. Frey, Hyperbranched polyglycerols: from the controlled synthesis of biocompatible polyether polyols to multipurpose applications. Acc. Chem. Res. **43**, 129–141 (2010). doi:10.1021/ar900158p

65. L. Zhou, C. Gao, W. Xu, X. Wang, Y. Xu, Enhanced biocompatibility and biostability of CdTe quantum dots by facile surface-initiated dendritic polymerization. Biomacromolecules **10**, 1865–1874 (2009). doi:10.1021/bm9002877
66. S. Sotoma, R. Igarashi, J. Iimura, Y. Kumiya, H. Tochio, Y. Harada, M. Shirakawa, Suppression of non-specific protein-nanodiamond adsorption enabling specific targeting of nanodiamonds to bio-molecules of interest. Chem. Lett. **44**, 354–356 (2015). doi:10.1246/cl.141036

Carbon Based Dots and Their Luminescent Properties and Analytical Applications

Yongqiang Dong, Jianhua Cai and Yuwu Chi

Abstract Carbon based dots (CDs) composed of sp^2 carbon structures and surface functional groups are a new kind of carbon nanomaterials, exhibiting unique luminescent properties due to the quantum confinement and edge effects. This chapter introduces CDs in detail from their synthetic strategies, morphological and structural characteristics, luminescent properties and mechanisms, and sensing applications. The synthesis methods are summarized as "top-down" and "bottom-up" approaches. Luminescent properties discussed include photoluminescence, upconversion luminescence, chemiluminescence, electrochemiluminescence. Sensing applications mainly refer to the chemical and biological sensors based on the luminescent properties of CDs. This chapter provides an overview of the research field and gives future perspectives for developing the exciting materials.

Keywords Carbon based dots · Synthesis · Morphology · Luminescent property · Mechanism · Sensing

1 What Are Carbon Based Dots?

Carbon based dots (CDs) are considered as a class of zero-dimensional carbon nanomaterials, which usually contain sp^2 carbon structures and functional groups. Due to the quantum confinement and edge effects, CDs exhibit many unique optical properties such as photoluminescence (PL), chemiluminescence (CL), and electrochemiluminescence (ECL) [1–7]. Like traditional semiconductor based quantum dots (QDs), CDs exhibit obvious advantages over organic dyes in photostability

Y. Dong · J. Cai · Y. Chi (✉)
Ministry of Education Key Laboratory of Analysis and Detection Technology for Food Safety, Fujian Provincial Key Laboratory of Analysis and Detection Technology for Food Safety, Fuzhou University, Fujian 350108, China
e-mail: y.w.chi@fzu.edu.cn

Y. Dong · J. Cai · Y. Chi
Department of Chemistry, Fuzhou University, Fujian 350108, China

N. Yang et al. (eds.), *Carbon Nanoparticles and Nanostructures*,
Carbon Nanostructures, DOI 10.1007/978-3-319-28782-9_6

against photobleaching and blinking. Furthermore, CDs usually have low toxicity and good biocompatibility, which overcomes the main drawbacks of QDs. Accordingly, CDs have attracted significant attention from researchers.

CDs were first discovered unexpectedly at 2004 [8]. The researchers used an eletrophoretic method to purify single-walled carbon nanotubes (SWCNTs), which have been prior chemically oxidized with concentration HNO_3 and extracted with basic water. To their surprise, some highly luminescent nanomaterials were obtained, which were considered as fragments of SWCNTs. Two years later, Sun's group obtained highly luminescent carbon nanomaterials again [9]. They prepared some small carbon particles (sub-10 nm) via laser ablation of graphite, followed by chemical oxidation and surface passivation. Bright and excitation-dependent PL emission was observed from those surface-passivated carbon nanoparticles, which were named as carbon dots. After that, these luminescent carbon nanoparticles of less than 10 nm in size have come to be known as carbon dots [10, 11], carbon quantum dots [12, 13], carbon nanodots [14], or carbogenic naodots [15, 16], and have attracted increasing attention [1, 2, 17–20]. In 2008, Ponomarenko et al. used ultra-high-resolution electron beam lithography to cut graphene to desired sizes. The obtained materials were first named as graphene quantum dots [21]. However, little attention has been paid to the graphene quantum dots since this process required very specialized equipment and led to a very low yield. Two years later, Pan's group found that graphene sheets could be cut hydrothermally into blue PL graphene quantum dots. The graphene quantum dots had diameters distributed in the range of 5–13 nm and heights between 1 and 2 nm [22]. After that, broad attention has been attracted by these fascinating luminescent materials [3–6, 23–31]. It seems that carbon dots and graphene quantum dots have their own definitions. Usually, carbon dots refer to carbon nanoparticles (either graphite nanocrystals or amorphous carbon nanoparticles) of less than 10 nm in size, while graphene quantum dots are single- or multi-layer graphene nanosheets of less than 100 nm in width. However, the two types of carbon nanomaterials have quite similar structural characteristics, including sp^2 carbon structures, edge states and functional groups. Furthermore, the two types of carbon nanomaterials have many common characteristics in luminescent properties [1, 25]. Therefore, carbon dots and graphene quantum dots should be the same in nature, and may be merged to CDs. In this chapter, the synthesis, luminescent properties and analytical applications of CDs will be introduced systematically.

2 Synthetic Strategies

Up to now, considerable attention has been focused on the synthesis of CDs, and many methods have been proposed to prepare different types of CDs. Usually, these methods can be classified into "bottom-up" and "top-down". Generally, the "bottom-up" strategies mean synthesizing CDs by carbonizing some special organic precursors while the "top-down" techniques refer to cutting some big-size carbon materials into small-size carbon particles.

2.1 "Top-Down" Approaches

"Top-down" approaches have been used widely in the preparation of CDs. These methods could be further divided into chemical oxidation releasing, chemical oxidation etching, electrochemical cutting, hydrothermal/solvothermal cutting, and Li^+/K^+ intercalation methods.

2.1.1 Chemical Oxidation Releasing

Chemical oxidation releasing methods mean oxidizing the aggregates of small sized carbon structures by some strong oxidative acid, producing soluble CDs. The chemical oxidation introduces a large quantities of hydrophilic oxygen-containing functional groups into some insoluble carbon aggregates. As a result, the nanosized carbon structures contained in the aggregates turn to water-soluble CDs, and are released into the solution. Accordingly, CDs could be obtained after suitable purification. Xu et al. [8] oxidized arc-discharged soot with 3.3 M HNO_3 to increase their hydrophilicity, followed by an extraction with a NaOH solution. Then, the obtained black suspension was separated by gel electrophoresis. Finally, luminescent carbon nanomaterials, which can now refer to CDs, were obtained. The CDs were characterized with atomic force microscopy (AFM) (Fig. 1). The results indicated that the nanoparticles have a narrow distribution of vertical sizes centered at about 1 nm, which fit well with the height of most single-layer or bilayer graphene quantum dots reported in the latter literatures. However the expensive carbon source and the complicated separation limited the further development of this method.

Sun's group refluxed carbon aggregates produced via laser ablation of a carbon target (graphite powder) with nitric acid solution (up to 2.6 M) to prepare CDs [9]. However, the obtained CDs must be surface-passivated with some organic

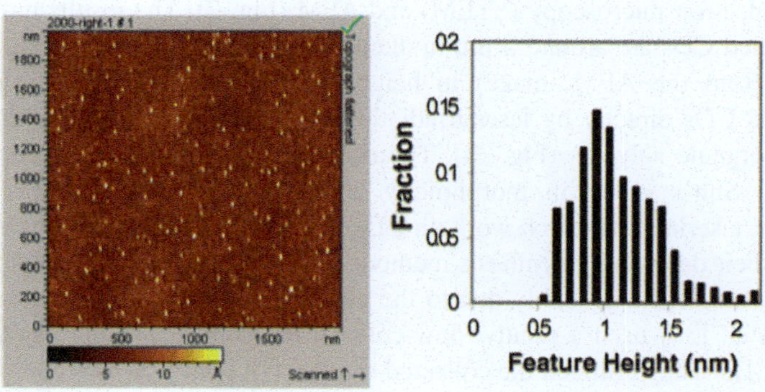

Fig. 1 A typical AFM topography image (*left*) and feature height distribution (*right*) for the orange fluorescent fraction deposited onto mica [8]

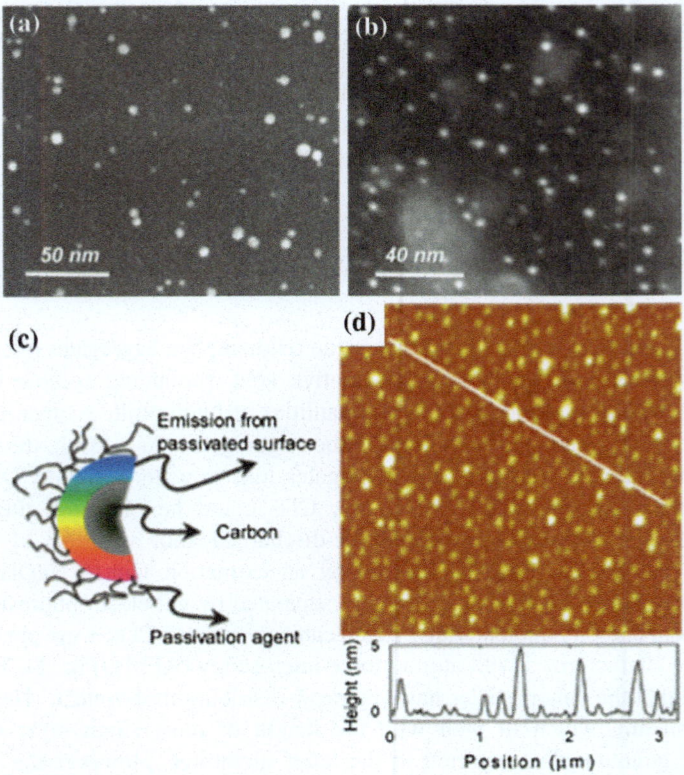

Fig. 2 Representative STEM images of CDs surface-passivated with **a** PEG1500 N and **b** PPEI-EI [9], **c** structure of the surface-passivated CDs and **d** AFM topography image of CDs on mica substrate, with the height profile along the *line* in the image [37]

molecules to produce bright PL emission, which will be further discussed in the following sections. They characterized the obtained PL CDs with scanning transmission electron microscopy (STEM) and AFM (Fig. 2). The results indicated that the obtained CDs are around 5 nm in diameter and less than 5 nm (may be mainly 1–2 nm from the AFM image) in height. Hu et al. [32] synthesized surface-passivated CDs directly by laser irradiation of carbon powders in surface ligands adopted organic solvents (Fig. 3a). The obtained CDs are similar with those prepared by Sun's group in morphology and showed bright blue PL emission. Although, a serial of research work on CDs has been carried out by Sun's group [6, 33–37], their developing synthesis methods for CDs have not been widely adopted by other researchers probably due to the need for high-cost equipments.

Liu et al. [38] used a relative low-cost carbon source, namely candle soot, to prepare CDs. They refluxed the collected candle soot with 5 M HNO_3 for 12 h. The formed suspension had been neutralized and dialyzed against water before being separated with polyacrylamide gel electrophoresis (PAGE). Finally, nine fractions

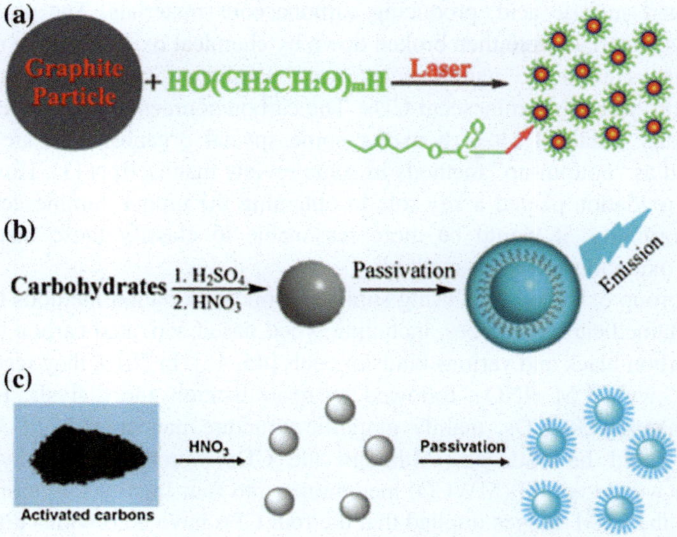

Fig. 3 **a** Schematic map of the one-step synthesis of luminescent CNPs from carbon powders in PEG200N solvent [32]. **b** Schematic illustration of preparation procedure of PL carbogenic dots [43]. **c** Schematic illustration of preparation procedure of PL CDs from AC [47]

of CDs emitting different PL were collected. However, the corresponding AFM analysis suggested that the different CDs might have no obvious difference in morphology, with an average height of about 1 nm. In a further study, Ray et al. [39] synthesized CDs using the similar method. However, proposed a new step-centrifugation to purify the obtained CDs. In brief, they removed the unreacted carbon soot by centrifugation at 3000 rpm firstly, obtaining light brownish-yellow supernatant. Subsequently, the aqueous supernatant was mixed with acetone (water/acetone volume ratio was 1:3) and centrifuged at 14,000 rpm for 10 min to collect the black precipitate, which was purified CDs. Then, the CDs were separated by stepwise centrifugation in a water/ethanol/chloroform (1:1:3) mixed solvent. Finally, CDs with particle sizes in the range of 2–6 nm could be obtained from the supernatant after the centrifugation at 8000 rpm. From the results of high resolution transmission electron microscopic (HRTEM), it can be found that the CDs should be graphitic in nature. Combining the results of both AFM and HRTEM in the two work mentioned above, it can be found that the lateral sizes of the CDs from candle soot should be much larger than the heights. In other words, the CDs should be layer like in morphology. After that, some other research groups used similar methods synthesized CDs from other kinds of soot, such as candle soot [40], natural gas soot and plant soot [41, 42]. The obtained CDs also exhibited similar morphologic and PL characteristics with those synthesized from candle soot. Peng et al. carbonized carbohydrates to prepare carbonaceous materials, which were used instead of candle soot to prepared CDs using the similar method (Fig. 3b) [43]. In brief, carbohydrates such as glucose, sucrose and starch were dehydrated using concentrated sulfuric acid, producing carbonaceous materials. The obtained carbonaceous materials were then broken down by chemical oxidation with nitric acid. Finally, the carbogenic nanoparticles were passivated using amine-terminated compounds, yielding luminescent CDs. The carbon sources used in these methods were usually obtained by carbonizing some special organics, and are therefore considered as "bottom-up" methods in some review manuscripts [1]. However, the chemical oxidation played a key role in obtaining the unique luminescent properties. Accordingly, it would be more reasonable to classify these methods into chemical oxidation releasing methods.

Chi's group extracted CDs using similar chemical oxidation methods from more easyilyobtained carbon sources, including wood based activated carbon (AC) [44], XC-72 carbon black and various kinds of coals [45, 46]. In 2010, they reacted wood based AC with 4 M HNO_3, followed by neutralization and dialysis. High yield (higher than 10 %) CDs, mainly graphitic structure nanocrystals of 3–4 nm in diameter could be obtained. Although the CDs were separated by different molecular weight cut off (MWCO) membranes into four fractions with different PL emission, the TEM images implied that the four CDs have no obvious difference in morphology. Nearly at the same time, another group also prepared CDs from AC by a similar method (Fig. 3c) [47]. In another work of Chi and coworkers, they chemically oxidized XC-72 by refluxing with 6 M HNO_3 for 24 h [45]. The resulted black suspension was centrifuged to obtain supernatant and sediment. On one hand, the supernatant was heated at 200 °C to evaporate the water and HNO_3, high yield (44.5 %) of reddish-brown solid was obtained (Fig. 4). The solid could be well dissolved in water, and exhibited bright green PL. AFM and TEM images indicated that the reddish-brown solid was mainly single layer CDs, which had a uniform lateral size of about 10 nm and an average height of 0.5 nm [48]. On the other hand, the sediment was washed, dissolved and adjusted to about pH 8 with ammonia water, then ultra-filtered through a centrifugal filter device with a 100 kDa MWCO membrane. The filtrate was collected and proved to be multi-layer CDs. The average lateral size and height were 18 and 2 nm, respectively. The production yield of this multi-layer CDs was 9.0 %. It should be pointed out that the residual HNO_3 after the chemical oxidation was usually neutralized with NaOH, thus a long time dialysis was required to remove the resultant salt. The authors proposed therein for the first time to remove the residual HNO_3 by evaporating it directly. Apparently, it is much simpler and more convenient. Zheng's group also used a similar method to synthesize CDs from lamp black, obtaining CDs of about 3–4 nm in lateral size [49]. In a subsequent study, Chi's group developed a more efficient method to extract single layer CDs in large-scale from coals. Coals were refluxed with 5 M HNO_3, followed by a serial of centrifugal separation procedures. Six coal samples of different ranks have been investigated. Every coal sample could be treated into two fractions, namely, nitric acid-soluble fraction and nitric acid-insoluble fraction. According to the characterization results of TEM, AFM and Raman spectra, the nitric acid-soluble fraction was revealed to be mainly composed of CDs, which had an average height of about 0.5 nm and an average lateral size of about 10 nm. The production yield of the single layer CDs was dependent on the rank of coal. The

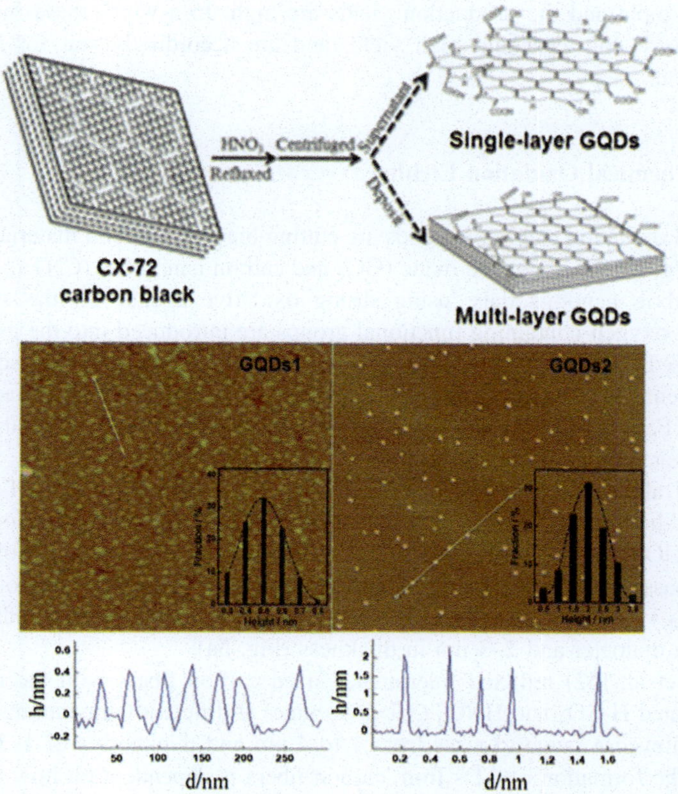

Fig. 4 Schematic illustration of preparation of single-layer CDs and multi-layer CDs from XC-72 carbon black, and the corresponding AFM images of the obtained CDs [45]

value decreased from 56.3 to 14.7 % in the six investigated coals when the coal rank increased gradually.

Generally, chemical oxidation releasing methods mean using oxidative acid (mainly concentrated HNO_3, usually not higher than 6 M) to react with some special carbon source (either artificial or natural obtained), combined with some separation treatments. HNO_3 at those concentrations are difficult to cut directly big sized carbon sources, such as graphite, carbon black, carbon nanotube or graphene, into small sized CDs [46]. However, it can introduce abundant hydrophilic function groups, including carboxyl and hydroxyl groups, into the carbon structures. Then, some small carbon structures present yet immobilized in the carbon sources could be released into the solution. That is to say, the morphology of the obtained CDs prepared by this kind of methods were mainly dependent on the nature of the carbon sources. These methods have some obvious advantages. For example, the chemical oxidation needs no any expensive equipment or agent, many very common carbon sources could be used as the precursor, the operation procedures are usually simple and the production yields are high. In a word, these methods are usually easy, low cost and high yield, and are accordingly suit for large-scale preparation.

2.1.2 Chemical Oxidation Etching

Chemical oxidation etching methods are cutting big sized carbon materials, such as graphite, graphene, graphene oxide (GO) and carbon nanotubes (CNTs), into small sized carbon nanomaterials, using strong oxidative agents. At the same time, abundant oxygen-containing functional groups are introduced into the as-produced small sized carbon materials, resulting in the formation of water-soluble and luminescent CDs. Some strong oxidants, such as the mixture of HNO_3 and H_2SO_4, $KMnO_4$, Fenton reagent (Fe^{2+}/Fe^{3+}/H_2O_2) have been used to react with big sized carbon materials to prepared small sized CDs.

Liu et al. [51] used artificial graphite produced by pyrolyzing self-assembled hexa-*peri*-hexabenzocoronene (HBC) columns at 1200 °C as the precursor [50], to prepare GO with a modified Hummers methods. The resultant GO solution was surface-passivated by refluxing with oligomeric poly(ethylene glycol) diamine (PEG_{1500N}) and reduced with hydrazine to obtain homogeneous nanodisks of ca. 60 nm in diameter and 2–3 nm in thickness (Fig. 5a).

Peng et al. [52] refluxed micrometer sized carbon fibers with the mixture of concentrated H_2SO_4 and HNO_3 (3:1 in volume) at different temperatures (Fig. 5b). They achieved a range of sizes from 1 to 4 nm and thicknesses of 1–3 graphene layers. The formation of CDs from carbon fibers is dependent on how the submicrometer domain structure of the sp^2 carbons were broken down. The authors considered that the broken down of the carbon fibers should share a similar reaction mechanism with the unzipping of carbon nanotubes into graphene nanoribbons by chemical oxidation [53]. In brief, the breakup of the structure and the planar graphitic domains was chemically initiated by the lining up of chemical functionalities, making the graphitic domains prone to fracture, preferably along the zigzag

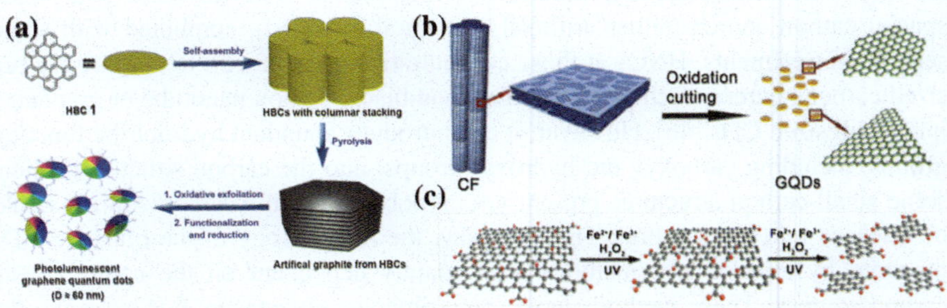

Fig. 5 **a** Processing diagram for the preparation of PL CDs by using HBC 1 as Carbon Source [50]. **b** Representation scheme of oxidation cutting of CF into GQDs [52]. **c** Schematic illustration of the synthesis of CDs by the photo-Fenton reaction method [64]

direction. After that, the mixture of H_2SO_4 and HNO_3 have been widely used to cut many big sized carbon sources, including graphene/graphene oxide (GO) [54–57], carbon fiber [58, 59], graphite powder [60, 61], coals [62], and petroleum coke [63] into small sized CDs. Although the obtained CDs by different authors were a little different in morphology, the main structural and PL characteristics are similar.

Zhou et al. [64] developed a photo-Fenton reaction method to cut GO into CDs (Fig. 5c). GO sheets reacted with Fenton reagent under UV irradiation, the reaction rate was strongly dependent on the extent of oxidization of the GO. Under the photo-assisted catalysis of Fe^{2+}/Fe^{3+} in water, H_2O_2 could be dissociated into hydroxyl radicals ($OH^{\bullet}$), which is considered as one of the most powerful oxidant [65]. The resultant $OH^{\bullet}$ and/or peroxide radical would attack carbon atoms connected with hydroxyl and epoxide groups, and break the C–C/C=C bonds. Meanwhile, the newly formed oxygen-containing groups such as the quinone group or radicals that might serve further as new photo-Fenton reaction sites. As a result, a mass scale production of CDs with periphery carboxylic groups could be obtained.

2.1.3 Electrochemical Cutting

The electrochemical cutting methods adopted high redox potentials, ranging from ± 1.5 to ± 3 V, which were high enough to either oxidize the C-C bonds or oxidize water to generate high oxidative $OH^{\bullet}$ and $O^{\bullet}$ radicals playing the role of an electrochemical “scissors” in their oxidative cleavage reactions. The first work about the electrochemical synthesis of CDs was reported by Ding’s group [66]. They used the multiwalled carbon nanotubes (MWCNTs), which were grown on a carbon paper by the chemical vapor deposition method, to serve as the working electrode in an electrochemical cell. Then the applied potential was cycled between −2.0 and +2.0 V in an acetonitrile solution containing 0.1 M tetrabutylammonium (TBA) perchlorate. Finally, CDs with a uniform spherical shape and a narrow size distribution being 2.8 ± 0.5 nm diameter were obtained (Fig. 6a). At the same time, the MWCNTs with straight and well isolated shapes became entangled together with swelling and curled features (Fig. 6b, c). Furthermore, the tube walls were opened (Fig. 6d). A parallel contrast experiment indicated that no CDs was formed when carbon paper without any MWCNTs was used. All the results suggest that the CDs should be electrochemically cut from the SWCNTs. Considering MWCNTs were formed with scrolled graphene layers, the authors proposed that TBA cations most probably intercalated into the gaps during electrochemical cycling and broke the tubes near the defects. Then, CDs were exfoliated and entered into the electrolyte solution.

After that, some researchers tried to synthesize CDs from some more readily available carbon sources, for example graphite material, by electrochemical cutting methods. Pang et al. [67] used a graphite column as the working electrode, and used aqueous electrolyte instead of organic electrolyte. The graphite column electrode was electrochemically oxidized at +3 V against a saturated calomel reference

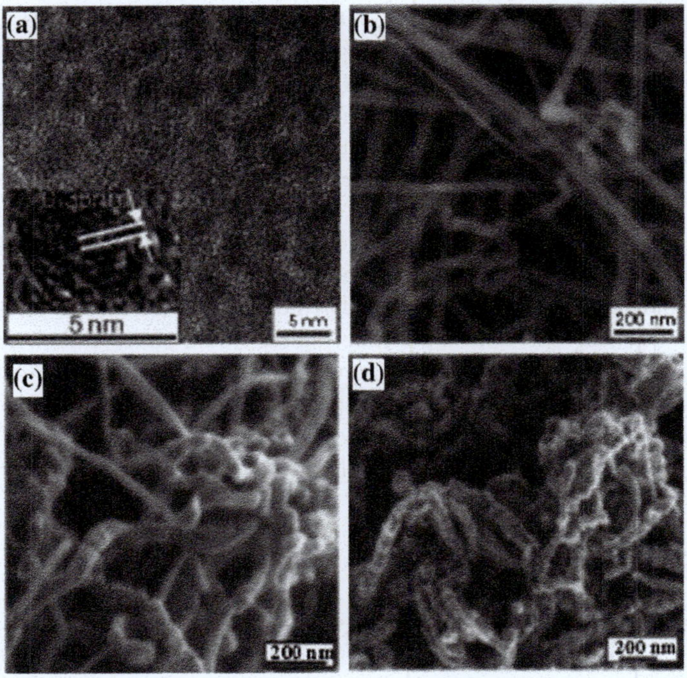

Fig. 6 HRTEM images of CDs (**a**); SEM images of pristine MWCNTs (**b**); MWCNTs after 100 cycles where the applied potential was scanned between 2.0 and −2.0 V at 0.5 V/s (**c**); and MWCNTs after 1000 cycles (**d**). *Inset* in panel **a** is the HRTEM image of a typical CD [66]

electrode. The obtained dark brown solution was separated by centrifugation and centrifugal filter devices. Finally, they obtained two kinds of CDs, the diameters of which were 1.9 ± 0.3 nm and 3.2 ± 0.5 nm, respectively. Chi's group cycled the graphite rod working electrode between −3 and +3 V instead (Fig. 7a) [68]. they obtained two kinds of spherical CDs respectively with average sizes of about 20 and 2 nm, which were separated using a 10 kDa MWCO membrane. The small CDs were found to exhibit bright blue PL and excellent ECL properties. Li et al. [69] used the mixture solution of NaOH/EtOH as electrolyte to oxidize graphite rods. Then CDs with diameters within 4 nm could be obtained. Loh's group developed an ionic liquid (IL) assisted electro-oxidation method to prepare CDs from graphite [70]. The water-soluble IL 1-butyl-3-methylimi-dazolium tetrafluoroborate [bmim][BF_4] was mixed with water at a mass ratio of 1:9, and used as the electrolyte. There were three stages during the electrochemical exfoliation, corresponding to the production of CDs with an average size of 8–10 nm, carbon nanoribbons and graphene respectively. It was considered that the formation of CDs was related to the $OH^{\bullet}$ and $O^{\bullet}$ radicals produced by the electro-oxidation of water, while the exfoliation of graphene nanoribbons should be caused by the intercalation of BF_4^- ions into the graphite anode (Fig. 7b).

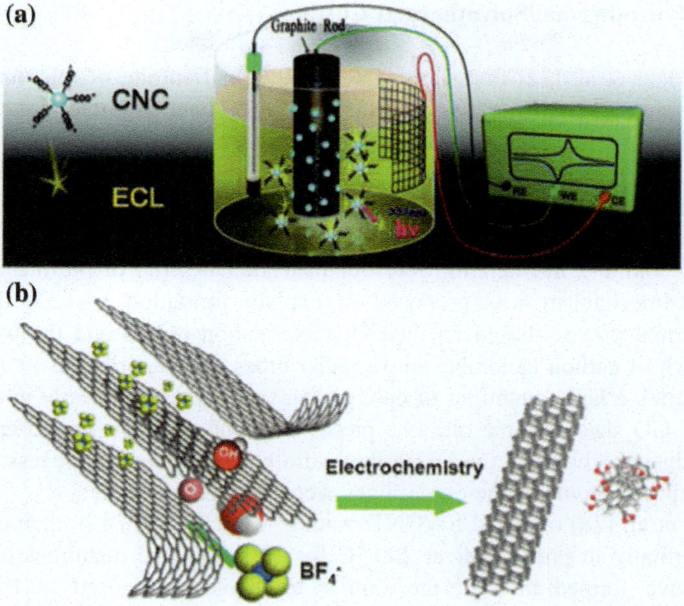

Fig. 7 **a** Electrochemical production of CDs from a graphite rod which are capable of electrochemiluminescence (ECL) [68]. **b** Illustration of the exfoliation process showing the attack of the graphite edge planes by hydroxyl and oxygen radicals, which facilitate the intercalation of BF_4^- anion [70]

Qu's group used a similar electrochemical method to synthesize CDs from a recently popular carbon nanomaterial, namely graphene [71]. They prepared a graphene film by direct filtration of the aqueous reduced graphene oxide colloidal suspensions through a filter membrane with pore size of 220 nm. After being mechanically peeled from the filter, the graphene film was treated with O_2 plasma for seconds prior to being electrochemically oxidized. Then CDs with uniform sizes of 3–5 nm could be obtained after the purification. In one of their follow-up reports, N-containing tetrabutylammonium perchlorate (TBAP) in acetonitrile was used as the electrolyte instead of PBS, producing nitrogen-doped CDs (N-CDs) whose morphologies were similar with those of N-free counterparts [72].

Some other similar works have been carried out to synthesize CDs from carbon fibers [73], graphite rod [74], and MWCNTs [75, 76]. Generally, these obtained CDs by electrochemical cutting have a typical crystallized carbon core featuring graphitic sp^2 carbon atoms and a surface with functional groups, mainly carboxyl and carboxyl groups. The sizes of the CDs are usually distributed in the range of 2–10 nm, depending on the nature of the carbon source, electrolyte and potential window.

In general, these electrochemical cutting methods are usually safe and convenient, however, the low production yields and the boring purification procedures may limit the development of these methods.

2.1.4 Hydrothermal/Solvothermal Cutting

In 2010, Pan et al. [22] first showed a hydrothermal cutting of micrometer sized graphene sheets into CDs. GO synthesized by a modified Hummer's method were thermally reduced first at 200–300 °C, then chemically oxidized using the mixture of H_2SO_4 and HNO_3 solution for 15–20 h under mild ultrasonication. The resultant graphene sheets were dispersed in weak alkaline solution (pH = 8), followed by a further hydrothermal treatment at 200 °C for 10 h. Finally, CDs with 5–13 nm diameters and 1–2 nm heights were obtained after a serial of purification procedures. A mechanism was proposed to explain formation of CDs during the hydrothermal process based on their characterization results and the well-known breakdown of carbon nanotubes into smaller tubes and nanoribbons in acid media [77]. In brief, a large quantities of epoxy chains were present linearly in the carbon lattice of GO sheets. Some ultrafine pieces surrounded by the mixed epoxy lines and/or edges might break up during the hydrothermal treatment process, by which the bridging O atoms in the epoxy lines were removed (Fig. 8a).

Dong et al. [78] oxidized SWCNTs with 8 M HNO_3 for 24 h, then deoxidized hydrothermally in pure water at 200 °C for 12 h. A large quantities of ultrafine pieces have formed through the similar mechanism proposed in Pan's work (Fig. 8b). However, the formed ultrafine pieces were fixed together with the SWCNTs matrixes due to the fact that most of the oxygen-containing functional groupswas removed during the hydrothermal treatment and the formed ultrafine piecesbecame water-insoluble. Therefore, a subsequent chemical oxidation was necessary to release the ultrafine pieces from the SWCNTs matrixes for obtaining CDs. Yang et al. [79] developed an ozonation pre-oxide method to extract CDs from reduced GO. GO dispersed in DI water with hydrogen peroxide at different pH was first treated with ozone for 1 h. The obtained ozonized GO (O–GO) was thermally deoxidized in argon atmosphere before being oxidized by ozone again. Then CDs of 2–5 nm in diameter could be obtained. The two synthesis methods included a similar three-step treatments, namely chemical oxidation, thermal deoxidation, and chemical oxidation. In other words, the two kinds of CDs should share a similar formation mechanism, although the oxidants were different.

Tetsuka et al. [80] developed an ammonia-assistant hydrothermal cutting method to synthesize CDs. GO sheets dispersed in an ammonia solution were heated hydrothermally at 70–150 °C for 5 h. Then amino-functionalized CDs were formed, with an average diameter of about 2.5 nm and an average height of about 1.13 nm. The method provided an efficient route to synthesize CDs of different PL wavelengths by tuning the temperature applied in the hydrothermal treatment. Li's group added directly a surface passivation agent, PEG 10,000, into the GO aqueous solution, which was heated hydrothermally at 200 °C for 24 h to synthesize surface-passivated CDs. Apparently, the two modified hydrothermal methods were both very simple.

Some other research groups used appropriate organic solvents instead of aqueous solution, and developed solvothermal methods to synthesize CDs. Yang's group dissolved GO sheets in DMF, and sonicated the DMF solution for 30 min,

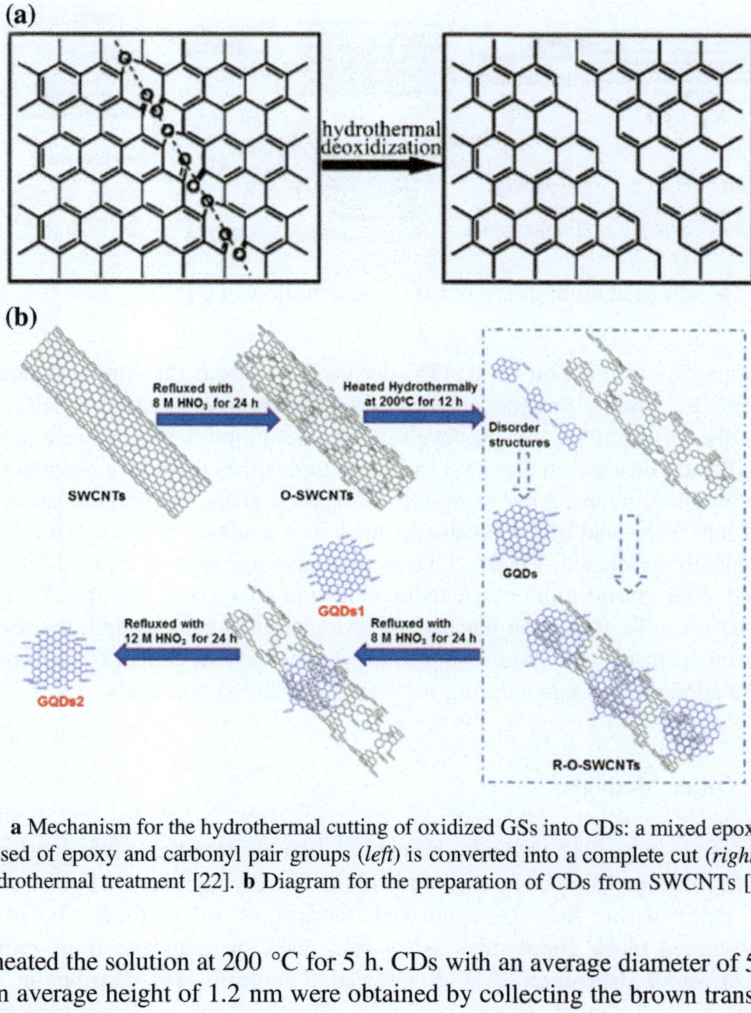

Fig. 8 **a** Mechanism for the hydrothermal cutting of oxidized GSs into CDs: a mixed epoxy chain composed of epoxy and carbonyl pair groups (*left*) is converted into a complete cut (*right*) under the hydrothermal treatment [22]. **b** Diagram for the preparation of CDs from SWCNTs [78]

then heated the solution at 200 °C for 5 h. CDs with an average diameter of 5.3 nm and an average height of 1.2 nm were obtained by collecting the brown transparent suspension [81].

2.1.5 Intercalation Methods

Lin et al. [82] developed an intercalation approach to prepare water-soluble CDs from MWCNTs and graphite flakes (Fig. 9). K atoms were intercalated into the covalently-bonded graphene sheets in MWCNTs to obtain a compound (K-GICs). It reacted with EtOH in an inert atmosphere to generate hydrogen gas and thus exfoliate the thin graphite sheets. Upon a short exposure of the K-GICs to air, many defects were formed on the graphene walls. Finally, with the assistance of ultrasonication, K-GICs continued to react violently with EtOH–H_2O, further exfoliating and disintegrating the walls of MWCNTs to yield monolayered CDs with an average

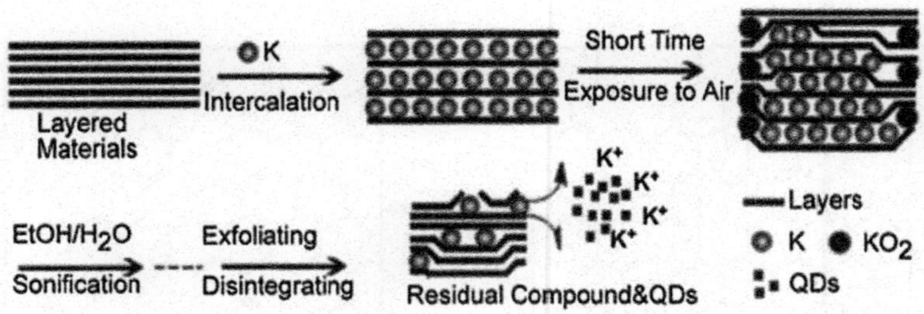

Fig. 9 The scheme of the formation of CDs by K intercalation [82]

size around 18.5 nm. Zhu et al. [83] derived CDs from the lithium-intercalated graphite. In brief, graphite anodes for lithium-ion batteries (LIBs) were charge-discharged at 1 C for 400 cycles. Then the graphite anodes were separated from LiBs and rinsed with dimethyl carbonate three times to remove residual solvent and partial lithium ions. After being briefly exposed to air, the graphite anodes were poured into water and ultrasonically cleaned. The resultant graphite particles were ultrasonically exfoliated. Finally, CDs with an average size of about 3.5 nm were obtained after purification of centrifugation and dialysis. Although the authors proposed a similar formation mechanism like that of K-intercalation, the real reaction mechanism may be more complicated since electrochemical oxidation of the graphite anode may happen during the charge-discharge processes.

2.1.6 Other Methods

Ponomarenko et al. [21] used ultra-high-resolution electron beam lithography to cut graphene to desired sizes. The method exhibited a very high level of precision. Lee et al. reported on the size-controlled fabrication of uniform GQDs using self-assembled block copolymers as an etch mask on graphene films grown by chemical vapor deposition [84]. It allowed a uniform size distribution of the obtained CDs. Li et al. synthesized CDs directly from activated carbon by a one-step hydrogen peroxide-assisted ultrasonic treatment [85]. The obtained CDs had diameters range from 5 to 10 nm. Apparently, the method is a green and convenient way to prepare CDs. Liu et al. [86] prepared CDs by shaking commercial graphite nanoparticles whose diameters were 4 nm in the ethanol/H_2O (1:1 v/v ratio) mixture on a vortex mixer. The resultant solution was centrifuged at 2000 rpm for 30 min and the supernatant was collected. From the supernatant, monolayer CDs of less than 4 nm in diameter could be obtained. The obtained CDs were considered to have pure sp^2 carbon crystalline structure without oxygenous defects, and may therefore be useful in investigating the PL mechanism of CDs. However, all the three methods mentioned above might be limited in applications by the very low production yield.

2.2 "Bottom-up" Methods

In general, bottom-up methods are those turning organic precursors into CDs. Therefore, carbonization processes of the organic precursors were involved. According to the way of carbonization, the "Bottom-up" methods could be further classified as solvothermal decomposition, direct thermal decomposition, microwave pyrolysis in solvent, supported thermal decomposition, refluxing pyrolysis in solvent, dehydrating organics with concentrated H_2SO_4 and other methods.

2.2.1 Solvothermal Decomposition

Giannelis's group developed a one-step hydrothermal decomposition of low-temperature-melting molecular precursors to synthesize hydrophilic CDs [16]. A mixture of citric acid monohydrate (CA) and an organic ammonium ($HOCH_2CH_2OCH_2CH_2NH_2$) was heated at 65 °C for 3 days, then heated hydrothermally in a Teflon equipped stainless steel autoclave at 300 °C for 2 h. After well washed the resultant solid with acetone, the residue was dispersed in water and filtered off. Then CDs with an average size of 7 nm was obtained from the filtrate. It should be pointed out that this is the first report about the synthesis of CDs through a "bottom-up" method. After that, many similar works have been carried out to prepare hydrophilic CDs (Fig. 10a), using the complex of CA and some organic ammonium molecules such as cysteine, glycine [87], ethylenediamine [88], ethanolamine [89], urea and thiourea [90], and polyethylene glycol (PEG) diamine [91]. More works have carried out to find other organics as the precursor to synthesize CDs (see Table 1). Furthermore, besides water, some organic solvents have also been used as the solvent. The morphology of the CDs

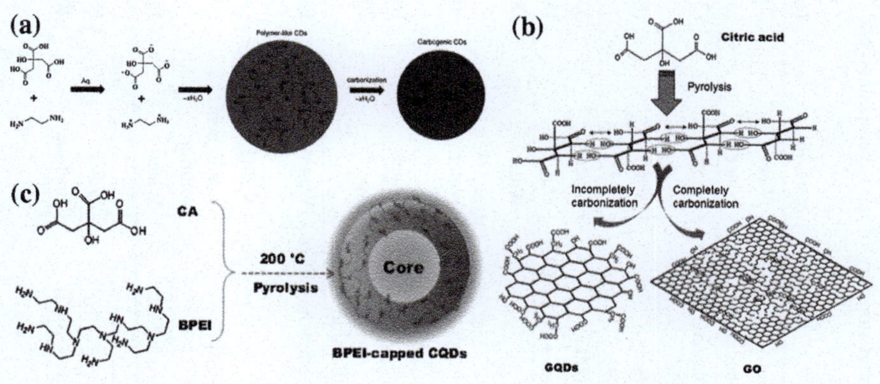

Fig. 10 **a** A synthetic route using citric acid and ethylenediamine: from ionization to condensation, polymerization, and carbonization [88]. **b** Diagram for the synthesis of CDs and GO by tuning the carbonization degree of CA [116]. **c** The scheme of the formation of BPEI-functionalized CDs [117]

Table 1 A brief summary of the experimental conditions and the morphologies of the CDs synthesized via solvothemal methods using different organic precursors

Precursor	Solvent	Temperature (°C)	Time (h)	Purification	Size (nm)	Height (nm)	Ref.
CA + cysteine	Water	200	3	–	5–9	0.5–3.5	[87]
CA + ethylenediamine	Water	150–300	5	Dialysis	2–6	2.81	[88]
CA + ethanolamine	Water	230 or 300	–	Dialysis	19 ± 1 4–10	–	[89]
CA + thiourea Or CA + urea	Water	160	4	Centrifugation	2.69 ± 0.42 3.10 ± 0.54	–	[90]
CA + PEG1000	Glycerin	220	12	Dialysis	5 ± 2	–	[92]
Sodium citrate + NH_4HCO_3	Water	180	4	Dialysis	1.59	–	[93]
Dopamine	Water	180	6	Centrifugation and dialysis	3.8	–	[94]
Phenylboronic acid	Water	160	8	Centrifugation and dialysis	2.5–6.5	–	[95]
Cornflour	Water	180	5	Centrifugation	2–6	–	[96]
CCl_4 and 1,2-ethylenediamine	No	200	0.5–2	Dialysis	5–15	–	[97]
Folic acid + ethylene glycol	Water	180	12	Centrifugation and dialysis	3.5–5.5	–	[98]
Histidine	NaOH solution	180	12	Dialysis	3–5	–	[99]
Chitosan + PEG or chitosan + PEI	Acetic acid solution	200	3	Centrifugation	3.4 ± 0.46/3.9 ± 0.48	–	[100]
EG400, PEG 1500 or PEG6000	Water	120	72	Dialysis	2–4	–	[101]
Milk	Water	180	2	Filtration	2–4	–	[102]
Hydroquinone + $SiCl_4$	Acetone	200	2	Dialysis	7 ± 2	4–6	[103]

(continued)

Table 1 (continued)

Precursor	Solvent	Temperature (°C)	Time (h)	Purification	Size (nm)	Height (nm)	Ref.
BSA	Water/ethanol (1:1)	180	12	Centrifugation and dialysis	6.8 ± 1.3	8	[104]
bPEI	Ammonium persulfate	100–200	5–10	Centrifugation and dialysis	3–4	–	[105]
Honey	H_2O_2	100	2	Dialysis	2	–	[106]
Orange juice	Ethanol	120	2.5	Centrifugation in acetone	1.5–4.5, 2.5	–	[107]
Streptomycin	Water	200	12	Centrifugation	2.97	–	[108]
Glucose	KH_2PO_4 solution	200	12	Centrifugation and dialysis	1.83 and 3.83	–	[109]
Glucosamine/HCl	Water	140	12	Dialysis	15–70	–	[110]
Cetylpyridinium chloride	NaOH, pH 11.5	150	2	Extraction	2	–	[111]
Xylan	NH_4OH	200	12	Centrifugation	6.99 ± 1.33	1.3–1.5	[112]
Glycine, Tris, EDTA, or cadaverine	Water	300	2	Centrifugation and filtration	2.6 ± 0.5 3.3 ± 0.4 3.0 ± 0.5 7.9 ± 0.8	–	[113]
Fructose + hydrochloric acid	Water	170	4	Dialysis	5.4	–	[114]
Apple juice	Water	180	12	Centrifugation and filtration	2.8 ± 0.4	–	[115]

obtained from the solvothermal methods are present in Table 1. Obviously, the size of the CDs are affected by both the precursor and the experimental conditions. In general, the average sizes of most reported CDs are below 10 nm. Unfortunately, the height data of most the obtained CDs was absent. Therefore, it is difficult to judge where the obtained CDs were nanoparticles or nanosheets.

It can be seen from the Table 1, although the experimental conditions were different when different precursors were used, all the the methods could be described as follow: Organic precursors and suitable solvent were put into a Teflon equipped stainless steel autoclave, and heated at 100–300 °C for some time (0.5–12 h). Generally, these hydrothermal methods are quite simple in operation, and the obtained CDs show good PL activities. Furthermore, the obtained CDs usually show high PL activity and excellent dispersity in aqueous solution. Therefore, solvothermal methods have been widely used in the synthesis of CDs.

2.2.2 Direct Thermal Decomposition Without Solvent

At 2008, Giannelis's group also developed a direct thermal decomposition route to synthesize CDs [16]. 4-aminoantipyrine (4AAP) was calcined in air at 300 °C for 2 h, and dissolved in $CF_3CH(OH)CF_3$, then precipitated by the addition of water. The obtained precipitate was washed thoroughly with water and ethanol. Then CDs of irregular in shape and 5–9 nm in size were obtained. However the particles appeared strongly aggregated in water, and could be only dispersed in organic solvents such as DMSO, DMF and $CF_3CH(OH)CF_3$. In a follow-up study, the group reported a similar thermal decomposition to synthesize hydrophilic CDs [15]. In brief, sodium 11-aminoundecanoate reacted with CA to form the corresponding ammonium carboxylate salt, which was subsequently heated directly in air at 300 °C for 2 h. After a serial of centrifugation, CDs with a core-shell structure were obtained. The total size of the obtained CDs ranges from 10 to 20 nm while that of the cores is between 5 and 10 nm. Chi's group developed an easy method for the preparation of blue emission CDs and GO by tuning the carbonization degree of CA (Fig. 10b) [116]. CA was heated at 200 °C for about 30 min. Then the formed orange liquid was added drop by drop into NaOH solution under vigorous stirring. As a result, CDs with 15 nm in width and 0.5–2.0 nm in thickness could be obtained. In another work of Chi's group, the mixture of CA and branched polyethylenimine (BPEI) was used as the precursor to synthesize BPEI-functionalized CDs via similar direct thermal decomposition route (Fig. 10c) [117]. The obtained BPEI-CDs were considered as spherical graphite nanocrystals with an average size of about 6.2 nm. Up to now, the direct thermal decomposition method have been widely used in synthesize various kinds of CDs (see Table 2). In general, the precursors were carbonized by being heated directly in the atmosphere of air or some inert gas. Then CDs could be extracted from the obtained carbon solids using some suitable solvents. Compared with the solvothermal decomposition methods, the direct thermal decomposition usually adopted higher temperatures and shorter heating time. Like CDs synthesized by hydrothermal decomposition

Table 2 A brief summary of the experimental conditions and the morphologies of the CDs synthesized via direct thermal decomposition methods using different organic precursors

Precursor	Temperature (°C)	Time	Atmosphere	Solvent	Purification	Size	Height (nm)	Ref.
Konjac flour	470	1.5 h	Air	Ethanol	Filtration	3.37	–	[118]
Shrimp eggs	180	25 min	Air	Water	Centrifugation and filtration	3	3.24 ± 1.06	[119]
Glutamic acid	200	5 min	Air	NaOH, pH 7	No	4.6	–	[120]
3-(3,4-dihydroxyphenyl)-l-Alanine	300	2 h	Air	DMF	Centrifugation	3.64	–	[121]
Lauryl gallate, propyl gallate, methyl gallate	270	2 h	Air	Ethanol	Washing with hexane	2 3 3.5	–	[122]
Glutamic acid	165	–	Air	Water	No	2.5–6.4	–	[123]
Paper	Burned	–	Air	Water	Filtration	2–5	–	[124]
Egg white or yolk	250	2 h	N_2	Water	Centrifugation and ultra-filtration	2.23 and 2.03	–	[125]
Watermelon peel	220	2 h	Air	Water	Filtration and centrifugation	2 ± 0.5	–	[126]
Ethanolamine	150	2 h	Air	Water	No	2.7	–	[127]
Ethanolamine + H_2O_2	250	7 min	Air	Water	No	2	–	
1,4-addition polymers	250	2 h	N_2	Water	Centrifugation	3–6 nm	–	[128]
EDTA-2Na·$2H_2O$	400	2 h	N_2	Acetone	Centrifugation	3.8 nm	–	[129]
CA + $LiNO_3$	280	3 h	Argon	Water	Dialysis	3–5	–	[130]
Poly(styrene-co-glycidylmethacrylate) photonic crystals	200 300 400	2 h	N_2	Water	Centrifugation	2.42 3.51 4.46	–	[131]
Tris(hydroxymethyl) aminomethane + betaine hydrochloride	250	2 h	Air	Water	Precipitated by adding acetone	7	–	[10]

methods, the CDs obtained by the direct thermal decomposition methods were usually considered as carbon nanoparticles rather than graphene nanosheets. TEM images indicated that the sizes of the obtained CDs were usually less than 10 nm. However, most of the authors didn't characterize the heights. Therefore, it is difficult to say accurately the morphologies of the obtained CDs. Anyway, these methods are usually very simple, and have been adopted widely.

2.2.3 Microwave Pyrolysis in Solvent

At 2009, a facile microwave pyrolysis approach to synthesize CDs was first carried out by Zhu and coworkers [132]. A saccharide, such as glucose and fructose, and PEG-200 were dissolved in water and heated in a 500 W microwave oven for 2–10 min. It was found that the size of the obtained CDs increased with the heating time. For example, the average diameters of the CDs were 2.75 ± 0.45 and 3.65 ± 0.6 nm for the heating times of 5 and 10 min. Qu's group used a carbohydrate (glycerol, glycol, glucose, sucrose, etc.) and a tiny amount of an inorganic ion as the precursor [133]. After being heated for several minutes, bright emission CDs could be obtained. The results imply that the surface passivation reagent may be unnecessary. Up to now, many CDs have been synthesized from different organic precursors by similar microwave pyrolysis methods (Table 3). These methods can be generally described as heating the solution of some soluble organic molecular by microwave at a certain power (300–900 W) for some time (usually from dozens of seconds to dozens of minutes), combining with some suitable purification such as dialysis, centrifugation and ultrafiltration. It can be seen from Table 3, the organic precursors were usually some small molecules which were easily to be dehydrated. Sometimes, some polymers acting as the surface passivation reagent or some heteroatom-containing molecules as the hetero atom sources for doping were also added into the precursors. The solvents could be water as well as some common organic solvents, dependent on the necessary. The quite simple operations and short reaction time have make this kind of methods very popular.

2.2.4 Supported Thermal Decomposition

Supported synthetic method has been widely adopted for the synthesis of monodisperse nanomaterials, involving molecular sieve, porous carbon and so on. Supported synthetic method was first applied in the synthesis of CDs by Liu and coworkers in 2009 [150]. Satellite-like polymer/F127/silica composites were prepared by an aqueous route using silica colloid spheres functionalized with amphiphilic triblock copolymer F127 ($EO_{106}PO_{70}EO_{106}$, Mw = 12,600; EO = ethylene oxide, PO = propylene oxide) as carriers and resols (phenol/formaldedyde resins, Mw < 500) as carbon precursors. Then the satellite-like polymer/F127/silica composites were heated at 900 °C in Ar for 2 h, leading to corresponding carbon/silica composites (Fig. 11a). The carbon dots were released by etcing the silica spheres

Table 3 A brief summary of the experimental conditions and the morphologies of the CDs synthesized via microwave pyrolysis methods using different organic precursors

Precursor	Solvent	Power (W)	Time (min)	Purification	Size	Height	Ref.
CA + 1,2-ethylenediamine, 1,4-butanediamine, diethylamine, or diethylamine	Water	700	2–4	Dialysis	2.2–3	–	[134]
PEG + serine	Glycerin	–	10	Dialysis	3–4	–	[135]
CA + PEI	Water	850	5	Ultrafiltration	12	–	[136]
[BMIM][Br] and [BMIM][BF4]	Water	650	10	Centrifugation and dialysis	2–6	–	[137]
CA + ethylenediamine	Water	700	4–15	Dialysis	2.5 ± 0.5	–	[138]
Glucose +tryptophan	Water	700	5	–	29	–	[139]
Phytic acid + ethylenediamine	Water	700	8	Centrifugation	6–11, 9	–	[140]
Glycerol + (3-amino-propyl)triethoxysilane	–	700	30	Dialysis	3.6 ± 1.6	–	[141]
Glycerol + PEI125 k	PBS, pH7	700	10	Dialysis	4–12, 7	–	[142]
Ascorbic acid	Water	900	1–5	Centrifugation	3.1 ± 0.3	–	[143]
Glucose + poly(acrylatesodium)	Water	300	4	–	3.5	–	[144]
Resorcinol	H_2SO_4	800	1/3–2	Extraction with butanol + DDW	0.5–6	–	[145]
DMF	Chlorosulfonic	700	2/3	Dialysis	1–6	–	[146]
Glycerol	Phosphate, pH7.4	750	14	–	–	2.1 ± 0.76	[133]
Glucose + PEG200	Water	500	5 10	–	2.75 ± 0.45 3.65 ± 0.6	–	[132]
Glycerol + 4,7,10-trioxa-1,13-tridecanediamine	Water	700	10	Dialysis	3.5	–	[147]
Glucose	Water	280–700	1–9	–	1.65–21	–	[148]
Glucose + ammonia	Water	280	1–9	Dialysis	1.7–5.8	–	[149]

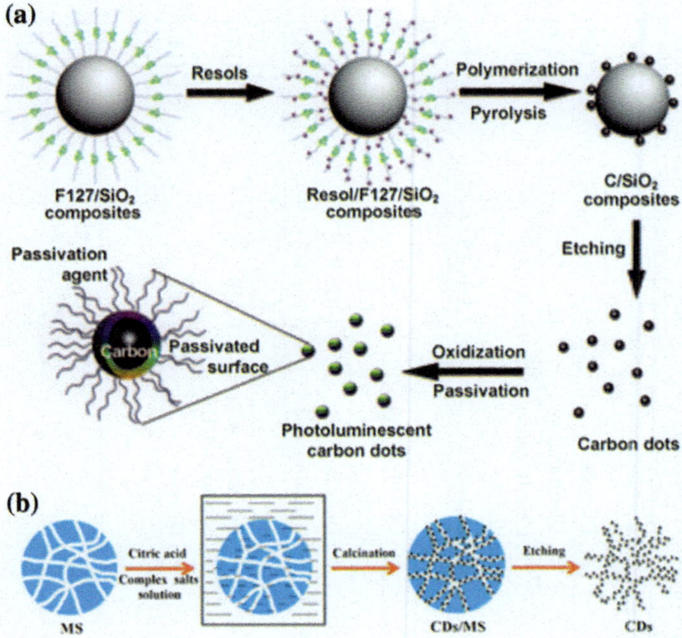

Fig. 11 **a** Processing diagram for the synthesis of PL CDs using silica colloid spheres as carriers [150]. **b** Processing diagram for the synthesis of PL CDs using mesoporous silica spheres [151]

with concentrated NaOH. High PL CDs could be finally obtained after neutralization, dialysis, centrafugation, chemical oxidation, neutralization, dialysis, and surface passivation. Apparently, the method is quite complicated. Li's group reported a relative simple approach for preparing hydrophilic CDs by using mesoporous silica spheres as nanoreactors in an impregnation method [151]. The resulting highly efficient photoluminescent CDs without any further treatment are monodisperse, photostable and of low toxicity, and show excellent luminescence properties (Fig. 11b). It should be pointed out that the abundant carboxyl groups in the precusor (CA) and the relative low temperture should be quite important for obtaining directly the soluble CDs. The method has also been adpoted by Zong et al. [152].

2.2.5 Refluxing Pyrolysis in Solvent

CDs prepared by heating under reflux were fist reported by Sun's group [153]. In a typical synthesis procedure, 0.69 mL of 1,2-ethylenediamine (EDA) was added to 1 mL of CCl_4, which acted as both the precursor and the solvent. The formed mixture was subsequently heated at 80 °C for 1 h. Then CDs with diameters ranging from 1 to 5 nm (called as carbon nitride dots in the paper due to the contained N atoms) could be obtained after a dialysis purification. Some follow-up works have been carried by other groups (Table 4). Organic precursors with or without surface passivation agents were dissolved in suitable solvents, such as—water, NaOH solution, H_2SO_4 solution and some organic solvents.

2.2.6 Stepwise Organic Synthesis

Li's group has made many efforts to develop stepwise organic synthesis method, which could tune accurately the conjugated carbon atoms of CDs [3, 161–165]. The CDs were made up of 168, 132, and 170 conjugated carbons produced by oxidation of polyphenylene dendritic precursors (Fig. 12). The CDs were stabilized by 2′,4′,6′-trialkyl phenyl groups. The stepwise synthesis method provides a versatile synthesis of large, stable colloidal CDs with desired sizes and structures. However, it has not been adopted by other groups probably due to the relative low efficiency.

2.2.7 Dehydrating Organics with Concentrated H_2SO_4

It has been well known that concentrated H_2SO_4 can make many organics be carbonized due to its strong dehydration ability. In 2010, Zhang et al. firstly synthesized CDs by dehydrating organics with concentrated H_2SO_4 [166]. In brief, 2.04 g sucrose was dissolved in 1 mL water, and dehydrated by adding 2 mL concentrated H_2SO_4 without any heating. The dehydration was finished by adding 40 mL water. The obtained uniformly brown solution was neutralized and purified by dialysis (retaining molecular weight higher than 1000 Da). Then the pure CDs were separated into two fractions by dialysis by a membrane with MWCO of 3500 Da. The outside fraction contained smaller CDs of about 2 nm in size, exhibiting green PL emission. The inside fraction were bigger CDs of about 5 nm in size. The bigger CDs exhibited no obvious PL emission, but emitted a bright PL after the surface was passivated with PEG2000 N (Fig. 13). Since then, ethylene glycol, bovine serum albumin, hair and ethylenediaminetetraacetic acid (EDTA) have also been dehydrated by concentrated H_2SO_4 to prepare various kinds of CDs [167–170]. However, the dehydration reactions were usually carried out with heating assistance (from 40 to 200 °C).

2.2.8 Other Methods

Kim et al. [171] used thermal plasma jet to etch graphene into CDs. A large amount of ethylene gas was injected continuously into Ar plasma to generate a carbon atomic beam (Fig. 14). The beam was then flowed through a carbon tube attached to the anode and then dispersed into a chamber. Carbon materials including CDs were obtained by a gas phase collision reaction. CDs could be extracted from

Table 4 A brief summary of the experimental conditions and the morphologies of the CDs synthesized *via* refluxing pyrolysis methods using different organic precursors

Precursor	Solvent	Temperature (°C)	Time	Purification	Size (nm)	Height (nm)	Ref.
CA + PEG-diamine	Glycerin	270	3 h	Dialysis	5–7	5–7	[154]
PEG200	NaOH solution	120	6 h	Dialysis	5	–	[155]
Diethylamine	$CHCl_3$	–	1 h or 60 h	Dialysis	1–3 or 2–4	–	[156]
PEG 200	NaOH solution	120	6 h	Dialysis	4	–	[157]
Glycerol + PEG-diamine or Glycerol	Glycerol	230	30 min	Dialysis	5.5 ± 1.1	–	[158]
Lactose + Tris	Water	100	24 h	Dialysis	1.5	–	[159]
Sucrose	Oil acid	215	5 min	Centrifugation	2	–	[12]
Waste frying oil	H_2SO_4 solution	100	5 min	Dialysis	1–4, 2.6 nm	–	[160]

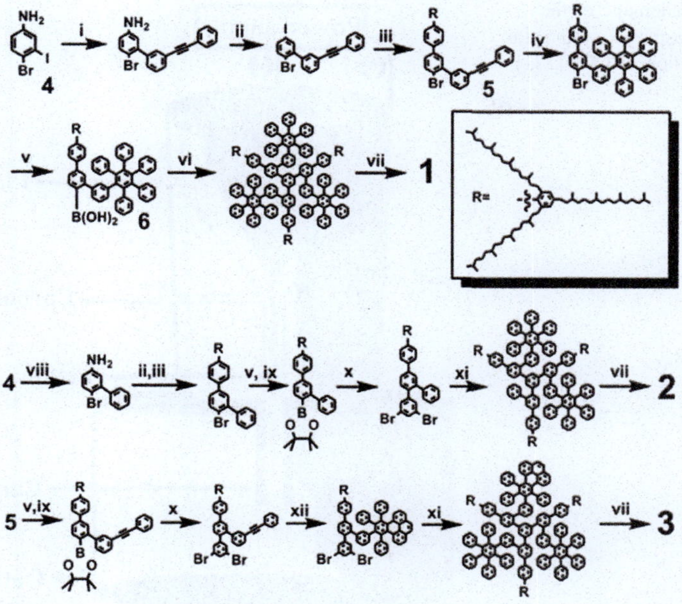

Fig. 12 Schematic illustration of the stepwise synthesis of CDs 1–3 [164]

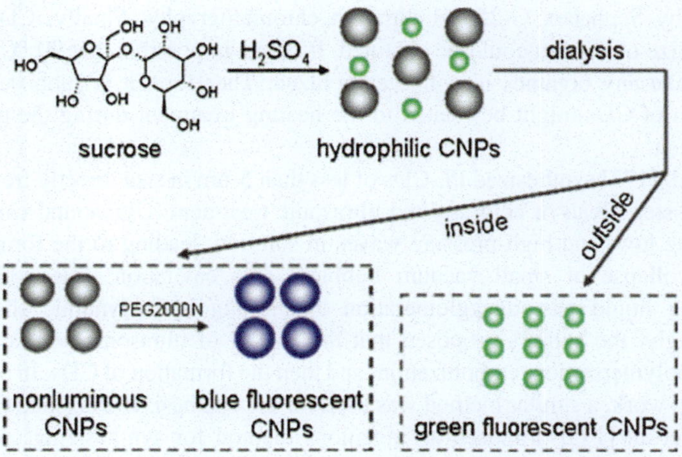

Fig. 13 The synthesis procedure of *green* and *blue* luminescent CDs by dehydration of glucose [166]

carbon soot simply by dispersing it in ethanol using a stirring rod. The size of the CDs could be tuned by adjusting the length of the carbon tube.

Jiang et al. [172] extracted CDs directly from commercial coffee. In brief, coffee powder was dissolved in hot water (90 °C), then centrifuged at 14,000 rpm. The

Fig. 14 Schematic of the thermal plasma jet system for the production of CDs [171]

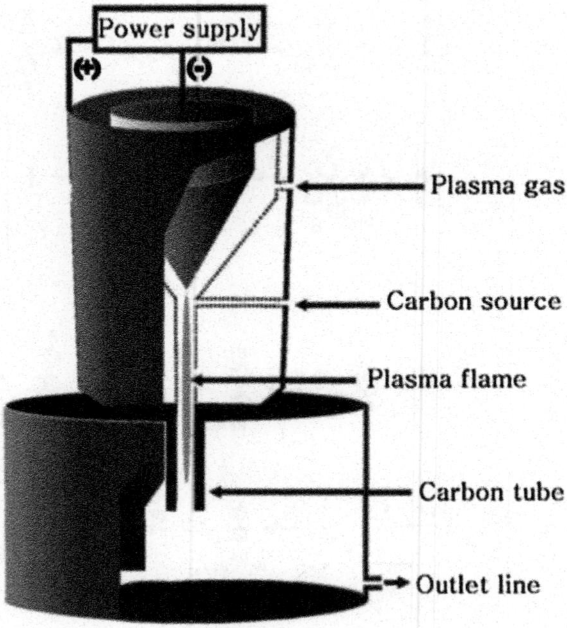

resulting supernatant was further filtered through a 0.22 μm membrane and further purified by Sephadex G-25 gel filtration chromatography. Finally, CDs with an average size of 4.4 nm could be obtained. It seems impossible that 90 °C hot water could make any organics in coffee carbonized. Therefore, it is supposed that the formation of CDs might be related to the heating treatment during the produce of coffee bean.

Li et al. [173] synthesized PL CDs of less than 5 nm in size directly from glucose by a one-step alkali or acid assisted ultrasonic treatment. Ultrasound can generate alternating low- and high-pressure waves in solution, leading to the formation and violent collapse of small vacuum bubbles. This cavitation causes high speed impinging liquid jets, deagglomeration and strong hydrodynamic shear-forces. Accordingly, the authors proposed that the energy of ultrasonic waves led to the glucose polymerization, carbonization, and then the formation of CDs. In one of their follow-up work, a similar method was used for the synthesis of N-doped CDs [174].

Fang et al. [175] proposed an ingenious method for synthesizing cross-linked hollow CDs by simply mixing acetic acid (AC), water and diphosphorus pentoxide. It was proposed that P_2O_5 could on one hand react with water to convert high chemical energy into thermal energy, and on the other hand remove composite water from many organic and inorganic compounds. Furthermore, the produced polyphosphoric acid (PPA) still maintains a strong dehydrating effect before its conversion into monophosphoric acid, promoting the carbonization reaction. Thus, highly sticky liquid PPA captured the hot gas of AC to suppress the heat loss. Due to the catalysis by both PPA and P_2O_5, the carbonization reactions of AC was started and continued at 117 °C.

Lu et al. [176] synthesized well-defined CDs on a ruthenium surface using C_{60} molecules as a precursor. The ruthenium surface interacted strongly with the C_{60} molecules, leading to surface vacancies on ruthenium, which helps the embedding of C_{60} molecules. The fragmentation of the embedded C_{60} molecules at elevated temperatures produced carbon clusters that undergo diffusion and aggregation to form CDs.

3 Morphology and Structural Characteristics

3.1 Morphology

For the CDs prepared by "top-down" methods, the morphology of CDs may firstly affected by the nature of the precursors. For example, chemical oxidation using concentrated HNO_3 could release nanoparticles with diameters around 5 nm and heights less than 5 nm from the carbon soot by laser ablation of graphite powder [37], and graphene nanosheets with heights centered at around 1 nm and lateral sizes of 2–6 nm from candle soot [38], and graphene nanosheets with an average height of about 0.5 nm and an average lateral size of about 10 nm from coals [46]. The morphology of CDs may also affected by the preparation method. For instance, GO could be etched into nanosheets with uniform lateral dimensions of ~40 nm by Fenton reagent under an UV irradiation for 15 min [64], into nanosheets with uniform sizes of 3–5 nm by electrochemically oxidation [71], and into CDs with uniform size of 2.5 nm by hydrothermal treatment in an ammonia solution [80]. Even a small change of the experimental conditions would also affect the morphology of the obtained CDs. For example, the lateral size of the CDs synthesized from GO by chemical oxidation with Fenton reagent decreased with the increase of reaction time [64]. Moreover, as carbon fibers was chemically oxidized with the mixture of concentrated H_2SO_4 and HNO_3 (3:1 in volume), the sizes of the obtained CDs increased as the reaction temperature was decreased gradually from 120, 100 to 80 °C [52]. However, in general, the average size of most reported CDs obtained from "top-down" methods were below 10 nm. The largest diameter of CDs reported so far is 60 nm [50]. CDs synthesized from "top-down" methods have been considered as either spherical nanoparticles (carbon quantum dots) or nanosheets (graphene quantum dots). The heights of the graphene quantum dots were usually less than 3 nm. Especially, the average height of some monolayer graphene quantum dots were as low as 0.5 nm. The heights of these CDs were usually less than the lateral sizes. However, the heights of most reported "carbon quantum dots" have not been characterized. Therefore, it is difficult to say that the "carbon quantum dots" were really spherical nanoparticles.

In general, most CDs synthesized by "bottom-up" methods were considered as carbon quantum dots. TEM images indicated that the lateral sizes of most of CDs were less than 10 nm. However, the heights of most CDs were usually absent (see Tables 1, 2, 3, 4). Therefore, it is impossible to say whether these CDs were layer like or spherical like. From the existing data, the heights of many CDs may be less than the lateral sizes [87, 88, 103, 112, 116]. In other words, many of the "bottom-up" obtained CDs, which were considered as carbon quantum dots may also be layer like.

3.2 Structural Characteristics

Usually, the structural characteristics of CDs were characterized by Fourier transform infrared (FTIR), X-ray photoelectron spectroscopy (XPS), X-ray diffraction (XRD) patterns, Raman spectroscopy and HRTEM. The XPS spectra indicated that most CDs were composed mainly of carbon, hydrogen, oxygen, (and nitrogen for N-doped CDs) (Fig. 15a, b), which were usually confirmed by the elemental analysis results. The high resolution spectra of C_{1s} (Fig. 15c, d) of CDs usually exhibited three main peaks associated with carbon atoms corresponding to banding energy of C–C sp^2, C–O, C=O bonds (and C–N bond for N-doped CDs). FTIR spectra (Fig. 15e) of CDs usually presented absorption peaks of C=C ($\nu_{C=C}$ at

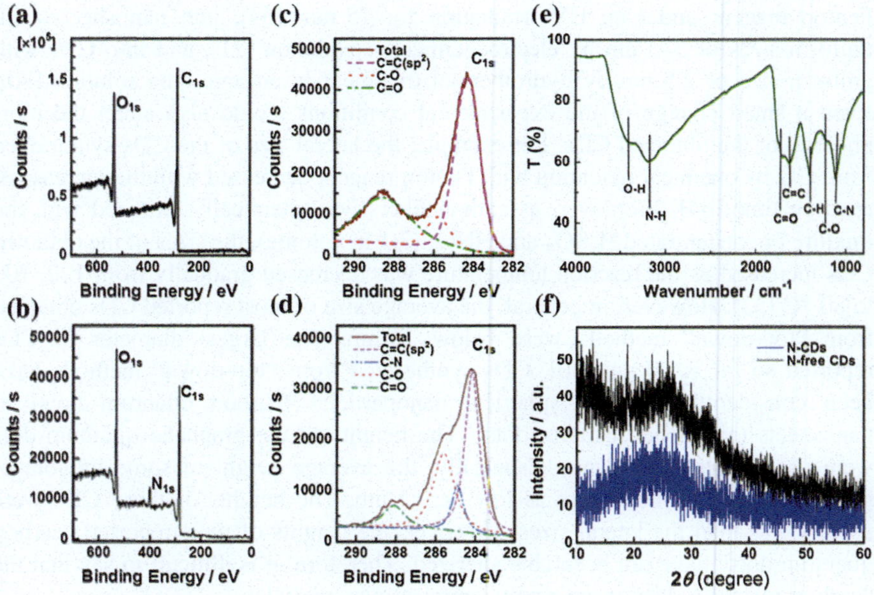

Fig. 15 XPS spectrum (**a**, **b**) and high-resolution C1s spectrum (**c**, **d**) of the N-free CDs (**a**, **c**) and N-CDs (**b**, **d**), FTIR spectrum of N-CDs (**e**), XRD pattern of N-CDs and N-free CDs (**f**) [168]

around 1615 cm^{-1}), C–O (ν_{C-O} at around 1230 cm^{-1}, ν_{O-H} at 3405 cm^{-1}), C=O ($\nu_{C=O}$ at around 1720 cm^{-1}), and C–N for N-doped CDs (ν_{C-N} at around 1110 cm^{-1}, ν_{N-H} at around 3180 cm^{-1}). The experimental results of XPS and FTIR indicated that most of the obtained CDs were composed of both sp^2 carbon structures and oxygen-containing functional groups including carboxyl, carbonyl, hydroxyl and epoxy groups.

XRD patterns of most reported CDs featured a broad (002) peak around 25° (0.34 nm), which is often broad due to the small size of GQDs (Fig. 15f). Sometimes, the broad peak would centered at $2\theta < 25°$, suggesting the interlayer spacing of larger than 0.34 nm. The larger interlayer spacing could be attributed to the abundant oxygen-containing groups on the edge or basal plane of CDs. In other words, the oxygen-containing groups could expand the space of graphene layers. Furthermore, the oxygen-containing groups distributed on the basal plane might lead to larger spacing than that at the edges of graphene sheets [29]. However, the N-doping seemed to have no effect on the interlayer spacing.

The Raman technique is also a powerful and non-destructive tool for the characterization of GQDs. Usually, CDs exhibited a D band around 1350 cm^{-1} and a G band at 1590 cm^{-1}, which are attributed to the first-order scattering of the E_{2g} vibration mode in the graphite sheet and structure defects, respectively (Fig. 16a). The intensity ratios (I_D/I_G) of CDs has been used to compare the structural order between crystalline and defect states. The I_D/I_G values of the CDs usually distributed in the range from ~0.5 to ~1.5, dependent on the synthesis method.

HRTEM images indicated that most CDs are crystalline (Fig. 16b, c, d). Two kinds of lattice fringes could be usually observed, namely (002) interlayer spacing around 0.34 nm and the (100) in-plane lattice spacing around 0.24 nm. The spacing of both the two kinds of lattice fringes in CDs synthesized by different methods would be a little different, depending upon the exact experimental conditions used. Usually the interlayer spacing mainly distributed from 0.31 to 0.34 nm, while the in-plane lattice spacing varied from 0.20 to 0.24 nm. However, not all CDs are crystalline, in particular those synthesized by "bottom-up" methods at relative low

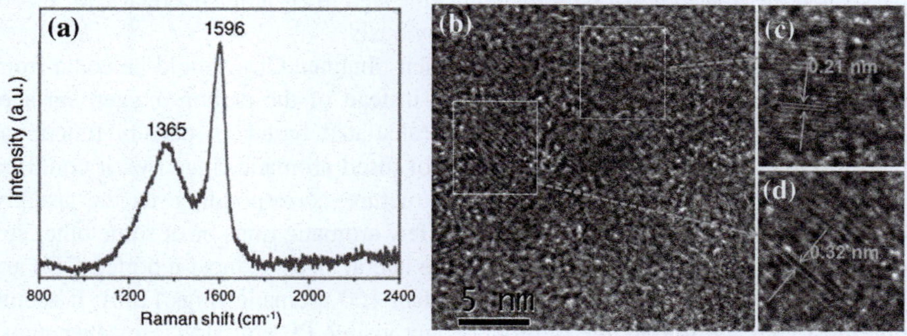

Fig. 16 Raman spectra (**a**) [71] and HRTEM images (**b**) of the CDs. Typical single CDs with lattice parameters of 0.21 nm (**c**) and 0.32 nm (**d**), respectively [87]

temperatures [101, 160, 177]. That may be caused by the incompletely carbonization of precursors.

4 Luminescent Properties

4.1 Photoluminescence

4.1.1 PL Mechanism

PL is one of the most fascinating features for all CDs. Up to now, CDs of different PL colors, ranging from the deep ultraviolet to near-infrared, could be obtained via various synthetic approaches. Although the exact mechanism of PL for CDs is still controversial, the PL of CDs is usually considered to be either intrinsic state emission or defect state (or surface state) emission. Herein, some popular and reasonable hypothesis should be introduced. Sun's group found the surface passivation with some organic polymers could make lowly luminescent CDs emit bright PL, and accordingly proposed that the PL of their obtained CDs should be mainly attributed to the radiative recombination of the excitons from the passivated defects on CD surface [37]. The viewpoint was subsequently accepted by Ding's group [66]. Up to now, many important evidences have further indicated that the surface states play important roles in the PL properties of CDs. First, the PL activities of many CDs could be improved greatly by surface-passivation with some polymers or organic molecules [37, 47, 150, 178, 179]. Secondly, the PL properties of many CDs could be tuned by modifying their surface functional groups or oxidation degree [13, 57, 80, 121, 122, 180–184]. Pan's group further proposed that the emissive defects of CDs should be the free zigzag sites with a carbene-like triplet ground state (Fig. 17a, b), according to their observed pH dependent PL behaviors of CDs (The PL intensity decreased greatly when the pH value was decreased gradually from pH 13 to pH 1) [22, 55, 129]. Pan's mechanism hypothesis was also accepted by Lingam and coworkers, based on their experimental results [185]. They passivated the free edges of CDs by being annealed in H_2 atmosphere at 250 °C. As a result, the PL emission was quenched obviously.

Lee's group considered that the PL of their obtained CDs should be come from the quantum-sized graphite structure itself instead of the carbon-oxygen surface. From the energy gap of π-π^* transitions calculated based on density functional theory (DFT) as a function of the number of fused aromatic rings (*N*), it could be found that the quantum-sized graphite structure corresponding to the usually observed blue emission should have only a few aromatic rings or of some other sp^2 configuration of similar size (Fig. 17c). The lateral sizes of most reported CDs are bigger than 3 nm, which contained more than 100 aromatic rings [186]. It seems that most CDs are impossible to produce any visible PL emission. However, most reported CDs have abundant oxygen-containing functional groups, such as

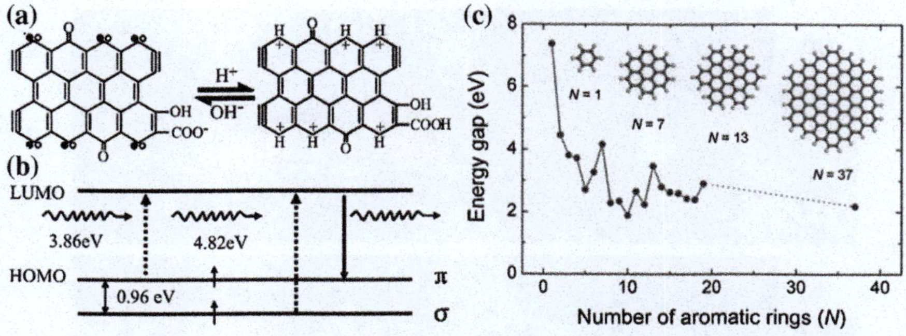

Fig. 17 **a** Models of the CDs in acidic (*right*) and alkali (*left*) media. The two models can be converted reversibly depending on pH. The pairing of σ(•) and π (o) localized electrons at carbene-like zigzag sites and the presence of triple bonds at the carbyne-like armchair sites are represented. **b** Typical electronic transitions of triple carbenes at zigzag sites observed in the optical spectra [22]. **c** Energy gap of p–p* transitions calculated based on DFT as a function of the number of fused aromatic rings (N). The *inset* shows the structures of the graphene molecules used for calculation [29]

hydroxyl, carboxyl groups. Then, some sp^2 clusters being isolated within the sp^3 C–O matrix would be formed at the edge of CDs [116, 187]. The radiative recombination of electron-hole pairs in such small sp^2 clusters could give rise to PL [188, 189]. Apparently, some defect states may be also present in these sp^2 clusters, and affect the energy levels. Then the PL properties of CDs should be a result of the combining effect of the sp^2 clusters and the surface states.

4.1.2 Effect of Excitation Wavelength on the PL of CDs

The excitation-dependent emission of CDs was firstly observed by Sun's group (Fig. 18a, b) [9]. They found that the obtained surface-passivated CDs could afford multi-PL colors, from blue to red, under different excitation wavelengths. Take the PPEI-EI coated CDs as an example, when the excitation wavelengths were progressively increased from 400 to 600 nm in 20 nm increment, the emission spectra red-shifted gradually, while the emission intensities decreased dramatically (Fig. 18c). Similar phenomenon could be also observed in other CDs, no matter synthesized by "top-down" or "Bottom-up" methods [11, 16, 44, 45, 47, 55, 61, 73, 129, 131–133, 146, 147, 150, 178, 179, 190]. The excitation-dependent emission seems to be a common phenomenon for most of CDs. However, in one following research of Sun's group, surface passivated CDs obtained from the similar method was further separated by an aqueous gel column. The most fluorescent fractions had a PL quantum yield (PLQY) close to 60 % [191]. What is more important, these CDs showed bandgap-like PL emission, namely the emission spectrum is nearly independent on the excitation wavelengths [191]. Similar excitation-independent PL could be also observed in their other works (Fig. 18d) [192, 193]. Then it seems

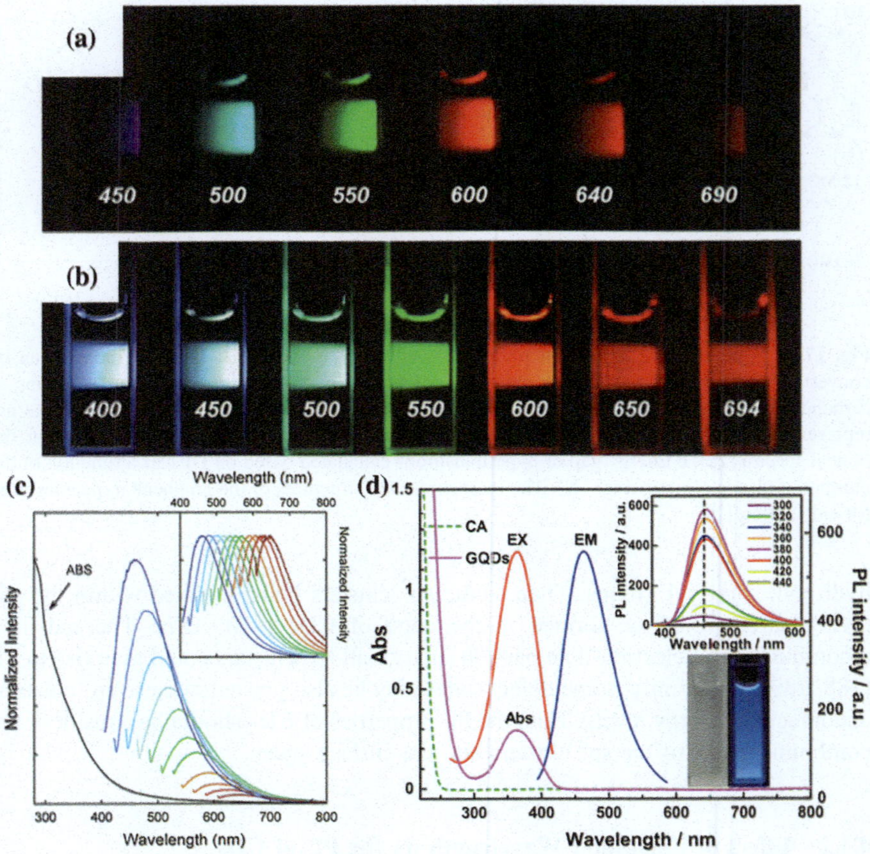

Fig. 18 **a** Aqueous solution of the PEG1500 N-attached CDs excited at 400 nm and photographed through band-pass filters of different wavelengths as indicated, **b** and excited at the indicated wavelengths and photographed directly. **c** The absorption (ABS) and luminescence emission spectra of PPEI-EI CDs in an aqueous solution. The emission spectral intensities are normalized to quantum yields (normalized to spectral peaks in the *inset* of **c**) [9]. **d** UV–Vis absorption of CA and the CDs, and PL spectra of the CDs. *Inset* in **d**: (*upper*) Emission spectra of the CDs with excitation of different wavelength; (*lower*) Photographs of the solution of CDs taken under visible light (*left*) and under 365 nm UV light (*right*) [116]

reasonable that the excitation-dependent PL behaviors may reflect not only effects from particles of different sizes in the sample, but also a distribution of different emissive sites on each CDs [9]. Actually, many CDs synthesized by "Bottom-up" methods also had excitation-independent PL emissions [87–89, 93, 116, 117]. Generally, these CDs exhibiting excitation-independent PL had higher PLQYs than those CDs exhibiting excitation-dependent PL. Then, the excitation-independent PL may reflect uniform size distribution and surface states.

4.1.3 Effect of PH Value of Solution on the PL of CDs

The pH value of solution is an important factor affecting the PL properties of CDs. Pan's group found that their obtained CDs could emitted strong PL in alkaline solutions, while the PL intensities decreased greatly in acidic solutions [22, 55, 129]. Furthermore, the pH effect on the PL intensities of the CDs is reversible. They considered the PL of their obtained CDs was produced by the free zigzag sites with a carbene-like triplet ground, and accordingly proposed a structural model to explain pH-dependent PL. In brief, under alkaline conditions, the zigzag sites in the CDs were free and PL active. However, under acidic solutions, the zigzag sites in the CDs were protonated and inactive. Some other research groups also noticed similar pH-dependent PL from their obtained CDs [52, 79, 160]. However, most CDs exhibited different responses toward the change of pH value. For example, the PL intensities of most reported CDs were relatively strong in neutral solutions, while the PL intensities decreased in strong acidic or/and strong alkaline solutions [44, 45, 54, 59, 63, 67, 87–89, 91, 93, 97, 101, 108, 120, 125, 159, 170, 172, 194]. The PL intensities of some other CDs were nearly independent on the pH value [83, 128, 133, 169]. Then, it can be noticed that CDs synthesized by different methods may have different pH-responses t. Presently, it is difficult to explain definitely the effecting mechanisms of pH value on the PL properties of most CDs, due to the fact that the exact PL mechanisms of CDs are still open questions. However, the effect of pH value on the PL of CDs should be related to their surface states.

4.1.4 Effect of Size on the PL of CDs

As discussed above, the PL of CDs should be attributed to either the surface states or the contained small sp^2 clusters. Then, it seems that the PL properties should be independent on the size of the CDs. However, many evidences showed that the PL properties of CDs could be affected by their size. First of all, the PL intensity of CDs could be affect by their size. Sun's group found that CDs of larger sizes (30–50 nm in average diameter) with the same surface passivation were much less luminescent than CDs of smaller sizes (less than 10 nm in diameter) [9]. They proposed that a large surface-to-volume ratio in CDs is necessary for the CDs upon surface passivation to exhibit strong PL. Secondly, the PL spectra of CDs could be also affected by their size. For example, Pang's group separated the CDs produced by electrochemical oxidation of graphite into two fractions (1.9 ± 0.3 and 3.2 ± 0.5 nm) by MWCO membranes [67]. It was found that the smaller CDs emitted blue PL, while the bigger ones emitted yellow PL. Lee's group separated the CDs obtained from alkali-assisted electrochemical oxidation of graphite by column chromatography [69]. The PL emission of the obtained CD fractions red-shifted from UV light to near-infrared light when the size of the CDs was increased gradually from 1.2 to 3.8 nm. Peng and coworkers synthesized CDs by chemically oxidizing carbon nanofibers [52]. Three kinds of CDs with different size distribution (1–4, 4–8 and 7–11 nm) were obtained by tuning the reaction

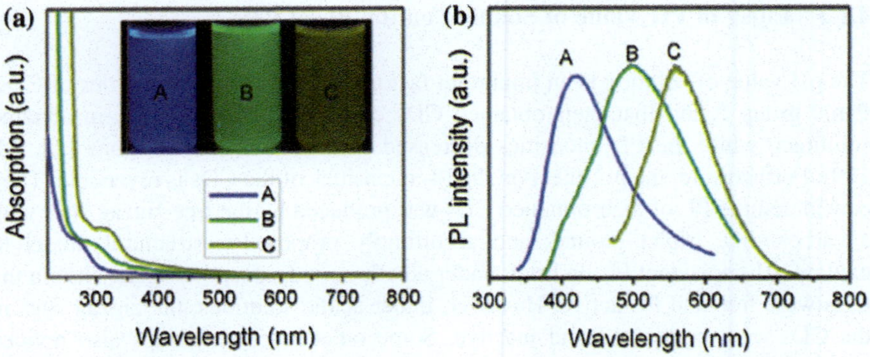

Fig. 19 **a** UV-vis spectra of CDs *A*, *B*, and *C*, corresponds to synthesized reaction temperature at 120, 100, and 80 °C, respectively. *Inset* of panel a is a photograph of the corresponding CDs under UV light with 365 nm excitation. **b** PL spectra of CDs with different emission color excited at 318, 331, and 429 nm, respectively [52]

temperature. With the increase of CD size, the PL changed from blue and green to yellow (Fig. 19). Similar size-dependent PL was also observed by some other research groups [131, 132]. It should be pointed out that the size-effects on the PL spectra of CDs were only in the CDs prepared by the same or a quite similar method. Considering the possible PL mechanisms of CDs discussed above, the PL spectra of CDs should be related to either the surface states or the contained sp^2 clusters. CDs synthesized from the same method should present similar surface states. Then, the size-dependent PL of CDs might reflect the effect of sp^2 clusters with different sizes. Apparently, the size-dependent PL is not a general law for all CDs. The size of the isolated sp^2 clusters and the surface states of CDs synthesized by different methods may be very different. Therefore, CDs prepared by different methods may have quite different PL properties even their size distributions are similar.

4.1.5 Enhancing PL of CDs

Surface Passivation

In general, “top-down” methods are quite easy for preparing CDs of different PL colors, ranging from blue to red. However, the QYs of these CDs without further treatment are relatively low, usually less than 3 % [8, 38, 39, 45, 67, 68, 83]. Therefore, many works have been carried out to improve the PLQY of these CDs. One of the most popular method is passivating the surface states of these CDs by some organic polymers. For example, Sun’s group passivated the chemical oxidized CDs with diamine-terminated oligomeric poly(ethylene glycol) (Fig. 20a) [9]. As a result, bright PL emissions were observed. The PLQY of the obtained surface-passivated CDs could be changed from about 4 % to more than 10 %. The

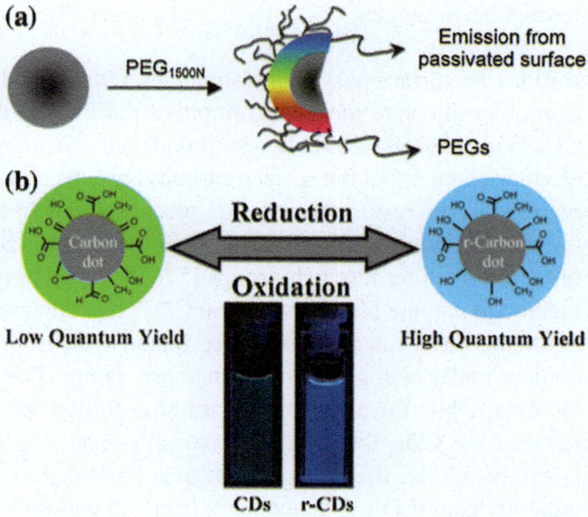

Fig. 20 **a** Schematic illustration of passivation of CDs [9]. **b** Graphical representation of the synthesis of reduced state carbon dots with *blue* luminescence from original carbon dots. *Inset*: photographs of aqueous solutions of the CDs (*left*) and the r-CDs (*right*) obtained under UV light (360 nm) [184]

variation probably depended on the effectiveness of the reaction for surface passivation. They also pointed out other molecules or polymers, such as poly(propionylethyleneimine-*co*-ethyleneimine), could also be used for the passivation of CDs. Since then, the post-passivation strategies have been widely used in improving the PLQY of CDs [37, 47, 150, 178, 179]. However, this strategy usually involves a long-time and boring refluxing procedure. Therefore, some researchers tried to develop a one-step procedure that integrated synthesis and passivation. Hu and coworkers used a pulsed Nd:YAG laser to irradiate graphite or carbon black dispersed in diamine hydrate, diethanolamine, or PEG200 under ultrasonication [32]. The obtained CDs could exhibit similar strong PL behaviors as those of post-passivated CDs. CDs obtained by "bottom-up" methods could also be post-passivated to improve the PL activity [166]. However, more surface-passivated CDs synthesized by "bottom-up" methods were obtained by adding directly the passivation agents into the organic precursors. For example, Gianelis's group synthesized a serial of ammonium carboxylate complexes through the acid-base combination between organic amines and organic acid [10, 15, 16, 89, 122]. During the thermal treatment, one serves as the carbon source (either the acid or the base part, dependent on their stability), the other part serves as the surface modifier. Other groups mixed the organic molecules used as the carbon source with some polymers together followed by various kinds of thermal treatments [92, 100, 132, 135, 136, 142, 154, 158]. Most of the obtained modified or passivated CDs showed good PL activities.

Reduction Strategies

Besides the surface passivation strategies, reducing CDs with some strong reductants is another effective method to improve the PLQY of CDs. Zheng's group reacted CDs with $NaBH_4$ [184]. It was found that the morphologies of the CDs had no obvious change, but the surface carbonyl groups reduced while the hydroxyl groups increased. As a result, the emission spectrum blue-shifted from 520 to 470 nm, while the PLQY increased from 2 to 24 % (Fig. 20b). Similar results have been also observed by other research groups [57, 195]. Qu's group reported a photo-reducing method to prepare blue luminescent CDs [180]. In brief, CDs obtained by chemically oxidizing GO or candle soot were mixed with isopropanol, followed by UV irradiation under a high-pressure mercury lamp. The PLQY could have 3.7 time increase, while the emission spectra blue-shifted accordingly. Dong and coworkers reduced the CDs obtained by chemically oxidizing XC-72 carbon with hydrazine [196]. Similarly, the PL spectrum blue-shifted from 525 to 455 nm, accompanied with obvious PLQY enhancement (from 1.3 to 4.8 %). The blue PL of the hydrazine reduced CDs may be related to the formed luminol-like units. Nearly at the same time, similar phenomenon was observed by Feng and coworkers [183].

Doping

Sun's group proposed that doping was a new strategy for obtaining highly PL CDs. They coated CDs with ZnO or ZnS, then passivated with PEG 1500. Compared with the undoped CDs, the PLQYs (50 and 45 %) of the doped CDs were much higher [193]. In their following-up study, CDs were doped with ZnS and TiO_2 using a similar method. The doped CDs were further processed by gel column fractionation to harvest the most PL dots, which exhibited PLQY of up to 78 % [192]. Other reseach groups have synthesized N-dopping CDs (N-CDs. For example, Qu's group obtained N-CDs by electrochemically oxidizing chemically reduced GO in acetontrile in the presence of tetrabutylammonium perchlorate (TBAP) as the electrolyte [72]. Liu et al. [197] prepared N-CDs by solvothermally treating GO in DMF at 200 °C for 4.5 h. Hu et al. [198] synthesized N-CDs by hydrothermally treating GO in the presence of ammonia. Their obtained N-CDs also exhibited bright PL emission.

"Bottom-up" methods seem much easier to obtain high PL heteroatom-doped CDs. The other popular method is synthesizing CDs by using a serial of heteroatom-containing organics as the precursor. Dong and coworkers synthesized three kinds of CDs, namely CDs without doping (labeled as O-CDs due to the contained oxygen-containing function groups), N-CDs, nitrogen and sulfur co-doped CDs (N,S-CDs) [87]. It was found that O-CDs show very weak PL, N-CDs show much stronger PL than that of O-CDs, and N,S-CDs show the strongest PL. Considering the PL should be mainly attributed to the surface states of CDs, a model was proposed for the PL of the doped CDs. In brief, the doping of nitrogen can introduce the CDs a new kind of surface state. Electrons trapped by the new formed surface states might be able to facilitate a high yield of radiative recombination. The density of the nitrogen-relative surface state could be increased dramatically by the co-doped sulfur atoms, resulting in the further enhancement of PL. Up to now, a large quantity of N-CDs or N,S-CDs have been synthesized using various organics as the precursor [90, 91, 97, 98, 102, 108, 114, 118, 120, 123, 127, 137, 149, 160, 168, 170, 174, 199–201]. Most of those doped CDs showed good PL, while others didn't exhibit any obvious advantage in PL over those CDs without any doping [127, 149, 168, 174, 200]. Accordingly, the exact enhancement mechanism of the heteroatom-doping may still need further investigation.

4.2 Upconversion Luminescence

Upconversion luminescence (UCL) is an anti-Stokes process whereby low-energy photons are converted to higher-energy ones. The UCL behaviors of CDs were first reported by Sun's group [37]. A small aliquot of the aqueous solution of surface-passivated CDs was dropped on a cover glass and evaporated the water, then measured by a confocal fluorescence microscope equipped with laser. The CDs emitted strongly emission when excited with laser of either 458 or 800 nm (Fig. 21a). They proposed that the luminescence excited by the laser of 800 nm is two-photon luminescence of the CDs.

The UCL of CDs was subsequently observed by Lee's group using a Fluorolog-TCSPC luminescence spectrometer [69]. The PL spectra of their obtained CDs excited by long-wavelength light (500–1000 nm) with the upconverted emissions centered in the range from 325 to 425 nm (Fig. 21b). Similar UCL

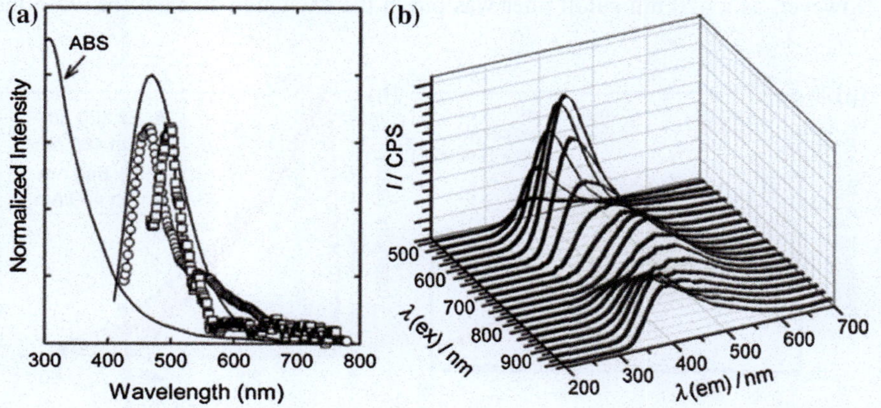

Fig. 21 **a** The one-photon (*squares*, 458 nm excitation) and two-photon (*circles*, 800 nm excitation) luminescence spectra of the CDs on glass substrate are compared with solution-phase absorption (ABS) and luminescence (*solid line*, 400 nm excitation) spectra [37]. **b** UCL spectra of CDs excited at different wavelength [69]

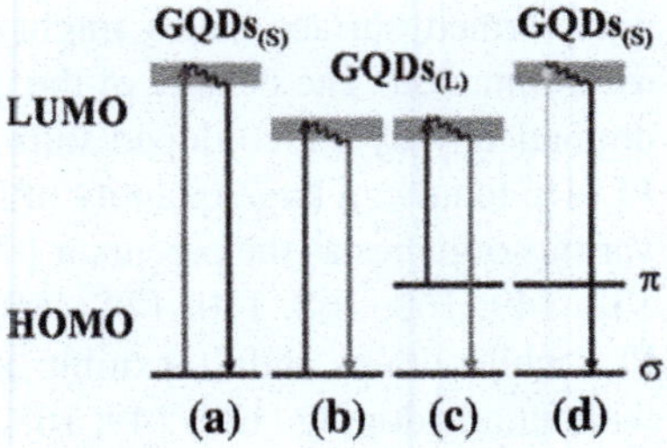

Fig. 22 A schematic illustration of various typical electronic transitions processes of CDs. Normal PL mechanisms in CDs for small size (**a**) and large size (**b**); Upconverted PL mechanisms in CDs for large size (**c**) and small size (**d**) [202]

was observed by Li's group from PEG passivated GQDs prepared via chemical oxidation [202]. When the excitation wavelength was increased gradually from 600 to 800 nm, the upconverted emissions peaks shifted accordingly from 390 to 468 nm. However, the shifting between the energy of upconverted emission light and excitation light seemed to be unchanged, about 1.1 eV. Accordingly, they considered that the anti-Stokes transition behaviors of the CDs were related to the energy levels of π and σ orbitals provided by the carbene ground-state multiplicity. A possible mechanism was proposed (Fig. 22). When excited with a bunch of low-energy photons, the electrons of the π orbital would transfer to the LUMO, then come back to σ orbital and emit high-energy photons.

Up to now, many CDs prepared by different methods have been claimed to have the UCL properties, and applied in many fields such as bioimaging and analytical detection [14, 54, 136, 177, 190, 195, 203–205]. However, Tan and coworkers argued that the so-called UCL of CDs under excitation of a xenon lamp should be artificial [206]. CDs synthesized by chemically oxidizing GO could produce luminescence under the excitation of either 800 or 400 nm (Fig. 23a). Although the luminescence intensity excited at 800 nm is about 1/100 of the normal luminescence intensity excited at 400 nm, the two emission spectra are quite similar. However, as a 420 nm cutoff filter was put in the excitation channel (between the

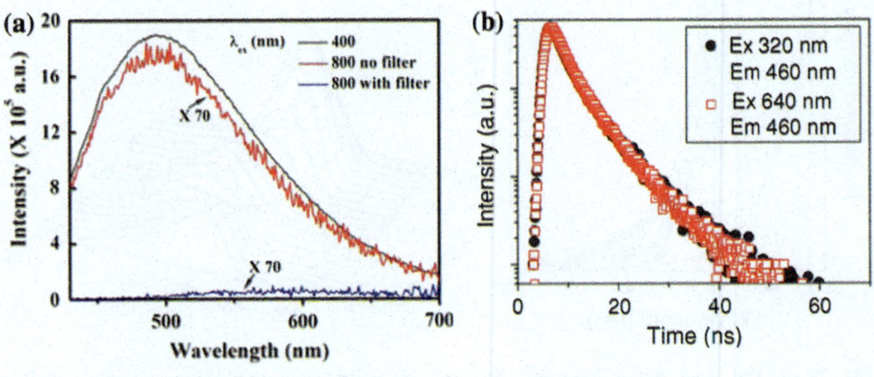

Fig. 23 **a** Emission spectra excited at 400 nm (*black*), 800 nm without (*red*) and with (*blue*) a 420 nm cutoff filter in the excitation channel [206]. **b** Time resolved PL monitored at 460 nm emission excited by the selected light of 320 (*circles*) and 640 nm (*squares*) [207]

excitation source and the sample in order to eliminate the light at 400 nm (800/2) from light source), nearly no luminescence could be detected when excited at 800 nm. Considering second-order diffraction light (λ/2) is always present in the selected excitation light (λ) due to limitation of diffraction gratings used in the monochromators of the spectrofluorimeter, the so-called UCL should be not real UCL but normal luminescence excited by the second-order diffraction light of wavelength λ/2. Gan and coworkers provided new evidences to support the conclusion of Tan [207]. The time resolved PL measurement of the CDs synthesized by hydrothermally treating GO was performed. It was found that the decay spectra of the normal PL and the "UCL" are nearly the same (Fig. 23b). Apparently, no real UCL takes place in CDs under excitation of a xenon lamp. However, the authors pointed out the excitation wavelength-dependent PL indicated that there are abundant energy levels in most CDs. The multi-energy levels in CDs may play the role of essential intermediate states. That is to say, CDs should be possible to produce UCL. The excitation conditions is crucial for the observation of UCL. Their experimental results indicated that both non-coherent photons and coherent photons from continuous wave laser with low power density cannot be up-converted by CDs to produce a UCL. However, UCL from CDs could be available under excitation of pulsed laser with enough high power density.

4.3 *Electrochemiluminescence*

4.3.1 ECL Mechanisms

ECL is a CL that triggered by electrochemical methods. The analytical methods based on the ECL signal combine the advantages of CL and electrochemical analysis, and accordingly show many distinct advantages including no optical background, easy reaction control, high sensitivity and selectivity, and wide response range. Accordingly, the ECL behaviors of CDs has been well applied in sensors [2]. What is more important, ECL has been proposed to be a powerful technic to study the surface states of various quantum dots [208–210]. Therefore, ECL is another important luminescence property of CDs.

The ECL properties of CDs were first discovered by Chi's group [68]. The ECL emission of CDs produced from the electrochemical oxidation of graphite was detected as the potential was cycled between +1.8 and−1.5 V (Fig. 24a). Similar ECL behaviors was subsequently found from the PEG passivated CDs produced by microwave heating glucose and PEG_{200} [132]. The ECL signals were considered to be caused by the electron-transfer annihilation of negatively charged CDs ($R^{\bullet -}$) and positively charged CDs ($R^{\bullet +}$). However, either $R^{\bullet -}$ or $R^{\bullet +}$ seems impossible to be stable for more than 33 s (calculated by dividing the potential window by the scan rate). Therefore, the ECL signals should be coreactant ECL signals produced by the reactions between $R^{\bullet-}$ (or $R^{\bullet+}$) and some unknown electro-generated free radical species from the solvent or impurities. In 2010, the real annihilation ECL signals between $R^{\bullet-}$ and $R^{\bullet+}$ were detected by applying 1 Hz potential steps between +1.8 and −1.5 V (Fig. 24b) [44]. The observation of cathodic and anionic ECL signals implied that both $R^{\bullet-}$ and $R^{\bullet+}$ radicals in solution are stable enough to transfer charge upon colliding and produce the excited stated CDs.

ECL spectra of the many different CDs have been measured and compared with their corresponding PL spectra (Fig. 24c). The maximum wavelengths of the ECL emission ranged from 535 to 600 nm, which were substantially red-shifted from those of the corresponding PL emission [44, 57, 68, 211, 212]. Accordingly, the ECL activities of CDs were considered to be mainly attributed to the surface states. Xu and coworkers studied the effect of oxidation degree on the ECL behaviors of CDs [121]. It was found that a high oxygen content could give the CDs a good ECL activity, which further confirming that the ECL of CDs should be mainly characterized by the surface states for the formation of radicals. Accordingly, a model has been proposed to explain the ECL behaviors of CDs (Fig. 24d) [68].

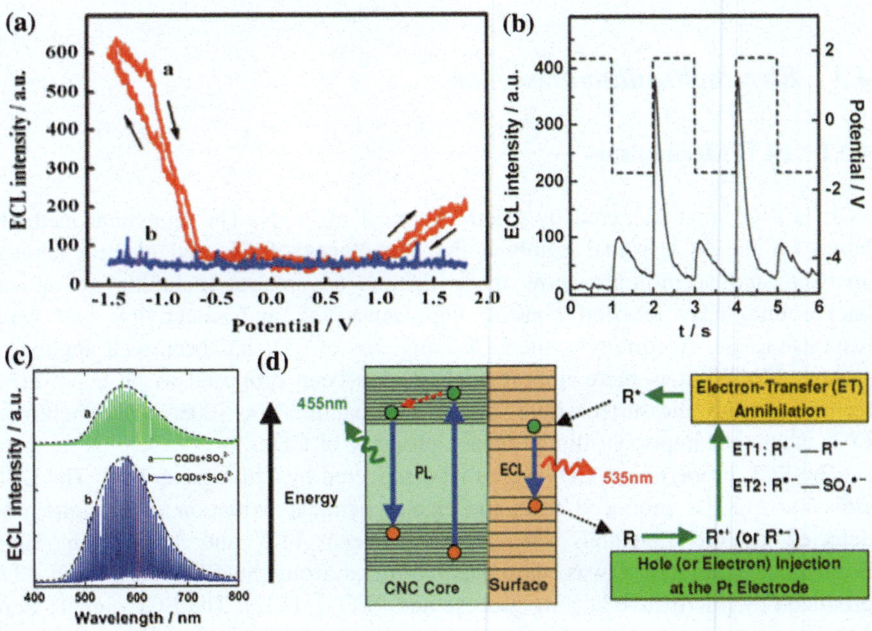

Fig. 24 **a** ECL responses of CDs (*red line*) at a Pt electrode in 0.1 M PBS (pH 7.0) [68]. **b** ECL transients of CDs by stepping potential (*upper curves*) between +1.8 and −1.5 V [44]. **c** ECL spectra of CDs [212]. **d** Schematic illustration of the ECL and PL mechanisms in CDs. $R^{\bullet+}$, $R^{\bullet-}$, and R^{*} represent negatively charged, positively charged, and excited-state CNCs, respectively [68]

4.3.2 Coreactant ECL Systems

In 2009, it was found that CDs prepared from the electrochemical oxidation of graphite could produce strong and stable cathodic ECL signal in the presence of peroxydisulfate ($S_2O_8^{2-}$), a coreactant commonly used in ECL systems (Fig. 25a) [68]. The ECL signal was considered to be caused by the electron transfer annihilation $R^{\bullet-}$ between and the electro-generated $SO_4^{\bullet-}$ (Fig. 24d). Then the first coreactant ECL system about CDs has been developed. Up to now, the coreactant ECL of CD–$S_2O_8^{2-}$ system has been well studied and used in sensors [57, 121, 213–219]. Besides $S_2O_8^{2-}$, SO_3^{2-} and L-cysteine (L–Cys) were subsequently found to be able to enhance the cathodic ECL signal of CDs [211, 212]. However, an anodic polarization was found to be essential for both SO_3^{2-} (Fig. 25b) and L–Cys to produce the cathodic ECL signal. For the CD–SO_3^{2-} system, the electrochemical oxidation of SO_3^{2-} produced sulfur trioxide anion free radicals ($SO_3^{\bullet-}$), which subsequent initiate a three-step auto-catalytic reaction in the presence of dissolved oxygen (O_2). In the following cathodic polarization process, the electron transfer between $SO_4^{\bullet-}$ generated in the autocatalytic reaction and the electro-generated $R^{\bullet-}$ finally produce the excited stated CDs, which give rise to the strong ECL emission when go back to their ground state (Fig. 25c). The ECL behaviors of CD–L–Cys system is quite similar with those of CD–SO_3^{2-} system. The electrochemical oxidation of cysteine (RSH) in the anodic polarization process produced ($RO^{\bullet}$), which

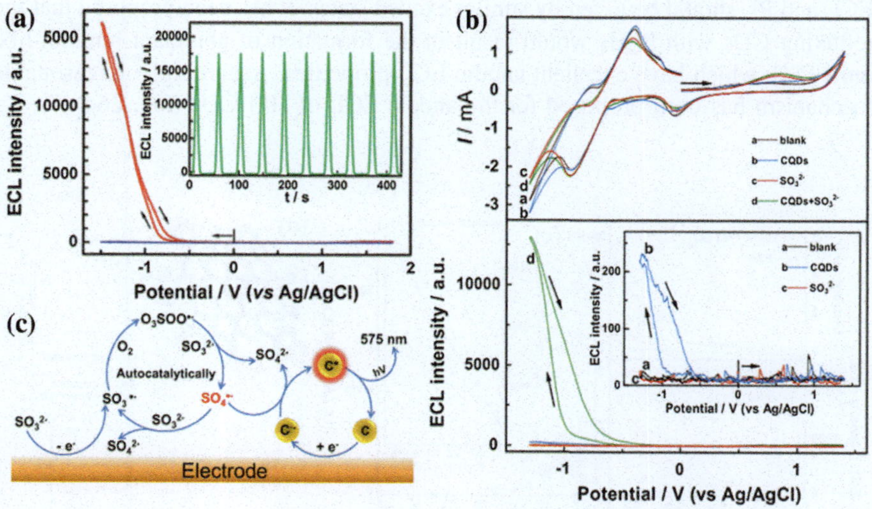

Fig. 25 **a** ECL behavior of CDs in the absence (*blue line*) and presence of 1 mM $K_2S_2O_8$. *Inset* in (**a**) shows the ECL response of CD–$S_2O_8^{2-}$ system obtained during a continuous potential scan [44]. **b** Cyclic voltammogram (*upper lines*) and the corresponding ECL potential curves (*lower lines*) of CDs in the absence (*blue lines*) and presence (*green lines*) of SO_3^2. *Inset* in (**b**) shows amplified ECL responses of CDs without SO_3^{2-} [212]. **c** Schematic diagram for the ECL reaction mechanism of CD–SO_3^{2-} system [212]

was further oxidized to $RSO_2^{\bullet}$ by O_2 dissolved in the solution. Subsequently, the formed $RSO_2^{\bullet}$ initiated a three-step autocatalytic reaction, in which three kinds of free radicals ($RSO^{\bullet}$, $RSO_2^{\bullet}$, and $RSO_3^{\bullet}$) were produced continuously. These free radicals reacted with electro-generated $R^{\bullet -}$ in the following cathodic polarization process, producing excited CDs.

As discussed above, most CDs exhibited good ECL activities. In particular, CDs could produce strong cathodic ECL signal in the presence of some sulfur-containing coreactants. However, much less attention has been paid to the anodic ECL of CDs. Yu's group observed that the CDs obtained from hydrothemal treatment of GO could produce strong anodic ECL signal in the presence of H_2O_2 as coreactant [218]. The authors argued that the anodic ECL signal was caused by the electron transfer between the electro-generated $R^{\bullet +}$ and $O_2^{\bullet -}$ formed by the electro-oxidation of H_2O_2. However, the observed anodic ECL seems to be a special case for CDs without any modification. Dong and coworkers observed a similar anodic ECL signal from hydrazide-modified CDs (HM-CDs) [196]. They refluxed the CDs from chemical oxidation of XC-72 carbon black with hydrazine. As a result, partial oxygen-containing groups on the CDs have been removed, while a lot of hydrazide groups have been introduced. In alkaline solutions, HM-CDs produced an obvious electro-oxidation peak at around +0.64 V and a strong ECL signal with peak at around +0.72 V (Fig. 26a, b). Furthermore, the anodic ECL signal could be enhanced greatly by H_2O_2 (Fig. 26c). The ECL spectrum of the HM-CD/H_2O_2 system was measured and compared with the main PL emission spectrum of the HM-CDs. It was found that two spectra were essentially the same, implying that the ECL and PL should be given by similar excited states. It has been proposed that the refluxing CDs with N_2H_4 would result in the formation of abundant luminol-like units [74], which have excellent anodic ECL properties. Accordingly, a reasonable mechanism has been proposed for the anodic ECL of HM-CDs (Fig. 26d).

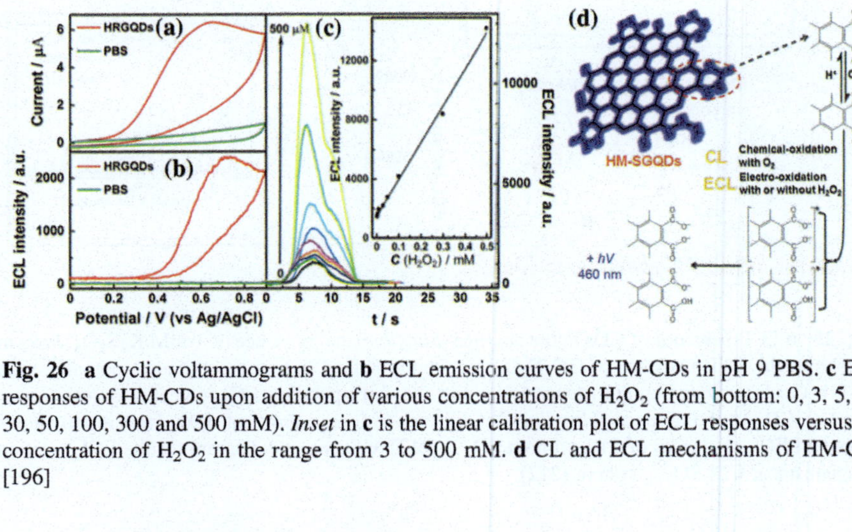

Fig. 26 **a** Cyclic voltammograms and **b** ECL emission curves of HM-CDs in pH 9 PBS. **c** ECL responses of HM-CDs upon addition of various concentrations of H_2O_2 (from bottom: 0, 3, 5, 10, 30, 50, 100, 300 and 500 mM). *Inset* in **c** is the linear calibration plot of ECL responses versus the concentration of H_2O_2 in the range from 3 to 500 mM. **d** CL and ECL mechanisms of HM-CDs [196]

4.4 Chemiluminescence

CL is the generation of electromagnetic radiation as light by the release of energy from a chemical reaction. CL has been proved to be important for both fundamental study and analytical application. Analytical methods based on CL usually show advantages including high sensitivity, wide response range, simple instrumentation and lack of background scattering light interference. The CL behaviors of CDs were first detected by Chi's group, during they were investigating the ECL behaviors of CDs synthesized by hydrothermally treating the mixture of $HOCH_2CH_2OCH_2CH_2NH_2$ and CA [44]. It was found that, in the presence of $K_2S_2O_8$, the CDs produced a strong background emission signal, which was supposed as CL. However, the authors didn't pay more attention to the CL behaviors of CDs.

Lin's group found that the CDs, synthesized from glycerine and PEG using a microwave method, could produce strong CL emission in the presence of $NaNO_2$ and H_2O_2 [135]. The CL spectrum of the CDs–$NaNO_2$–H_2O_2 system located in the wide range of 400–600 nm and centered at 520 nm. The range of the CL spectra was similar to that of the PL spectra of the CDs, implying that CDs should act as the emitters. However, the maximum wavelength of the CL spectrum was red-shifted in comparison to that of the main PL spectrum, indicating that the CL of the CDs should be attributed to their surface states. A possible CL mechanism has been proposed for the CDs–$NaNO_2$–H_2O_2 system (Fig. 27a). On one hand, the mixing of $NaNO_2$ and acidified H_2O_2 produced ONOOH and $OH^{\bullet}$, which could serve as the hole injector and convert CDs to $R^{\bullet +}$. On the other hand, ONOOH could react with H_2O_2 in acidic solution tor from $O_2^{\bullet -}$, which could react CDs into $R^{\bullet -}$. Finally, electron-hole annihilation in or between $R^{\bullet +}$ and $R^{\bullet -}$ resulted in the energy release with CL emission. In another work of Lin's group, similar annihilation ECL behaviors of CDs were also observed (Fig. 27b). The formation of $R^{\bullet +}$ and $R^{\bullet -}$ are related to the intermediate radicals generated during the reaction of H_2O_2 and $NaHSO_3$, such as hydroxide radical ($OH^{\bullet}$), sulfate anionic radical ($SO_4^{\bullet -}$), superoxide anionic radical ($O_2^{\bullet}$), and sulfur trioxide anionic radical ($SO_3^{\bullet}$) [220]. Lin's group further found that CDs could emit strong CL signal in the presence of classical oxidants, such as acidic potassium permanganate and cerium (IV) [221]. It

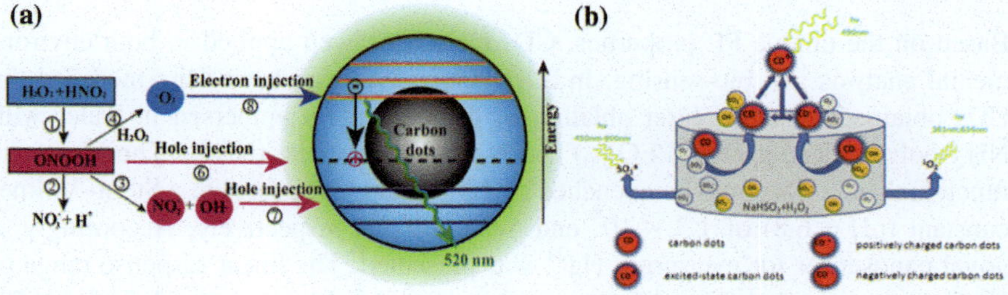

Fig. 27 **a** Schematic illustration of the CL mechanism of CDs–$NaNO_2$–H_2O_2 system [135]. **b** CL reaction mechanism for the H_2O_2-$NaHSO_3$-CDs system [220]

was proposed that oxidants could inject holes into the CDs. The radiative recombination of oxidant-injected holes and the electrons in CDs results in energy release in the form of CL emission. Teng and coworkers further investigated the CL behavior and mechanism of CDs in the presence of acidic $KMnO_4$ [49]. The results indicated that reduced CDs, containing a large amount of hydroxyl groups, could produce much stronger CL intensity than CDs. Accordingly, the surface groups of CDs should play an important role on their CL properties.

All the CL behaviors mentioned above were observed in acidic solutions. However, some CDs could also produce CL emission in alkaline solution. Zhao and coworkers observed the CL behaviors of CDs in the presence of a strong alkaline solution. The CL intensity was dependent on the concentrations of CDs and hydroxyl ion (OH^-), but independent on the concentration of dissolved oxygen. The authors claimed that the CL was caused by the radiative recombination of the injected electrons by "chemical reduction" of CDs with thermally excited generated holes [222]. Amjadi and coworkers reported that CDs could be oxidized directly by $K_3Fe(CN)_6$ in an alkaline solution, giving a strong CL emission [92]. The CL intensity was related to the dissolved oxygen. It was proposed that CDs could be on one hand oxidized by $K_3Fe(CN)_6$ to $R^{\bullet +}$, and on the other hand reduced by the present superoxide anion radical ($O_2^{\bullet -}$) to $R^{\bullet -}$. The electron-transfer annihilation between $R^{\bullet +}$ and $R^{\bullet -}$ resulted in the CL emission. The HM-CDs reported by Dong and coworkers also produced strong CL signal in alkaline solution [196]. The CL intensity was proportional to the concentration of OH^- in the pH range of 8–11. Furthermore, the concentration of dissolved oxygen would also affect the FL intensity. The CL spectrum was found to be similar to the ECL spectrum and the main PL spectrum of HM-CDs. The CL behaviors of HM-CDs were believed to be related to their contained luminol-like units. A reasonable CL mechanism has been proposed accordingly (Fig. 26b).

5 Sensing Applications

5.1 PL Sensors

5.1.1 Response to Hg^{2+} and Relevant Sensing Applications

Based on the unique PL properties, CDs have been well applied in both environmental analysis and bio-sensing. In 2010, Gonçalves and coworkers modified the CDs obtained by direct laser ablation of carbon targets immersed in water with NH_2–polyethylene-glycol (PEG_{200}) and N-acetyl-l-cysteine (NAC). The PL of the functionalized CDs could be quenched by Hg^{2+} and Cu^{2+} ions with a Stern–Volmer constant (pH = 6.8) of 1.3×10^5 and 5.6×10^4 M, respectively. Accordingly, a novel nanosensor for measuring Hg^{2+} was presented. The linear response range of Hg^{2+} ranged from 1.00×10^{-7} to 2.69×10^{-6} M. It was proved that the PL quenching was related to the complexing between the metal ions and the functional groups (sulfydryl) [223]. In one of their follow-up work, an optical fiber sensor for Hg^{2+} in aqueous solution based on sol-gel immobilized the modified CDs mentioned above was described [224]. The nanosensor showed a fast (less than 10 s), reversible and stable response. Also, the nanosensor allowed the detection of sub-μmole concentrations of Hg^{2+} in aqueous solution.

Gao et al. [225] combined blue-emission CDs with red-emission hydrophilic CdSe@ZnS QDs to form a nanohybride, which was used as a ratiometric PL probe for the detection of Hg^{2+}. Due to the strong chelating ability of carboxylmethyldithiocarbamate modified on the surface of QDs to Hg^{2+}, the red-emission of QDs was quenched by Hg^{2+} while the blue-emission of CDs remained constant. As a result, an obviously distinguishable PL color evolution (from red to blue) could be observed (Fig. 28). The detection limit of the method was 0.1 μM. Up to now, various sensors for Hg^{2+} based on CDs have been fabricated with signal-off process [93, 98, 115, 138, 155, 167, 223–227]. Most of the experiments were designed reasonably, and the results were good. However, some of the detection were performed in pure water, without any buffer solution. That is to say, the interferences from solution pH vand some commonly existing anions were omitted. Additionally, the quenching mechanism was not so clear, although most of the experimental results seemed good in terms of both the sensitivity and the selectivity.

Some research groups combined the nano-technic with DNA technic to develop novel PL signal-on system for highly sensitive and selective detection of Hg^{2+}.

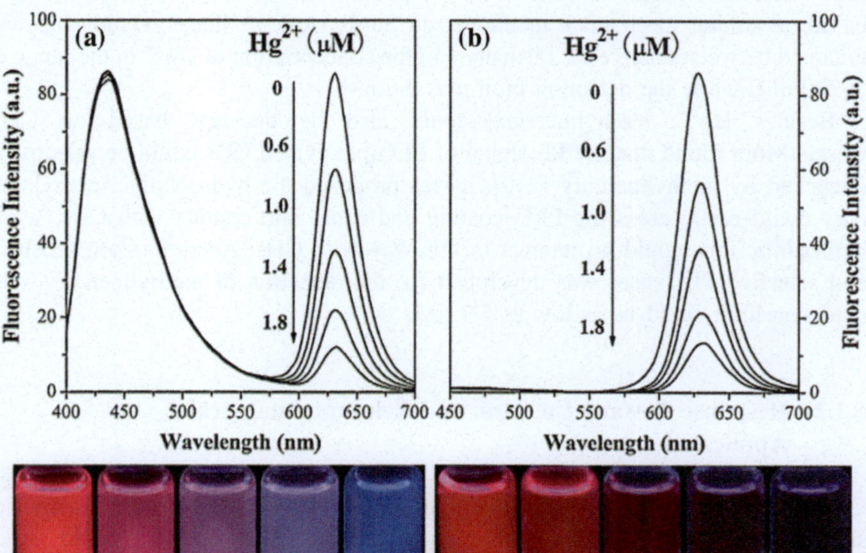

Fig. 28 PL responses of the ratiometric probe solution (**a**) and the single GDTC-QDs solution (**b**) upon the addition of different amounts of Hg^{2+}. The images in the *bottom* are their corresponding fluorescence images [225]

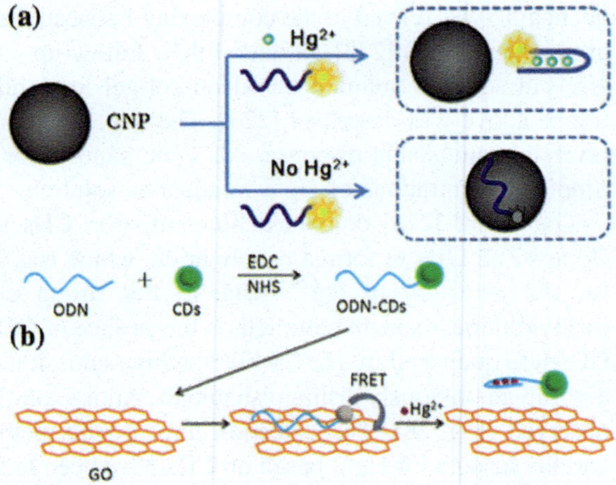

Fig. 29 **a** A schematic (not to scale) illustrating the CNP-based fluorescent Hg^{2+} detection based on conformational change of a Hg^{2+}-specific T-rich OND (PH). PH: a FAM-labeled Hg^{2+}-specific OND probe [228]. **b** Schematic illustration of the GO-based sensor system for Hg^{2+} detection [229]

Sun's group used the poor PL CDs from candle soot as a PL quencher to inhibit the PL of dye-labeled T-rich single-stranded DNA via π-π stacking interactions between DNA bases and CDs [228]. In the presence of Hg^{2+}, the PL signal of the dye was recovered due to the formation of T–Hg^{2+}–T induced hairpin. Accordingly, a sensitive and selective sensor for Hg^{2+} was developed (Fig. 29a). The detection limit could be as low as 10 nM. Different from Sun's work, Cui and coworkers used high PL CDs as the PL emitters to label oligodeoxyribonucleotides (Fig. 29b) [229]. The PL signal was quenched by GO, and subsequently recovered by Hg^{2+} based on similar mechanism mentioned in Sun's work. A linear relationship was obtained between relative PL intensity and the concentration of Hg^{2+} in the range of 5–200 nM, while the detection limit was 2.6 nM.

Besides Hg^{2+}, methylmercury could also be detected based on CDs. Costas-Mora found that the PL signal of PEG-passivated CDs could be selectively quenched by methylmercury [230]. It was proposed the hydrophobic methylmercury could easily cross the PEG coating and came into contact with CDs, while hydrophilic ions could no interact in this way with CDs. Accordingly a sensitive and selective PL sensor was developed for the detection of methylmercury. The detection limit could be as low as 5.9 nM.

5.1.2 Response Toward Cu^{2+} Ion and Relevant Analytical Applications

Chi's group reported that poly(ethylenimine) (BPEI)-functionalized CDs synthesized by thermally treating the mixture of branched BPEI and CA could act as probe for Cu^{2+} [231]. It was found that Cu^{2+} ions could be captured by the amino groups of the BPEI-CDs to form an absorbent complex at the surface of CDs, resulted in a strong quenching of PL via an inner filter effect (Fig. 30a). Accordingly, a novel sensing system was designed. In pH 4 PBS, the sensing

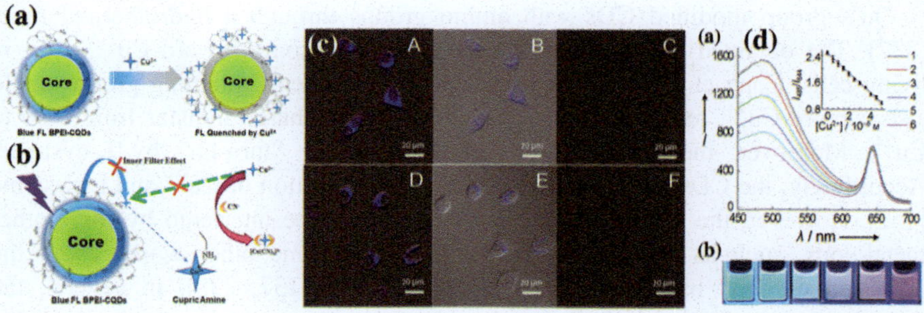

Fig. 30 **a** Schematic diagram for the PL of the BPEI-CDs quenched by Cu^{2+} [231]. **b** Diagram for the "recovery" effect of cyanide on the PL of BPEI-CDs/Cu^{2+} system [233]. **c** Confocal PL images of HepG$_2$ (*A*) and HeLa cells (*D*) incubated with CD-TPEA in PBS (pH 7.4), overlay of confocal PL and bright-field images of HepG$_2$ (*B*) and HeLa cells (*E*) stained with CD-TPEA, confocal fluorescence images of HepG$_2$ (*C*) and HeLa cells (*F*) which were first stained with CD-TPEA and then incubated with $CuCl_2$ and 100 mM PDTC [234]. **d** PL spectra of the (*upper figure*), and PL colors (*lower figure*) of CdSe@C-TPEA ratiometric probe solutions after exposure to $CuCl_2$ [235]

system offered a rapid, reliable, and selective detection of Cu^{2+} ion with a detection limit as low as 6 nM and a dynamic range from 10 to 1100 nM. In one of their following work, BPEI-CDs were encapsulated into the zeolitic imidazolate framework materials (ZIF-8), which could strongly and selectively accumulate target analytes due to the adsorption property. Then the sensitivity toward Cu^{2+} was improved greatly, with a wide response range of 2–1000 nM and a low detection limit of 80 pM [232]. In their another work, the BPEI-CDs/Cu^{2+} system was used for the detection of cyanide based on the PL "turn on" effect (Fig. 30b) [233].

Tian's group also synthesized amino-functionalized CDs for Cu^{2+} detection. CDs synthesized from electrochemical oxidation of graphite was capped with amino TPEA ([N-(2-aminoethyl)-*N,N,N'*-tris(pyridin-2-ylmethyl) ethane-1,2-diamine], AE-TPEA), and used for Cu^{2+} detection. The developed sensing system showed broad linear response range ($\sim 10^{-6}$–10^{-4} M) and low detection limit (~ 10 nM) [234]. Furthermore, the PL probe was successfully applied for intracellular sensing and imaging of Cu^{2+} (Fig. 30c). In one of their follow-up work, AE-TPEA modified CDs were hybridized with CdSe/ZnS QDs as dual-emission fluorophore, in which red-emission CdSe/ZnS QDs embedded in silica shells were inserted to Cu^{2+} [235]. Therefore, Upon the addition of Cu^{2+}, the intensity of blue emission from CDs showed continuous quenching, whereas that of red emission from the QDs still remained constant. The changes in the intensities of the two emission peaks resulted in continuous PL color changes (Fig. 30d). The nanohybrid was successfully applied for imaging and biosensing of Cu^{2+} ions in living cells. A similar work was carried out by Liu and coworkers [236]. CDs were coated on the surface of Rhodamine B-doped silica nanoparticles by the silylation reaction. The PL of blue emission CDs was quenched, resulting in the ratiometric PL response of the dual-emission silica nanoparticles.

Qu's group modified CDs with amino groups through a hydrothermal route [237]. The obtained CDs showed a satisfactory selectivity toward Cu^{2+}, even in living cells. Zong et al, proposed that CDs with oxygen-containing groups could also capture Cu^{2+}, resulting in PL quenching due to charge transfer from CDs to Cu^{2+}. Moreover, the quenched PL signal could be "turn-on" by L-cysteine. Accordingly, the CDs might be used for "off-on" detection of Cu^{2+} and L-cysteine [152]. However, the quenching effect by Cu^{2+} might be interfered by some other metal ions, such as Fe^{3+}, Fe^{2+}, Hg^{2+}. There are some other works about the detection of Cu^{2+} based on various CDs [96, 136, 137, 238]. In general, the detection mechanisms are similar to those mentioned above.

5.1.3 Response Toward Fe^{3+} Ion and Relevant Sensing Applications

Fe^{3+} ion has been found to be able to quench the PL signals of various CDs. For example, the PL intensity of CDs synthesized by hydrothermally treating candle soot in sodium hydroxide solution could be inhibited obviously by some metal ions, such as Hg^{2+}, Cr^{3+}, Al^{3+}, and particularly Fe^{3+} [239]. Accordingly, the authors proposed that the CDs could be used to measure Cr^{3+}, Al^{3+}, and Fe^{3+} in human body fluids, where the content of Hg^{2+} is extremely low. For instance, in the detection of Cr^{3+} in solution, F^- was used as the masking agent to eliminate the interferences of Al^{3+} and Fe^{3+}. The other two metal ions could also be detected in the similar way. The PL signal of N-CDs synthesized by the pyrolysis of konjac flour could be quenched by Fe^{3+} through the electron transfer mechanism [118]. Although the quenching signal would be interfered by some other metal ions, such as Co^{2+}, Ni^{2+}, Ag^+, and Cu^{2+}, only the PL signal quenched by Fe^{3+} could be recovered by L-lysine, implying a potential application in the detection of Fe^{3+} and L-lysine. Qian and coworkers found that the bright blue PL of N-CDs by solvothermally treating the mixture of CCl_4 and diamines could be selectively quenched by Fe^{3+}. It was proposed that the PL quenching might originate from excited-state electron transfer between Fe^{3+} and CDs because the contained carboxylic acid in CDs can form complexes with Fe^{3+} [97]. Accordingly, in the presence of Fe^{2+}, the CDs were used to detect H_2O_2, which was considered to be able to oxidize Fe^{2+} into Fe^{3+}. Similar PL quenching behaviors caused by Fe^{3+} or Fe^{2+}/H_2O_2 were also observed by Yang's group [240]. However, they proposed different mechanisms (Fig. 31a). Firstly, Fe^{3+} quenched the PL of CDs through a dynamic way, which was proved by the Stern-Volmer equation, temperature dependent quenching and PL lifetime measurements. Then, a hemin sensor was achieved based on the Fe^{3+}/CDs system. Secondly, Fe^{2+}/H_2O_2 quenched the PL signal of CDs due to the formation of hydroxyl radicals ($OH^{\bullet}$), which is strongly oxidative. Thus an H_2O_2 sensor with a low detection limit (0.9 ppb) was realized. Actually, the dynamic quenching mechanism between CDs and Fe^{3+} was also accepted by other researchers [241].

Qu's group reported that Fe^{3+} could anaerobically oxidized the hydroquinone of CDs synthesized by hydrothermal treatment of dopamine, resulting in the PL

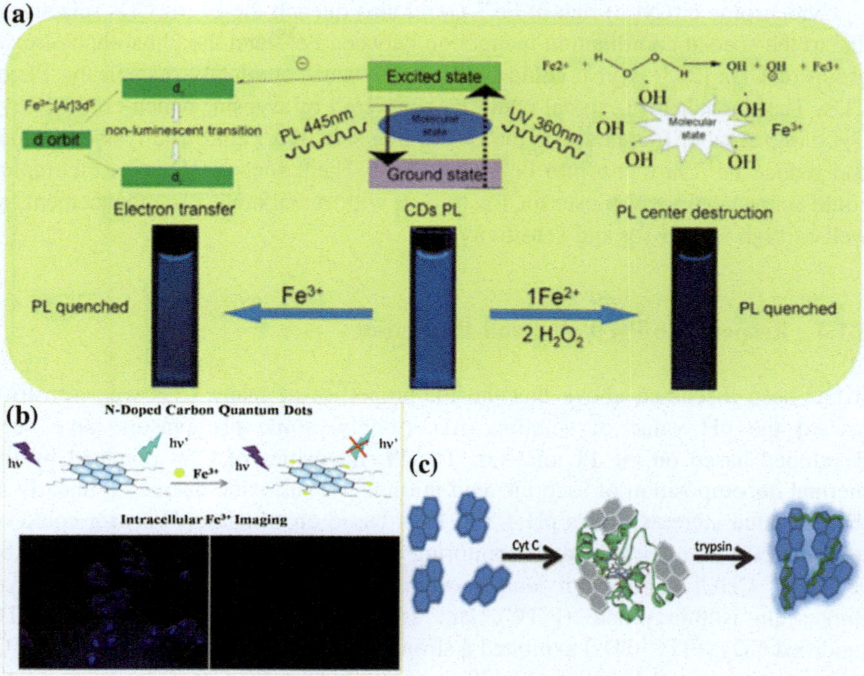

Fig. 31 **a** Fe^{3+} and H_2O_2 sensors based on visual fluorescent detection were achieved, utilizing a special kind of CDs. The origin of the PL of these CDs may be their molecular state, and the PL can be quenched through electron transfer or PL center destruction [240]. **b** Sensing Principle of the N-Doped C-Dots based probe for Fe^{3+} (*upper figure*) and PL microscopy images of Hela cells first incubated with N-CDs for 6 h and then incubated with $FeCl_3$ (*lower figure*) [91]. **c** Schematic illustration of the fluorescent biosensor for trypsin based-on self-assembled GQDs [243]

quenching due to formation of quinone molecule [94]. The CDs were used as an effective PL sensing platform for label-free sensitive and selective detection of Fe^{3+}. Furthermore, based on a competition mechanism, the CDs/Fe^{3+} platform could also be used for the detection of dopamine. In another work of the research group, two different ionic liquids were used to synthesize surface-different N-CDs [137]. The PL signals of the two obtained N-CDs could be selectively quenched by Cu^{2+} and Fe^{3+}, respectively. The authors considered that the PL quenching behaviors should be caused by electron or energy transfer. Zhang and coworkers proposed that the strong coordination of oxygen-rich groups on CDs to Fe^{3+} caused PL quenching via nonradiative electron-transfer [91]. Accordingly, a sensitive label-free sensing platform for Fe^{3+} was developed. Furthermore, the CDs were successfully applied for the PL imaging of intracellular Fe^{3+} (Fig. 31b). Hu and coworkers proposed that Fe^{3+} could act as an energy donor to inhibit the fluorescence resonance energy transfer between CDs and Rhodamine B (RhB), resulting in the PL quenching of RhB [242]. Therefore, the RhB@CDs sensing system could be applied in the sensitive and selective detection of Fe^{3+}.

Cytochrome c (Cyt c) rich in Fe^{3+} could also quench the PL of CDs selectively due to the special coordination interaction between Fe^{3+} and the phenolic hydroxyl group of CDs [243]. Cyt c could bind to CDs and completely quench the PL of CDs. The quenched PL signal could be recovered by trypsin, which could cleave Cyt c into smaller fragments on the C-terminal side of arginine and lysine residues and reduce Fe^{3+} in Cyt c into Fe^{2+} (Fig. 31c). Then, such a CDs–Cyt c complex could serve as a novel biosensor for trypsin with remarkable PL enhancement, as well as high selectivity and sensitivity.

5.1.4 Response to PH Value and PH Sensor

It has been discussed above that the PL properties of many CDs were sensitive toward the pH value of solution. Accordingly, some pH sensors have been developed based on the PL of CDs. The PL intensities of CDs obtained by the thermal decomposition of ascorbic acid in dimethyl sulfoxide decreased linearly as the pH value increased from pH 4.5 to 11.5, based on which a pH sensor could be used directly for environmental monitoring applications [244]. CDs synthesized by refluxing $CHCl_3$ and diethylanime were modified with a pH-sensitive dye, fluorescein isothiocyanate (FITC), and used as a pH probe [156]. The FITC modified CDs (FITC-CDs) exhibited a strong emission peak at 524 nm from FITC, and two emission peaks at around 470 and 566 nm from the CDs when respectively excited by 405 and 534 nm light beams. The emission intensity at 524 nm increased with the increase of pH value, while those at 470 and 565 nm decreased gradually. Measuring the ratio of FITC FL intensity at 524 nm to CD FL intensity at 470 nm (or 565 nm) at various pH showed that the ratio of FL intensity increased linearly with pH in the range from pH 5 to 8. There were some other CDs might be used as pH probes. For example, the PL intensity of N-CDs synthesized by solvothermal reaction of CCl_4 and 1,2-ethylenediamine was inversely proportional to pH values from 5.0 to 13.5 [97]. The PL intensity of CDs obtained by electrochemically oxidizing graphite decreased linearly as the pH was increased gradually over the range from 7 to 14 [67].

5.1.5 K^+ Sensor

Qu's group synthesized a kind of covalently aminated CDs, whose PL signal could be quenched by 18-crown-6 (18C6E) functionalized graphene [245]. The tight binding of primary alkyl-ammonium with 18C6E brought CDs and graphene into appropriate proximity and hence induced energy transfer, resulted in the PL quenching. However, the energy transfer process was inhibited because of competition between K^+ and ammonium for 18C6E, lead to the PL recovery (Fig. 32).

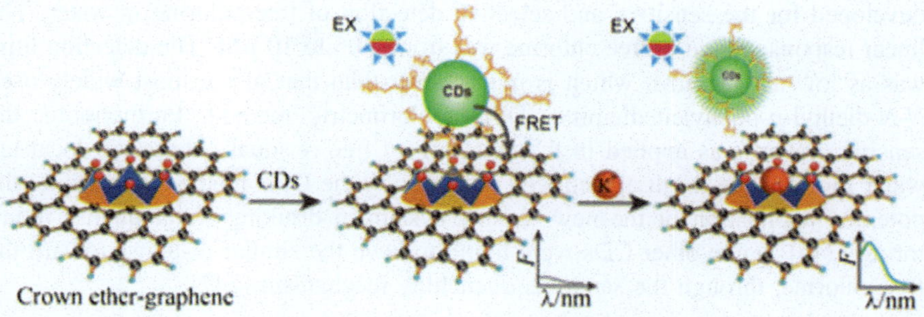

Fig. 32 Schematic illustration of the FRET model based on CDs–graphene and the mechanism of K^+ determination [245]

Accordingly, a PL resonance energy transfer sensor was constructed for measuring the concentration of K^+.

5.1.6 Pb^{2+} Sensor

Wee et al. dehydrated Bovine serum albumin using concentrated H_2SO_4 to synthesize high PL CDs, whose PL signal could be selectively quenched by Pb^{2+} [169]. It was supposed that the PL quenching should be caused by the electron transfer between the CDs and Pb^{2+}. Then a sensor for Pb^{2+} has been developed with a limit of detection of 5.05 μM.

5.1.7 Response to Free Chlorine and Corresponding Sensors

Chi's group found that free chlorine was found to be able to destroy the passivated surface of the CDs synthesized by pyrolysis of CA. As a result, the bright blue PL signal was quenched greatly (Fig. 33). Thereby, a green sensing system has been

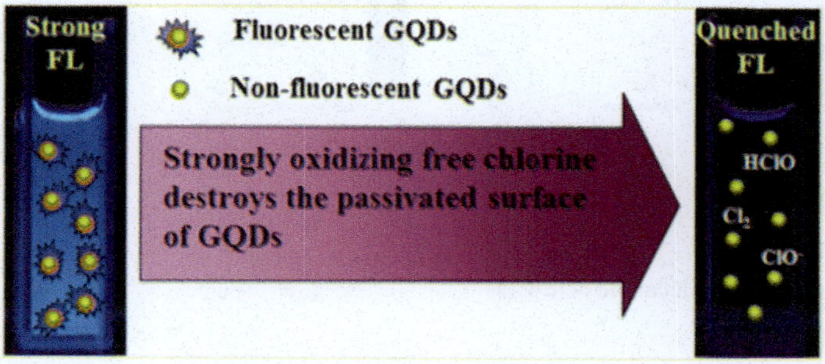

Fig. 33 Schematic illustration of quenching effect of free chlorine on the PL of CDs

developed for the sensitive and selective detection of free chlorine in water. The linear response range of free chlorine was from 0.05 to 10 μM. The detection limit was as low as 0.05 μM, which is much lower than that of the most widely used *N*-*N*-diethyl-*p*-phenylenediamine (DPD) colorimetric method. Furthermore, the sensing system was applied in the detection of free residual chlorine in local tap water sample. The result agreed well with that by the DPD method, suggesting the potential application of the new sensing system in drinking water quality monitoring [246]. Some other CDs were found to have the similar response toward the free chlorine, through the same PL quenching mechanism [247].

5.1.8 Phosphate and Phosphate-Containing Metabolites Detection

Huang's group developed an off-on PL probe of europium (Eu^{3+})-adjusted CDs for the detection of phosphate (Pi) [248]. In principle, Eu^{3+} displays a certain affinity to the oxygen-donor atoms and thus can coordinate with the carboxylate groups on the CDs surface, resulting in the aggregation of CDs. Accordingly, the PL signal of CDs was quenched (turn-off) by Eu^{3+}. However, Eu^{3+} exhibits higher affinity to the oxygen-donor atoms originated from phosphates than that from carboxylate groups. Therefore, the CDs aggregation species could be disassociated in the presence of Pi, leading to "turn-on" PL signal (Fig. 34). The off-on PL probe showed high selectivity, and could be applied in the detection of Pi in very complicated matrixes such as artificial wetlands system. Similar work has been carried out by Qiu's group, using another kind of CDs as the PL emitter [249]. The off-on model has also been applied in the detection of some phosphate-containing metabolites, such as adenosine triphosphate (ATP). CDs synthesized by thermally treating the mixture of CA and glutathione (GSH) were used as the PL emitter, whose PL signal was quenched by Fe^{3+} and then recovered by ATP [250]. Accordingly, a nanosensor for ATP detection was developed, which was successfully applied to estimate the ATP level in cell lysates and human blood serum. Actually, the PL signal of CDs quenched by Fe^{3+} could be also recovered by Pi [251]. Apparently,

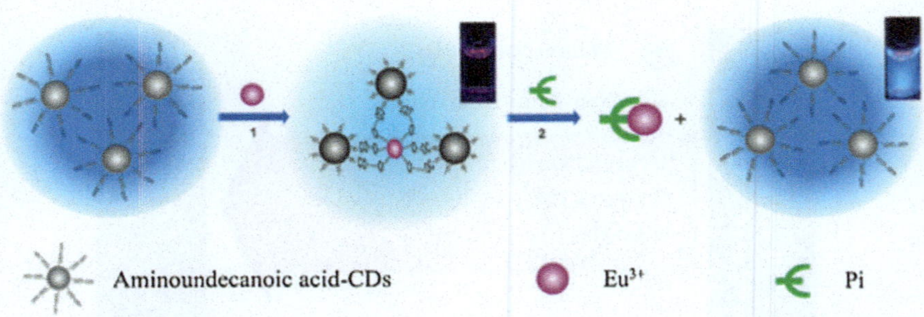

Fig. 34 Schematic representation of Pi detection based on the off–on fluorescence probe of carbon dots adjusted by Eu^{3+} [248]

the mechanisms of PL quenching and recovery were quite similar with those described in Huang's work.

5.1.9 Glucose Detection

Qiu's group developed a label-free signal on sensor for the detection of glucose based on the PL recovery of CDs. It was proposed that cationic boronic acid-substituted bipyridinium salt (BBV) could quench the PL signal of anionic CDs due to the fact that the electron could transfer from the excited CDs to the bipyridinium. However, in the presence of glucose, the boronic acids were converted to more tetrahedral anionic glucoboronate esters, which could effectively neutralize the net charge of the cationic bipyridinium and remove the quencher from the immediate vicinity of the CDs (Fig. 35a). Accordingly, PL signal of CDs quenched by BBV could be recovered by glucose. There was a linear relationship between the recovered PL intensity of CDs and the concentration of glucose from 1 to 60 mM [252].

Besides the signal on sensor, some signal off sensors have also developed for the detection of glucose. Qu et al., functionalized CDs with 3-Aminobenzeneboronic acid, which would covalently bind with the cis-diols of glucose. Therefore, glucose could selectively lead to the assembly of the boronic acid functionalized CDs, resulting in the PL quenching of CDs (Fig. 35b). Then the nanohybrid offered a facile and low cost detection method of glucose with high sensitivity and selectivity [253]. There was a good linear FL response in a wide range from 0.1 to 10 mM. The detection limit for glucose was 0.5 μM or lower. After that, Shen et al. [95] synthesized boronic acid modified CDs by hydrothermal carbonization of phenylboronic acid. The obtained boronic acid modified CDs showed similar PL response towards glucose as mentioned in Qu's work. Moreover, the sensitivity was further improved (with a linear response range of 9–900 μM and a detection limit of 1.5 μM). Furthermore, the proposed sensing system was successfully used for the detection of glucose in human serum.

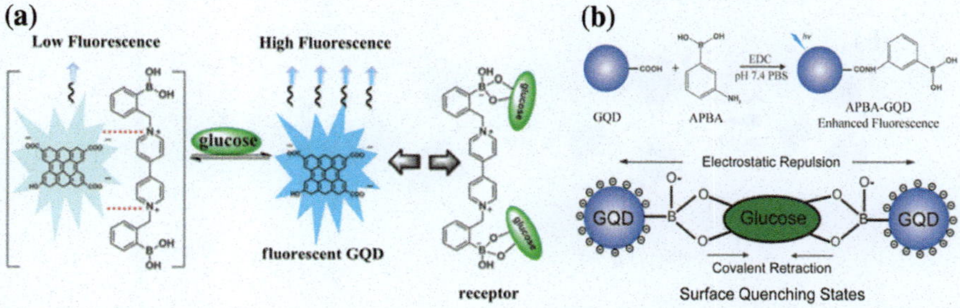

Fig. 35 **a** Proposed glucose-sensing mechanism based on BBV receptor and PL CDs [252]. **b** Schematic representation of the functionalization of CDs with APBA (*upper*) and proposed mechanism of surface quenching states (SQS) for glucose recognition [253]

5.1.10 Immunosensor

Zhu et al. [254] developed a PL immunoassay based on CDs for the detection of human immunoglobulin G (IgG, antigen). CDs synthesized by hydrothermally treating CA were firstly carboxylated by bath sonication with NaOH and $ClCH_2COONa$, then conjugated with goat antihuman IgG (gIgG, antibody). The obtained CDs–gIgG conjugate exhibited the same PL properties as CDs–COOH. However, in the presence of IgG, the PL intensity of the CDs–gIgG conjugate was enhanced obviously due to the specific interaction between gIgG and IgG (Fig. 36a). The antigen-induced PL enhancement was explained by the decrease of surface defects of CDs–COOH. Another signal-on sensor for IgG based on the PL of CDs was developed by Zhao's group. CDs were functionalized covalently with mouse antihuman immunoglobulinG (mIgG, antibody), without any obvious change in the PL properties. The PL intensity of the mIgG modified CDs was quenched by GO due to the fact that both π-π stacking interaction between CDs and GO, and the nonspecific binding interaction between mIgG and the GO surface would bring GO and CDs into FRET proximity to facilitate the PL quenching of CDs. However, IgG (antigen) would bind the mIgG due to the specific antibody-antigen interaction, which effectively increased the distance between mIgG modified CDs and GO surface. Then the quenched PL signal was recovered.

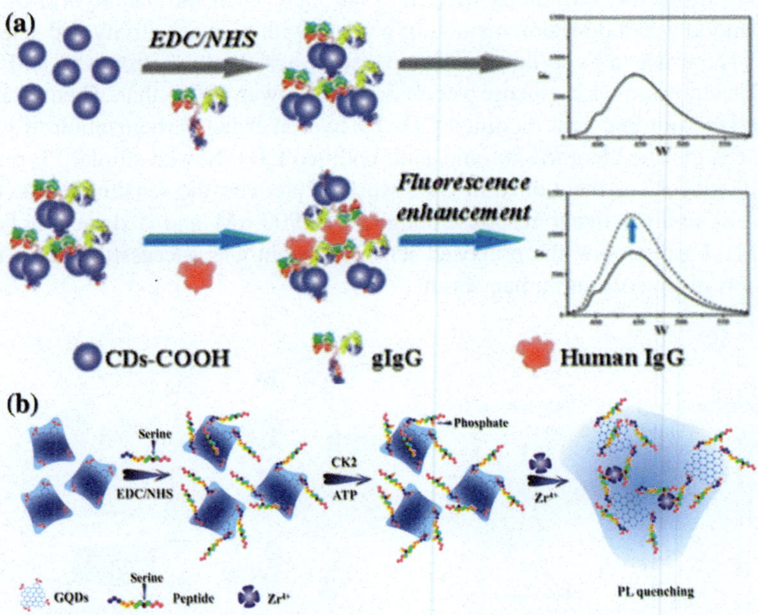

Fig. 36 **a** Schematic illustration for the detection mechanism of human IgG using the carboxyl-functionalized carbon dots [254]. **b** Schematic representation of the CK2 kinase assay based on the aggregation and PL quenching of phosphorylated peptide—CD conjugates via Zr^{4+} ion linkages [257]

Accordingly, the concentration of IgG could be detected by determine the recovered PL intensity [255]. Du and coworkers labeled PL CDs with 4,4-dibrominated biphenyl antigen (PBBAg), and functionalized AuNPs with anti-PBB15 antibody (antiPBBAb) [256]. Thus the obtained PBBAg-CDs were combined with the antigen-binding site of the antiPBBAb-AuNPs. As a result, the PL of CDs was quenched through FRET. However, in the presence of 4,4-dibrominated biphenyl (PBB15), PBBAg-CDs were partly displaced from antiPBBAb-AuNPs due to the competitive immunoreaction, leading to a PL recovery. The increased PL intensity was proportional to the concentration of PBB15. Hence, the concentration of PBB15 could be quantified. Qiu's group modified CDs with serine-containing peptide, which was subsequently phosphorylated by Ser/Thr-specific protein kinase casein kinase II (CK2) in the presence of adenosine 5′-triphosphate (ATP) [257]. The phosphorylated peptide-CDs would be aggregated as Zr^{4+} ions were introduced into the system as linkages between the phosphorylated sites of phosphopeptides via the multicoordinative interactions between Zr^{4+} ions and phosphate groups (Fig. 36b). As a result, the PL of CDs was quenched. Then the PL sensing system could be used for evaluating the activity of protein kinase.

5.1.11 DNA Sensor

CDs have also been applied in the detection of DNA. Peng's group reduced CDs with $NaBH_4$ to prepare rCDs, which were subsequently connected with single-stranded DNA. The obtained ssDNA-rCDs probe was adsorbed on the surface of GO through electrostatic attractions and π-π stacking interactions, leading to substantial PL quenching due to FRET. In the presence of target DNA, ssDNA-rCDs were hybridized into double-stranded DNA-rCDs, which were then detached and liberated from GO (Fig. 37a). As a result, PL recovery was observed. The established method for DNA detection has a broad linear range of 6.7–46.0 nM with a detection limit of 75.0 pM [60]. At the same time, they tried to replace GO by carbon nanotubes as the PL quencher [258]. As expected, similar results were obtained.

5.1.12 Glutathione Detection

Shi and coworkers reported a dual-mode nanosensor with both colorimetric and fluorometric readuot based on CDs and gold nanoparticles (AuNPs) for discriminative detection of GSH in the presence of cysteine and homocysteine [259]. The used CDs had plenty of amine groups on the surfaces, while AuNPs have a large quantity of carboxylic groups. Accordingly, the addition of CDs into the AuNPs solution caused the aggregation of AuNPs and CDs, leading to an obvious color change of AuNPs (from red to blue) and PL quenching of CDs. However, in the presence of GSH, which showed strong affinity to AuNPs due to the multi-denstate anchor together with the specific steric structure existing in GSH, AuNPs could be

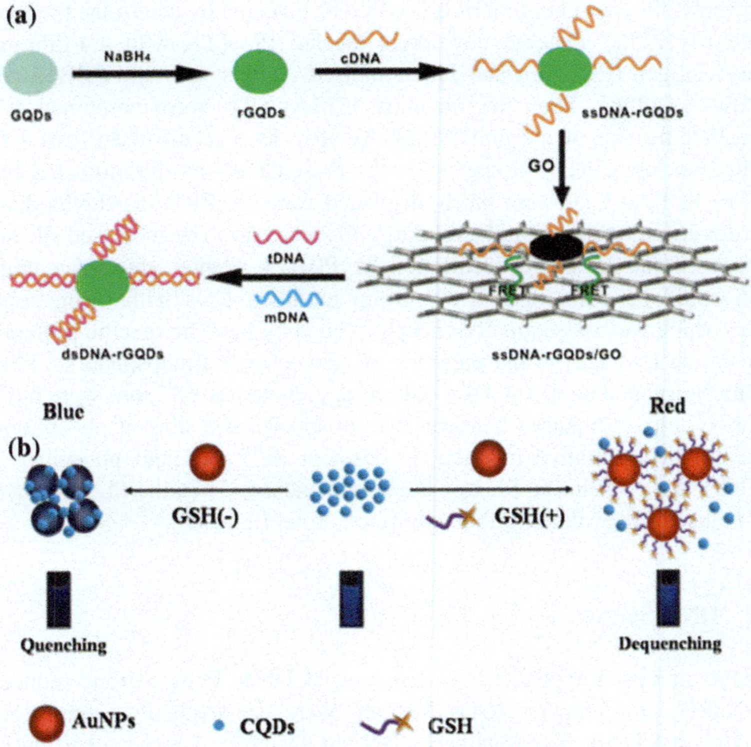

Fig. 37 **a** Schematic illustration of a universal fluorescence sensing platform for the detection of DNA based on fluorescence resonance energy transfer (FRET) between CDs and graphene oxide [60]. **b** Schematic principle of GSH detection by using the dual-mode nanosensor with both colorimetric and fluorometric readout [259]

protected from being aggregated and enlarge the inter-particle distance (Fig. 37b). Then, the color change and PL signal recovery happened.

5.1.13 Organic Pollutants Detection

Gayuela et al. [194], passivated the CDs synthesized by chemical oxidation of MWCNTs with acetone. The obtained passivated CDs were used for the detection of 2,4-dinitrophenol (DNP) at pH 3.5 and 2-amino-3,4,8-trimethyl-3H-imidazo [4,5-f]quinoxaline (4,8-DiMeIQx) at physiological pH. It was proposed that the protonated DNP might interact with oxygenated surfaces of CDs by hydrogen bond, causing PL quenching. Li et al. [260] developed a molecularly imprinted PL sensor for the determination of dimethoate based on the FRET between methyl red (MR) and CDs (Fig. 38). In brief, a molecularly templated polymer was prepared by electropolymerization of MR-doped *o*-phenylenediamine as a functional

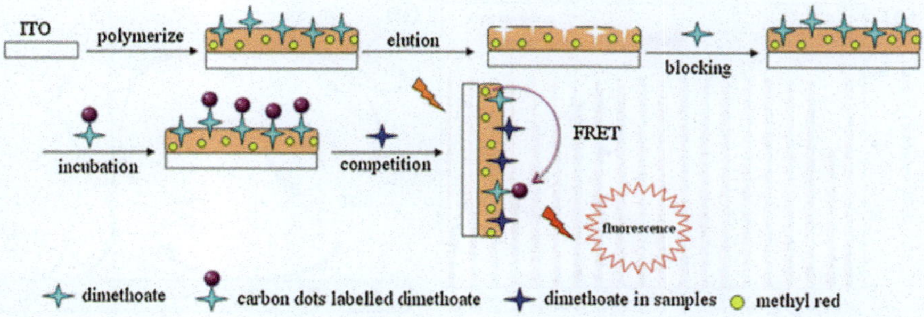

Fig. 38 Outline of the procedure followed to construct the MIP sensor and determination of dimethoate [260]

monomer and dimethoate as the template. After a careful elution with methanol, the molecularly imprinted polymer was subsequently immersed in dimethoate solution to mask all the vacant binding cavities and incubated in CDs-labeled dimethoate solution. FRET effect between CDs-labeled dimethoate and methyl red enhanced the PL signal of CDs. During dimethoate detection, a competitive reaction occurred between dimethoate and CDs-labeled dimethoate. Then the PL intensity decreased as the CDs-labeled dimethoate molecules were replaced by dimethoate molecules [260]. Dai et al. [99], developed a "turn-on" PL sensor for melamine based on melamine-induce decrease of the FRET efficiency between AuNPs and CDs. It was found that CDs would be prone to get close to the surfaces of AuNPs, resulting in the PL quenching. However, melamine could compete with CDs due to the contained amino groups, leading to the restoration of the PL.

5.2 *ECL Sensor*

5.2.1 Metal Ions Sensors

Zhu's group found that the ECL signal of CDs-$S_2O_8{}^{2-}$ could be quenched obviously by some metal ions, including Ni^{2+},Pb^{2+}, Cu^{2+}, Co^{2+}, Fe^{2+} and Cd^{2+}. It was proposed that the interaction between metal ions and the oxygen-containing functional groups such as hydroxyl and carboxyl groups of CDs led to the aggregation of CDs, and thereby the decrease of ECL emission [57]. The hypothesis was proved by the fact that the quenched ECL signal could be recovered completely by ethylenediaminetetraacetic acid (EDTA), which is a strong metal ion chelator. Subsequently, they used cysteine, a weaker chelator, as an effective masking agent for the determination of Cd^{2+} with acceptable selectivity. Dong et al. [219] have also carried a similar work. It was found that the ECL signal could be quenched by not only the metal ions mentioned in Zhu's work, but also Cr(VI) anion. The quenching effect of all the metal cations could be eliminated by the added EDTA (Fig. 39a). As a result,

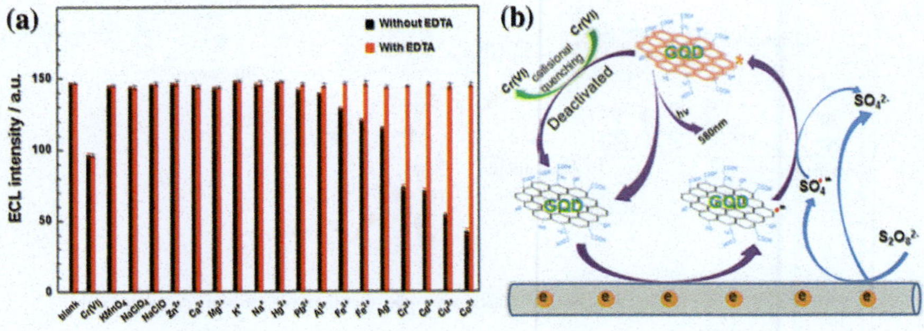

Fig. 39 **a** ECL responses of $CD/S_2O_8^{2-}$ system to Cr(VI) and other metal ions in the absence and presence of EDTA, and **b** ECL reaction mechanism of the $CD/S_2O_8^{2-}$ system in the presence of Cr (VI) [219]

a selective and sensitive ECL sensor was developed for the detection of Cr(VI) anion. As an anion, Cr(VI) seems impossible to make the negative charged CDs aggregated. Apparently, the quenching mechanism should be different from that proposed by Zhu. Therefore, the quenching mechanism was investigated. It was found that the quenching efficient dependent only on the concentration of Cr(VI), but independent on the concentrations of CDs and $S_2O_8^{2-}$. Therefore, the ECL emission should be quenched by Cr(VI) through a dynamic way rather than a static way (Fig. 39b). That is to say, the excited stated CDs (CD^*) were deactivated upon contact with Cr(VI), leading to the decrease of ECL emission. The hypothesis was further proved by the fact that the relationship between the quenching ratio and the concentration of Cr(VI) fits well with Stern-Volmer equation. Dong and coworkers developed another coreactant ECL system, namely CDs/L–Cys system, whose ECL emission could be selectively quenched by Pb^{2+} [211]. It was proposed that Pb^{2+} or its complexes with L–Cys could inhibit the formation of the free radicals ($RSO^{\bullet}$, $RSO_2^{\bullet}$, and $RSO_3^{\bullet}$) from the oxidation of L–Cys, then block the formation of CD^*. Accordingly, a sensitive and selective ECL sensor was developed for sensing Pb^{2+}, and have been applied in the dection of Pb^{2+} in a river water sample.

5.2.2 Biosensing

Yu's group developed a sensitive sandwich-type immunosensor for prostate protein antigen (PSA) detection based on the ECL of CDs [261]. First, 3D graphene modified with AuNPs (3D-GR@AuNPs) were fabricated on the surface of a glassy carbon electrode, then incubated with primary anti-PSA (Ab1). Second, CDs were mixed with nanoporous silver (NPS) and sonicated to form NPS@CDs composite, which was subsequently conjugated with the secondary anti-PSA (Ab_2). In the presence of PSA, NPS@CDs could be bind on the surface of the modified GCE due to the specific antibody-antigen interaction (Fig. 40a). Thus the concentration of PSA could be detect by measuring the ECL intensity of the modified GCE in the

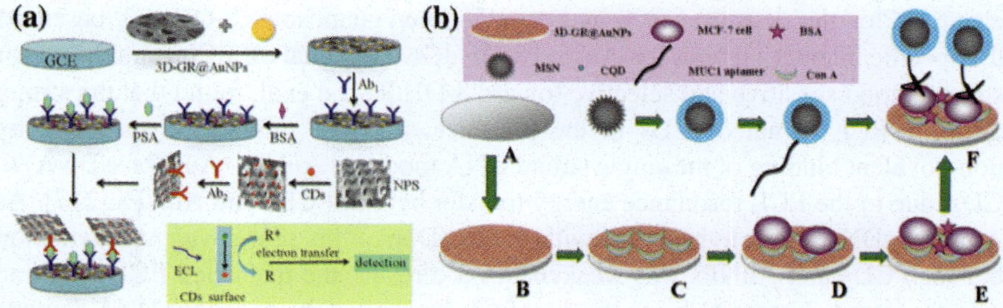

Fig. 40 **a** Schematic representation of the fabrication process of the PSA immunosensor [261]. **b** The fabrication process of the ECL cyto-sensor [216]

presence of $S_2O_8^{2-}$ as the coreactant. In one of their following-up works, a similar strategy was proposed for the detection of Michigan cancer foundation-7 (MCF-7) human breast cancer cells [216]. In brief, concanavalin A (Con-A, a lectin) was immobilized on the surface of the 3D-GR@AuNPs modified GCE, while CDs functionalized mesoporous silica nanoparticles (MSNs) was conjugated with DNA aptamers targeting mucin1 (MUC1). In the presence of MCF-7 cells, CDs@MSNs as the ECL luminophore were fabricated on the surface of the modified GCE, producing strong ECL emission in the presence of $S_2O_8^{2-}$ (Fig. 40b). The proposed method showed a good analytical performance for the detection of MCF-7 cancer cells ranging from 500 to 2×10^7 cells mL^{-1} with a detection limit of 230 cells mL^{-1}. In their another work, CDs were coated on ZnO nanoshpere to form ZnO@CDs hybrid and used for the detection leukemia cells, accordingly to a similar principle used in the detection of MCF-7 cells [214]. They have also proposed an ECL aptamer sensor for adenosine triphosphate (ATP) determination based on the ECL behaviors of CDs [218]. CDs contained abundant carboxyl groups were first covalently bind to the amino-functionalized SiO_2 nanoparticles to obtain SiO_2/CDs composites, which were subsequently incubated with an amino group functionalized single strand DNA (ssDNA2). At the same time, another thiol-functionalized single strand DNA (ssDNA1) was modified on the gold working electrode. Thus, SiO_2/CDs were captured by the modified electrode through the interaction among ssDNA1, ATP and ssDNA2, producing strong anodic ECL emission in the presence of H_2O_2. The developed signal on ECL aptamer sensor exhibited excellent analytical performance for ATP determination, ranging from 5.0×10^{-12} to 5.0×10^{-9} mol L^{-1} with the detection limit of 1.5×10^{-9} mol L^{-1}.

Zhang et al. immobilized carboxyl groups-containing CDs on the surface of amine-functionalized Au/SiO_2 core-shell nanoparticles to form Au/SiO_2/CDs hybrid, which was easily to be immobilized on the electrode surface, producing strong ECL signal in the presence of $S_2O_8^{2-}$ [213]. The ECL intensity would be decreased as the electron transfer on the electrode surface was hinder. Thus, rabbit anti-8-OHdG (Ab) was conjugated on the Au/SiO_2/CDs modified Pt working electrode for the detection of 8-hydroxyl-2′-deoxyguanosine (8-OHdG), based on the specific interaction between Ab and 8-OHdG. The signal off ECL immunosensor showed good sensitive and selective toward 8-OHdG. Lu et al. found that the strong cathodic ECL signal of CDs in the presence of $S_2O_8^{2-}$ could be quenched by non-covalent binding of the single strand DNA modified AuNPs (AuNPs-ssDNA) to CDs, due to the ECL resonance energy transfer between CDs and AuNPs [217]. As AuNPs-ssDNA was hybridized with target DNA, the non-covalent interaction between CDs and AuNPs was weakened. As a result, the quenched ECL emission was recovered. Accordingly, a signal on ECL sensor was proposed for DNA damage detection. Lou and coworker developed a signal off ECL sensor for DNA detection based on site-specific cleavage of *Bam*HI endonuclease [262]. Bidentate chelation of the dithiocarbamate DNA (DTC-DNA) was assembled on the clean gold electrode (GE) via bidentate chelation (S–Au–S bonds). Then CDs were bound to the DTC-DNA modified GE using ethylenediamine as the binder. After being hybridized with target DNA, the modified GE produced strong cathodic ECL emission, which would be decreased obviously as the double strand DNA was cleaved by *Bam*HI endonuclease. The signal off ECL sensor exhibited a linear range from 5 fM to 100 pM with a detection limit of 0.45 fM.

5.2.3 Pentachlorophenol Detection

Yang et al. [215] immobilized CDs on graphene to form CDs/GR hybrid, which was readily to be fabricated on titanium ribbon working electrode, produced a strong cathodic ECL signal in the presence $S_2O_8^{2-}$. The ECL system was used for the detection of pentachlorophenol (PCP), which was considered to be able to react with the excited-state CDs (Fig. 41). The sensing performance were excellent, with

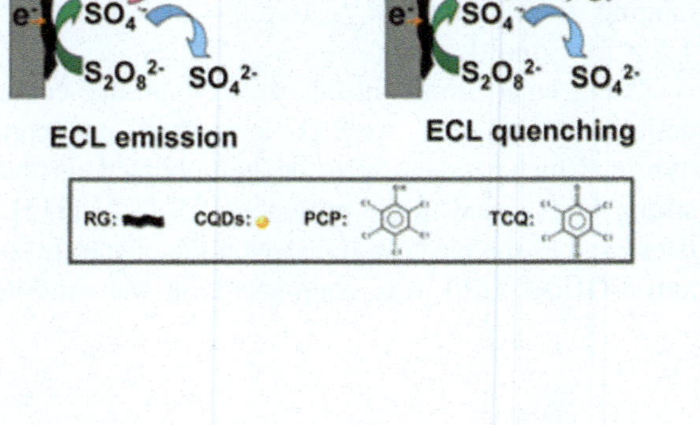

Fig. 41 Illustrative ECL Detection Mechanism for PCP with CDs/GR in $S_2O_8^{2-}$ Solution [215]

a wide linear range from 1.0×10^{-12} to 1.0×10^{-8} mol L^{-1}, and a detection limit of 1.0×10^{-12} mol L^{-1}.

5.3 CL Sensor

Lin's group observed that the mixing of nitrite and acidified hydrogen peroxide could produce peroxynitrous acid, which could further react with CDs to produce strong CL emission [135]. Accordingly, a sensitive CL sensor has been developed for the detection of nitrites. The CL intensity of the developed CL sensor increased linearly with nitrite concentration in the range from 1.0×10^{-7} to 1.0×10^{-5} M, with a detection limit of 5.3×10^{-8} M. Furthermore, the established method has been successfully applied for the determination of nitrite in pond water, river water, and milk with good recovery and high reproducibility.

Shi et al. [263] developed a selective CL sensing system for the detection of Co^{2+}. It was proposed that the surface functional groups (PEG and, then cetyltrimethylammonium bromide, CTAB) in the used CDs could adsorb the Co^{2+} ions, which could catalyze the dissociation of H_2O_2 through a Fenton-like reaction. As a result, abundant $OH^{\bullet}$ radical was produced around the surface of CDs. In alkaline solutions, $OH^{\bullet}$ radical was changed into $O_2^{\bullet -}$ radical, which would be recombined into energy-rich excited singlet oxygen dimol species $(O_2)_2^*$. The energy transfer between $(O_2)_2^*$ and CDs resulted in the formation of CDs^*. Then a strong CL emission could be observed during CDs^* returned to the ground state (Fig. 42). The proposed CL

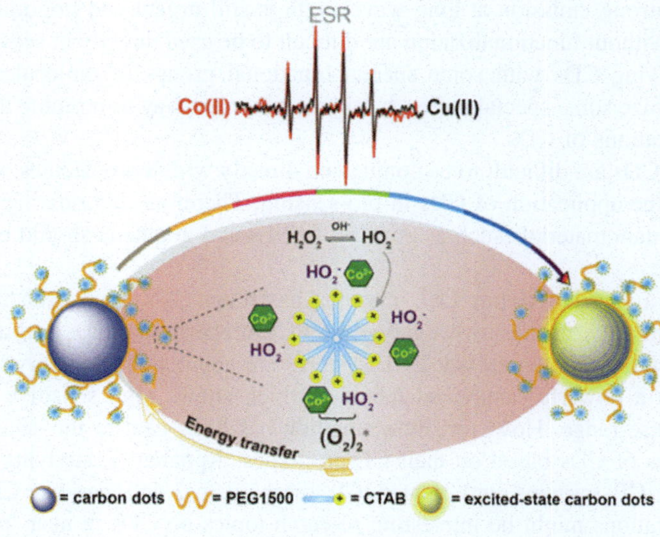

Fig. 42 Possible CL mechanism for CTAB@CD–Co(II)–H_2O_2–OH—system [263]

system exhibited a stable response to Co(II) over a concentration range from 1.0 to 1000 nM with a detection limit as low as 0.67 nM.

Amjadi observed the CL of CDs caused by the oxidation of Ce(IV) [264]. Furthermore, they found that uric acid could quench the CL of CDs/Ce(IV) system due to the competition of uric acid with CDs for Ce(IV). The CL intensity decreased linear as the concentration of uric acid was increased in the range of 1.0×10^{-6} to 1.0×10^{-4} M.

Dong et al. [196] synthesized hydrazide-modified CDs (HM-CDs), which could produce strong CL emission in alkaline solutions. The CL intensity increased dramatically with the pH value in the range from pH 8–11. There was a good semilogarithmic correlation between the CL intensity and the pH value, suggesting that the CL intensity is proportional to the concentration of OH^-. Furthermore, the CL intensity was related to the concentration of the dissolved oxygen (O_2), suggesting that the HM-CDs could be applied in O_2 sensing.

5.4 Summary

Although great attention has been paid to the sensing applications based on the luminescent properties of CDs, the applications of CDs in sensors are still on the initial stage. To further promote the sensing applications of CDs, efforts can be made from the following aspects:

(a) Most of the reported high luminescent CDs emit only blue light, which is quite unfavorable for intracellular sensing of CDs. Therefore, synthesizing CDs with strong emission at long wavelength is still urgent and important.
(b) CDs without functionalization are difficult to be used directly in sensing. Thus, modifying CDs with some special functional groups or molecules that can recognize some special targets is still an important way to promote the sensing applications of CDs.
(c) Most CDs are difficult to be conjugated directly with biomolecules, which will limit the application of CDs in bio-sensing. Therefore, hybridizing CDs with other nanomaterials such as AuNPs, AgNPs and graphene should be an ideal choice.
(d) Only a few coreactant ECL systems based on CDs have been developed. Accordingly, further investigating the ECL properties of CDs and developing more CD-based coreactant ECL systems are apparently necessary.
(e) CL is a powerful analytical method, which shows high sensitive and wide response range. However, little attention has been paid to the sensing applications of CDs based on their CL properties. Apparently, studying new CDs based CL systems and related CL mechanisms, and various CL sensing applications might be interesting research topics on CDs in near future.

References

1. S.N. Baker, G.A. Baker, Luminescent carbon nanodots: emergent nanolights. Angew. Chem. Int. Ed. **49**(38), 6726–6744 (2010). doi:10.1002/anie.200906623
2. Y. Xu, J. Liu, C. Gao, E. Wang, Applications of carbon quantum dots in electrochemiluminescence: a mini review. Electrochem. Commun. **48**, 151–154 (2014). doi:10.1016/j.elecom.2014.08.032
3. X. Yan, B.S. Li, L.S. Li, Colloidal graphene quantum dots with well-defined structures. Acc. Chem. Res. **46**(10), 2254–2262 (2013). doi:10.1021/ar300137p
4. H.J. Sun, L. Wu, W.L. Wei, X.G. Qu, Recent advances in graphene quantum dots for sensing. Mater. Today **16**(11), 433–442 (2013). doi:10.1016/j.mattod.2013.10.020
5. M. Bacon, S.J. Bradley, T. Nann, Graphene quantum dots. Part. Part. Syst. Char. **31**(4), 415–428 (2014). doi:10.1002/ppsc.201300252
6. L. Cao, M.J. Meziani, S. Sahu, Y.P. Sun, Photoluminescence properties of graphene versus other carbon nanomaterials. Acc. Chem. Res. **46**(1), 171–180 (2012). doi:10.1021/ar300128j
7. H.T. Li, Z.H. Kang, Y. Liu, S.T. Lee, Carbon nanodots: synthesis, properties and applications. J. Mater. Chem. **22**(46), 24230–24253 (2012). doi:10.1039/c2jm34690g
8. X.Y. Xu, R. Ray, Y.L. Gu, H.J. Ploehn, L. Gearheart, K. Raker, W.A. Scrivens, Electrophoretic analysis and purification of fluorescent single-walled carbon nanotube fragments. J. Am. Chem. Soc. **126**(40), 12736–12737 (2004). doi:10.1021/ja040082h
9. Y.P. Sun, B. Zhou, Y. Lin, W. Wang, K.A.S. Fernando, P. Pathak, M.J. Meziani, B.A. Harruff, X. Wang, H.F. Wang, Quantum-sized carbon dots for bright and colorful photoluminescence. J. Am. Chem. Soc. **128**(24), 7756–7757 (2006). doi:10.1021/ja062677d
10. A.B. Bourlinos, R. Zbořil, J. Petr, A. Bakandritsos, M. Krysmann, E.P. Giannelis, Luminescent surface quaternized carbon dots. Chem. Mater. **24**(1), 6–8 (2012). doi:10.1021/cm2026637
11. F. Wang, S. Pang, L. Wang, Q. Li, M. Kreiter, C.Y. Liu, One-step synthesis of highly luminescent carbon dots innoncoordinating solvents. Chem. Mater. **22**(16), 4528–4530 (2010). doi:10.1021/cm101350u
12. X. Chen, W. Zhang, Q. Wang, J. Fan, C8-structured carbon quantum dots: synthesis, blue and green double luminescence, and origins of surface defects. Carbon **79**, 165–173 (2014). doi:10.1016/j.carbon.2014.07.056
13. S. Chandra, S.H. Pathan, S. Mitra, B.H. Modha, A. Goswami, P. Pramanik, Tuning of photoluminescence on different surface functionalized carbon quantum dots. RSC Adv. **2**(9), 3602–3606 (2012). doi:10.1039/c2ra00030j
14. X.M. Wen, P. Yu, Y.R. Toh, X.Q. Ma, J. Tang, On the upconversion fluorescence in carbon nanodots and graphene quantum dots. Chem. Commun. **50**(36), 4703–4706 (2014). doi:10.1039/c4cc01213e
15. A.B. Bourlinos, A. Stassinopoulos, D. Anglos, R. Zboril, V. Georgakilas, E.P. Giannelis, Photoluminescent carbogenic dots. Chem. Mater. **20**(14), 4539–4541 (2008). doi:10.1021/cm800506r
16. A.B. Bourlinos, A. Stassinopoulos, D. Anglos, R. Zboril, M. Karakassides, E.P. Giannelis, Surface functionalized carbogenic quantum dots. Small **4**(4), 455–458 (2008). doi:10.1002/smll.200700578
17. Y.F. Wang, A.G. Hu, Carbon quantum dots: synthesis, properties and applications. J. Mater. Chem. C **2**(34), 6921–6939 (2014). doi:10.1039/c4tc00988f
18. Y.B. Song, S.J. Zhu, B. Yang, Bioimaging based on fluorescent carbon dots. RSC Adv. **4**(52), 27184–27200 (2014). doi:10.1039/c3ra47994c
19. K. Hola, Y. Zhang, Y. Wang, E.P. Giannelis, R. Zboril, A.L. Rogach, Carbon dots—emerging light emitters for bioimaging, cancer therapy and optoelectronics. Nano Today **9**(5), 590–603 (2014). doi:10.1016/j.nantod.2014.09.004

20. J.C.G. Esteves da Silva, H.M.R. Gonçalves, Analytical and bioanalytical applications of carbon dots. TrAC Trends Anal. Chem. **30**(8), 1327–1336 (2011). doi:10.1016/j.trac.2011.04.009
21. L.A. Ponomarenko, F. Schedin, M.I. Katsnelson, R. Yang, E.W. Hill, K.S. Novoselov, A.K. Geim, Chaotic dirac billiard in graphene quantum dots. Science **320**(5874), 356–358 (2008). doi:10.1126/science.1154663
22. D. Pan, J. Zhang, Z. Li, M. Wu, Hydrothermal route for cutting graphene sheets into blue-luminescent graphene quantum dots. Adv. Mater. **22**(6), 734–738 (2010). doi:10.1002/adma.200902825
23. S. Zhu, S. Tang, J. Zhang, B. Yang, Control the size and surface chemistry of graphene for the rising fluorescent materials. Chem. Commun. **48**(38), 4527–4539 (2012). doi:10.1039/c2cc31201h
24. X. Zhou, S. Guo, J. Zhang, Solution-processable graphene quantum dots. ChemPhysChem **14**(12), 2627–2640 (2013). doi:10.1002/cphc.201300111
25. Z. Zhang, J. Zhang, N. Chen, L. Qu, Graphene quantum dots: an emerging material for energy-related applications and beyond. Energ. Environ. Sci. **5**(10), 8869–8895 (2012). doi:10.1039/c2ee22982j
26. X. Wang, G. Sun, P. Routh, D.H. Kim, W. Huang, P. Chen, Heteroatom-doped graphene materials: syntheses, properties and applications. Chem. Soc. Rev. **43**(20), 7067–7098 (2014). doi:10.1039/c4cs00141a
27. J. Shen, Y. Zhu, X. Yang, C. Li, Graphene quantum dots: emergent nanolights for bioimaging, sensors, catalysis and photovoltaic devices. Chem. Commun. **48**(31), 3686–3699 (2012). doi:10.1039/c2cc00110a
28. L. Lin, M. Rong, F. Luo, D. Chen, Y. Wang, X. Chen, Luminescent graphene quantum dots as new fluorescent materials for environmental and biological applications. TrAC Trends Anal. Chem. **54**, 83–102 (2014). doi:10.1016/j.trac.2013.11.001
29. L. Li, G. Wu, G. Yang, J. Peng, J. Zhao, J.J. Zhu, Focusing on luminescent graphene quantum dots: current status and future perspectives. Nanoscale **5**(10), 4015–4039 (2013). doi:10.1039/c3nr33849e
30. J.L. Li, B. Tang, B. Yuan, L. Sun, X.G. Wang, A review of optical imaging and therapy using nanosized graphene and graphene oxide. Biomaterials **34**(37), 9519–9534 (2013). doi:10.1016/j.biomaterials.2013.08.066
31. J. Guttinger, F. Molitor, C. Stampfer, S. Schnez, A. Jacobsen, S. Droscher, T. Ihn, K. Ensslin, Transport through graphene quantum dots. Rep. Prog. Phys. **75**(12), 126502 (2012). doi:10.1088/0034-4885/75/12/126502
32. S.L. Hu, K.Y. Niu, J. Sun, J. Yang, N.Q. Zhao, X.W. Du, One-step synthesis of fluorescent carbon nanoparticles by laser irradiation. J. Mater. Chem. **19**(4), 484–488 (2009). doi:10.1039/b812943f
33. X. Wang, L. Cao, F. Lu, M.J. Meziani, H. Li, G. Qi, B. Zhou, B.A. Harruff, F. Kermarrec, Y. P. Sun, Photoinduced electron transfers with carbon dots. Chem. Commun. **25**, 3774–3776 (2009). doi:10.1039/b906252a
34. S.T. Yang, X. Wang, H.F. Wang, F.S. Lu, P.J.G. Luo, L. Cao, M.J. Meziani, J.H. Liu, Y.F. Liu, M. Chen, Carbon dots as nontoxic and high-performance fluorescence imaging agents. J. Phys. Chem. C **113**(42), 18110–18114 (2009). doi:10.1021/jp9085969
35. S.T. Yang, L. Cao, P.G.J. Luo, F.S. Lu, X. Wang, H.F. Wang, M.J. Meziani, Y.F. Liu, G. Qi, Y.P. Sun, Carbon dots for optical imaging in vivo. J. Am. Chem. Soc. **131**(32), 11308–11309 (2009). doi:10.1021/ja904843x
36. Q. Li, T.Y. Qhulchanskyy, R.L. Liu, K. Koynov, D.Q. Wu, A. Best, R. Kumar, A. Bonoiu, P. N. Prasad, Photoluminescent carbon dots as biocompatible nanoprobes for targeting cancer cells. J. Phys. Chem. C **114**(28), 12062–12068 (2010). doi:10.1021/jp911539r
37. L. Cao, X. Wang, M.J. Meziani, F.S. Lu, H.F. Wang, P.J.G. Luo, Y. Lin, B.A. Harruff, L.M. Veca, D. Murray, Carbon dots for nultiphoton bioimaging. J. Am. Chem. Soc. **129**(37), 11318–11319 (2007). doi:10.1021/ja073527l

38. H. Liu, T. Ye, C. Mao, Fluorescent carbon nanoparticles derived from candle soot. Angew. Chem. Int. Ed. **46**(34), 6473–6475 (2007). doi:10.1002/anie.200701271
39. S.C. Ray, A. Saha, N.R. Jana, R. Sarkar, Fluorescent carbon nanoparticles: synthesis, characterization, and bioimaging application. J. Phys. Chem. C **113**(43), 18546–18551 (2009). doi:10.1021/jp905912n
40. X.H. Wang, K.G. Qu, B.L. Xu, J.S. Ren, X.G. Qu, Multicolor luminescent carbon nanoparticles: synthesis, supramolecular assembly with porphyrin, intrinsic peroxidase-like catalytic activity and applications. Nano Res. **4**(9), 908–920 (2011). doi:10.1007/s12274-011-0147-4
41. M. Tan, L. Zhang, R. Tang, X. Song, Y. Li, H. Wu, Y. Wang, G. Lv, W. Liu, X. Ma, Enhanced photoluminescence and characterization of multicolor carbon dots using plant soot as a carbon source. Talanta **115**, 950–956 (2013). doi:10.1016/j.talanta.2013.06.061
42. L. Tian, D. Ghosh, W. Chen, S. Pradhan, X.J. Chang, S.W. Chen, Nanosized carbon particles from natural gas soot. Chem. Mater. **21**(13), 2803–2809 (2009). doi:10.1021/cm900709w
43. J.M. Berlin, T.T. Pham, D. Sano, K.A. Mohamedali, D.C. Marcano, J.N. Myers, J.M. Tour, Noncovalent functionalization of carbon nanovectors with an antibody enables targeted drug delivery. ACS Nano **5**(8), 6643–6650 (2011). doi:10.1021/nn2021293
44. Y. Dong, N. Zhou, X. Lin, J. Lin, Y. Chi, G. Chen, Extraction of electrochemiluminescent oxidized carbon quantum dots from activated carbon. Chem. Mater. **22**(21), 5895–5899 (2010). doi:10.1021/cm1018844
45. Y. Dong, C. Chen, X. Zheng, L. Gao, Z. Cui, H. Yang, C. Guo, Y. Chi, C.M. Li, One-step and high yield simultaneous preparation of single- and multi-layer graphene quantum dots from CX-72 carbon black. J Mater Chem C **22**(18), 8764–8766 (2012). doi:10.1039/c2jm30658a
46. Y. Dong, J. Lin, Y. Chen, F. Fu, Y. Chi, G. Chen, Graphene quantum dots, graphene oxide, carbon quantum dots and graphite nanocrystals in coals. Nanoscale **6**(13), 7410–7415 (2014). doi:10.1039/c4nr01482k
47. Z.A. Qiao, Y. Wang, Y. Gao, H. Li, T. Dai, Y. Liu, Q. Huo, Commercially activated carbon as the source for producing multicolor photoluminescent carbon dots by chemical oxidation. Chem. Commun. **46**(46), 8812–8814 (2010). doi:10.1039/c0cc02724c
48. Y. Dong, C.X. Guo, Y. Chi, C.M. Li, Reply to comment on "one-step and high yield simultaneous preparation of single- and multi-layer graphene quantum dots from CX-72 carbon black". J. Mater. Chem. **22**(40), 21777–21778 (2012). doi:10.1039/c2jm34130a
49. P. Teng, J. Xie, Y. Long, X. Huang, R. Zhu, X. Wang, L. Liang, Y. Huang, H. Zheng, Chemiluminescence behavior of the carbon dots and the reduced state carbon dots. J. Lumin. **146**, 464–469 (2014). doi:10.1016/j.jlumin.2013.09.036
50. R. Liu, D. Wu, X. Feng, K. Mullen, Bottom-up fabrication of photoluminescent graphene quantum dots with uniform morphology. J. Am. Chem. Soc. **133**(39), 15221–15223 (2011). doi:10.1021/ja204953k
51. W.S. Hummers, R.E. Offeman, Preparation of graphitic oxide. J. Am. Chem. Soc. **80**(6), 1339 (1958). doi:10.1021/ja01539a017
52. J. Peng, W. Gao, B.K. Gupta, Z. Liu, R. Romero-Aburto, L. Ge, L. Song, L.B. Alemany, X. Zhan, G. Gao, S.A. Vithayathil, B.A. Kaipparettu, A.A. Marti, T. Hayashi, J.J. Zhu, P.M. Ajayan, Graphene quantum dots derived from carbon fibers. Nano Lett. **12**(2), 844–849 (2012). doi:10.1021/nl2038979
53. D.V. Kosynkin, A.L. Higginbotham, A. Sinitskii, J.R. Lomeda, A. Dimiev, B.K. Price, J.M. Tour, Longitudinal unzipping of carbon nanotubes to form graphene nanoribbons. Nature **458**(7240), 872–876 (2009). doi:10.1038/nature07872
54. S.J. Zhuo, M.W. Shao, S.T. Lee, Upconversion and downconversion gluorescent graphene quantum dots. ACS Nano **6**(2), 1059–1064 (2012). doi:10.1021/nn2040395
55. D. Pan, L. Guo, J. Zhang, C. Xi, Q. Xue, H. Huang, J. Li, Z. Zhang, W. Yu, Z. Chen, Z. Li, M. Wu, Cutting sp^2 clusters in graphene sheets into colloidal graphene quantum dots with strong green fluorescence. J. Mater. Chem. **22**(8), 3314–3318 (2012). doi:10.1039/c2jm16005f

56. S. Kim, S.W. Hwang, M.K. Kim, D.Y. Shin, D.H. Shin, C.O. Kim, S.B. Yang, J.H. Park, E. Hwang, S.H. Choi, Anomalous behaviors of visible luminescence from graphene quantum dots: interplay between size and shape. ACS Nano **6**(9), 8203–8208 (2012). doi:10.1021/nn302878r
57. L.L. Li, J. Ji, R. Fei, C.Z. Wang, Q. Lu, J.R. Zhang, L.P. Jiang, J.J. Zhu, A facile microwave avenue to electrochemiluminescent two-color graphene quantum dots. Adv. Funct. Mater. **22**(14), 2971–2979 (2012). doi:10.1002/adfm.201200166
58. M. Nurunnabi, Z. Khatun, K.M. Huh, S.Y. Park, D.Y. Lee, K.J. Cho, Y.K. Lee, In vivo biodistribution and toxicology of carboxylated graphene quantum dots. ACS Nano **7**(8), 6858–6867 (2013). doi:10.1021/nn402043c
59. M. Nurunnabi, Z. Khatun, M. Nafiujjaman, D.G. Lee, Y.K. Lee, Surface coating of graphene quantum dots using mussel-inspired polydopamine for biomedical optical imaging. ACS Appl. Mater. Inter. **5**(16), 8246–8253 (2013). doi:10.1021/am4023863
60. Z.S. Qian, X.Y. Shan, L.J. Chai, J.J. Ma, J.R. Chen, H. Feng, A universal fluorescence sensing strategy based on biocompatible graphene quantum dots and graphene oxide for the detection of DNA. Nanoscale **6**(11), 5671–5674 (2014). doi:10.1039/c3nr06583a
61. Y. Sun, S. Wang, C. Li, P. Luo, L. Tao, Y. Wei, G. Shi, Large scale preparation of graphene quantum dots from graphite with tunable fluorescence properties. Phys. Chem. Chem. Phys. **15**(24), 9907–9913 (2013). doi:10.1039/c3cp50691f
62. R. Ye, C. Xiang, J. Lin, Z. Peng, K. Huang, Z. Yan, N.P. Cook, E.L. Samuel, C.C. Hwang, G. Ruan, G. Ceriotti, A.R. Raji, A.A. Marti, J.M. Tour, Coal as an abundant source of graphene quantum dots. Nat. Commun. **4**, 2943 (2013). doi:10.1038/ncomms3943
63. M. Wu, Y. Wang, W. Wu, C. Hu, X. Wang, J. Zheng, Z. Li, B. Jiang, J. Qiu, Preparation of functionalized water-soluble photoluminescent carbon quantum dots from petroleum coke. Carbon **78**, 480–489 (2014). doi:10.1016/j.carbon.2014.07.029
64. X.J. Zhou, Y. Zhang, C. Wang, X.C. Wu, Y.Q. Yang, B. Zheng, H.X. Wu, S.W. Guo, J.Y. Zhang, Photo-fenton reaction of graphene oxide: a new strategy to prepare graphene quantum dots for DNA cleavage. ACS Nano **6**(8), 6592–6599 (2012)
65. K. Ikehata, M.G. EI-Din, Aqueous pesticide degradation by hydrogen peroxide ultraviolet irradiation and Fenton-type advanced oxidation processes: a review. Environ. Eng. Sci. **5**(2), 81–135 (2006). doi:10.1139/S05-046
66. J.G. Zhou, C. Booker, R.Y. Li, X.T. Zhou, T.K. Sham, X.L. Sun, Z.F. Ding, An electrochemical avenue to blue luminescent nanocrystals from multiwalled carbon nanotubes (MWCNTs). J. Am. Chem. Soc. **129**(4), 744–745 (2007). doi:10.1021/ja0669070
67. Q.L. Zhao, Z.L. Zhang, B.H. Huang, J. Peng, M. Zhang, D.W. Pang, Facile preparation of low cytotoxicity fluorescent carbon nanocrystals by electrooxidation of graphite. Chem. Commun. **41**, 5116–5118 (2008). doi:10.1039/b812420e
68. L. Zheng, Y. Chi, Y. Dong, J. Lin, B. Wang, Electrochemiluminescence of water-soluble carbon nanocrystals released electrochemically from graphite. J. Am. Chem. Soc. **131**(13), 4564–4565 (2009). doi:10.1021/ja809073f
69. H. Li, X. He, Z. Kang, H. Huang, Y. Liu, J. Liu, S. Lian, C.H. Tsang, X. Yang, S.T. Lee, Water-soluble fluorescent carbon quantum dots and photocatalyst design. Angew. Chem. Int. Ed. **49**(26), 4430–4434 (2010). doi:10.1002/anie.200906154
70. J. Lu, J.X. Yang, J.Z. Wang, A.L. Lim, S. Wang, K.P. Loh, One-pot synthesis of fluorescent carbon nanoribbons, nanoparticles, and graphene by the exfoliation of graphite in ionic liquids. ACS Nano **3**(8), 2367–2375 (2009). doi:10.1021/nn900546b
71. Y. Li, Y. Hu, Y. Zhao, G. Shi, L. Deng, Y. Hou, L. Qu, An electrochemical avenue to green-luminescent graphene quantum dots as potential electron-acceptors for photovoltaics. Adv. Mater. **23**(6), 776–780 (2011). doi:10.1002/adma.201003819
72. Y. Li, Y. Zhao, H. Cheng, Y. Hu, G. Shi, L. Dai, L. Qu, Nitrogen-doped graphene quantum dots with oxygen-rich functional groups. J. Am. Chem. Soc. **134**(1), 15–18 (2012). doi:10.1021/ja206030c

73. L. Bao, Z.L. Zhang, Z.Q. Tian, L. Zhang, C. Liu, Y. Lin, B. Qi, D.W. Pang, Electrochemical tuning of luminescent carbon nanodots: from preparation to luminescence mechanism. Adv. Mater. **23**(48), 5801–5806 (2011). doi:10.1002/adma.201102866
74. M. Zhang, L. Bai, W. Shang, W. Xie, H. Ma, Y. Fu, D. Fang, H. Sun, L. Fan, M. Han, C. Liu, S. Yang, Facile synthesis of water-soluble, highly fluorescent graphene quantum dots as a robust biological label for stem cells. J. Am. Chem. Soc. **22**(15), 7461–7467 (2012). doi:10.1039/c2jm16835a
75. D.B. Shinde, V.K. Pillai, Electrochemical resolution of multiple redox events for graphene quantum dots. Angew. Chem. Int. Ed. **52**(9), 2482–2485 (2013). doi:10.1002/anie.201208904
76. D.B. Shinde, V.K. Pillai, Electrochemical preparation of luminescent graphene quantum dots from multiwalled carbon nanotubes. Chem. Eur. J. **18**(39), 12522–12528 (2012). doi:10.1002/chem.201201043
77. D.V. Kosynkin, A.L. Higginboham, A. Sinitskii, J.R. Lomeda, A. Dimiev, B.K. Price, J.M. Tour, Longitudinal unzipping of carbon nanotubes to form graphene nanoribbons. Nature **458**(7240), 872–876 (2009). doi:10.1038/nature07872
78. Y. Dong, H. Pang, S. Ren, C. Chen, Y. Chi, T. Yu, Etching single-wall carbon nanotubes into green and yellow single-layer graphene quantum dots. Carbon **64**, 245–251 (2013). doi:10.1016/j.carbon.2013.07.059
79. F. Yang, M. Zhao, B. Zheng, D. Xiao, L. Wu, Y. Guo, Influence of pH on the fluorescence properties of graphene quantum dots using ozonation pre-oxide hydrothermal synthesis. J. Mater. Chem. **22**(48), 25471–25479 (2012). doi:10.1039/c2jm35471c
80. H. Tetsuka, R. Asahi, A. Nagoya, K. Okamoto, I. Tajima, R. Ohta, A. Okamoto, Optically tunable amino-functionalized graphene quantum dots. Adv. Mater. **24**(39), 5333–5338 (2012). doi:10.1002/adma.201201930
81. S. Zhu, J. Zhang, C. Qiao, S. Tang, Y. Li, W. Yuan, B. Li, L. Tian, F. Liu, R. Hu, H. Gao, H. Wei, H. Zhang, H. Sun, B. Yang, Strongly green-photoluminescent graphene quantum dots for bioimaging applications. Chem. Commun. **47**(24), 6858–6860 (2011). doi:10.1039/c1cc11122a
82. L. Lin, S. Zhang, Creating high yield water soluble luminescent graphene quantum dots via exfoliating and disintegrating carbon nanotubes and graphite flakes. Chem. Commun. **48**(82), 10177–10179 (2012). doi:10.1039/c2cc35559k
83. X. Zhu, H. Wang, Q. Jiao, X. Xiao, X. Zuo, Y. Liang, J. Nan, J. Wang, L. Wang, Preparation and characterization of the fluorescent carbon dots derived from the lithium-Intercalated graphite used for cell Imaging. Part. Part. Syst. Char. **31**(7), 771–777 (2014). doi:10.1002/ppsc.201300327
84. J. Lee, K. Kim, W.I. Park, B.H. Kim, J.H. Park, T.H. Kim, S. Bong, C.H. Kim, G. Chae, M. Jun, Y. Hwang, Y.S. Jung, S. Jeon, Uniform graphene quantum dots patterned from self-assembled silica nanodots. Nano Lett. **12**(12), 6078–6083 (2012). doi:10.1021/nl302520m
85. H. Li, X. He, Y. Liu, H. Yu, Z. Kang, S.-T. Lee, Synthesis of fluorescent carbon nanoparticles directly from active carbon via a one-step ultrasonic treatment. Mater. Res. Bull. **46**(1), 147–151 (2011). doi:10.1016/j.materresbull.2010.10.013
86. F. Liu, M.H. Jang, H.D. Ha, J.H. Kim, Y.H. Cho, T.S. Seo, Facile synthetic method for pristine graphene quantum dots and graphene oxide quantum dots: origin of blue and green luminescence. Adv. Mater. **25**(27), 3657–3662 (2013). doi:10.1002/adma.201300233doi:10.1002/adma.201300233
87. Y. Dong, H. Pang, H.B. Yang, C. Guo, J. Shao, Y. Chi, C.M. Li, T. Yu, Carbon-based dots co-doped with nitrogen and sulfur for high quantum yield and excitation-independent emission. Angew. Chem. Int. Ed. **52**(30), 7800–7804 (2013). doi:10.1002/anie.201301114
88. S. Zhu, Q. Meng, L. Wang, J. Zhang, Y. Song, H. Jin, K. Zhang, H. Sun, H. Wang, B. Yang, Highly photoluminescent carbon dots for multicolor patterning, sensors, and bioimaging. Angew. Chem. Int. Ed. **52**(14), 3953–3957 (2013). doi:10.1002/anie.201300519

89. M.J. Krysmann, A. Kelarakis, P. Dallas, E.P. Giannelis, Formation mechanism of carbogenic nanoparticles with dual photoluminescence emission. J. Am. Chem. Soc. **134**(2), 747–750 (2012). doi:10.1021/ja204661r
90. D. Qu, M. Zheng, P. Du, Y. Zhou, L. Zhang, D. Li, H. Tan, Z. Zhao, Z. Xie, Z. Sun, Highly luminescent S, N co-doped graphene quantum dots with broad visible absorption bands for visible light photocatalysts. Nanoscale **5**(24), 12272–12277 (2013). doi:10.1039/c3nr04402e
91. H. Zhang, Y. Chen, M. Liang, L. Xu, S. Qi, H. Chen, X. Chen, Solid-phase synthesis of highly fluorescent nitrogen-doped carbon dots for sensitive and selective probing ferric ions in living cells. Anal. Chem. **86**(19), 9846–9852 (2014). doi:10.1021/ac502446m
92. M. Amjadi, J.L. Manzoori, T. Hallaj, M.H. Sorouraddin, Direct chemiluminescence of carbon dots induced by potassium ferricyanide and its analytical application. Spectrochimi. Acta A **122**, 715–720 (2014). doi:10.1016/j.saa.2013.11.097
93. Y. Guo, Z. Wang, H. Shao, X. Jiang, Hydrothermal synthesis of highly fluorescent carbon nanoparticles from sodium citrate and their use for the detection of mercury ions. Carbon **52**, 583–589 (2013). doi:10.1016/j.carbon.2012.10.028
94. K. Qu, J. Wang, J. Ren, X. Qu, Carbon dots prepared by hydrothermal treatment of dopamine as an effective fluorescent sensing platform for the label-free detection of iron(III) ions and dopamine. Chem. Eur. J. **19**(22), 7243–7249 (2013). doi:10.1002/chem.20130004295
95. P. Shen, Y. Xia, Synthesis-modification integration: one-step fabrication of boronic acid functionalized carbon dots for fluorescent blood sugar sensing. Anal. Chem. **86**(11), 5323–5329 (2014). doi:10.1021/ac5001338
96. J. Wei, X. Zhang, Y. Sheng, J. Shen, P. Huang, S. Guo, J. Pan, B. Feng, Dual functional carbon dots derived from cornflour via a simple one-pot hydrothermal route. Mater. Lett. **123**, 107–111 (2014). doi:10.1016/j.matlet.2014.02.090
97. Z. Qian, J. Ma, X. Shan, H. Feng, L. Shao, J. Chen, Highly luminescent N-doped carbon quantum dots as an effective multifunctional fluorescence sensing platform. Chem. Eur. J. **20** (8), 2254–2263 (2014). doi:10.1002/chem.201304374
98. R. Zhang, W. Chen, Nitrogen-doped carbon quantum dots: facile synthesis and application as a “turn-off” fluorescent probe for detection of Hg^{2+} ions. Biosens. Bioelecton. **55**, 83–90 (2014). doi:10.1016/j.bios.2013.11.074
99. H. Dai, Y. Shi, Y. Wang, Y. Sun, J. Hu, P. Ni, Z. Li, A carbon dot based biosensor for melamine detection by fluorescence resonance energy transfer. Sens. Actuat. B Chem. **202**, 201–208 (2014). doi:10.1016/j.snb.2014.05.058
100. A. Sachdev, I. Matai, P. Gopinath, Implications of surface passivation on physicochemical and bioimaging properties of carbon dots. RSC Adv. **4**(40), 20915 (2014). doi:10.1039/c4ra02017k
101. R.-J. Fan, Q. Sun, L. Zhang, Y. Zhang, A.-H. Lu, Photoluminescent carbon dots directly derived from polyethylene glycol and their application for cellular imaging. Carbon **71**, 87–93 (2014). doi:10.1016/j.carbon.2014.01.016
102. L. Wang, H.S. Zhou, Green synthesis of luminescent nitrogen-doped carbon dots from milk and its imaging application. Anal. Chem. **86**(18), 8902–8905 (2014). doi:10.1021/ac502646x
103. Z. Qian, X. Shan, L. Chai, J. Ma, J. Chen, H. Feng, Si-doped carbon quantum dots: a facile and general preparation strategy, bioimaging application, and multifunctional sensor. ACS Appl. Mater. Inter. **6**(9), 6797–6805 (2014). doi:10.1021/am500403n
104. Q. Wang, X. Huang, Y. Long, X. Wang, H. Zhang, R. Zhu, L. Liang, P. Teng, H. Zheng, Hollow luminescent carbon dots for drug delivery. Carbon **59**, 192–199 (2013). doi:10.1016/j.carbon.2013.03.009
105. L. Hu, Y. Sun, S. Li, X. Wang, K. Hu, L. Wang, WuY Liang X-j, Multifunctional carbon dots with high quantum yield for imaging and gene delivery. Carbon **67**, 508–513 (2014). doi:10.1016/j.carbon.2013.10.023
106. X. Yang, Y. Zhuo, S. Zhu, Y. Luo, Y. Feng, Y. Dou, Novel and green synthesis of high-fluorescent carbon dots originated from honey for sensing and imaging. Biosens. Bioelectron. **60**, 292–298 (2014). doi:10.1016/j.bios.2014.04.046

107. S. Sahu, B. Behera, T.K. Maiti, S. Mohapatra, Simple one-step synthesis of highly luminescent carbon dots from orange juice: application as excellent bio-imaging agents. Chem. Commun. **48**(70), 8835–8837 (2012). doi:10.1039/c2cc33796g
108. W. Wang, Y.C. Lu, H. Huang, J.J. Feng, J.R. Chen, A.J. Wang, Facile synthesis of water-soluble and biocompatible fluorescent nitrogen-doped carbon dots for cell imaging. Analyst **139**(7), 1692–1696 (2014). doi:10.1039/c3an02098c
109. Z.C. Yang, M. Wang, A.M. Yong, S.Y. Wong, X.H. Zhang, H. Tan, A.Y. Chang, X. Li, J. Wang, Intrinsically fluorescent carbon dots with tunable emission derived from hydrothermal treatment of glucose in the presence of monopotassium phosphate. Chem. Commun. **47**(42), 11615–11617 (2011). doi:10.1039/c1cc14860e
110. Z.C. Yang, X. Li, J. Wang, Intrinsically fluorescent nitrogen-containing carbon nanoparticles synthesized by a hydrothermal process. Carbon **49**(15), 5207–5212 (2011). doi:10.1016/j.carbon.2011.07.038
111. O. Kozák, K.K.R. Datta, M. Greplová, V. Ranc, J. Kašlík, R. Zbořil, Surfactant-derived amphiphilic carbon dots with tunable photoluminescence. J. Phys. Chem. C **117**(47), 24991–24996 (2013). doi:10.1021/jp4040166
112. Z.C. Liang, L. Zeng, X.D. Cao, Q. Wang, X.H. Wang, R.C. Sun, Sustainable carbon quantum dots from forestry and agricultural biomass with amplified photoluminescence by simple NH4OH passivation. J. Mater. Chem. C **2**, 9760–9766. (2014). doi:10.1039/C4TC01714E
113. P.C. Hsu, H.T. Chang, Synthesis of high-quality carbon nanodots from hydrophilic compounds: role of functional groups. Chem. Commun. **48**(33), 3984–3986 (2012). doi:10.1039/c2cc30188a
114. X.M. Li, S.P. Lau, L.B. Tang, R.B. Ji, P.Z. Yang, Multicolour light emission from chlorine-doped graphene quantum dots. J. Mater. Chem. C **1**, 7308–7313 (2014). doi:10.1039/C3TC31473A
115. Y. Xu, C.J. Tang, H. Huang, C.Q. Sun, Y.K. Zhang, Q.F. Ye, A.J. Wang, Green synthesis of fluorescent carbon quantum dots for detection of Hg^{2+}. Chine. J. Anal. Chem. **42**(9), 1252–1258 (2014). doi:10.1016/s1872-2040(14)60765-9
116. Y. Dong, J. Shao, C. Chen, H. Li, R. Wang, Y. Chi, X. Lin, G. Chen, Blue luminescent graphene quantum dots and graphene oxide prepared by tuning the carbonization degree of citric acid. Carbon **50**(12), 4738–4743 (2012). doi:10.1016/j.carbon.2012.06.002
117. Y. Dong, R. Wang, H. Li, J. Shao, Y. Chi, X. Lin, G. Chen, Polyamine-functionalized carbon quantum dots for chemical sensing. Carbon **50**(8), 2810–2815 (2012). doi:10.1016/j.carbon.2012.02.046
118. X. Teng, C. Ma, C. Ge, M. Yan, J. Yang, Y. Zhang, P.C. Morais, H. Bi, Green synthesis of nitrogen-doped carbon dots from konjac flour with "off–on" fluorescence by Fe^{3+} and l-lysine for bioimaging. J. Mater. Chem. B **2**(29), 4631–4639 (2014). doi:10.1039/c4tb00368c
119. P.Y. Lin, C.W. Hsieh, M.L. Kung, L.Y. Chu, H.J. Huang, H.T. Chen, D.C. Wu, C.H. Kuo, S. L. Hsieh, S. Hsieh, Eco-friendly synthesis of shrimp egg-derived carbon dots for fluorescent bioimaging. J. Biotechnol. **189**, 114–119 (2014). doi:10.1016/j.jbiotec.2014.08.043
120. J. Niu, H. Gao, Synthesis and drug detection performance of nitrogen-doped carbon dots. J. Lumin. **149**, 159–162 (2014). doi:10.1016/j.jlumin.2014.01.026
121. Y. Xu, M. Wu, X.Z. Feng, X.B. Yin, X.W. He, Y.K. Zhang, Reduced carbon dots versus oxidized carbon dots: photo- and electrochemiluminescence investigations for selected applications. Chem. Eur. J. **19**(20), 6282–6288 (2013). doi:10.1002/chem.201204372
122. K. Hola, A.B. Bourlinos, O. Kozak, K. Berka, K.M. Siskova, M. Havrdova, J. Tucek, K. Safarova, M. Otyepka, E.P. Giannelis, R. Zboril, Photoluminescence effects of graphitic core size and surface functional groups in carbon dots: COO^{-} induced red-shift emission. Carbon **70**, 279–286 (2014). doi:10.1016/j.carbon.2014.01.008
123. J. Niu, H. Gao, L. Wang, S. Xin, G. Zhang, Q. Wang, L. Guo, W. Liu, X. Gao, Y. Wang, Facile synthesis and optical properties of nitrogen-doped carbon dots. New J. Chem. **38**(4), 1522–1527 (2014). doi:10.1039/c3nj01068f

124. J. Wei, J. Shen, X. Zhang, S. Guo, J. Pan, X. Hou, H. Zhang, L. Wang, B. Feng, Simple one-step synthesis of water-soluble fluorescent carbon dots derived from paper ash. RSC Adv. **3**(32), 13119–13122 (2013). doi:10.1039/c3ra41751d
125. J. Wang, C.F. Wang, S. Chen, Amphiphilic egg-derived carbon dots: rapid plasma fabrication, pyrolysis process, and multicolor printing patterns. Angew. Chem. Int. Ed. **51** (37), 9297–9301 (2012). doi:10.1002/anie
126. J. Zhou, Z. Sheng, H. Han, M. Zou, C. Li, Facile synthesis of fluorescent carbon dots using watermelon peel as a carbon source. Mater. Lett. **66**(1), 222–224 (2012). doi:10.1016/j.matlet.2011.08.081
127. X. Dong, Y. Su, H. Geng, Z. Li, C. Yang, X. Li, Y. Zhang, Fast one-step synthesis of N-doped carbon dots by pyrolyzing ethanolamine. J. Mater. Chem. C **2**(36), 7477–7481 (2014). doi:10.1039/c4tc01139b
128. Z. Jiang, A. Nolan, J.G. Walton, A. Lilienkampf, R. Zhang, M. Bradley, Photoluminescent carbon dots from 1,4-addition polymers. Chem. Eur. J. **20**(35), 10926–10931 (2014). doi:10.1002/chem.201403076
129. D. Pan, J. Zhang, Z. Li, C. Wu, X. Yan, M. Wu, Observation of pH-, solvent-, spin-, and excitation-dependent blue photoluminescence from carbon nanoparticles. Chem. Commun. **46**(21), 3681–3683 (2010). doi:10.1039/c000114g
130. F. Wang, M. Kreiter, B. He, S. Pang, C.Y. Liu, Synthesis of direct white-light emitting carbogenic quantum dots. Chem. Commun. **46**(19), 3309–3311 (2010). doi:10.1039/c002206c
131. X. Guo, C.F. Wang, Z.Y. Yu, L. Chen, S. Chen, Facile access to versatile fluorescent carbon dots toward light-emitting diodes. Chem. Commun. **48**(21), 2692–2694 (2012). doi:10.1039/c2cc17769b
132. H. Zhu, X. Wang, Y. Li, Z. Wang, F. Yang, X. Yang, Microwave synthesis of fluorescent carbon nanoparticles with electrochemiluminescence properties. Chem. Commun. **34**, 5118–5120 (2009). doi:10.1039/b907612c
133. X. Wang, K. Qu, B. Xu, J. Ren, X. Qu, Microwave assisted one-step green synthesis of cell-permeable multicolor photoluminescent carbon dots without surface passivation reagents. J. Mater. Chem. **21**(8), 2445–2450 (2011). doi:10.1039/c0jm02963g
134. X. Zhai, P. Zhang, C. Liu, T. Bai, W. Li, L. Dai, W. Liu, Highly luminescent carbon nanodots by microwave-assisted pyrolysis. Chem. Commun. **48**(64), 7955–7957 (2012). doi:10.1039/c2cc33869f
135. Z. Lin, W. Xue, H. Chen, J.M. Lin, Peroxynitrous-acid-induced chemiluminescence of fluorescent carbon dots for nitrite sensing. Anal. Chem. **83**(21), 8245–8251 (2011). doi:10.1021/ac202039h
136. A. Salinas-Castillo, M. Ariza-Avidad, C. Pritz, M. Camprubi-Robles, B. Fernandez, M. J. Ruedas-Rama, A. Megia-Fernandez, A. Lapresta-Fernandez, F. Santoyo-Gonzalez, A. Schrott-Fischer, L.F. Capitan-Vallvey, Carbon dots for copper detection with down and upconversion fluorescent properties as excitation sources. Chem. Commun. **49**(11), 1103–1105 (2013). doi:10.1039/c2cc36450f
137. A. Zhao, C. Zhao, M. Li, J. Ren, X. Qu, Ionic liquids as precursors for highly luminescent, surface-different nitrogen-doped carbon dots used for label-free detection of Cu^{2+}/Fe^{3+} and cell imaging. Anal. Chem. Acta **809**, 128–133 (2014). doi:10.1016/j.aca.2013.10.046
138. Y. Zhai, Z. Zhu, C. Zhu, J. Ren, E. Wang, S. Dong, Multifunctional water-soluble luminescent carbon dots for imaging and Hg^{2+} sensing. J. Mater. Chem. B **2**(40), 6995–6999 (2014). doi:10.1039/c4tb01035c
139. E.F. Simoes, J.C. da Silva, J.M. Leitao, Carbon dots from tryptophan doped glucose for peroxynitrite sensing. Anal. Chem. Acta **852**, 174–180 (2014). doi:10.1016/j.aca.2014.08.050
140. W. Wang, Y. Li, L. Cheng, Z. Cao, W. Liu, Water-soluble and phosphorus-containing carbon dots with strong green fluorescence for cell labeling. J. Mater. Chem. B **2**(1), 46–48 (2014). doi:10.1039/c3tb21370f

141. Y.F. Huang, X. Zhou, R. Zhou, H. Zhang, K.B. Kang, M. Zhao, Y. Peng, Q. Wang, H.L. Zhang, W.Y. Qiu, One-pot synthesis of highly luminescent carbon quantum dots and their nontoxic ingestion by zebrafish for in vivo imaging. Chem. Eur. J. **20**(19), 5640–5648 (2014). doi:10.1002/chem.201400011
142. C. Liu, P. Zhang, X. Zhai, F. Tian, W. Li, J. Yang, Y. Liu, H. Wang, W. Wang, W. Liu, Nano-carrier for gene delivery and bioimaging based on carbon dots with PEI-passivation enhanced fluorescence. Biomaterials **33**(13), 3604–3613 (2012). doi:10.1016/j.biomaterials.2012.01.052
143. J. Gong, X. An, X. Yan, A novel rapid and green synthesis of highly luminescent carbon dots with good biocompatibility for cell imaging. New J. Chem. **38**(4), 1376–1379 (2014). doi:10.1039/c3nj01320k
144. Q. Liu, S. Xu, C. Niu, M. Li, D. He, Z. Lu, L. Ma, N. Na, F. Huang, H. Jiang, J. Ouyang, Distinguish cancer cells based on targeting turn-on fluorescence imaging by folate functionalized green emitting carbon dots. Biosens. Bioelectron. **64**, 119–125 (2015). doi:10.1016/j.bios.2014.08.052
145. J. Wang, C. Cheng, Y. Huang, B. Zheng, H. Yuan, L. Bo, M.-W. Zheng, S.-Y. Yang, Y. Guo, D. Xiao, A facile large-scale microwave synthesis of highly fluorescent carbon dots from benzenediol isomers. J. Mater. Chem. C **2**(25), 5028–5035 (2014). doi:10.1039/c3tc32131b
146. S. Liu, L. Wang, J. Tian, J. Zhai, Y. Luo, W. Lu, X. Sun, Acid-driven, microwave-assisted production of photoluminescent carbon nitride dots from N. N-dimethylformamide. RSC Adv. **1**(6), 951–953 (2011). doi:10.1039/c1ra00249j
147. C. Liu, P. Zhang, F. Tian, W. Li, F. Li, W. Liu, One-step synthesis of surface passivated carbon nanodots by microwave assisted pyrolysis for enhanced multicolor photoluminescence and bioimaging. J. Mater. Chem. **21**(35), 13163–13167 (2011). doi:10.1039/c1jm12744f
148. R.J. Libin Tang, Xiangke Cao, Jingyu Lin, Hongxing Jiang, Xueming Li, Kar Seng Teng, Chi Man Luk, Songjun Zeng, Jianhua Hao, Shu Ping Lau, Deep ultraviolet photoluminescence of water-soluble self-passivated graphene quantum dots. ACS Nano **6**, 5102–5110 (2012). doi:10.1021/nn300760g
149. L.B. Tang, R.B. Ji, X.M. Li, G.X. Bai, C.P. Liu, J.H. Hao, J.Y. Lin, H.X. Jiang, K.S. Teng, Z.B. Yang, Deep ultraviolet to near-infrared emission and photoresponse in layered N-doped graphene quantum dots. ACS Nano **8**(6), 6312–6320 (2014). doi:10.1021/nn501796r
150. R. Liu, D. Wu, S. Liu, K. Koynov, W. Knoll, Q. Li, An aqueous route to multicolor photoluminescent carbon dots using silica spheres as carriers. Angew. Chem. Int. Ed. **48**(25), 4598–4601 (2009). doi:10.1002/anie.200900652
151. J. Zong, Y. Zhu, X. Yang, J. Shen, C. Li, Synthesis of photoluminescent carbogenic dots using mesoporous silica spheres as nanoreactors. Chem. Commun. (Camb.) **47**(2), 764–766 (2011). doi:10.1039/c0cc03092a
152. J. Zong, X. Yang, A. Trinchi, S. Hardin, I. Cole, Y. Zhu, C. Li, T. Muster, G. Wei, Carbon dots as fluorescent probes for "off-on" detection of Cu^{2+} and L-cysteine in aqueous solution. Biosens. Bioelectron. **51**, 330–335 (2014). doi:10.1016/j.bios.2013.07.042
153. S. Liu, J. Tian, L. Wang, Y. Luo, J. Zhai, X. Sun, Preparation of photoluminescent carbon nitride dots from CCl4 and 1,2-ethylenediamine: a heat-treatment-based strategy. J. Mater. Chem. **21**(32), 11726–11729 (2011). doi:10.1039/c1jm12149a
154. E.J. Goh, K.S. Kim, Y.R. Kim, H.S. Jung, S. Beack, W.H. Kong, G. Scarcelli, S.H. Yun, S.K. Hahn, Bioimaging of hyaluronic acid derivatives using nanosized carbon dots. Biomacromolecules **13**(8), 2554–2561 (2012). doi:10.1021/bm300796q
155. R. Liu, H. Li, W. Kong, J. Liu, Y. Liu, C. Tong, X. Zhang, Z. Kang, Ultra-sensitive and selective Hg^{2+} detection based on fluorescent carbon dots. Mater. Res. Bull. **48**(7), 2529–2534 (2013). doi:10.1016/j.materresbull.2013.03.015
156. H. Nie, M. Li, Q. Li, S. Liang, Y. Tan, L. Sheng, W. Shi, S.X.-A. Zhang, Carbon dots with continuously tunable full-color emission and their application in ratiometric pH sensing. Chem. Mater. **26**(10), 3104–3112 (2014). doi:10.1021/cm5003669

157. W. Kong, R. Liu, H. Li, J. Liu, H. Huang, Y. Liu, Z. Kang, High-bright fluorescent carbon dots and their application in selective nucleoli staining. J. Mater. Chem. B **2**(31), 5077–5082 (2014). doi:10.1039/c4tb00579a
158. C.-W. Lai, Y.-H. Hsiao, Y.-K. Peng, P.-T. Chou, Facile synthesis of highly emissive carbon dots from pyrolysis of glycerol; gram scale production of carbon dots/mSiO_2 for cell imaging and drug release. J. Mater. Chem. **22**(29), 14403–14409 (2012). doi:10.1039/c2jm32206d
159. Y.Y. Zhang, M. Wu, Y.Q. Wang, X.W. He, W.Y. Li, X.Z. Feng, A new hydrothermal refluxing route to strong fluorescent carbon dots and its application as fluorescent imaging agent. Talanta **117**, 196–202 (2013). doi:10.1016/j.talanta.2013.09.003
160. Y. Hu, J. Yang, J. Tian, L. Jia, J.-S. Yu, Waste frying oil as a precursor for one-step synthesis of sulfur-doped carbon dots with pH-sensitive photoluminescence. Carbon **77**, 775–782 (2014). doi:10.1016/j.carbon.2014.05.081
161. X. Yan, B. Li, X. Cui, Q. Wei, K. Tajima, Li L-s, Independent tuning of the band gap and redox potential of graphene quantum dots. J. Phys. Chem. Lett. **2**(10), 1119–1124 (2011). doi:10.1021/jz200450r
162. X. Yan, X. Cui, B. Li, L.S. Li, Large, solution-processable graphene quantum dots as light absorbers for photovoltaics. Nano Lett. **10**(5), 1869–1873 (2010). doi:10.1021/nl101060h
163. Yan L-sLX, Colloidal graphene quantum dots. J. Phys. Chem. Lett. **1**, 2572–2576 (2010). doi:10.1021/jz100862f|J
164. X.C. Xin Yan, Liang-shi Li, Synthesis of large, stable colloidal graphene quantum dots with tunable size. J. Am. Chem. Soc. **132**, 5944–5945 (2010). doi:10.1021/ja1009376
165. M.L. Mueller, X. Yan, J.A. McGuire, L.S. Li, Triplet states and electronic relaxation in photoexcited graphene quantum dots. Nano Lett. **10**(7), 2679–2682 (2010). doi:10.1021/nl101474d
166. J. Zhang, W. Shen, D. Pan, Z. Zhang, Y. Fang, M. Wu, Controlled synthesis of green and blue luminescent carbon nanoparticles with high yields by the carbonization of sucrose. New J. Chem. **34**(4), 591–593 (2010). doi:10.1039/b9nj00662a
167. Y. Liu, Liu C-y, Zhang Z-y, Synthesis of highly luminescent graphitized carbon dots and the application in the Hg^{2+} detection. Appl. Surf. Sci. **263**, 481–485 (2012). doi:10.1016/j.apsusc.2012.09.088
168. Y. Dong, H. Pang, H. Yang, J. Jiang, Y. Chi, T. Yu, Nitrogen-doped carbon-based dots prepared by dehydrating EDTA with hot sulfuric acid and their electrocatalysis for oxygen reduction reaction. RSC Adv. **4**(62), 32791–32795 (2014). doi:10.1039/c4ra06594h
169. S.S. Wee, Y.H. Ng, S.M. Ng, Synthesis of fluorescent carbon dots via simple acid hydrolysis of bovine serum albumin and its potential as sensitive sensing probe for lead (II) ions. Talanta **116**, 71–76 (2013). doi:10.1016/j.talanta.2013.04.081
170. D. Sun, R. Ban, P.-H. Zhang, G.-H. Wu, J.-R. Zhang, J.-J. Zhu, Hair fiber as a precursor for synthesizing of sulfur- and nitrogen-co-doped carbon dots with tunable luminescence properties. Carbon **64**, 424–434 (2013). doi:10.1016/j.carbon.2013.07.095
171. J.S. Suh JKa, Size-controllable and low-cost fabrication of graphene quantum dots using thermal plasma jet. ACS Nano **8**, 4190–4196 (2014)
172. C. Jiang, H. Wu, X. Song, X. Ma, J. Wang, M. Tan, Presence of photoluminescent carbon dots in Nescafe(R) original instant coffee: applications to bioimaging. Talanta **127**, 68–74 (2014). doi:10.1016/j.talanta.2014.01.046
173. H. Li, X. He, Y. Liu, H. Huang, S. Lian, S.-T. Lee, Z. Kang, One-step ultrasonic synthesis of water-soluble carbon nanoparticles with excellent photoluminescent properties. Carbon **49**(2), 605–609 (2011). doi:10.1016/j.carbon.2010.10.004
174. Z. Ma, H. Ming, H. Huang, Y. Liu, Z. Kang, One-step ultrasonic synthesis of fluorescent N-doped carbon dots from glucose and their visible-light sensitive photocatalytic ability. New J. Chem. **36**(4), 861–864 (2012). doi:10.1039/c2nj20942j
175. Y. Fang, S. Guo, D. Li, C. Zhu, W. Ren, S. Dong, E. Wang, Easy synthesis and imaging applications of cross-linked green fluorescent hollow carbon nanoparticles. ACS Nano **6**, 400–409 (2012). doi:10.1021/nn2046373

176. J. Lu, P.S. Yeo, C.K. Gan, P. Wu, K.P. Loh, Transforming C_{60} molecules into graphene quantum dots. Nat. Nanotechnol. **6**(4), 247–252 (2011). doi:10.1038/nnano.2011.30
177. J. Shen, Y. Zhu, X. Yang, J. Zong, J. Zhang, C. Li, One-pot hydrothermal synthesis of graphene quantum dots surface-passivated by polyethylene glycol and their photoelectric conversion under near-infrared light. New J. Chem. **36**(1), 97–101 (2012). doi:10.1039/c1nj20658c
178. H. Peng, J. Travas-Sejdic, Simple aqueous solution route to luminescent carbogenic dots from carbohydrates. Chem. Mater. **21**(23), 5563–5565 (2009). doi:10.1021/cm901593y
179. Y. Liu, Liu C-y, Zhang Z-y, Synthesis and surface photochemistry of graphitized carbon quantum dots. J. Colloid Interface Sci. **356**(2), 416–421 (2011). doi:10.1016/j.jcis.2011.01.065
180. H. Sun, L. Wu, N. Gao, J. Ren, X. Qu, Improvement of photoluminescence of graphene quantum dots with a biocompatible photochemical reduction pathway and its bioimaging application. ACS Appl. Mater. Interfaces **5**(3), 1174–1179 (2013). doi:10.1021/am3030849
181. G.S. Kumar, R. Roy, D. Sen, U.K. Ghorai, R. Thapa, N. Mazumder, S. Saha, K.K. Chattopadhyay, Amino-functionalized graphene quantum dots: origin of tunable heterogeneous photoluminescence. Nanoscale **6**(6), 3384–3391 (2014). doi:10.1039/c3nr05376h
182. S.H. Jin, D.H. Kim, G.H. Jun, S.H. Hong, S. Jeon, Tuning the photoluminescence of graphene quantum dots through the charge transfer effect of functional groups. ACS Nano **7**(2), 1239–1245 (2012). doi:10.1021/nn3046759
183. Y. Feng, J. Zhao, X. Yan, F. Tang, Q. Xue, Enhancement in the fluorescence of graphene quantum dots by hydrazine hydrate reduction. Carbon **66**, 334–339 (2014). doi:10.1016/j.carbon.2013.09.008
184. H. Zheng, Q. Wang, Y. Long, H. Zhang, X. Huang, R. Zhu, Enhancing the luminescence of carbon dots with a reduction pathway. Chem. Commun. **47**(38), 10650–10652 (2011). doi:10.1039/c1cc14741b
185. K. Lingam, R. Podila, H. Qian, S. Serkiz, A.M. Rao, Evidence for edge-state photoluminescence in graphene quantum dots. Adv. Funct. Mater. **23**(40), 5062–5065 (2013). doi:10.1002/adfm.201203441
186. G. Eda, Y.Y. Lin, C. Mattevi, H. Yamaguchi, H.A. Chen, I.S. Chen, C.W. Chen, M. Chhowalla, Blue photoluminescence from chemically derived graphene oxide. Adv. Mater. **22**(4), 505–509 (2010). doi:10.1002/adma.200901996
187. K.P. Loh, Q. Bao, G. Eda, M. Chhowalla, Graphene oxide as a chemically tunable platform for optical applications. Nat. Chem. **2**(12), 1015–1024 (2010). doi:10.1038/nchem.907
188. J. Robertson, G.A.J. Amaratunga, Photoluminescence behavior of hydrogenated amorphous carbon. J. Appl. Phys. **80**(5), 2998–3003 (1996). doi:10.1063/1.363158
189. F. Demichelis, S. Schreiter, A. Tagliaferro, Photoluminescence in a-C-H films. Phys. Rev. B **51**(4), 2143–2147 (1995). doi:10.1103/PhysRevB.51.2143
190. M. Li, W. Wu, W. Ren, H.-M. Cheng, N. Tang, W. Zhong, Y. Du, Synthesis and upconversion luminescence of N-doped graphene quantum dots. Appl. Phys. Lett. **101**(10), 103107 (2012). doi:10.1063/1.4750065
191. X. Wang, L. Cao, S.T. Yang, F. Lu, M.J. Meziani, L. Tian, K.W. Sun, M.A. Bloodgood, Y. P. Sun, Bandgap-like strong fluorescence in functionalized carbon nanoparticles. Angew. Chem. Int. Ed. **49**(31), 5310–5314 (2010). doi:10.1002/anie.201000982
192. P. Anilkumar, X. Wang, L. Cao, S. Sahu, J.H. Liu, P. Wang, K. Korch, K.N. Tackett 2nd, A. Parenzan, Y.P. Sun, Toward quantitatively fluorescent carbon-based "quantum" dots. Nanoscale **3**(5), 2023–2027 (2011). doi:10.1039/c0nr00962h
193. Y.P. Sun, X. Wang, F.S. Lu, L. Cao, M.J. Meziani, P.J.G. Luo, L.R. Gu, L.M. Veca, Doped carbon nanoparticles as a new platform for highly photoluminescent dots. J. Phys. Chem. C **112**(47), 18295–18298 (2008). doi:10.1021/jp8076485
194. A. Cayuela, M.L. Soriano, M. Valcarcel, Strong luminescence of carbon dots induced by acetone passivation: efficient sensor for a rapid analysis of two different pollutants. Anal. Chim. Acta **804**, 246–251 (2013). doi:10.1016/j.aca.2013.10.031

195. S. Zhu, J. Zhang, S. Tang, C. Qiao, L. Wang, H. Wang, X. Liu, B. Li, Y. Li, W. Yu, X. Wang, H. Sun, B. Yang, Surface chemistry routes to modulate the photoluminescence of graphene quantum dots: from fluorescence mechanism to up-conversion bioimaging applications. Adv. Func. Mater. **22**(22), 4732–4740 (2012). doi:10.1002/adfm.201201499
196. Y. Dong, R. Dai, T. Dong, Y. Chi, G. Chen, Photoluminescence, chemiluminescence and anodic electrochemiluminescence of hydrazide-modified graphene quantum dots. Nanoscale **6**(19), 11240–11245 (2014). doi:10.1039/c4nr02539c
197. Q. Liu, B. Guo, Z. Rao, B. Zhang, J.R. Gong, Strong two-photon-induced fluorescence from photostable, biocompatible nitrogen-doped graphene quantum dots for cellular and deep-tissue imaging. Nano Lett. **13**(6), 2436–2441 (2013). doi:10.1021/nl400368v
198. C. Hu, Y. Liu, Y. Yang, J. Cui, Z. Huang, Y. Wang, L. Yang, H. Wang, Y. Xiao, J. Rong, One-step preparation of nitrogen-doped graphene quantum dots from oxidized debris of graphene oxide. J. Mater. Chem. B **1**(1), 39–42 (2013). doi:10.1039/c2tb00189f
199. X.-M. Wei, Y. Xu, Y.-H. Li, X.-B. Yin, X.-W. He, Ultrafast synthesis of nitrogen-doped carbon dots via neutralization heat for bioimaging and sensing applications. RSC Adv. **4**(84), 44504–44508 (2014). doi:10.1039/c4ra08523j
200. L. Tang, R. Ji, X. Li, K.S. Teng, S.P. Lau, Energy-level structure of nitrogen-doped graphene quantum dots. J. Mater. Chem. C **1**(32), 4908–4915 (2013). doi:10.1039/c3tc30877d
201. Q.-Q. Shi, Y.-H. Li, Y. Xu, Y. Wang, X.-B. Yin, X.-W. He, Y.-K. Zhang, High-yield and high-solubility nitrogen-doped carbon dots: formation, fluorescence mechanism and imaging application. RSC Adv. **4**(4), 1563–1566 (2014). doi:10.1039/c3ra45762a
202. J. Shen, Y. Zhu, X. Yang, J. Zong, J. Zhang, C. Li, One-pot hydrothermal synthesis of graphene quantum dots surface-passivated by polyethylene glycol and their photoelectric conversion under near-infrared light. New J. Chem. **36**(1), 97–101 (2012). doi:10.1039/c1nj20658c
203. S.J. Zhuo, M.W. Shao, S.T. Lee, Upconversion and downconversion fluorescent graphene quantum dots: ultrasonic preparation and photocatalysis. ACS Nano **6**(2), 1059–1064 (2012). doi:10.1021/nn2040395
204. S. Zhu, J. Zhang, X. Liu, B. Li, X. Wang, S. Tang, Q. Meng, Y. Li, C. Shi, R. Hu, B. Yang, Graphene quantum dots with controllable surface oxidation, tunable fluorescence and up-conversion emission. RSC Adv. **2**(7), 2717–2720 (2012). doi:10.1039/c2ra20182h
205. Y. Su, M. Xie, X. Lu, H. Wei, H. Geng, Z. Yang, Y. Zhang, Facile synthesis and photoelectric properties of carbon dots with upconversion fluorescence using arc-synthesized carbon by-products. RSC Adv. **4**(10), 4839–4842 (2014). doi:10.1039/c3ra45453c
206. D.Z. Tan, S.F. Zhou, J.R. Qiu, Comment on "upconversion and downconversion fluorescent graphene quantum dots: ultrasonic preparation and photocatalysis". ACS Nano **6**(8), 6530–6531 (2012). doi:10.1021/nn3016822
207. Z. Gan, X. Wu, G. Zhou, J. Shen, P.K. Chu, Is there real upconversion photoluminescence from graphene quantum dots. Adv. Opt. Mater. **1**(8), 554–558 (2013). doi:10.1002/adom.201300152
208. S.K. Poznyak, D.V. Talapin, E.V. Shevchenko, H. Weller, Quantum dot chemiluminescence. Nano Lett. **4**(4), 693–698 (2004). doi:10.1021/nl049713w
209. Z. Ding, B.M. Quinn, S.K. Haram, L.E. Pell, B.A. Korgel, A.J. Bard, Electrochemistry and electrogenerated chemiluminescence from silicon nanocrystal quantum dots. Science **296**(5571), 1293–1297 (2002). doi:10.1126/science.1069336
210. Y. Bae, N. Myung, A.J. Bard, Electrochemistry and electrogenerated chemiluminescence of CdTe nanoparticles. Nano Lett. **4**(6), 1153–1161 (2004). doi:10.1021/nl049516x
211. Y. Dong, W. Tian, S. Ren, R. Dai, Y. Chi, G. Chen, Graphene quantum dots/L-cysteine coreactant electrochemiluminescence system and its application in sensing lead(II) ions. ACS Appl. Mater. Interface **6**(3), 1646–1651 (2014). doi:10.1021/am404552s
212. Y. Dong, C. Chen, J. Lin, N. Zhou, Y. Chi, G. Chen, Electrochemiluminescence emission from carbon quantum dot-sulfite coreactant system. Carbon **56**, 12–17 (2013). doi:10.1016/j.carbon.2012.12.086

213. T.T. Zhang, H.M. Zhao, X.F. Fan, S. Chen, X. Quan, Electrochemiluminescence immunosensor for highly sensitive detection of 8-hydroxy-2′-deoxyguanosine based on carbon quantum dot coated Au/SiO_2 core-shell nanoparticles. Talanta **131**, 379–385 (2015). doi:10.1016/j.talanta.2014.08.024
214. M. Zhang, H. Liu, L. Chen, M. Yan, L. Ge, S. Ge, J. Yu, A disposable electrochemiluminescence device for ultrasensitive monitoring of K562 leukemia cells based on aptamers and ZnO@carbon quantum dots. Biosens. Bioelectron. **49**, 79–85 (2013). doi:10.1016/j.bios.2013.05.003
215. S. Yang, J. Liang, S. Luo, C. Liu, Y. Tang, Supersensitive detection of chlorinated phenols by multiple amplification electrochemiluminescence sensing based on carbon quantum dots/graphene. Anal. Chem. **85**(16), 7720–7725 (2013). doi:10.1021/ac400874h
216. M. Su, H. Liu, L. Ge, Y. Wang, S. Ge, J. Yu, M. Yan, Aptamer-Based electrochemiluminescent detection of MCF-7 cancer cells based on carbon quantum dots coated mesoporous silica nanoparticles. Electrochim. Acta **146**, 262–269 (2014). doi:10.1016/j.electacta.2014.08.129
217. Q. Lu, W. Wei, Z. Zhou, Z. Zhou, Y. Zhang, S. Liu, Electrochemiluminescence resonance energy transfer between graphene quantum dots and gold nanoparticles for DNA damage detection. Analyst **139**(10), 2404–2410 (2014). doi:10.1039/c4an00020j
218. J. Lu, M. Yan, L. Ge, S. Ge, S. Wang, J. Yan, J. Yu, Electrochemiluminescence of blue-luminescent graphene quantum dots and its application in ultrasensitive aptasensor for adenosine triphosphate detection. Biosens. Bioelectron. **47**, 271–277 (2013). doi:10.1016/j.bios.2013.03.039
219. Y. Chen, Y. Dong, H. Wu, C. Chen, Y. Chi, G. Chen, Electrochemiluminescence sensor for hexavalent chromium based on the graphene quantum dots/peroxodisulfate system. Electrochim. Acta **151**, 552–557 (2015). doi:10.1016/j.electacta.2014.11.068
220. W. Xue, Z. Lin, H. Chen, C. Lu, J.-M. Lin, Enhancement of ultraweak chemiluminescence from reaction of hydrogen peroxide and bisulfite by water-soluble carbon nanodots. J. Phys. Chem. C **115**(44), 21707–21714 (2011). doi:10.1021/jp207554t
221. Z. Lin, W. Xue, H. Chen, J.M. Lin, Classical oxidant induced chemiluminescence of fluorescent carbon dots. Chem. Commun. **48**(7), 1051–1053 (2012). doi:10.1039/c1cc15290d
222. L. Zhao, F. Di, D. Wang, L.H. Guo, Y. Yang, B. Wan, H. Zhang, Chemiluminescence of carbon dots under strong alkaline solutions: a novel insight into carbon dot optical properties. Nanoscale **5**(7), 2655–2658 (2013). doi:10.1039/c3nr00358b
223. H. Gonçalves, P.A.S. Jorge, J.R.A. Fernandes, J.C.G. Esteves da Silva, Hg(II) sensing based on functionalized carbon dots obtained by direct laser ablation. Sensor. Actuat. B-Chem. **145** (2), 702–707 (2010). doi:10.1016/j.snb.2010.01.031
224. H.M. Goncalves, A.J. Duarte, J.C. Esteves da Silva, Optical fiber sensor for Hg(II) based on carbon dots. Biosens. Bioelectron. **26**(4), 1302–1306 (2010). doi:10.1016/j.bios.2010.07.018
225. B. Cao, C. Yuan, B. Liu, C. Jiang, G. Guan, M.Y. Han, Ratiometric fluorescence detection of mercuric ion based on the nanohybrid of fluorescence carbon dots and quantum dots. Anal. Chim. Acta **786**, 146–152 (2013). doi:10.1016/j.aca.2013.05.015
226. W. Wang, T. Kim, Z. Yan, G. Zhu, I. Cole, N.T. Nguyen, Q. Li, Carbon dots functionalized by organosilane with double-sided anchoring for nanomolar Hg^{2+} detection. J. Colloid Interface Sci. **437**, 28–34 (2015). doi:10.1016/j.jcis.2014.09.013
227. F. Yan, Y. Zou, M. Wang, X. Mu, N. Yang, L. Chen, Highly photoluminescent carbon dots-based fluorescent chemosensors for sensitive and selective detection of mercury ions and application of imaging in living cells. Sens. Actuat. B Chem. **192**, 488–495 (2014). doi:10.1016/j.snb.2013.11.041
228. H. Li, J. Zhai, J. Tian, Y. Luo, X. Sun, Carbon nanoparticle for highly sensitive and selective fluorescent detection of mercury(II) ion in aqueous solution. Biosens. Bioelectron. **26**(12), 4656–4660 (2011). doi:10.1016/j.bios.2011.03.026

229. X. Cui, L. Zhu, J. Wu, Y. Hou, P. Wang, Z. Wang, M. Yang, A fluorescent biosensor based on carbon dots-labeled oligodeoxyribonucleotide and graphene oxide for mercury(II) detection. Biosens. Bioelectron. **63**, 506–512 (2015). doi:10.1016/j.bios.2014.07.085
230. I. Costas-Mora, V. Romero, I. Lavilla, C. Bendicho, In situ building of a nanoprobe based on fluorescent carbon dots for methylmercury detection. Anal. Chem. **86**(9), 4536–4543 (2014). doi:10.1021/ac500517h
231. Y. Dong, R. Wang, G. Li, C. Chen, Y. Chi, G. Chen, Polyamine-functionalized carbon quantum dots as fluorescent probes for selective and sensitive detection of copper ions. Anal. Chem. **84**(14), 6220–6224 (2012). doi:10.1021/ac3012126
232. X. Lin, G. Gao, L. Zheng, Y. Chi, G. Chen, Encapsulation of strongly fluorescent carbon quantum dots in metal-organic frameworks for enhancing chemical sensing. Anal. Chem. **86** (2), 1223–1228 (2014). doi:10.1021/ac403536a
233. Y. Dong, R. Wang, W. Tian, Y. Chi, G. Chen, "Turn-on" fluorescent detection of cyanide based on polyamine-functionalized carbon quantum dots. RSC Adv. **4**(8), 3701–3705 (2014). doi:10.1039/c3ra45893h
234. Q. Qu, A. Zhu, X. Shao, G. Shi, Y. Tian, Development of a carbon quantum dots-based fluorescent Cu^{2+} probe suitable for living cell imaging. Chem. Commun. **48**(44), 5473–5475 (2012). doi:10.1039/c2cc31000g
235. A. Zhu, Q. Qu, X. Shao, B. Kong, Y. Tian, Carbon-dot-based dual-emission nanohybrid produces a ratiometric fluorescent sensor for in vivo imaging of cellular copper ions. Angew. Chem. Int. Ed. **51**(29), 7185–7189 (2012). doi:10.1002/anie.201109089
236. X. Liu, N. Zhang, T. Bing, D. Shangguan, Carbon dots based dual-emission silica nanoparticles as a ratiometric nanosensor for Cu^{2+}. Anal. Chem. **86**(5), 2289–2296 (2014). doi:10.1021/ac404236y
237. H. Sun, N. Gao, L. Wu, J. Ren, W. Wei, X. Qu, Highly photoluminescent amino-functionalized graphene quantum dots used for sensing copper ions. Chem. Eur. J. **19**(40), 13362–13368 (2013). doi:10.1002/chem.201302268
238. S. Zhang, Q. Wang, G. Tian, H. Ge, A fluorescent turn-off/on method for detection of Cu^{2+} and oxalate using carbon dots as fluorescent probes in aqueous solution. Mater. Lett. **115**, 233–236 (2014). doi:10.1016/j.matlet.2013.10.086
239. L. Liu, Y. Li, L. Zhan, Y. Liu, C. Huang, One-step synthesis of fluorescent hydroxyls-coated carbon dots with hydrothermal reaction and its application to optical sensing of metal ions. Sci. China Chem. **54**(8), 1342–1347 (2011). doi:10.1007/s11426-011-4351-6
240. Y. Song, S. Zhu, S. Xiang, X. Zhao, J. Zhang, H. Zhang, Y. Fu, B. Yang, Investigation into the fluorescence quenching behaviors and applications of carbon dots. Nanoscale **6**(9), 4676–4682 (2014). doi:10.1039/c4nr00029c
241. Y. Liu, N. Xiao, N. Gong, H. Wang, X. Shi, W. Gu, L. Ye, One-step microwave-assisted polyol synthesis of green luminescent carbon dots as optical nanoprobes. Carbon **68**, 258–264 (2014). doi:10.1016/j.carbon.2013.10.086
242. S. Hu, Q. Zhao, Q. Chang, J. Yang, J. Liu, Enhanced performance of Fe^{3+} detection via fluorescence resonance energy transfer between carbon quantum dots and Rhodamine B. RSC Adv. **4**(77), 41069–41075 (2014). doi:10.1039/c4ra06371f
243. X. Li, S. Zhu, B. Xu, K. Ma, J. Zhang, B. Yang, W. Tian, Self-assembled graphene quantum dots induced by cytochrome c: a novel biosensor for trypsin with remarkable fluorescence enhancement. Nanoscale **5**(17), 7776–7779 (2013). doi:10.1039/c3nr00006k
244. S. Gomez-de Pedro, A. Salinas-Castillo, M. Ariza-Avidad, A. Lapresta-Fernandez, C. Sanchez-Gonzalez, C.S. Martinez-Cisneros, M. Puyol, L.F. Capitan-Vallvey, J. Alonso-Chamarro, Microsystem-assisted synthesis of carbon dots with fluorescent and colorimetric properties for pH detection. Nanoscale **6**(11), 6018–6024 (2014). doi:10.1039/c4nr00573b
245. W. Wei, C. Xu, J. Ren, B. Xu, X. Qu, Sensing metal ions with ion selectivity of a crown ether and fluorescence resonance energy transfer between carbon dots and graphene. Chem. Commun. **48**(9), 1284–1286 (2012). doi:10.1039/c2cc16481g

246. Y. Dong, G. Li, N. Zhou, R. Wang, Y. Chi, G. Chen, Graphene quantum dot as a green and facile sensor for free chlorine in drinking water. Anal. Chem. **84**(19), 8378–8382 (2012). doi:10.1021/ac301945z
247. Z. Huang, F. Lin, M. Hu, C. Li, T. Xu, C. Chen, X. Guo, Carbon dots with tunable emission, controllable size and their application for sensing hypochlorous acid. J. Lumin. **151**, 100–105 (2014). doi:10.1016/j.jlumin.2014.02.013
248. H.X. Zhao, L.Q. Liu, Z.D. Liu, Y. Wang, X.J. Zhao, C.Z. Huang, Highly selective detection of phosphate in very complicated matrixes with an off-on fluorescent probe of europium-adjusted carbon dots. Chem. Commun. **47**(9), 2604–2606 (2011). doi:10.1039/c0cc04399k
249. J.M. Bai, L. Zhang, R.P. Liang, J.D. Qiu, Graphene quantum dots combined with europium ions as photoluminescent probes for phosphate sensing. Chem. Eur. J. **19**(12), 3822–3826 (2013). doi:10.1002/chem.201204295
250. J.J. Liu, X.L. Zhang, Z.X. Cong, Z.T. Chen, H.H. Yang, G.N. Chen, Glutathione-functionalized graphene quantum dots as selective fluorescent probes for phosphate-containing metabolites. Nanoscale **5**(5), 1810–1815 (2013). doi:10.1039/c3nr33794d
251. J. Xu, Y. Zhou, G. Cheng, M. Dong, S. Liu, C. Huang, Carbon dots as a luminescence sensor for ultrasensitive detection of phosphate and their bioimaging properties. Luminescence **30** (4), 411–415 (2015). doi:10.1002/bio.2752
252. Y.H. Li, L. Zhang, J. Huang, R.P. Liang, J.D. Qiu, Fluorescent graphene quantum dots with a boronic acid appended bipyridinium salt to sense monosaccharides in aqueous solution. Chem. Commun. **49**(45), 5180–5182 (2013). doi:10.1039/c3cc40652k
253. Z.B. Qu, X.G. Zhou, L. Gu, R.M. Lan, D.D. Sun, D.J. Yu, G.Y. Shi, Boronic acid functionalized Graphene quantum dots as fluorescent. Chem. Comm. **49**(84), 9830–9832 (2013). doi:10.1039/C3CC44393K
254. L. Zhu, X. Cui, J. Wu, Z. Wang, P. Wang, Y. Hou, M. Yang, Fluorescence immunoassay based on carbon dots as labels for the detection of human immunoglobulin G. Anal. Methods **6**(12), 4430–4436 (2014). doi:10.1039/c4ay00717d
255. H. Zhao, Y. Chang, M. Liu, S. Gao, H. Yu, X. Quan, A universal immunosensing strategy based on regulation of the interaction between graphene and graphene quantum dots. Chem. Commun. **49**(3), 234–236 (2013). doi:10.1039/c2cc35503e
256. D. Bu, H. Zhuang, G. Yang, X. Ping, An immunosensor designed for polybrominated biphenyl detection based on fluorescence resonance energy transfer (FRET) between carbon dots and gold nanoparticles. Sens. Actuat. B Chem. **195**, 540–548 (2014). doi:10.1016/j.snb.2014.01.079
257. Y. Wang, L. Zhang, R.P. Liang, J.M. Bai, J.D. Qiu, Using graphene quantum dots as photoluminescent probes for protein kinase sensing. Anal. Chem. **85**(19), 9148–9155 (2013). doi:10.1021/ac401807b
258. Z.S. Qian, X.Y. Shan, L.J. Chai, J.J. Ma, J.R. Chen, H. Feng, DNA nanosensor based on biocompatible graphene quantum dots and carbon nanotubes. Biosens. Bioelectron. **60**, 64–70 (2014). doi:10.1016/j.bios.2014.04.006
259. Y. Shi, Y. Pan, H. Zhang, Z. Zhang, M.J. Li, C. Yi, M. Yang, A dual-mode nanosensor based on carbon quantum dots and gold nanoparticles for discriminative detection of glutathione in human plasma. Biosens. Bioelectron. **56**, 39–45 (2014). doi:10.1016/j.bios.2013.12.038
260. S. Li, J. Luo, G. Yin, Z. Xu, Y. Le, X. Wu, N. Wu, Q. Zhang, Selective determination of dimethoate via fluorescence resonance energy transfer between carbon dots and a dye-doped molecularly imprinted polymer. Sens. Actuat. B Chem. **206**, 14–21 (2015). doi:10.1016/j.snb.2014.09.038
261. L. Wu, M. Li, M. Zhang, M. Yan, S. Ge, J. Yu, Ultrasensitive electrochemiluminescence immunosensor for tumor marker detection based on nanoporous sliver@carbon dots as labels. Sens. Actuat. B Chem. **186**, 761–767 (2013). doi:10.1016/j.snb.2013.06.092

262. J. Lou, S. Liu, W. Tu, Z. Dai, Graphene quantums dots combined with endonuclease cleavage and bidentate chelation for highly sensitive electrochemiluminescent DNA biosensing. Anal. Chem. **87**(2), 1145–1151 (2015). doi:10.1021/ac5037318
263. J. Shi, C. Lu, D. Yan, L. Ma, High selectivity sensing of cobalt in $HepG_2$ cells based on necklace model microenvironment-modulated carbon dot-improved chemiluminescence in Fenton-like system. Biosens. Bioelectron. **45**, 58–64 (2013). doi:10.1016/j.bios.2013.01.056
264. M. Amjadi, J.L. Manzoori, T. Hallaj, Chemiluminescence of graphene quantum dots and its application to the determination of uric acid. J. Lumin. **153**, 73–78 (2014). doi:10.1016/j.jlumin.2014.03.020

Photoluminescent Properties of Carbon Nanodots

Bao-Ping Qi, Guo-Jun Zhang, Zhi-Ling Zhang and Dai-Wen Pang

Abstract With unique and tunable photoluminescent properties, carbon nanodots (CNDs), as a new class of optical tags, have been extensively studied. In this chapter, we introduce the basic knowledge with respect to CNDs, including their structures and compositions, optical properties and applications in the bioimaging and biosensors. In particular, the photoluminescence (PL) mechanisms of CNDs, which are able to instructively improve its optical properties, have been emphasized and discussed in details. We hope to inspire research into the origins of the unique properties of CNDs and intrigue the researchers with different research backgrounds to participate in this field and explore the PL mechanisms of CNDs.

Keywords Carbon nanodots · Photoluminescence · Mechanism · Density functional theory · Electrochemiluminescence

1 Introduction

Since carbon nanodots (CNDs) were accidentally discovered by Scrivens et al. in the process of purifying single-walled carbon nanotubes (SWCNTs) through arc-discharge methods [1], luminous CNDs have attracted a great deal of interests from the viewpoints of both fundamental studies and applications. Typically, CNDs are smaller than 10 nm in size, and show size- and excitation-dependent photoluminescence (PL) behaviors. CNDs are mainly composed of carbon, oxygen, and

B.-P. Qi · Z.-L. Zhang · D.-W. Pang (✉)
Key Laboratory of Analytical Chemistry for Biology and Medicine (Ministry of Education), College of Chemistry and Molecular Sciences, State Key Laboratory of Virology, The Institute for Advanced Studies, Wuhan Institute of Biotechnology, Wuhan University, Wuhan 430072, People's Republic of China
e-mail: dwpang@whu.edu.cn

G.-J. Zhang
School of Laboratory Medicine, Hubei University of Chinese Medicine, Huangjia Lake West Road, Wuhan 430065, People's Republic of China

N. Yang et al. (eds.), *Carbon Nanoparticles and Nanostructures*,
Carbon Nanostructures, DOI 10.1007/978-3-319-28782-9_7

other heteroatom with many oxygenous functional groups at their surface [2, 3], thus imparting them with ease to be functionalized and excellent water solubility [4–6]. In addition, CNDs are characteristic of small sizes, low cytotoxicity, excellent photo-stability and chemical inertness [7]. Due to these features, CNDs have already displayed their potentials in biolabeling, in vivo, and dynamic tracer applications [8, 9].

One of the most notable issues is whether the emerging optical materials CNDs are able to displace the commonly used semiconductor quantum dots (QDs), having the unknown environmental and biological hazards for their heavy metal [10, 11]. The luminous carbonaceous CNDs exhibit low-toxicity and eco-friendly properties. However, compared with the QDs, the quantum yield of CNDs is much less, which would hinder their further developments and applications. Hence, it is necessary to make clear the PL mechanisms of CNDs to improve the quantum yield of CNDs. Both the complicated carbon/oxygen chemical bonds and the non-stoichiometric nature of CNDs make it difficult to study the intrinsic PL properties. Up to now, several PL mechanisms have been mainly suggested to originate from surface state, conjugated structures, special structure sites, etc. [12–15]. The viewpoint that the radiative recombinations of the CNDs surface-confined electrons and holes are responsible for the PL phenomenon is the surface state. The π-electron systems, zigzag sites and luminol-like units on the CNDs are also deemed to be the center of PL. The surface state is able to explain the excitation-dependent PL behaviors of CNDs, but the free zigzag sites are more suitable for the pH-dependent properties of CNDs. Up to now, none of them could account for all the PL phenomena of CNDs.

Herein, we introduce the CNDs regarding the structures and compositions, the basic optical properties and their applications in bioimaging and biosensors. In addition, we pay much attention to the PL mechanisms of CNDs, which is a significant aspect. We look forward to obtaining the better understanding into the origins of their PL behaviors, achieving higher quantum yield, and developing novel applications in further.

2 Structures and Compositions

CNDs are very small in size, and usually less than 10 nm (Fig. 1a). They are composed of a carbon backbone with sp^2 carbon or sp^3 carbon (Fig. 1c), and abundant in oxygen-containing groups at the edge or the base plane (Fig. 1b). The special structures of CNDs determine their special properties, among which the PL performance has caused widely public concern. Mostly, CNDs are divided into two members, carbon dots (CDs) and graphene quantum dots (GQDs). CDs are always spherical, and they are divided into carbon nanoparticles (CNPs) without crystal-lattice and carbon quantum dots (CQDs) with obvious crystal lattice. Another carbon-based nanomaterial, whose size and surface functionality are similar to the CDs, is GQDs. GODs have the lateral dimension larger than their height, which can be regarded as small pieces of graphene. The UV-visible absorptions and PL spectra

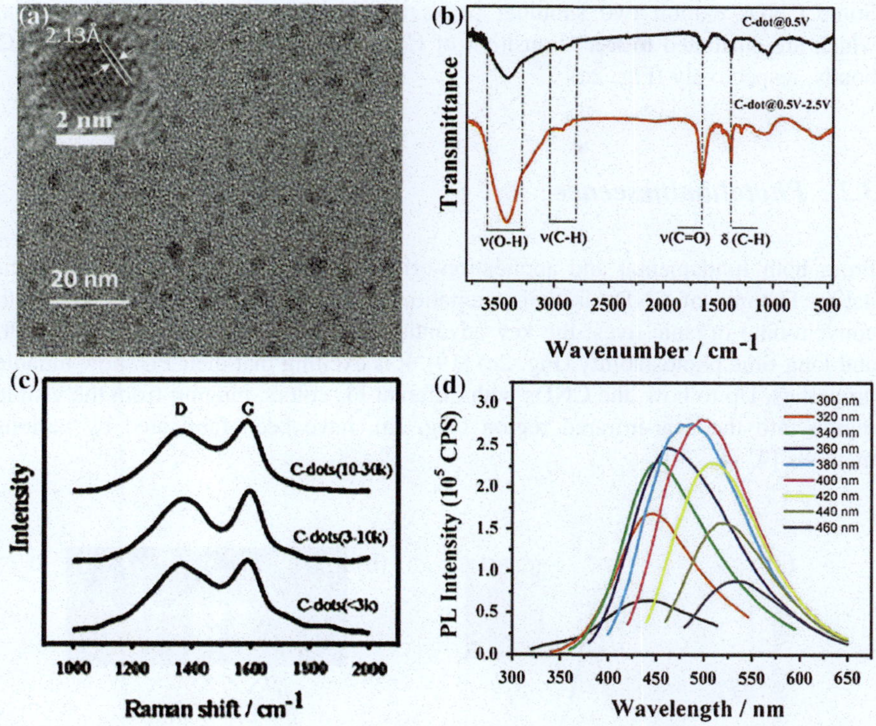

Fig. 1 **a** The high-resolution transmission electron microscopy images of CNDs [16]. **b** FT-IR spectra of CNDs [17]. **c** Raman spectra of CNDs [18]. **d** The excitation-dependent PL spectra of CNDs in water [17]

of the carbonaceous QDs (CQDs and GQDs) from the quantum confinement effect are not so clear, which are obviously different from those of the semiconductor QDs. So both GQDs and CQDs are not proper to be brought into the system of QDs. Typically, both CDs and GQDs display many parallel performances such as excitation-dependent PL behaviors (Fig. 1d). Based on these similar structures and properties, herein CDs and GQDs are also categorized as CNDs.

3 Optical Properties

3.1 Absorbance

CNDs typically show strong optical absorption in the UV region (200–320 nm) due to the effective photon-harvesting, with a tail extending out into the visible range.

Some CNDs exhibit two shoulder peaks in the strong background absorption, which are attributed to π-π* transition of C=C bonds and n-π* transition of C=O bonds, respectively (Fig. 2a).

3.2 Photoluminescence

From both fundamental and application-oriented stances, one of the most fascinating features of CNDs is their excitation dependence [7]. As compared with conventional organic dyes, the key advantage of CNDs include non-blinking PL and long-time photostability (Fig. 2c) [19]. It is exciting that their PL show tunable properties. Up to now, the CNDs with different PL colors, ranging from the visible region into the near-infrared region (Fig. 2b), have been fabricated by various methods [4, 18, 20].

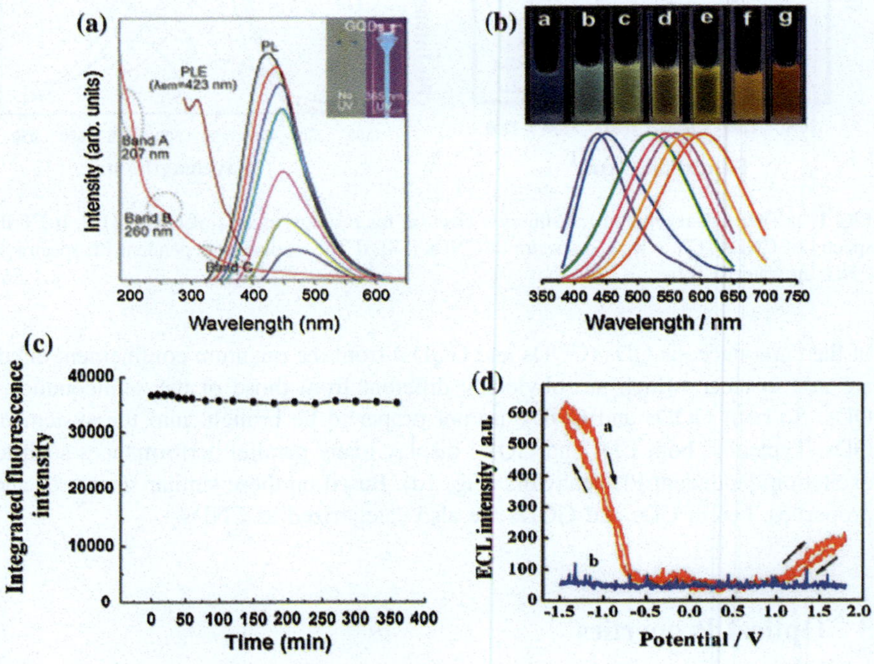

Fig. 2 **a** UV-visible absorption spectra of CNDs in water [35]. **b** *Top* Optical images of as-prepared multicolor fluorescent CNDs. *Bottom* PL spectra of CNDs [18]. **c** Dependence of fluorescence intensity on excitation time for CNDs in water [19]. **d** ECL responses: (*a*) with and (*b*) without CNDs at a Pt electrode in 0.1 M PBS (pH 7.0) [36]

3.3 *Upconversion Photoluminescence*

Upon simultaneous absorption of two or sequential absorption of multiple longer wavelength photons, the emission wavelength is shorter than the excitation wavelength, which is called upconversion PL. Some CNDs exhibit this PL feature [21–24]. Li et al. have reported that the CNDs prepared by the hydrazine hydrate reduction of graphene oxide with surface-passivated by a polyethylene glycol showed the upconverted emissions [25]. When the excitation wavelength changed from 600 to 800 nm, the upconverted emissions peaks shifted from 390 to 468 nm, respectively. The shifting between the energy of the upconverted emission wavelength and the excitation wavelength was almost unchanged and stayed around 1.1 eV, which was close to the δE between the σ and π orbitals. Based on this conclusion, they speculate on the upconversion PL due to the anti-Stokes PL. Since the long excitation wavelength owns the traits of deep-tissue penetration, low photon-induced toxicity, low back ground interference, etc., upconversion PL will enable CNDs to be desirable materials in in vivo imaging.

3.4 *Electrochemiluminescence (ECL)*

Similar to QDs [26, 27] and Si nanodots [28], CNDs exhibit ECL properties as well. For the first time, Chi et al. have reported the ECL phenomenon of the CNDs solution at a Pt disk working electrode (Fig. 2d). The ECL mechanism of the CNDs was suggested to involve the formation of excited-state CNDs (R*) via electron-transfer annihilation of negatively charged ($R^{\cdot -}$) and positively charged ($R^{\cdot +}$) CNDs. The intensity of cathodic ECL was larger than that of anodic ECL, indicating that $R^{\cdot +}$ was more stable than $R^{\cdot -}$. From then on, the ECL properties of CNDs have progressively drawn attention [29–32].

ECL was demonstrated to be more sensitive than PL to surface chemistry, which makes it a powerful tool to study surface states of nanocrystals [33]. Especially, the current signal and the light signal are obtained simultaneously, facilitating the investigation of light emission mechanism of the luminophor [34]. It is of great significance to compare the ECL spectrum with PL spectrum of CNDs, which will be discussed in the following.

4 Photoluminescence Mechanisms

Although CNDs have been studied for ten years, knowledge into the origins of their PL is still an open question and requires further clarification. The PL phenomena of CNDs have been mainly suggested to originate from surface state, conjugated structures, special structure sites, etc. On one hand, the quantum yields of CNDs are lower than those of QDs. On the other hand, the CNDs prepared by various methods from different raw materials are mainly blue or green in color. In order to effectively improve the optical properties of CNDs, a generally recognized PL mechanism capable of explaining all the PL phenomena of CNDs is urgently needed.

4.1 Surface State

The periodic lattices in the crystals are destroyed in some directions, resulting in a new state near the surface, which is the so-called surface state. The surface state in CNDs has been deemed to be the hybridization of the carbon backbone and the linked functional groups [12]. The characteristics of the excitation dependence from CNDs may reflect that there is a distribution of different emissive sites in each CNDs [7]. Sun et al. have attributed the PL from CNDs to the presence of surface energy traps that became emissive upon stabilization as a result of the surface passivation. The electrons and holes were generated likely by efficient photoinduced charge separations in the CNDs, and the roles of surface passivation by the organic or other functionalization was able to make the surface sites more stable to facilitate more effective radiative recombinations. At the same time, they draw a conclusion that there must be a quantum confinement of emissive energy traps to the particle surface, since the larger CNDs with the same surface passivation were found to be much less luminescent [15].

The CNDs prepared by different kinds of methods also exhibited the behaviors of excitation dependence [37–40], but no additional passivation step was required for PL to occur. In common, there are more or less oxygenous functional groups on the surface of CNDs. The oxygen-based groups on the carbon core could be regarded as the primary surface state of CNDs. From then on, lots of researches have been focused on the effect of oxygenous groups on the PL of CNDs. Zheng et al. [41] have reported that the CNDs with green emission could be changed to the blue ones through $NaBH_4$ reduction. Liu et al. [42] have prepared blue-color emissions of CNDs without oxygenous defects and their oxidized form with green-color emissions. They reveal that the green PL of CNDs originates from defect states with oxygenous functional groups, whereas the blue luminescence of CNDs is dominated by intrinsic states in the high-crystalline structure.

Our group has investigated the surface oxidation degree on the PL of CNDs in detail and proposed a surface oxidation-related PL mechanism for CNDs [17, 18, 43]. Bao et al. [17] have developed an electrochemical method to prepare luminescent CNDs with controllable sizes. The higher the applied potentials, the smaller the resulting CNDs. Importantly, once the CNDs were exfoliated into the solution, the size of the as-prepared CNDs would not change further, making it possible for the further electrochemical oxidation. The as-prepared CNDs at 0.5 V were further electrochemically oxidized at a home-made platinum cup electrode at 2.5 V. After that, the optimal emission wavelength of the further oxidized CNDs was red-shifted

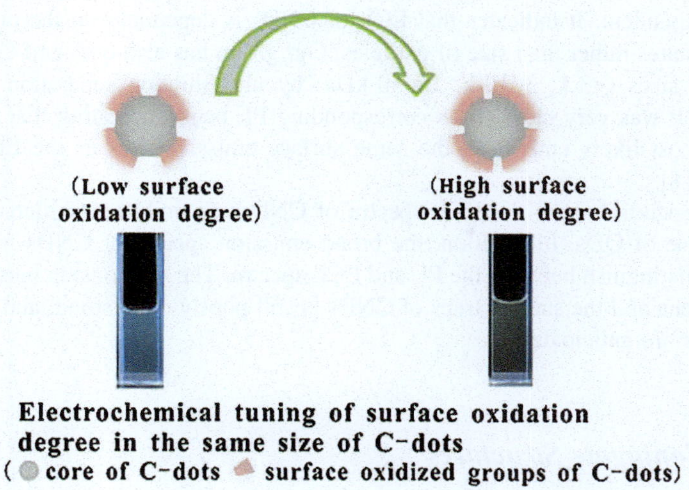

Fig. 3 Illustration of emission from CNDs along with the variation of surface oxidation degree [17]

by ca. 45 nm. In other words, the high surface oxidation degree resulted in the red-shifted PL of CNDs (Fig. 3). Surface states are the key to the PL of the CNDs. This work opens a new window not only to controllably prepare CNDs with small sizes and long emission wavelengths, but also to understand the PL mechanisms of CNDs. Our group [43] has also obtained two kinds of CNDs with the shifting or non-shifting PL at varied excitations by electro-oxidizing carbon paste electrodes with different compositions. CNDs with more complex surface states, related to a higher degree of surface oxidation, afford fluorescence emissions with varied energies at different excitations. However, it is demonstrated that CNDs with less surface states have the non-shifting fluorescence properties. Therefore, the emissions are proposed to be mainly attributed to the surface states caused by the surface oxidation of CNDs.

In order to confirm the surface state on the surface of CNDs, ECL, which is a highly sensitive technique for probing the surface of nanoparticles [44], has been adopted to probe the surface-state electronic transitions of the CNDs. Most ECL phenomena from semiconductor QDs have been observed to originate from surface states, which are often significantly red shifted from the PL peaks by as much as hundreds of nanometers, since these defect states are located in the band gap [27, 44, 45]. Ding et al. have discovered that the ECL spectrum of CNDs prepared by electrochemical etching was red shifted about 50 nm from the PL spectrum in organic solution [30]. The authors attributed the origin of both the ECL and PL to the surface traps in CNDs. Dong et al. [46] have obtained a series of the oxidized CNDs with the same oxidation degree by chemical oxidation. The PL spectra of the oxidized CNDs with different sizes (1–3, 3–5, 5–10, 10–30 kDa) were particle sized-dependent, but interestingly, ECL spectra of CNDs were all 600 nm, showing size-independent. It indicates that ECL of CNDs is dependent on the presence of surface states rather than size of particles. Our group has also obtained CNDs with different sizes (<3 k, 3–10 k, 10–30 kDa) by ultrafiltration separation and each ECL peak was very close to its corresponding PL peak, indicating that the PL of CNDs most likely emit from the same surface emissive sites as the ECL of the CNDs [18].

These studies show the ECL spectra of CNDs are tinily red-shifted compared with those of QDs. In addition, the broad emission spectra of CNDs make them hard to distinguish between the PL and ECL spectra. The relationship between ECL phenomena and the surface traps of CNDs is still poorly understood, and the better materials are anticipated.

4.2 Conjugate Structures

The sp^2 domain CNDs with hardly oxygenous defects, achieved by both physical exfoliation of graphite nanoparticles and intercalation into graphite through special compounds, show PL properties [42, 47]. Some studies have demonstrated the possibility of tailoring the sp^2 domains and designing the edge structures of CNDs. Li et al. have achieved the full sp^2 domain CNDs with different sizes through chemical synthesis, clearly demonstrating that a large sp^2 domain results in the decrease of band gap [48–51]. However, the complex synthetic steps and the strong tendency to aggregate hinder its future development. By electrooxidation of graphite, our group has obtained two kinds of CNDs with blue and yellow emissions, respectively [19]. The red shift in the emission wavelength was attributed to the increase of the conjugation system, which was testified by the peak position of ν(C=O) in FTIR.

The density functional theory (DFT) has been widely used in the study of CNDs with single layer, which can provide a convincing evidence for the conjugated structures-related PL mechanism. Theoretical modelling and calculations allow precise isolation of the influence of each factor while fixing the others. Eda et al. [52] have reported that sp^2 domains isolated within sp^3 carbon matrix were responsible for the observed blue fluorescence from graphene oxide, which was firstly supported by calculations based on DFT. To confirm that the strong emission comes from the quantum-sized graphite fragment of CNDs, Li et al. [53] have performed theoretical calculations to investigate the relationship between luminescence and cluster sizes. As the size of the fragment increases, the gap decreases gradually, which is in good agreement with Eda's report. In order to study the effect of conjugated structures on PL properties of CNDs, our group has developed an efficient strategy edge-functionalization method [16]. The added conjugated structures through the formation of aromatic pyrazine led to the red-shifted emissions of CNDs, which was testified by DFT-based calculation (Fig. 4). Both the experimental results and the DFT-based calculations suggest that the mechanism for conjugated structures in CNDs to tune the band gap of CNDs. Sk et al. [54] have

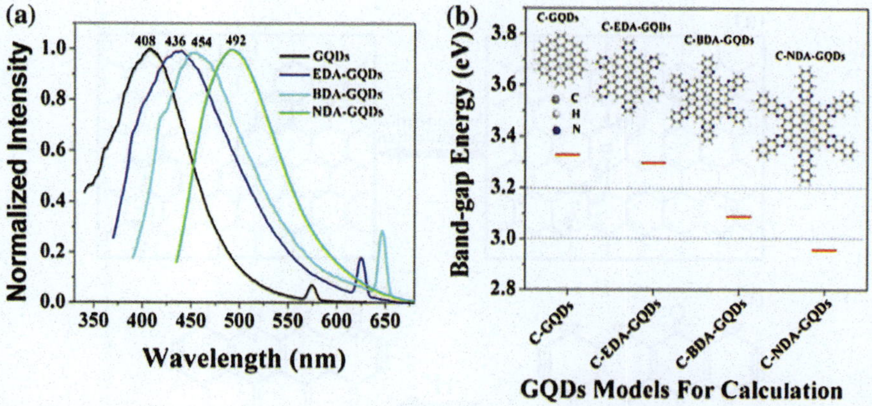

Fig. 4 **a** The PL emission spectra of CNDs dispersed in ethanol. **b** The calculated band-gap energy of the corresponding CNDs dispersed in ethanol based on DFT. The *inset* shows the structures of CNDs used for calculations [16]

revealed that the PL of CNDs is sensitively affected by their size and covers the entire visible light spectrum by varying the size of sp^2 domains from 0.89 to 1.80 nm based on the DFT and time-dependent DFT calculations.

4.3 Special Structure Sites

Owing to the small diameter of CNDs, Pan et al. have proposed that there is a high concentration of free zigzag sites at the edge of CNDs, leading to the strong PL [55]. It may originate from free zigzag sites with a carbene-like triplet ground state (Fig. 5). The energy difference (δE) between the two electronic transitions observed in the PL emission of CNDs is determined to be 0.96 eV, within the value (<1.5 eV) for triple carbenes defined by Hoffmann. The proposed PL mechanism based on the emissive free zigzag sites could explain pH-dependent PL phenomena of CNDs. Under acidic conditions, the free zigzag sites of CNDs are protonated, forming a reversible complex between the zigzag sites and H^+. Thus the emissive triple carbene state is broken and becomes inactive in PL. However, under alkaline conditions, the free zigzag sites are restored, thereby leading to the restoration of PL. If pH is switched repeatedly between 13 and 1, the PL intensity varies reversibly.

The main PL spectrum, chemiluminescence (CL) spectrum and ECL spectrum of hydrazide-modified CNDs are essentially the same, implying that the three spectra are given by similar excited states [56]. It has also been reported that refluxing CNDs with N_2H_4 would result in the formation of abundant luminol-like units at the edges of CNDs, which is a well-known luminophore, exhibiting excellent PL, CL and ECL properties. The main PL spectrum of hydrazide-modified CNDs is quite

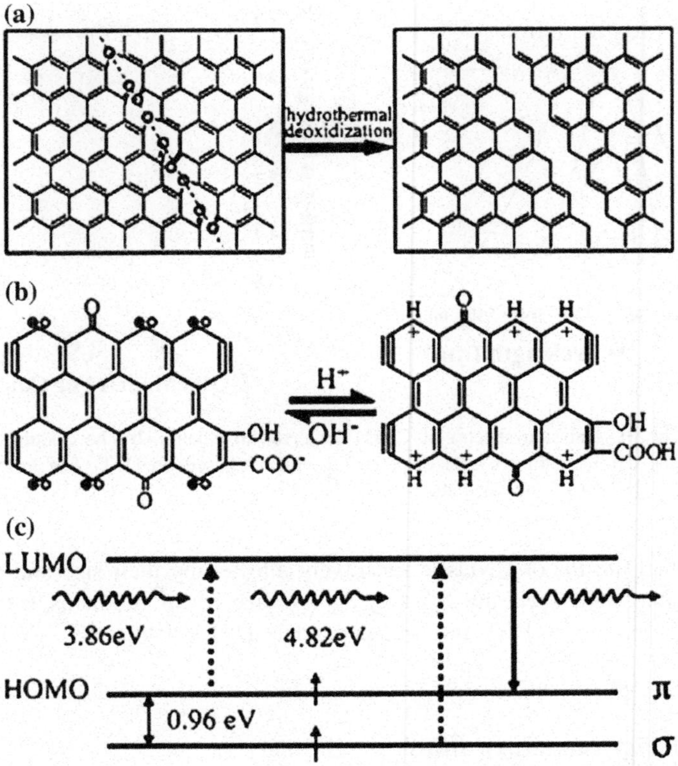

Fig. 5 **a** Mechanism for the preparation of CNDs. **b** Models of the CNDs in acidic and alkali media. The two models can be converted reversibly depending on pH. The pairing of σ(•) and π(o) localized electrons at carbene-like zigzag sites and the presence of triple bonds at the carbyne-like armchair sites are represented. **c** Typical electronic transitions of triple carbenes at zigzag sites [55]

similar to that of luminol, suggesting that the PL of CNDs might be derived from the luminol-like units.

PL mechanisms of CNDs may be attributed to either combining effect or competition between the intrinsic and defect state emission. Our group has revealed that the PL properties of CNDs were mainly influenced by both their size and degree of surface oxidation through surface analytical techniques time-resolved photoluminescence spectroscopy and the ECL technique [18]. Increasing the degree of surface oxidation leads to a narrowing of the energy gap of the surface; meanwhile, larger CNDs with an extensive π-electron system, which couple with surface electronic states, can also lead to a narrowing of the energy gap of the surface states. Sun et al. have classified the emissions of CNDs into two primary categories: emissions that originate from created or induced energy band gaps in a single graphene sheet and emissions that are associated with defects in single- and/or multiple-layer graphene [13]. They have proposed that the band gap fluorescence in graphene is associated with the conjugated structures on a single sheet, which is quenched by π-domains in neighboring sheets in a few-layer configuration. However, the defect-derived emissions are associated with the defect site across several sheets in a similar few-layer configuration, and thus little affected by the interlayer interactions.

Though the above PL mechanisms have been raised, none of them is able to explain all the PL phenomena of CNDs. Obviously, PL mechanism of CNDs is currently ambiguous. To resolve it, researchers in different fields are needed to participate in this study.

5 Applications

5.1 Bioimaging

The luminous CNDs combined with the characteristics of excellent photostability, ease to be functionalized, low cytotoxicity, chemical inertness, and small size, render them applicable in bioimaging, biolabeling and dynamic tracer [8]. Our group has reported that the multicolor CNDs showed low cytotoxicity to the VERO cells through 3-(4,5-dimethylthiazol-2-yl)-2,5-diphenyltetrazolium bromide (MTT) and finally realized the multicolor imaging of cells [18]. The cells efficiently took up CNDs, which was regarded as an endocytosis mechanism [57–59]. Lee et al. have first studied the in vivo bio-distribution and the potential toxicity of CNDs (Fig. 6) [60]. Four kinds of cancer cells and one kind of normal cell were chosen as in vitro cell culture models to examine the possible adverse effects of CNDs. At the tested exposure levels, no acute toxicity or morphological changes were noted in KB, MDA-MB231, A549 cancer cells, and MDCK normal cells, respectively. A long-term in vivo study revealed that CNDs mainly accumulated in liver, spleen, lung, kidney, and tumor sites after intravenous injection. The serum biochemical analysis and histological evaluation study revealed that CNDs did not cause appreciable toxicity to the treated mice. With adequate studies of toxicity, both in vitro and in vivo, luminous CNDs may be considered for potential biological applications. Chen et al., have demonstrated that CNDs could be used to specifically label and track molecular targets involved in dynamic cellular processes in live cells [61]. The internalization, trafficking, and recycling of insulin receptors in adipocytes have been monitored in real-time by using insulin-conjugated CNDs, which were synthesized through amidation reaction. They reveal for the first time that the insulin receptor dynamics are stimulated by apelin and inhibited by TNFα, providing evidence for the molecular mechanisms underlying the regulation of these cytokines in insulin sensitivity. These activities have been successfully proceeded, which can be attributed to the excellent photostability of CNDs. This study demonstrates the great potentials of CNDs in live-cell imaging, particularly for investigating dynamic cellular processes.

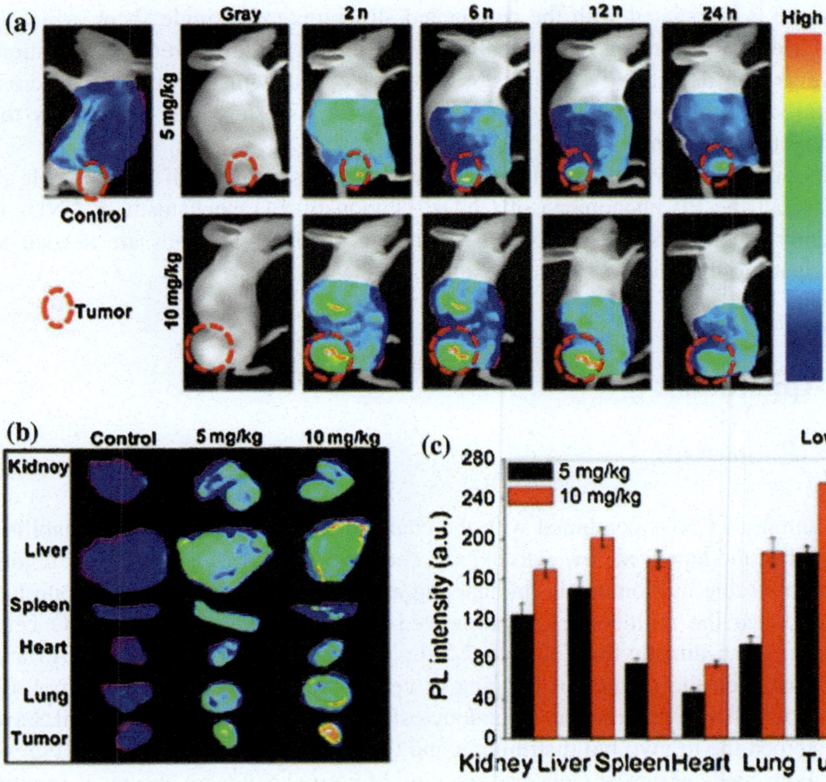

Fig. 6 In vivo imaging and biodistribution of CNDs. **a** The in vivo imaging of KB tumor bearing mice after intravenous injection of CNDs, **b** ex vivo images of isolated organs of mice at 24 h after injection of CNDs, and **c** PL intensities of the carboxylated CNDs from isolated organs at different dosages [60]

5.2 *Sensors*

Based on PL and ECL properties of CNDs discussed above, various sensors have been designed as signal-off or signal-on processes to detect the target. Zhao et al. have provided a novel and versatile signaling transduction strategy in the fluoroimmunoassay through regulating the interaction between graphene and CNDs (Fig. 7) [62]. When adding graphene to the mIgG-CNDs solution, the π-π stacking interaction between graphene and CNDs could bring graphene and CNDs into resonance energy transfer (RET) proximity to facilitate the PL quenching of CNDs. In the sensing process, the addition of IgG specifically linked with the mIgG increases the distance between CNDs and graphene surface, resulting in restoration the PL of CNDs. Based on fluorescence resonance energy transfer (FRET) from upconverting phosphors (UCPs) to CNDs, our group has also presented a new aptamer biosensor for thrombin [63].

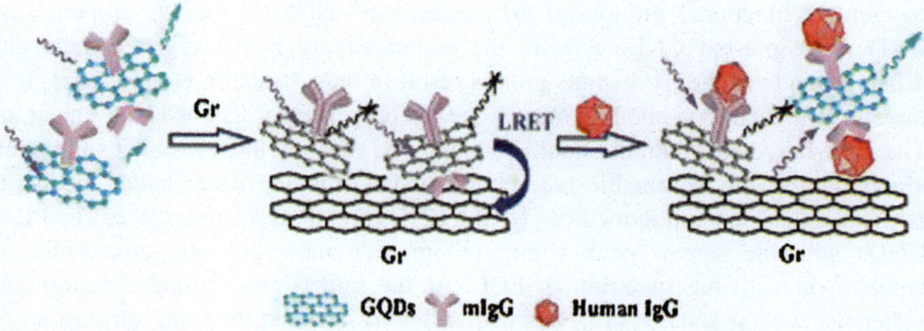

Fig. 7 Schematic illustration of a universal immunosensing strategy based on regulation of the interaction between graphene and CNDs [62]

Since Chi et al. have reported that CNDs can generate the ECL signal, CNDs has been widely used in the fabrication of the ECL-related sensors [64–71]. Zhu et al. have reported that the co-reaction reagent $S_2O_8{}^{2-}$ could significantly increase the ECL signal of the CNDs [31], and a novel ECL sensor for Cd^{2+} has been proposed based on the competitive coordination between the chelator cysteine and CNDs for metal ions. For the first time, our group has demonstrated that CNDs prepared by electrochemical etching could act as the coreactants for the anodic ECL of $Ru(bpy)_3^{2+}$ [72]. It has been suggested that the benzylic alcohol units on the CNDs are responsible for the capability as core actants in the anodic ECL process of $Ru(bpy)_3^{2+}$/CNDs. For the low biotoxicity, the system $Ru(bpy)_3^{2+}$/CNDs would be much promising in bioanalysis, which was exemplified by the quantitative detection of dopamine molecules. This work has deepened and broadened the knowledge of CNDs.

In addition, CNDs has also been applied in the fields of photocatalyst [53], organic photovoltaic (OPV) devices [73], oxygen reduction reaction [49, 74], etc. Due to the special structures, CNDs shows the peroxidase-like activities, realizing the great performance and stability in H_2O_2 detection [75, 76].

6 In Conclusion

Based on the excellent photostability, ease to be functionalized, low cytotoxicity, and chemical inertness, the luminous CNDs with small sizes have, no doubt, brought about tremendous interest. They would have the potentials to partially take the place of QDs. However, the research on CNDs is still in its early stage, and the PL from most CNDs with low quantum yields is blue or green. It is particularly important that the explicit PL mechanism of CNDs will facilitate the improvement of the quantum yields and the red-shifted emission.

Obviously, the ambiguous PL mechanism of CNDs hindered their developments. Herein, we would like to share our viewpoints on some critical issues on the PL mechanism of CNDs. On one hand, it is necessary to make clear the effect of the oxygenous functional groups on the surface of CNDs on the PL properties of CNDs. The present CNDs contain the comparatively high oxygen content and different kinds of the oxygenous groups result in their structure complexities. But, the effect of each oxygenous group on the PL properties of CNDs is not uncertain. The DFT-based calculations should be employed to study these problems in details via designing the reasonable theoretical models. On the other hand, the better fabrication and separation methods for the CNDs are of great importance. The PL of CNDs with the narrow peak width will provide more accurate information in comparison with the spectrum of ECL. At the same time, the more appropriate materials, such as a series of CNDs with different sizes but the same surface, are in urgent need for the ECL technique. For the complexity of the CNDs, more methods and more researchers in different fields are anticipated to participate in this domain.

CNDs have shown primary potential in imaging and related biosensor applications. Once the low toxic CNDs possess high quantum yield and tunable PL, they will have a huge impact in both health and environmental applications.

Acknowledgments This work was supported by the National Basic Research Program of China (973 Program, Grant No. 2011CB933600), the National Natural Science Foundation of China (Grant No. 21535005), the 111 Project (111-2-10), and Collaborative Innovation Center for Chemistry and Molecular Medicine.

References

1. X. Xu, R. Ray, Y. Gu, H.J. Ploehn, L. Gearheart, K. Raker, W.A. Scrivens, Electrophoretic analysis and purification of fluorescent single-walled carbon nanotube fragments. J. Am. Chem. Soc. **126**(40), 12736–12737 (2004). doi:10.1021/ja040082h
2. W. Kwon, J. Lim, J. Lee, T. Park, S.-W. Rhee, Sulfur-incorporated carbon quantum dots with a strong long-wavelength absorption band. J. Mater. Chem. C **1**(10), 2002–2008 (2013). doi:10.1039/c3tc00683b
3. Y.P. Sun, X. Wang, F.S. Lu, L. Cao, M.J. Meziani, P.J.G. Luo, L.R. Gu, L.M. Veca, Doped carbon nanoparticles as a new platform for highly photoluminescent dots. J. Phys. Chem. C **112**(47), 18295–18298 (2008). doi:10.1021/jp8076485
4. H. Tetsuka, R. Asahi, A. Nagoya, K. Okamoto, I. Tajima, R. Ohta, A. Okamoto, Optically tunable amino-functionalized graphene quantum dots. Adv. Mater. **24**(39), 5333–5338 (2012). doi:10.1002/adma.201201930
5. F. Jiang, D. Chen, R. Li, Y. Wang, G. Zhang, S. Li, J. Zheng, N. Huang, Y. Gu, C. Wang, C. Shu, Eco-friendly synthesis of size-controllable amine-functionalized graphene quantum dots with antimycoplasma properties. Nanoscale **5**(3), 1137–1142 (2013). doi:10.1039/c2nr33191h
6. S.J. Zhu, Q.N. Meng, L. Wang, J.H. Zhang, Y.B. Song, H. Jin, K. Zhang, H.C. Sun, H.Y. Wang, B. Yang, Highly photoluminescent carbon dots for multicolor patterning, sensors, and bioimaging. Angew. Chem. Int. Ed. **52**(14), 3953–3957 (2013). doi:10.1002/anie.201300519
7. S.N. Baker, G.A. Baker, Luminescent carbon nanodots: emergent nanolights. Angew. Chem. Int. Ed. **49**(38), 6726–6744 (2010). doi:10.1002/anie.200906623
8. C.Q. Ding, A.W. Zhu, Y. Tian, Functional surface engineering of C-dots for fluorescent biosensing and in vivo bioimaging. Acc. Chem. Res. **47**(1), 20–30 (2014). doi:10.1021/ar400023s

9. J. Shen, Y. Zhu, X. Yang, C. Li, Graphene quantum dots: emergent nanolights for bioimaging, sensors, catalysis and photovoltaic devices. Chem. Commun. **48**(31), 3686–3699 (2012). doi:10.1039/c2cc00110a
10. Y.Y. Su, Y. He, H.T. Lu, L.M. Sai, Q.N. Li, W.X. Li, L.H. Wang, P.P. Shen, Q. Huang, C.H. Fan, The cytotoxicity of cadmium based, aqueous phase-synthesized, quantum dots and its modulation by surface coating. Biomaterials **30**(1), 19–25 (2009). doi:10.1016/j.biomaterials.2008.09.029
11. Y.Y. Su, M. Hu, C.H. Fan, Y. He, Q.N. Li, W.X. Li, L.H. Wang, P.P. Shen, Q. Huang, The cytotoxicity of CdTe quantum dots and the relative contributions from released cadmium ions and nanoparticle properties. Biomaterials **31**(18), 4829–4834 (2010). doi:10.1016/j.biomaterials.2010.02.074
12. S. Zhu, Y. Song, X. Zhao, J. Shao, J. Zhang, B. Yang, The photoluminescence mechanism in carbon dots (graphene quantum dots, carbon nanodots, and polymer dots): current state and future perspective. Nano Res. **8**(2), 355–381 (2015). doi:10.1007/s12274-014-0644-3
13. L. Cao, M.J. Meziani, S. Sahu, Y.P. Sun, Photoluminescence properties of graphene versus other carbon nanomaterials. Acc. Chem. Res. **46**(1), 171–180 (2013). doi:10.1021/ar300128j
14. L. Li, G. Wu, G. Yang, J. Peng, J. Zhao, J.-J. Zhu, Focusing on luminescent graphene quantum dots: current status and future perspectives. Nanoscale **5**(10), 4015–4039 (2013). doi:10.1039/c3nr33849e
15. Y.P. Sun, B. Zhou, Y. Lin, W. Wang, K.A. Fernando, P. Pathak, M.J. Meziani, B.A. Harruff, X. Wang, H. Wang, P.G. Luo, H. Yang, M.E. Kose, B. Chen, L.M. Veca, S.Y. Xie, Quantum-sized carbon dots for bright and colorful photoluminescence. J. Am. Chem. Soc. **128**(24), 7756–7757 (2006). doi:10.1021/ja062677d
16. B.P. Qi, H. Hu, L. Bao, Z.L. Zhang, B. Tang, Y. Peng, B.S. Wang, D.W. Pang, An efficient edge-functionalization method to tune the photoluminescence of graphene quantum dots. Nanoscale **7**(14), 5969–5973 (2015). doi:10.1039/C5nr00842e
17. L. Bao, Z.L. Zhang, Z.Q. Tian, L. Zhang, C. Liu, Y. Lin, B. Qi, D.W. Pang, Electrochemical tuning of luminescent carbon nanodots: from preparation to luminescence mechanism. Adv. Mater. **23**(48), 5801–5806 (2011). doi:10.1002/adma.201102866
18. L. Bao, C. Liu, Z.L. Zhang, D.W. Pang, Photoluminescence-tunable carbon nanodots: surface-state energy-gap tuning. Adv. Mater. **27**(10), 1663–1667 (2015). doi:10.1002/adma.201405070
19. Q.L. Zhao, Z.L. Zhang, B.H. Huang, J. Peng, M. Zhang, D.W. Pang, Facile preparation of low cytotoxicity fluorescent carbon nanocrystals by electrooxidation of graphite. Chem. Commun. **41**, 5116–5118 (2008). doi:10.1039/b812420e
20. X.M. Li, S.P. Lau, L.B. Tang, R.B. Ji, P.Z. Yang, Multicolour light emission from chlorine-doped graphene quantum dots. J. Mater. Chem. C **1**(44), 7308–7313 (2013). doi:10.1039/c3tc31473a
21. M. Li, W. Wu, W. Ren, H.-M. Cheng, N. Tang, W. Zhong, Y. Du, Synthesis and upconversion luminescence of N-doped graphene quantum dots. Appl. Phys. Lett. **101**(10), 103107 (2012). doi:10.1063/1.4750065
22. A. Salinas-Castillo, M. Ariza-Avidad, C. Pritz, M. Camprubi-Robles, B. Fernandez, M. J. Ruedas-Rama, A. Megia-Fernandez, A. Lapresta-Fernandez, F. Santoyo-Gonzalez, A. Schrott-Fischer, L.F. Capitan-Vallvey, Carbon dots for copper detection with down and upconversion fluorescent properties as excitation sources. Chem. Commun. **49**(11), 1103–1105 (2013). doi:10.1039/c2cc36450f
23. C. Wang, X. Wu, X. Li, W. Wang, L. Wang, M. Gu, Q. Li, Upconversion fluorescent carbon nanodots enriched with nitrogen for light harvesting. J. Mater. Chem. **22**(31), 15522 (2012). doi:10.1039/c2jm30935a
24. X.M. Wen, P. Yu, Y.R. Toh, X.Q. Ma, J. Tang, On the upconversion fluorescence in carbon nanodots and graphene quantum dots. Chem. Commun. **50**(36), 4703–4706 (2014). doi:10.1039/C4cc01213e

25. J. Shen, Y. Zhu, C. Chen, X. Yang, C. Li, Facile preparation and upconversion luminescence of graphene quantum dots. Chem. Commun. **47**(9), 2580–2582 (2011). doi:10.1039/c0cc04812g
26. L. Bao, L. Sun, Z.-L. Zhang, P. Jiang, F.W. Wise, H.D. Abruña, D.-W. Pang, Energy-level-related response of cathodic electrogenerated-chemiluminescence of self-assembled CdSe/ZnS quantum dot films. J. Phys. Chem. C **115**(38), 18822–18828 (2011). doi:10.1021/jp205419z
27. L. Sun, L. Bao, B.R. Hyun, A.C. Bartnik, Y.W. Zhcng, J.C. Reed, D.W. Pang, H.D. Abruna, G.G. Malliaras, F.W. Wise, Electrogenerated chemiluminescence from PbS quantum dots. Nano Lett. **9**(2), 789–793 (2009). doi:10.1021/nl803459b
28. Z. Ding, B.M. Quinn, S.K. Haram, L.E. Pell, B.A. Korgel, A.J. Bard, Electrochemistry and electrogenerated chemiluminescence from silicon nanocrystal quantum dots. Science **296** (5571), 1293–1297 (2002). doi:10.1126/science.1069336
29. H. Zhu, X. Wang, Y. Li, Z. Wang, F. Yang, X. Yang, Microwave synthesis of fluorescent carbon nanoparticles with electrochemiluminescence properties. Chem. Commun. **34**, 5118–5120 (2009). doi:10.1039/b907612c
30. J. Zhou, C. Booker, R. Li, X. Sun, T.-K. Sham, Z. Ding, Electrochemistry and electrochemiluminescence study of blue luminescent carbon nanocrystals. Chem. Phys. Lett. **493**(4–6), 296–298 (2010). doi:10.1016/j.cplett.2010.05.030
31. L.-L. Li, J. Ji, R. Fei, C.-Z. Wang, Q. Lu, J.-R. Zhang, L.-P. Jiang, J.-J. Zhu, A facile microwave avenue to electrochemiluminescent two-color graphene quantum dots. Adv. Funct. Mater. **22**(14), 2971–2979 (2012). doi:10.1002/adfm.201200166
32. J. Lu, M. Yan, L. Ge, S. Ge, S. Wang, J. Yan, J. Yu, Electrochemiluminescence of blue-luminescent graphene quantum dots and its application in ultrasensitive aptasensor for adenosine triphosphate detection. Biosens. Bioelectron. **47**, 271–277 (2013). doi:10.1016/j.bios.2013.03.039
33. M.M. Richter, Electrochemiluminescence (ECL). Chem. Rev. **104**(6), 3003–3036 (2004). doi:10.1021/cr020373d
34. L. Hu, G. Xu, Applications and trends in electrochemiluminescence. Chem. Soc. Rev. **39**(8), 3275–3304 (2010). doi:10.1039/b923679c
35. L. Lin, S. Zhang, Creating high yield water soluble luminescent graphene quantum dots via exfoliating and disintegrating carbon nanotubes and graphite flakes. Chem. Commun. **48**(82), 10177–10179 (2012). doi:10.1039/C2CC35559K
36. L. Zheng, Y. Chi, Y. Dong, J. Lin, B. Wang, Electrochemiluminescence of water-soluble carbon nanocrystals released electrochemically from graphite. J. Am. Chem. Soc. **131**(13), 4564–4565 (2009). doi:10.1021/ja809073f
37. K. Jiang, S. Sun, L. Zhang, Y. Lu, A. Wu, C. Cai, H. Lin, Red, Green, and blue luminescence by carbon dots: full-color emission tuning and multicolor cellular imaging. Angew. Chem. Int. Ed. **54**(18), 5360–5363 (2015). doi:10.1002/anie.201501193
38. M.J. Krysmann, A. Kelarakis, P. Dallas, E.P. Giannelis, Formation mechanism of carbogenic nanoparticles with dual photoluminescence emission. J. Am. Chem. Soc. **134**(2), 747–750 (2012). doi:10.1021/ja204661r
39. J. Zhou, C. Booker, R. Li, X. Zhou, T.K. Sham, X. Sun, Z. Ding, An electrochemical avenue to blue luminescent nanocrystals from multiwalled carbon nanotubes (MWCNTs). J. Am. Chem. Soc. **129**(4), 744–745 (2007). doi:10.1021/ja0669070
40. C.S. Lim, K. Hola, A. Ambrosi, R. Zboril, M. Pumera, Graphene and carbon quantum dots electrochemistry. Electrochem. Commun. **52**, 75–79 (2015). doi:10.1016/j.elecom.2015.01.023
41. H. Zheng, Q. Wang, Y. Long, H. Zhang, X. Huang, R. Zhu, Enhancing the luminescence of carbon dots with a reduction pathway. Chem. Commun. **47**(38), 10650–10652 (2011). doi:10.1039/c1cc14741b
42. F. Liu, M.H. Jang, H.D. Ha, J.H. Kim, Y.H. Cho, T.S. Seo, Facile synthetic method for pristine graphene quantum dots and graphene oxide quantum dots: origin of blue and green luminescence. Adv. Mater. **25**(27), 3657–3662 (2013). doi:10.1002/adma.201300233

43. Y.-M. Long, C.-H. Zhou, Z.-L. Zhang, Z.-Q. Tian, L. Bao, Y. Lin, D.-W. Pang, Shifting and non-shifting fluorescence emitted by carbon nanodots. J. Mater. Chem. **22**(13), 5917–5920 (2012). doi:10.1039/c2jm30639e
44. N. Myung, Z.F. Ding, A.J. Bard, Electrogenerated chemiluminescence of CdSe nanocrystals. Nano Lett. **2**(11), 1315–1319 (2002). doi:10.1021/nl0257824
45. N. Myung, Y. Bae, A.J. Bard, Effect of surface passivation on the electrogenerated chemiluminescence of CdSe/ZnSe nanocrystals. Nano Lett. **3**(8), 1053–1055 (2003). doi:10.1021/nl034354a
46. Y. Dong, N. Zhou, X. Lin, J. Lin, Y. Chi, G. Chen, Extraction of electrochemiluminescent oxidized carbon quantum dots from activated carbon. Chem. Mater. **22**(21), 5895–5899 (2010). doi:10.1021/cm1018844
47. S.H. Song, M.H. Jang, J. Chung, S.H. Jin, B.H. Kim, S.H. Hur, S. Yoo, Y.H. Cho, S. Jeon, Highly efficient light-emitting diode of graphene quantum dots fabricated from graphite intercalation compounds. Adv. Opt. Mater. **2**(11), 1016–1023 (2014). doi:10.1002/adom.201400184
48. X. Yan, B. Li, L.S. Li, Colloidal graphene quantum dots with well-defined structures. Acc. Chem. Res. **46**(10), 2254–2262 (2013). doi:10.1021/ar300137p
49. Q. Li, S. Zhang, L. Dai, L.S. Li, Nitrogen-doped colloidal graphene quantum dots and their size-dependent electrocatalytic activity for the oxygen reduction reaction. J. Am. Chem. Soc. **134**(46), 18932–18935 (2012). doi:10.1021/ja309270h
50. X. Yan, X. Cui, L.S. Li, Synthesis of large, stable colloidal graphene quantum dots with tunable size. J. Am. Chem. Soc. **132**(17), 5944–5945 (2010). doi:10.1021/ja1009376
51. X. Yan, B.S. Li, X. Cui, Q.S. Wei, K. Tajima, L.S. Li, Independent tuning of the band gap and redox potential of graphene quantum dots. J. Phys. Chem. Lett. **2**(10), 1119–1124 (2011). doi:10.1021/jz200450r
52. G. Eda, Y.Y. Lin, C. Mattevi, H. Yamaguchi, H.A. Chen, I.S. Chen, C.W. Chen, M. Chhowalla, Blue photoluminescence from chemically derived graphene oxide. Adv. Mater. **22** (4), 505–509 (2010). doi:10.1002/adma.200901996
53. H. Li, X. He, Z. Kang, H. Huang, Y. Liu, J. Liu, S. Lian, C.H. Tsang, X. Yang, S.T. Lee, Water-soluble fluorescent carbon quantum dots and photocatalyst design. Angew. Chem. Int. Ed. **49**(26), 4430–4434 (2010). doi:10.1002/anie.200906154
54. M.A. Sk, A. Ananthanarayanan, L. Huang, K.H. Lim, P. Chen, Revealing the tunable photoluminescence properties of graphene quantum dots. J. Mater. Chem. C **2**(34), 6954–6960 (2014). doi:10.1039/c4tc01191k
55. D. Pan, J. Zhang, Z. Li, M. Wu, Hydrothermal route for cutting graphene sheets into blue-luminescent graphene quantum dots. Adv. Mater. **22**(6), 734–738 (2010). doi:10.1002/adma.200902825
56. Y. Dong, R. Dai, T. Dong, Y. Chi, G. Chen, Photoluminescence, chemiluminescence and anodic electrochemiluminescence of hydrazide-modified graphene quantum dots. Nanoscale **6** (19), 11240–11245 (2014). doi:10.1039/c4nr02539c
57. P. Anilkumar, L. Cao, J.J. Yu, K.N. Tackett II, P. Wang, M.J. Meziani, Y.P. Sun, Crosslinked carbon dots as ultra-bright fluorescence probes. Small **9**(4), 545–551 (2013). doi:10.1002/smll.201202000
58. S.K. Bhunia, A. Saha, A.R. Maity, S.C. Ray, N.R. Jana, Carbon nanoparticle-based fluorescent bioimaging probes. Sci. Rep. **3**, 1473–1749 (2013). doi:10.1038/srep01473
59. C. Liu, P. Zhang, X. Zhai, F. Tian, W. Li, J. Yang, Y. Liu, H. Wang, W. Wang, W. Liu, Nano-carrier for gene delivery and bioimaging based on carbon dots with PEI-passivation enhanced fluorescence. Biomaterials **33**(13), 3604–3613 (2012). doi:10.1016/j.biomaterials.2012.01.052
60. M. Nurunnabi, Z. Khatun, K.M. Huh, S.Y. Park, D.Y. Lee, K.J. Cho, Y.K. Lee, In vivo biodistribution and toxicology of carboxylated graphene quantum dots. ACS Nano **7**(8), 6858–6867 (2013). doi:10.1021/nn402043c

61. X.T. Zheng, A. Than, A. Ananthanaraya, D.H. Kim, P. Chen, Graphene quantum dots as universal fluorophores and their use in revealing regulated trafficking of insulin receptors in adipocytes. ACS Nano **7**(7), 6278–6286 (2013). doi:10.1021/Nn4023137
62. H. Zhao, Y. Chang, M. Liu, S. Gao, H. Yu, X. Quan, A universal immunosensing strategy based on regulation of the interaction between graphene and graphene quantum dots. Chem. Commun. **49**(3), 234–236 (2013). doi:10.1039/C2CC35503E
63. Y. Wang, L. Bao, Z. Liu, D.W. Pang, Aptamer biosensor based on fluorescence resonance energy transfer from upconverting phosphors to carbon nanoparticles for thrombin detection in human plasma. Anal. Chem. **83**(21), 8130–8137 (2011). doi:10.1021/ac201631b
64. W. Deng, F. Liu, S. Ge, J. Yu, M. Yan, X. Song, A dual amplification strategy for ultrasensitive electrochemiluminescence immunoassay based on a Pt nanoparticles dotted graphene-carbon nanotubes composite and carbon dots functionalized mesoporous Pt/Fe. Analyst **139**(7), 1713–1720 (2014). doi:10.1039/C3AN02084C
65. Z.-X. Wang, C.-L. Zheng, Q.-L. Li, S.-N. Ding, Electrochemiluminescence of a nanoAg-carbon nanodot composite and its application to detect sulfide ions. Analyst **139** (7), 1751–1755 (2014). doi:10.1039/C3AN02097E
66. S. Yang, J. Liang, S. Luo, C. Liu, Y. Tang, Supersensitive detection of chlorinated phenols by multiple amplification electrochemiluminescence sensing based on carbon quantum dots/graphene. Anal. Chem. **85**(16), 7720–7725 (2013). doi:10.1021/ac400874h
67. P.J. Zhang, Z.J. Xue, D. Luo, W. Yu, Z.H. Guo, T. Wang, Dual-peak electrogenerated chemiluminescence of carbon dots for iron ions detection. Anal. Chem. **86**(12), 5620–5623 (2014). doi:10.1021/Ac5011734
68. M. Zhang, H. Liu, L. Chen, M. Yan, L. Ge, S. Ge, J. Yu, A disposable electrochemiluminescence device for ultrasensitive monitoring of K562 leukemia cells based on aptamers and ZnO@carbon quantum dots. Biosens. Bioelectron. **49**, 79–85 (2013). doi:10.1016/j.bios.2013.05.003
69. Y.Q. Dong, C.Q. Chen, J.P. Lin, N.N. Zhou, Y.W. Chi, G.N. Chen, Electrochemiluminescence emission from carbon quantum dot-sulfite coreactant system. Carbon **56**, 12–17 (2013). doi:10.1016/j.carbon.2012.12.086
70. H. Dai, C. Yang, Y. Tong, G. Xu, X. Ma, Y. Lin, G. Chen, Label-free electrochemiluminescent immunosensor for [small alpha]-fetoprotein: performance of Nafion-carbon nanodots nanocomposite films as antibody carriers. Chem. Commun. **48**(25), 3055–3057 (2012). doi:10.1039/C1CC16571B
71. S. Li, J. Luo, X. Yang, Y. Wan, C. Liu, A novel immunosensor for squamous cell carcinoma antigen determination based on CdTe@Carbon dots nanocomposite electrochemiluminescence resonance energy transfer. Sensor Actuat. B-Chem. **197**, 43–49 (2014). doi:10.1016/j.snb.2014.02.066
72. Y.M. Long, L. Bao, J.Y. Zhao, Z.L. Zhang, D.W. Pang, Revealing carbon nanodots as coreactants of the anodic electrochemiluminescence of Ru(bpy)(3)(2+). Anal. Chem. **86**(15), 7224–7228 (2014). doi:10.1021/ac502405p
73. V. Gupta, N. Chaudhary, R. Srivastava, G.D. Sharma, R. Bhardwaj, S. Chand, Luminscent graphene quantum dots for organic photovoltaic devices. J. Am. Chem. Soc. **133**(26), 9960–9963 (2011). doi:10.1021/ja2036749
74. Y. Li, Y. Zhao, H. Cheng, Y. Hu, G. Shi, L. Dai, L. Qu, Nitrogen-doped graphene quantum dots with oxygen-rich functional groups. J. Am. Chem. Soc. **134**(1), 15–18 (2012). doi:10.1021/ja206030c
75. Y. Zhang, C. Wu, X. Zhou, X. Wu, Y. Yang, H. Wu, S. Guo, J. Zhang, Graphene quantum dots/gold electrode and its application in living cell H_2O_2 detection. Nanoscale **5**(5), 1816–1819 (2013). doi:10.1039/c3nr33954h
76. A. Muthurasu, V. Ganesh, Horseradish Peroxidase Enzyme immobilized graphene quantum dots as electrochemical biosensors. Appl. Biochem. Biotechnol. **174**(3), 945–959 (2014). doi:10.1007/s12010-014-1019-7

Catalytic Applications of Carbon Dots

Zhenhui Kang and Yang Liu

Abstract Carbon materials have been used for a long time in heterogeneous catalysis, which can satisfy most of the desirable properties required for a suitable catalyst support. Based on their significant advantages, such as low cost, huge amount, easy accessibility, high surface area, diverse porous structure, and resistance to acidic or basic environments, the carbon nanostructures are hungered for using as proper catalysts directly. Carbon dots (CDs), a new class of carbon nanomaterials with sizes below 10 nm, were also demonstrated to be efficient catalysts, such as, photocatalysts for selective oxidation, light-driven acid-catalysis and hydrogen bond catalysis. In this chapter, we will highlight the preparative methods, which have been used successfully to produce active, selective and durable CDs catalysts and look at their properties and the reactions which they promote. We then consider the catalysts design based on these new CDs and look into the future.

1 Preparative Methods

Here we just simple summarize the typical synthesis methods, and will not give a very detailed discussion. The detailed describe about the synthesis of CDs can be found in the recent reviews [1–5].

Tremendous efforts have been made to develop the synthetic methods for CDs. The methods have been proposed during the last decade can be roughly classified into "Top-down" and "Bottom-up" approaches. The top-down routes are implemented via either physical or chemical techniques, among which the latter is in the

Z. Kang (✉) · Y. Liu
Jiangsu Key Laboratory for Carbon-Based Functional Materials & Devices,
Institute of Functional Nano & Soft Materials (FUNSOM), Soochow University,
199 Ren'ai Road, Suzhou 215123, Jiangsu, China
e-mail: zhkang@suda.edu.cn

Y. Liu
e-mail: yangl@suda.edu.cn

N. Yang et al. (eds.), *Carbon Nanoparticles and Nanostructures*,
Carbon Nanostructures, DOI 10.1007/978-3-319-28782-9_8

majority. Further, the mechanism of chemical type top-down routes can be described as defect-mediated fragmentation processes. Namely, the oxygen-containing functional groups (epoxy and hydroxyl groups) could create defects on graphite sheets and serve as chemically reactive sites, thus allowing graphite to be cleaved into smaller sheets. Typically, the top-down methods include electron beam lithography, Laser ablation, acidic exfoliation, electrochemical oxidation, microwave-assisted hydrothermal synthesis, and so on. The top-down routes for the preparation of CDs have the advantages of abundant raw materials, large scale production and simple operation. CDs can also be prepared through bottom-up routes, including the solution chemistry, cyclodehydrogenation of polyphenylene precursors, carbonizing some special organic precursors or, the fragmentation of suitable precursors, for example, the C60. Compared with the up-down routes, the reports concerning the bottom-up routes are relatively scarce. The bottom-up methods offers us exciting opportunities to control the CDs with well-defined molecular size, shape, and thus properties. Nevertheless, these methods always involve complex synthetic procedures, and the special organic precursors may be difficult to obtain. Anyway, three problems facing CDs fabrication need to be noticed: (i) carbonaceous aggregation during carbonization, which can be avoided by using electrochemical synthesis, confined pyrolysis or solution chemistry methods, (ii) size control and uniformity, which is important for uniform properties and mechanistic study, and can be optimized via post-treatment, such as gel electrophoresis, centrifugation, and dialysis and (iii) surface properties that are critical for solubility and selected applications, which can be tuned during preparation or post-treatment. For particular applications explore, it is important to control the sizes of CDs to get uniform properties. Many approaches have been proposed to obtain uniform CDs during preparation or post-treatment. In most of the reports, the as-synthesized CDs fragments were purified via post treatments like filtration, dialysis, centrifugation, column chromatography and gel-electrophoresis. Surface modification is a powerful method to tune the surface properties of materials for selected applications. There are many approaches for functionalizing the surface of CDs through the surface chemistry or interactions, such as covalent bonding, coordination, p–p interactions, and sol–gel technology. Doping is a widely used approach to tune the PL properties of photoluminescent (PL) materials. Various doping methods with dozens of elements such as N, S, and P have been reported to tune the properties of CDs.

2 Structures and Properties of CDs

Here we just simple summarize the typical structure and properties of CDs, and will not give a very detailed discussion. The detailed describe about the structure and properties of CDs can be found in the recent reviews [1–5].

2.1 Component and Structure

In general, the average sizes of CDs are mostly below 10 nm, which is usually depended on the preparation methods. Technically speaking, CDs are only composed of elemental C and H. However, limit by their strong tendency for aggregation due to the face-to-face attraction and the preparation methods such as oxidation derived exfoliation, CDs reported so far are always partially oxidized, with hydroxyl, epoxy/ether, carbonyl and carboxylic acid groups on the surfaces. Fourier transform infrared (FTIR) and X-ray photoelectron spectroscopy (XPS) spectra are commonly adopted to analyze their component. The crystalline nature of CDs can be investigated through X-ray diffraction (XRD) patterns, Raman spectroscopy and high resolution transmission electron microscopy (HRTEM) observations. Both (002) interlayer spacing and (100) in-plane lattice spacing exist in CDs, and the former has been widely studied. The interlayer spacing of CDs depends strongly on their oxidation degree, that is, the attached hydroxyl, epoxy/ether, carbonyl and carboxylic acid groups can increase the interlayer spacing of CDs. The Raman technique is also a powerful and non-destructive tool for the characterization of CDs. The G band is assigned to the E2g vibrational modes of the aromatic domains, whereas the D band arises from the breathing modes of the graphitic domains. Traditionally, the intensity ratio of "disorder" D to crystalline G (I_D/I_G) is used to compare the structural order between crystalline and amorphous graphitic systems. The I_D/I_G values of CDs vary significantly depending on the preparation methods. The HRTEM images of CDs features two kinds of lattice fringes, namely (002) interlayer spacing and (110) in-plane lattice spacing. Similar to the XRD pattern, the former centered at about 0.34 nm has been observed from CDs prepared by acidic oxidation from carbon black, microwave-assisted method, and electrochemical cutting method. The in-plane lattice spacing mostly centered at about 0.24 nm has been observed from CDS synthesized via microwave-hydrothermal protocol, amino-hydrothermal method, K intercalation, acidic oxidation from carbon fibers, and photo-Fenton reaction, excepting that valued at about 0.21 nm via hydrothermal cutting strategy and glucose carbonization method. Besides, these two kinds of lattice fringes have been simultaneously been observed in the case of electrochemical method from MWCNTs and acidic oxidation from natural graphite. Moreover, CDs are not always crystalline, amorphous CDs has also been prepared via hydrothermal method and citric acid carbonization.

2.2 Properties

Absorption. CDs typically show strong optical absorption in the UV region, with a tail extending out into the visible range. There may be some absorption shoulders attributed to the π–π^* transition of the C=C bonds, the n–π^* transition of C=O bonds and/or others. Moreover, CDs prepared via different methods also showed different absorption behaviors. For instance, the citric acid carbonized CDs had an absorption band of 362 nm, but the unsubstituted HBC derived CDs with size of about 60 nm only showed a weak shoulder at 280 nm. Thus the absorption peak position was also dependent on the preparation method. Besides, the variation of oxygen content was reported to play important role in deciding the absorption peak position of CDs.

Photoluminescence. For both CDs, one of the most fascinating features is their photo-luminescence. Till now, variously sized CDs with different PL colors, ranging from the visible into the near-infrared region, have been prepared via various synthetic approaches. CDs prepared via different approaches can emit PL with different colors, including DUV, blue, green, yellow, and red. Typically, the luminescence mechanism may derive from intrinsic state emission and defect state emission. However, the exact mechanism of PL for CDs remains unsettled. The luminescence has been tentatively suggested to arise from excitons of carbon, emissive traps, quantum confinement effect, aromatic structures, oxygen-containing groups, free zigzag sites and edge defects. A widely accepted mechanism for luminescence emission from CDs needs systematic investigation. Anyway, the PL of CDs is should be attributed to either combining effect or competition between intrinsic state emission and defect state emission. CDs prepared via various methods probably exhibit distinct PL mechanism, which leads to different dependences of their PL on size, excitation wavelength, pH, solvent, and concentration etc. The quantum yield (QY) of CDs varies with the fabrication method and the surface chemistry involved. As for the unpassivated CDs, their QYs ranged between 2 and 30 %, which are observed in CDs prepared via stepwise solution chemistry and microwave-assisted acidic oxidation, respectively. The CDs commonly contains carboxylic and epoxide groups, which can act as the non-radiative electron–hole recombination centers. Therefore, the removal of these oxygen-containing groups maybe improves the QY, either by reduction or surface passivation. More importantly, significantly enhanced QY of ~72 % is the highest value for CDs reported so far [6].

Photoinduced electron transfer property. For the utilization of PL compounds in light-energy conversion and related areas, there have been extensive investigations on their photoresponse, photoinduced charge separation and electron transfer processes. It was found that the PL from a CDs solution could be efficiently quenched in the presence of either electron acceptors such as 4-nitrotoluene and 2,4-dinitrotoluene or electron donors such as N,N-diethylaniline. Namely, the photoexcited CDs are excellent as both electron donors and electron acceptors. They also found efficient PL quenching in CDs by surface-doped metals through disrupting the excited state redox processes. Electron transfer in nanocomposites of CDs–GO (graphene oxide), CDs–MWNTs (multi-wall carbon nanotubes) and CDs–TiO_2 NPs without linker molecules was also studied. Significant PL quenching was observed in the CD–GO system, which was attributed to the ultrafast electron transfer from CDs to GO with a time constant of 400 fs. In comparison, addition of carbon nanotubes resulted in static quenching of fluorescence in CDs. No charge transfer was observed in either CD–MWNT or CD–TiO_2 nanocomposites. These interesting photoinduced electron transfer properties of CDs as an electron donor/acceptor may offer new opportunities for light energy conversion, catalysis and related applications, as well as mechanistic elucidation.

Bandgap engineering. CDs feature tunable bandgap due to their pronounced quantum confinement and edge effects. The bandgap in graphene-based materials can be tuned from 0 eV to that of benzene by changing size and/or surface chemistry, making it a rising carbon-based fluorescent material. Li et al. reported that the band gaps and redox potentials of CDs could be independently controlled, the former by size and the latter by functionalization. Based on the two major strategies, currently, increasing efforts have been made to customize their optical properties, among which the tunability based on the variation of size has already been discussed in the above section, Size dependence. Other strategies can be classified into three main groups: tuning oxidation degree, surface functionalization, and chemical doping, showing as below. The oxygen-containing functional groups on CDs can lead to radiative recombination of localized electron–hole pairs and surface emissive traps. And the variation of the degree of oxidation (or reduction) can induce the alteration of localized sp^2 clusters and structural defects, thus changing the PL. Chemical modifications using molecules with strong electron-donating or—accepting ability could also result in appreciable impact on electronic characteristics of grapheme. The electron-donating groups generally raised the HOMO levels, and electron-withdrawing groups lowered the LUMO levels. Doping carbon nanomaterials with heteroatoms can effectively tune their intrinsic properties, including electronic properties, surface and local chemical reactivity. In view of the remarkable quantum-confinement and edge effects of CDs, doping CDs with chemically bonded N atoms could drastically alter their electronic characteristics and offer more active sites, thus producing new phenomena and unique properties.

3 Catalytic Properties of CDs

3.1 Photoctalytic Activity for Selective Oxidation Reaction [7]

Selective oxidation of alcohols to carbonyls is a fundamental and significant transformation for the large-scale production of fine chemicals [8, 9]. Several UV and visible light driven photocatalytic systems for alcohols oxidation have been developed, such as Ru-$(bipy)_3^{2+}$ and its derivatives, TiO_2 or TiO_2-based materials (dye/TiO_2), Pd@CeO_2, and mesoporous carbon nitride (mpg-C_3N_4) [9–18]. The photocatalytic application of carbon based materials has focused primarily on that such catalyst could or not catalyze the reaction by UV and/or visible light [19–22]. In general, the UV and/or short wavelength visible light may damage or change the structure of organics [11–13]. From a broader perspective, exploiting long wavelength light driven photocatalysts is a direct and effective strategy to realize the photo-assisted selective organic transformations with high conversion and selectivity. 1–4 nm CDs are effective NIR light driven photocatalysts, which catalyze the efficient oxidation of benzyl alcohol to benzaldehyde with high conversion (92 %) and selectivity (100 %) in the presence of H_2O_2 as oxidant. The cooperation of CDs and NIR can prevent the overoxidation of the product to benzoic acid, and hold the reaction resting on the phase of benzaldelyde. Control catalytic experiments confirm that the catalytic activity of CDs is dependent on their photocatalytic activity for H_2O_2 decomposition and NIR light induced electron transfer property. Such metal-free photocatalytic system also selectively converts other alcohol substrates to their corresponding aldehydes with high conversion, demonstrating a potential pathway of accessing traditional alcohols oxidation chemistry. Details of the conversion and selectivity of benzyl alcohol oxidation catalyzed by CDs and contrast samples are shown in Table 1. After 12 h reaction under NIR light irradiation, high conversion efficiency of 92 % and high selectivity up to 100 % catalyzed by CDs (1–4 nm) were achieved simultaneously (Entry 1, Table 1). In contrast, CNPs (100–150 nm) and graphite (100–2000 nm) catalysts gave conversion of 71, 51 % and selectivity of only 78, 59 % under the same conditions, respectively (Entries 3 and 5). In the absence of NIR light, the selectivity yielded by three different catalysts all exhibited obvious decrease (Entries 2, 4 and 6), which means that the NIR light

Table 1 Selective oxidation of benzyl alcohol to benzaldehyde under the different light irradiation or not

Entry	Catalysts	Size (nm)	NIR light	Conv. (%)[a]	Sel. (%)
1	CDs	1–4	+	92	100
2	CDs	1–4	/[b]	93	54
3	CNPs	100–150	+	71	78
4	CNPs	100–150	/	64	57
5	Graphite	100–2000	+	51	59
6	Graphite	100–2000	/	46	60
7[c]	/	/	+	21	52
8	CDs	1–4	Vis[d]	93	84
9	CDs	1–4	UV[e]	95	75

All reactions were carried out applying a quartz three-neck flask. Typically, benzyl alcohol, CDs or other kinds of catalysts were added into a three-neck flask equipped with a condenser. The reaction mixture was vigorously stirred at 60 °C, followed by NIR irradiation for 12 h (450 W Xe arc lamp with an NIR-CUT filter to cut off light of wavelength <700 nm, remove the light when used control samples). H_2O_2 (30 wt% in water, 1.0 mL, 10.0 mmol) was added continuously with a syringe pump in 12 h

[a]Conversion rates were determined by GC with an FID detector

[b]Reference experiment without NIR light

[c]Reference experiment without catalyst

[d]Reference experiments with visible light

[e]Reference experiments with UV light

can enhance the catalytic ability of CDs for the selective oxidation. Here, the low selectivity yielded by contrast samples (Entries 2–6) are resulted from the formation of benzoic acid (overoxidation product of benzaldehyde) in the obtained products as measured by Gas Chromatography (GC) analysis. Additionally, in the absence of any catalyst, a thermocatalytic reaction of benzel alcohol was observed for the same reaction system with low conversion (21 %) and poor selectivity (52 %) to benzaldehyde (Entry 8). When using CDs as catalyst and visible or UV light as the irradiation light (Entries 8 and 9), the conversion of benzyl alcohol (≤95 %) was similar to that using NIR light, while the selectivity to benzaldehyde (84, 75 %) was lower than that achieved under NIR light (100 %). These results reveal that the conversion and selectivity of such catalytic reaction are influenced not only by the carbon catalysts (CDs, CNPs, and graphite) but also by irradiation lights (UV, Vis, and NIR). The cooperation of CDs and NIR irradiation can promote the oxidation of benzyl alcohol to benzaldehyde achieving a high conversion and selectivity. Compared to the other catalysts with high efficiency (such as TiO_2 with a conversion and selectivity ca. 99 % under UV and visible light) [16], the catalytic process obtained the similar conversion (92 %) and selectivity (100 %) under long wavelength NIR light. Although the oxidants are different in these two systems, the long wavelength near infrared (NIR) and infrared (IR) light are more moderate and environmentally friendly than the UV light. The active oxygen species in the present process appear to have a radical character. The further experiments also strongly suggest that a higher amount of $HO^{\cdot}$ would be generated by the unirradiated sample than that of the irradiated one. Without NIR irradiation, the mainly catalytic product is benzoic acid with only a little amount of benzaldehyde, indicating that during the reaction, benzyl alcohol is firstly oxidized into benzaldehyde, and then further oxidized into benzoic acid. Significantly, under the NIR light irradiation, the CDs' photo-induced electron transfer ability (especially as strong electron donors) protects the first step product (benzaldehyde) from overoxidation by the photoelectron reductive environment [23], thus yielding the high selectivity (100 %) to benzaldehyde. The scope of reactivity of the aforementioned CDs catalyst was further explored by Kang et al. using a variety of benzyl alcohol's derivatives under the same conditions under NIR light irradiations. Those benzyl alcohol's derivatives could also resulted into their oxidized aldehydes products in high conversions and selectivities with H_2O_2 as oxidant and NIR light irradiation.

3.2 Visible Light Induced Acid Catalyst [24]

The traditional acid catalysts (mineral acids, etc.) suffer from serious drawbacks that they necessitate costly and inefficient separation of catalysts from homogeneous reaction mixtures, resulting in a huge waste of energy and products, equipment corrosion, uncontrolled reaction, and demanding manufacturing conditions [25–33]. The use of solid acid catalysts offers advantages over liquid acids, such as reduced equipment corrosion, ease of product separation and catalyst recyclability [28–33].

Light-driven catalytic reactions can potentially provide a sustainable pathway for green controlled synthesis, and have attracted tremendous interest [34–37]. Carbon-based nanomaterials show great potential in photo-catalysis, redox-catalysis and acid-catalysis. As acid catalysts, carbon nanostructures (such as sulfated-graphene/-tube/-active carbons) typically suffer from lack of proper surface functionalization, low efficiency, and uncontrolled process [37–39]. 5–10 nm CDs have strong photo-induced proton generating capacity in solution under visible light irradiation. The catalytic activity of 5–10 nm CDs is strongly dependent on illumination intensity and reaction temperature. As light-driven acid-catalyst, they show CDs can catalyze a series of organic reactions (Esterification, Beckmann rearrangement and Aldol condensation) achieving high conversion efficiency in water solution under visible light irradiation. 5–10 nm CDs were produced from a graphite rod by electrochemical method in pure water. The reversible temperature- and concentration-dependent photo-proton generating ability should be due to the light-induced structure interconversion of the functional groups on CDs surface. And the proton generation mechanism of CDs under visible light irradiation was shown as follows. First, the hydroxyl group of CDs dissolved in water would release a free proton (or H_3O^+) under visible light irradiation. Then the intermediate would produce another proton (hydrolyzed to H_3O^+) from reaction of C=O group with H_3O^+. After the O in C=O links with HO^- by supramolecular interaction, the intermediate with the epoxy group was obtained under visible light irradiation. A series of reactions (Esterification, Beckmann rearrangement, Aldol condensation) needing acid as a catalyst was performed by Kang et al. to study the photo-induced proton generating property of CDs and reference samples (small-sized CDs with 1–4 nm, graphite particles) [40]. The reaction products were monitored by GC. Details of different reactions (Esterification, Beckmann rearrangement, Aldol condensation) catalyzed by CDs and reference samples (Entry 1–10). After 10 h reaction under visible light irradiation, high conversion efficiencies of 34.7–46.2 % were achieved in all reactions (Esterification reactions: Entry 1–8; Beckmann rearrangement: Entry 9; Aldol condensation: Entry 10, Table 2) catalyzed by CDs. Under the same conditions but without visible light the, conversion efficiencies are substantially smaller at <5 % (Table 2), showing visible light can enhance the catalytic ability of CDs for acid-catalyzed reactions. To further investigate the role of light in the reaction, Kang et al. also measured conversion efficiency versus reaction time (Esterification, Beckmann rearrangement, and Aldol condensation corresponding to Entry 1, 9, and 10, respectively, were selected for study) in alternating dark and light environment. While, the conversion efficiencies of all three reactions increase sharply (4.1–6.8 %) under light irradiation (time at 0–30 min, 60–90 min, 120–150 min, and 180–210 min), but only a little (only 0.2–0.6 %) without irradiation. It shows that light irradiation indeed has an important effect on the acid catalytic reactions. In the absence of catalysts, a thermocatalytic reaction with a low conversion efficiency of <1 % was only observed in the reaction system. Moreover, the filtrate solution test, i.e. 2 h reaction after removing the CDs from the reaction medium, showed no catalytic activity. All above-mentioned experiments collectively show that CDs indeed can act as a light-controlled acid catalyst in the present catalytic reactions.

Table 2 Photocatalytic activities for different reactions (Esterification, Beckmann and Aldol condensation) by 5–10 nm CDs photocatalyst with (+) or without (−) light (λ > 420 nm) irradiation

Entry	Reactants 1	Reactants 2	Products	Conv.[%] (light +/−)[a]
1	CH_3OH	CH_3COOH	CH_3COOCH_3	45.3/4.8
2	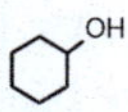	CH_3COOH	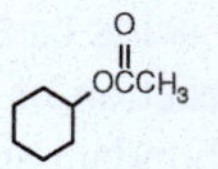	42.6/4.5
3	C_2H_5OH	CH_3COOH	$CH_3COOC_2H_5$	43.7/4.2
4	C_2H_5OH	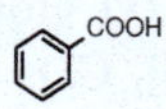	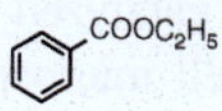	40.3/4.1
5	CH_3OH	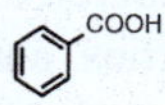	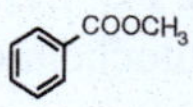	38.9/3.7
6	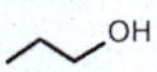	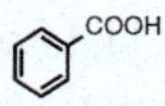	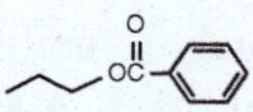	45.2/4.3
7	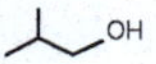	CH_3COOH	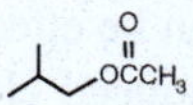	39.1/4.0
8		CH_3COOH	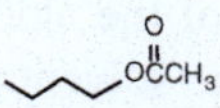	35.5/3.9
9	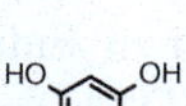	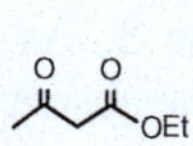	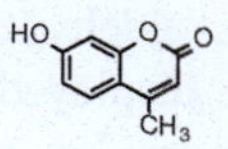	46.2/4.5
10	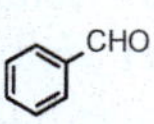	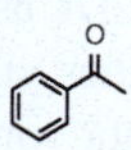	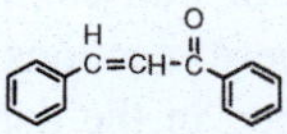	34.7/3.8

All reactions were performed in a quartz three-neck flask. Reactants and CDs were added into a three-neck flask equipped with a condenser. The reaction mixture was vigorously stirred at low temperatures, followed by visible light irradiation for 10 h (450 W Xe arc lamp with a CUT filter to cut off light of wavelength <420 nm, but removed for experiments without irradiation)

[a]Conversion efficiencies were determined by GC with an FID detector

The catalytic activities of CDs in aqueous solution are attributed to the remarkable visible light-induced proton generating capability associated with the oxygen-containing functional groups of CDs.

3.3 *Photoenhanced Hydrogen-Bond Catalytic Activity [41]*

Hydrogen bonding (H-bonding) plays a crucial role in enzyme-catalyzed reactions by orienting the substrate molecules and lowering barriers to reactions [42]. Recently, this kind of noncovalent interaction (H-bonding) was introduced into many organic chemical reactions [43–50], including the Diels-Alder, Aldol reaction, aza-Henry, Mannich, Michael, Morita-Baylis-Hillman, and Strecker reactions [51–58]. As a typical organocatalysis, hydrogen bond (H-bond) catalysis uses H-bonding interactions to accelerate organic reactions and stabilize anionic intermediates and transition states [59]. Several organic molecules with hydroxyl groups can be used as homogeneous H-bond catalysts [60–62]. For example, Chiral BINOL-derived Brønsted acids catalyze the enantioselective, asymmetric Morita–Baylis–Hillman (MBH) reaction of cyclohexenone with aldehydes [63]. Inorganic solid nanoparticles with abundant surface hydroxyl groups have been developed as alternative heterogeneous catalysts for H-bonding reactions [64, 65]. For example, aldol reactions could be catalyzed by hydroxyl groups on the $Fe(OH)_3$ shell of $Fe_3O_4@Fe(OH)_3$ core-shell microspheres [66]. CDs are stable and biocompatible and have strong and tunable PL. CDs are considered to be good H-bonding catalysts because of their rich photochemical properties and functional carboxylic and hydroxyl groups. As heterogeneous nanocatalysts for H-bond catalysis, CDs showed good photoenhanced catalytic abilities (89 % yield when 4-cyanobenzaldehyde is used) in the aldol condensation. A series of catalytic experiments confirmed that the catalytic activity of CDs can be effectively enhanced by visible light, which may be attributed to their photo-induced electron accepting properties. The aldol condensation with a high yield of the aldol condensation product under CDs as catalyst. Typically, acetone was the solvent and reactant in the first set of trials and reacted with a series of aromatic aldehydes at room temperature. Detailed yields of the aromatic compounds are displayed in Table 3. As shown in Entry 8, the aldol condensation exhibited a very low yield in the absence of CDs as catalysts whether under light irradiation or not, confirming the CDs as catalyst is necessary for the reaction. In general, these reactions yielded higher yields with visible light irradiation than those produced in the dark. These data confirm that visible light is necessary for good conversion. When benzaldehyde (Entry 1) is used, the lowest yield in visible light (19 %) with CDs as catalyst (Table 3) was obtained. Other aromatic aldehydes (Table 3, Entries 2–7) showed higher conversion. Among them, when 4-cyanobenzaldehyde (circled by the red frame) is used, the aldol condensation showed the highest yield (89 %, Entry 5) with CDs as catalyst. These catalytic results were similar to the results reported by Niu et al. [66] It would be attributed to the weak interaction between the para-position of aldehyde group may have with and the groups on the CDs surface, which make the reactant molecules closer to the catalytic active center. In contrast, without visible light irradiation, yields were reduced to 18 % for 4-cyanobenzaldehyde (Entry 5). When the reaction temperature was elevated to 50 °C, a total conversion of 4-cyanobenzaldehyde in aldol condensation was achieved in 4 h (Table 3, Entry 9). When 4-chlorobenzaldehyde (Entry 2) and 4-bromobenzaldehyde (Entry 3) were used, the yields reached 68 and 63 %, respectively, with visible light irradiation, whereas the yields fell to 15 and 13 % in the dark. In the present system, the reaction products were dehydrated products, and the typical one (Table 3, Entry 1) was (E)-4-phenylbut-3-en-2-one. Different solvents (Toluene, Ethanol, THF, and $CHCl_3$) and ketones (cyclopentanone, cyclohexanone, and acetone) were selected to evaluate the catalytic abilities of CDs (about

Table 3 Room-temperature aldol condensation between acetone and aromatic aldehydes in the presence of CDs with or without visible light

Entry[a]	Ar-	Light[b]	Dark[c]
		Yields (%)[d]	Yields (%)
1	C_6H_5	19	3
2	4-Cl–C_6H_4	68	15
3	4-Br–C_6H_4	63	13
4	4-NO_2–C_6H_4	65	8
5	4-CN–C_6H_4	89	18
6	4-CH_3–C_6H_4	67	9
7	4-CH_3O–C_6H_4	65	8
8[e]	4-CN–C_6H_4	Trace	Trace
9[f]	4-CN–C_6H_4	99	32

[a]Reaction conditions: 0.2 mmol aromatic aldehydes, 2 mL acetone, 60 mg CDs, room temperature, 24 h
[b,c]Reference experiments with and without visible light irradiation, respectively (Xenon lamp, 300 W, $\lambda \geq 420$ nm)
[d]The reaction products were isolated through silica column chromatography, and analyzed by GC
[e]Reference experiment without catalyst
[f]Heated to 50 °C, 4 h

5 nm) for aldol condensations with 4-cyanobenzaldehyde. When ethanol, THF, and $CHCl_3$ were used as solvents, the highest yield when 4-cyanobenzaldehyde is used (32 %) was obtained with THF as solvent. The lowest yield (12 %) was for $CHCl_3$. In the catalytic system, toluene was the most suitable solvent for the aldol condensation with CDs as catalysts. CDs with three different average particle sizes (5, >10, and <4 nm) were tested. The highest yield of 89 % was achieved for 5 nm CDs when 4-cyanobenzaldehyde is used, whereas yields of 40 and 25 % were obtained for CDs >10 and <4 nm, respectively. 5 nm CDs had the most satisfying photocatalytic performance for this H-bond catalysis reaction, which should attributes to the highest electron accepting ability when compared with other CDs (with size >10 and <4 nm). Further, control catalytic experiments confirmed that the efficient electron-accepting properties of CDs strengthened the O–H bond and activated the C=O bond of the 4-cyanobenzaldehyde, accelerating the aldol condensation.

3.4 *Visible-Light Photocatalysts for CO_2 Conversion [67]*

The significant rise in atmospheric CO_2 levels due to the combustion of hydrocarbon fuels has generated much concern. Among various CO_2 sequestration options, a compelling approach is photocatalytic conversion to recycle CO_2 back to hydrocarbon fuels, for which the use of solar irradiation may represent an ultimate solution. However, there are major challenges in finding potent photocatalysts

[68, 69]. Nanoscale wide-band-gap semiconductors such as titanium dioxide (TiO_2) and cadmium sulfide (CdS) were originally used and have since been quite popular in CO_2 photoreduction and related photocatalytic reactions, but their limitations in terms of the requirement for UV excitation and generally low conversion efficiencies have also become evident [68–70]. The functionalized carbon nanoparticles with gold or platinum coating were used as photocatalysts for the reduction of CO_2. In a typical experiment, an aqueous solution of the goldcoated particles was added to an optical cell in the photolysis setup, after which the solution was purged with CO_2 gas toward saturation at ambient temperature. The resulting solution in the optical cell was photoirradiated with visible light for 5 h. The photoreduction of CO_2 generally yields formic acid as a significant product. The quantification of the photoproduct was used to estimate quantum yields for photocatalytic reactions of CO_2 under the specific experimental conditions [71]. For the reduction to formic acid only, the estimated quantum yield was ~0.3 %, which was likely lower than the overall quantum yield for the CO_2 conversion. Nevertheless, even for the production of formic acid alone, the observed quantum yield is already at the higher end of the available literature values for reactions with a variety of photocatalysts under many different conditions. As a more direct comparison, the use of suspended TiO_2 nanoparticles (Degussa P25) as photocatalysts with UV light irradiation (through a 350 nm cutoff filter) in the same photolysis setup resulted in quantum yields that were an order of magnitude lower (also for formic acid only). For the platinum-coated photocatalysts in the CO_2 conversion reaction under otherwise identical experimental conditions, the results were generally similar to those with the gold-coated ones described above, though more quantitative comparisons, including those for effects of varying amounts of metal coating, are still being pursued. Mechanistically, photoexcitation of the surface-passivated small carbon nanoparticles likely results in charge separation to form surface-confined electrons and holes, on which significant experimental evidence has already become available [72, 73]. For example, the bright fluorescence emissions, which can be attributed to radiative recombinations on the particle surface, could be quenched by both electron donors and acceptors in an equally efficient fashion [72]. The emissions were also essentially diminished by the gold or platinum coating, as the coated metal was designed to soak up the surface-confined electrons, disrupting the radiative recombinations. Therefore, the functionalized carbon nanoparticles apparently served the function of harvesting visible photons to drive the photoreduction process, with the particle surface defects facilitating the charge separation by trapping the separated electrons and holes, phenomenologically similar to what occurs for metal-coated semiconductor nanoparticles (e.g., platinum-coated CdS or TiO_2) [74–76]. In addition to the obviously important advantage of strong visible absorption, other distinctive features of the nanoscale carbon-based photocatalysts include their aqueous solubility for the photoconversion under homogeneous reaction conditions and the confinement of the photoinduced charge separation to the particle surface, thus facilitating more efficient electron harvesting by the coated gold or platinum as the cocatalyst in the CO_2 conversion. Beyond CO_2 photoreduction, the same functionalized carbon nanoparticles with gold or platinum coating

could be used as photocatalysts for H_2 generation from water [77, 78]. In present system, the photoconversion efficiencies likely benefit from the solubility of the catalysts, surface confinement of the photoinduced charge separation and trapping, and the straightforward doping of metal cocatalysts.

3.5 *Photocatalyst for Overall Water-Splitting [79]*

Electronic structural analysis revealed that graphene oxide (GO) materials have conduction band minimum (CBM) and valence band maximum (VBM) levels suitable for generating H_2 and O_2, respectively, under visible-light irradiation [80]. Size modulation and chemical modification readily tune the electronic properties of graphene. The size effect results from quantum confinement, which becomes prominent when the sp^2 domain size is less than 10 nm. Quantum confinement causes the separation of the π and π^* orbitals, and creates a band gap in graphene [81–86]. Modifying graphene by oxygen adsorption forms C–O covalent bonds that damage the original orbitals and confine π electrons because of the reduction in sp^2 domain size. This modification renders the quantized discrete levels to be dictated by the nature of the sp^2 domains and associated functional groups. GO is a p-doped material because oxygen atoms are more electronegative than carbon atoms. Replacing oxygen functional groups on the GO sheet edge with nitrogen-containing groups transforms GO into an n-type semiconductor. In addition to surface modification by addition of functionalities, direct substitution with heteroatoms in the graphene lattice induces the modulation of optical and electronic properties. Teng et al. synthesized nitrogen-doped grapheme oxide-quantum dots (NGO-QDs) as the catalyst [79]. The NGO-QDs exhibited both p- and n-type conductivities. The diode configuration resulted in an internal Z-scheme charge transfer for effective reaction at the QD interface. Visible light (>420 nm) irradiation on the NGO-QDs resulted in simultaneous H_2 and O_2 evolution from pure water at an H_2:O_2 molar ratio of 2:1. An energetic band bending was present at the interface between semiconductor and solution, and a p–n type photochemical diode configuration, mimicking the biological photosynthesis system, provided a favorable situation to accomplish vectorial charge displacement for overall water-splitting. These developed NGO-QD photocatalyst consisted of nitrogen-doped graphene sheets stacked into crystals, with oxygen functional groups on the crystal surface. The band gap of the NGO-QDs was approximately 2.2 eV, and was capable of absorbing visible light to generate excitons. This NGO-QD construction resulted in the formation of p–n type photochemical diodes, in which the n-conductivity was caused by embedding nitrogen atoms in the graphene frame, and the p-conductivity by grafting oxygen functionalities on the graphene surface. Visible-light illumination on NGO-QDs suspended in pure water resulted in the evolution of H_2 and O_2 at a molar ratio of approximately 2:1. The p- and n-domains were responsible for the production of H_2 and O_2 gases, respectively. Nitrogen-free QDs with p-type conductivity catalyzed only H_2 evolution under irradiation, proving that the band bending in the p-type domains was favorable for electron injection to produce H_2. Likewise, NH_3-treated NGO-QDs showed n-type conductivity and catalyzed only O_2 evolution. The sp^2 clusters serve as the junction between the p- and n-domains and are the recombination sites for majority carriers from the two domains. The strong PL emission from the NGO-QDs with visible-light irradiation might be associated with the presence of the interfacial junction for recombination. The photochemical diode-type mechanism for water-splitting over NGO-QDs showed a remarkable similarity to that of biological photosynthesis [79].

3.6 CDs as Electrocatalysts [5]

Nitrogen doping has been a powerful way to modify the properties of carbon materials. N-doping was also demonstrated to significantly affect the properties of the CDs, including the emergence of size-dependent electrocatalytic activity for the oxygen reduction reaction. For technological relevance in clear energy production like fuel cells and clear fuel production, oxygen reduction reaction (ORR) and its reverse reaction—oxygen evolution reaction (OER) are at the centre of intensive research. Because of the sluggish kinetics of ORR, electrocatalysts are usually used to improve the kinetics of ORR and of which platinum is the "state-of-the-art". Unfortunately, the formidably high cost of platinum-based electrocatalysts has prompted researchers to look for non-platinum-based electrocatalysts for ORR, aiming at achieving comparable or even better electrocatalytic efficiency than that of platinum-based electrocatalysts. The ultra-small size of CDs along with their high stability and good electrical conductivity makes them interesting contenders as electrocatalytic materials for ORR [87]. Previous investigations on graphene have indicated that doped nitrogen atoms in carbon materials, especially in the form of pyridinium moieties, play a critical role in enhancing their electrocatalytic activities toward ORR. One of the pioneering reports on the use of CDs as electrocatalysts for ORR was by Li et al. [88]. They demonstrated that N-CDs with oxygen-rich functional groups prepared via an electrochemical procedure are electrocatalytically active toward electrochemical reduction of oxygen. The onset potential of ORR was found to be −0.16 V (vs. Ag/AgCl), which is close to that of commercial platinum-based electrocatalysts. Similar results were later obtained by Yan and co-workers and Liu et al. With N-CDs synthesised by totally different procedures [89, 90]. A comparison between nitrogen-free CDs and the N-CDs suggested that the electrocatalytic activity of the N-CDs is indeed closely associated with the N-doping effect. In addition, the N-CDs exhibited excellent tolerance to a possible crossover effect from methanol. First-principles investigations of the N-CDs suggested that pyridinic and graphitic nitrogen are responsible for the observed electrocatalytic activity [91]. In another report, Zhu and colleagues investigated the electrocatalytic activity of CDs prepared from natural biomass—soy milk [92].

Similar to the N-CDs, a much enhanced electrochemical reduction profile of oxygen was obtained. Likewise, OER also suffers from sluggish kinetics and a high over-potential is required in order to drive OER at a reasonably high rate. Currently, the best electrocatalysts for OER are ruthenium- and iridium-based materials. Again, the formidably high cost of these materials has urged researchers to search for alternative electrocatalysts that can offer high efficiency in OER and yet readily available at low cost. Unfortunately, reasonably high electrocatalytic activity of CDs toward OER has yet to be reported.

4 Catalysts Design Based on Carbon Dots

4.1 From CDs to Mesoporous Carbons Catalyst [93]

The design, fabrication and control of nanostructured materials are the core issues in catalysis and nanotechnology [94–96]. Carbon nanostructures, as typical inorganic materials, such as carbon nanotube, graphene, mesoporous carbon, have received a great deal of attention due to its wide applications [97–100]. Especially, mesoporous carbons (MCs) are widely used for gas separation, adsorbents, catalyst supports, and electrodes for electrochemical double layer capacitors and fuel cells [101–108]. Although many kinds of rigid and designed inorganic templates have been employed to obtain MCs with uniform pore sizes, the synthesis process still suffers from complex steps, harsh conditions, and a series post-processing [109, 110]. Therefore, seeking a facile and non-template method to obtain MCs is still a huge challenge for nanochemistry and nanotechnology. In light of their tinny size, high crystalline nature (fragments of graphite) and functionalized surface, CDs should be regarded as the construction units of carbon materials, and then realize the construction of mesoporous carbon without templates. The obtain MCs from CDs have a high specific surface area (183.6 m^2 g^{-1}) and uniform pore size distribution (5 nm). Typically, during the synthetic process of MCs, firstly, with the temperature increased, the water molecule would lose by β-elimination/condensation reactions from the hydroxyl/carboxyl groups on the surface of CDs, which makes the graphitic structure developed as well as the formation of carbon–oxygen-carbon bonds. After the further dehydration and/or carbonization, many CDs were connected with each other and finally lead to the formation of MCs. Also, MCs as catalyst exhibit high catalytic activity for selective oxidation of cyclooctene (32.43 % conversion based on cyclooctene and 87.53 % selectivity for epoxycyclooctane) with tert-Butyl hydroperoxide (TBHP) as initiator and air as oxidant at 80 °C. The high conversion of cyclooctene and selectivity to epoxycyclooctane was achieved simultaneously after 48 h. The conversion efficiency of cyclooctene increased from 7.81 to 32.43 % after 48 h. The selectivity of epoxycyclooctane also increased from 75.53 up to 87.53 %. But the selectivity to 2-cyclooctenone fell to 9.23 from 24.47 % with increasing reaction time. MCs catalyst really have good catalytic activity over the elective oxidation of cyclooctene. Moreover, the present selective oxidation of cyclooctene was carried out under a mild condition with air as oxidant, which suggested that the present process is energy-saving, environment-friendliness, and low cost. The oxygen-containing surface groups indeed are active sites for the elective oxidation of cyclooctene and have an effect on the catalytic activity of MCs.

4.2 *Metal Nanoparticle/CD Complex Photocatalyst for Hydrocarbon Selective Oxidation [111]*

The selective oxidation of cyclohexane to cyclohexanone and cyclohexanol (about 10^6 tons per year) is the key step of the commercial production of nylon-6 and nylon-66 polymers. Despite extensive development efforts for cyclohexane oxidization catalysts, today's commercial processes still suffer from high temperature (150–160 °C), high pressure (1–2 MPa), low yield (<15 %), low selectivity (<80 %), and excessive production of wastes [112, 113]. Catalysts enabling cyclohexane oxidation under mild conditions would alleviate the potential hazard of harsh reaction conditions, as well as allow for more selective and controlled reactions. As a preferred oxidant in liquid phase, hydrogen peroxide can work at low temperature, produces only water as a by-product and has a high oxygen uptake (47 %) [114–116]. However, cyclohexane oxidation with H_2O_2 at low temperature using the current catalysts (such as weak acid resin, iron bispidine complexes, Ru complex, Cr/Ti/Si oxides, Na-GeX, mesoporous TS-1, Ti-MCM-41, metal aluminophosphates, biomimetic systems and Cu–Cr_2O_3) invariably suffers from large amount of organic solvent (i.e. acetone and acetonitrile), poor selectivity, and low H_2O_2 efficiency, thus making its industrialization difficult [117–126]. CDs can function not only as an efficient photocatalyst (for highly selective oxidation), but also as a multi-functional component in photocatalyst design to promote wider spectrum absorption and separation of electron-hole, as well as to stabilize photolysis semiconductors [40]. Metal nanoparticles (Au, Cu, etc.) are known to possess catalytic activities for selective oxidation reactions [127, 128]. Additionally, surface plasma resonance absorption can also induce good photocatalytic abilities of metal nanoparticles (Au, Ag, etc.) [129, 130]. Kang et al. report the design of a tunable photocatalyst based on the composite of CDs and metal nanoparticles for the selective oxidation of cyclohexane. Significantly, the Au nanoparticles/CDs (Au/CDs) composite catalyst yielded oxidation of cyclohexane to cyclohexanone with 63.8 % efficiency and >99.9 % selectivity in the presence of H_2O_2 under visible light at room temperature (see Fig. 1). The totality of the experiments shows that the high conversion and selectivity in the present photocatalyst system should be attributed to the collective contribution and interactions of CDs and AuNPs. In present photocatalytic system, the reaction happened in the interface between Au NPs and CDs. Although the hydroxy radicals have strong oxidation abilities, the synergic effect of the functions of CDs and the SPR (surface

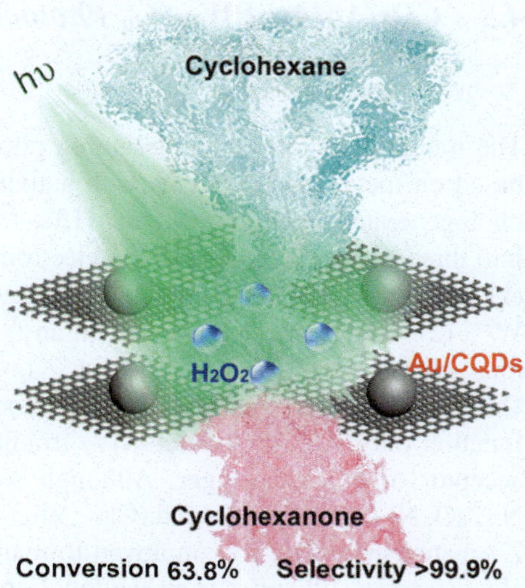

Fig. 1 Au/CQDs composites as a photocatalyst for selective oxidation of cyclohexane in the presence of H_2O_2 under visible light (reproduced from Liu et al. [111, p. 328])

plasmonic resonance) of AuNPs, still can protect the products (cyclohexanone and cyclohexanol) from the over oxidation. The totality of the experiment results suggests the following mechanistic processes in the photocatalytic oxidation of cyclohexane. The surface plasma resonance of AuNPs enhances the light absorption of Au/CDs composites. Under visible light, H_2O_2 is decomposed to hydroxyl radical ($HO^{\cdot}$), which serves as a strong oxidant for conversion of cyclohexane to cyclohexanone. In the process, the interaction between CDs and AuNPs under visible light plays a key role in the eventual high conversion and selectivity. Based on the proposed mechanism, Ag nanoparticles/CDs (Ag/CDs) and Cu nanoparticles/CDs (Cu/CDs) composite photocatalysts were synthesized, and used to catalyze cyclohexane oxidation under the same condition with Au/CDs system. Like the Au/CDs system, Ag/CDs and Cu/CDs composites exhibited similarly good catalytic performance under purple light (conversion 54.0 %, selectivity 84.1 % for Ag) and red light (conversion 46.7 %, selectivity 75.3 % for Cu), which correspond respectively to the surface plasma resonance of Ag and Cu nanoparticles. The results were consistent with the proposed catalytic mechanism discussed above. It is therefore possible to tune the wavelength response of the present photocatalyst system based on the surface plasma resonance of metal nanoparticles to achieve high-efficiency and high-selectivity oxidation of cyclohexane. Given its diversity and versatility of structural and composition design, metal nanoparticles/CDs composites may provide a powerful pathway for the development of high-performance catalysts and production processes for green chemical industry.

4.3 $CDs/Ag/Ag_3PW_{12}O_{40}$ Photocatalysts for Overall Water Splitting [131]

The traditional strategy for designing photocatalysts for overall water splitting is based on metal oxide systems with activities that are related to the band gap and photogenerated electron-hole pairs [132–138]. Typically, photocatalysts are divided into three groups according to the electronic configuration of their core metal ions. (1) Transition metal ions with d0 configurations are Ti^{4+}, Zr^{4+}, Ta^{5+}, Nb^{5+}, V^{5+}, and W^{6+}. (2) A rare-earth metal ion with an f^0 configuration is Ce^{4+}. (3) Typical metal ions with d^{10} configurations are Ga^{3+}, In^{3+}, Ge^{4+}, Sn^{4+}, and Sb^{4+} [70, 139–146]. Unfortunately, the majority of the photocatalysts are complex and expensive or function only with ultraviolet (UV) irradiation in need of an appropriate electron acceptor or hole scavenger. Although some semiconductors, such as La-doped NaTaO, Ni-doped $InTaO_4$ and $(Ga_{1-x}Zn_x)$-$(N_{1-x}O_x)$ solid solution (or to construct a Z-scheme structure), are employed to realize the overall water-splitting, these catalyst systems still suffer from the low light absorption (<510 nm), poor separation efficiency of electron-hole pairs, and the formidable complexity of the oxidative half-reaction [147–152]. Polyoxometalates (POMs), with unique photoelectric chemical properties, are potentially useful for photocatalytic H_2 and O_2 generation [153–158]. The utilization of surface plasmon resonance (SPR) has offered a new opportunity to overcome the limited efficiency of photocatalysts. SPR improves the solar-energy-conversion efficiency by (i) extending light absorption to longer wavelengths, (ii) increasing light scattering, and (iii) exciting electron–hole pairs in the semiconductor by transferring the plasmonic energy from the metal to the semiconductor [159]. In light of the remarkable photocatalytic properties of CDs, the SPR effect of metal Ag, and the photocatalytic hydrogen generation of POMs, the combination of CDs, Ag, and POMs may be a unique approach to construct stable and efficient complex photocatalyst for solar water splitting. Kang and Liu et al. report the design and fabrication of $CDs/Ag/Ag_3PW_{12}O_{40}$ nanocomposites, which served as photocatalysts for overall water splitting in visible light (light absorption extend to 650 nm) without any electron acceptors or hole scavengers. The estimated apparent quantum yield (AQY) was 4.9 % at 480 nm [143, 160]. A reaction mechanism was proposed to explain the photocatalytic water splitting with this composite photocatalyst in visible light. In present system, POMs play a key role for the water splitting. PB could be formed under visible light irradiation in a complex system [161, 162]. The absorption of visible light happened at Ag nanoparticles surface for the SPR effect, and then the absorbed photons would be efficiently separated into electrons and holes. Given the dipolar character of the surface plasmonic state of Ag nanoparticles, such those electrons made POM transfer to PB which resulted from the one- and two-electron reduced, and the holes will act as positive charge centers on the $Ag_3PW_{12}O_{40}$ surface [163, 164]. Also, PB could be further excited by visible light irradiation and transferred electrons to conductive band of POM as literatures reported [165–167]. These non-tight-binding electrons in the intermediate energy levels act like as "color center" in $Ag_3PW_{12}O_{40}$, which can be transiently stabilized and further photoexcited to the conduction band of $Ag_3PW_{12}O_{40}$ by photons in visible regions. On the other hand, the photoexcited electrons in the conduction band of $Ag_3PW_{12}O_{40}$ are thermodynamics feasible for water reductions. The Ag nanoparticles here played two vital roles for the enhanced photocatalytic water splitting efficiency: (1) the strong SPR-induced electric fields localized nearby at the $Ag/Ag_3PW_{12}O_{40}$ interfaces which can cause the electrons generation and enhance separation efficiencies of electron-hole pairs in $Ag_3PW_{12}O_{40}$ [168]; and (2) The electrons in the conduction band of $Ag_3PW_{12}O_{40}$ can inject into the contractile Ag nanoparticles which act as electron buffer and catalytic site for hydrogen generation. The insoluble CDs layer on the surface of $Ag_3PW_{12}O_{40}$ effectively protects $Ag_3PW_{12}O_{40}$ from dissolution in aqueous solution, thus, enhancing the structural stability of $CDs/Ag/Ag_3PW_{12}O_{40}$ during the photocatalytic processes. Also, the CDs with excellent storing-charges ability can also acted as like an electron buffer which can promote the electron-extraction from the conduction band of $Ag_3PW_{12}O_{40}$ and subsequently decrease the electron-hole recombination rate in $Ag_3PW_{12}O_{40}$ and increase the optical absorption of $Ag_3PW_{12}O_{40}$ for the increased unoccupied occupiable-states in the conduction band of $Ag_3PW_{12}O_{40}$ [168, 169]. Finally, CDs can enhance the electron transport due to their photo-induced electron transfer property. All of above mentioned positive roles of Ag, CDs, and $Ag_3PW_{12}O_{40}$ are responsible for the excellent photocatalytic water splitting properties of the composited $CDs/Ag/Ag_3PW_{12}O_{40}$ photocatalysts in visible light.

4.4 CDs Sensitized TiO_2 Nanotube Arrays for Photoelectrochemical Hydrogen Generation Under Visible Light [170]

TiO_2 is regarded as one of the most popular photoanode materials of photoelectrochemical (PEC) water splitting devices owing to its high resistance to photocorrosion, physical and chemical stability, easy availability and low cost. While, both the bulk and nanostructured photoanodes of pure TiO_2 materials still suffer the major limitation of weak response (or nonresponse) to the visible spectrum owning to their large band gap (about 3.2 eV). To date, a variety of strategies have been utilized to improve TiO_2 PEC performance, such as doping with C, N, Sn, Sr, Nb, coupling with secondary semiconductors and photosensitization of dyes [171–176]. Compared with the doping strategy, another approach with dye-based compounds or narrow band gap semiconductors as a sensitizer is convenient, and has structure and properties designable abilities [173–181]. This system also has achieved a certain degree of success in broadening the absorption spectrum to visible and near infrared (NIR) region, and therefore enhancing the solar-to-hydrogen (STH) efficiency of the PEC cells [174, 177–181]. For the efficient hydrogen generation under visible light, a good sensitizer should meet several criteria as described in the following. First, it must be able to enhance the absorption of the solar spectrum for the host materials. Second, it should own suited energy levels so that the photoexcited electrons in the HOMO level or conduction band can be efficiently injected into the semiconductor acceptor's conduction band, and simultaneously the holes in the LUMO level or valence band can oxidize the reducing substances in the electrolytes. Third, it is favorable to contain some specific functional groups on their surfaces which can make them easy and stable to attach to the semiconductor acceptors' surfaces. Fourth, it should be resistant to corrosion or degradation in the practical operation conditions of PEC cells for a long term. CDs were used as an alternative sensitizer for the PEC cells based on TiO_2 nanotube arrays (TiO_2 NTs). Under simulated sunlight illumination (AM 1.5G, 100 mW/cm^2), the photocurrent density of the CDs sensitized PEC cell is four times larger than the unsensitized one at 0 V versus Ag/AgCl. The corresponding hydrogen production rate was determinated to be about 14.1 μmol/h for a CDs sensitized TiO_2 nanotube arrays (CDs/TiO_2 NTs) photoanode (about 0.78 cm^2) with Faradaic efficiency nearly 100 %. The enhanced photocurrent after CDs deposited on the surface of TiO_2 is not due to the TiO_2 were excited by the CDs' upconverted emission. This conclusion was confirmed by the different IPCE results obtained by performing tests in different electrolytes. No positive IPCE value can be observed with excited wavelength over 410 nm for CDs/TiO_2 NTs photoelectrode in 1 M Na_2SO_4 aqueous electrolyte. The reason is that the holes in the HOMO level of CDs are not able to realize the oxygen evolution reaction (OER) while they can be consumed by the hole-scavenger in a sacrificial electrolyte such as Na_2S/Na_2SO_3 aqueous solution. And if the positive IPCE values over 410 nm observed by performing the measurement in the Na_2S/Na_2SO_3 aqueous solution were due to the upconverted fluorescence of CDs, there should also be able to observe a positive IPCE value over 410 nm in the 1 M Na_2SO_4 electrolyte. Therefore, a sensibilization mechanism was proposed to illustrate the role of CDs in the PEC cells. A photon with adequate energy can excite the electron in the HOMO level transferring to the LUMO level of CDs. Afterward, the excited electrons in the LUMO level of the CDs are transferred to the conduction band of the contacted TiO_2 NTs and then transported to the counter electrode for hydrogen evolution reaction along the TiO_2 NTs axial direction. While the holes left in the HOMO level of the CDs can oxidize the sacrificial reagent to complete a whole galvanic circle. It can be expected that several further works could be done to optimize the PEC hydrogen production efficiency of the CDs/TiO_2 system: (a) one can design more reasonable structures of TiO_2 to improve the absorbance of CDs and the charge carrier transport; (b) CDs can be modified by doping with S, N and P to further enhance the light-havesting ability, quantum efficiency and the chemical reaction kinetics; (c) Surface treatments of CDs may be favorable for stronger anchoring on TiO_2 and higher photogenerated charge carrier separation efficiency.

4.5 Modulation of Electron/Energy Transfer States at TiO_2-CDs Interface [182]

In light of their respective unique properties, CDs and TiO_2 nanotubes (TNTs) arrays may be integrated into a model nanosystem for the studies of catalysis, photoelectrical devices, and energy transfer systems in chemistry and nanoscience. Different energy and electron transfer states have been observed in the CDs/TNTs system due to the up-conversion photoluminescence and the electron donation/acceptance properties of the CDs decorated on TNTs. Kang and wang et al. show a bias-mediated electron/energy transfer process at the CDs/TiO_2 interface for the dynamic modulation of opto-electronic properties [182]. Different energy and electron transfer states have been observed in the CDs/TNTs system due to the up-conversion PL and the electron donation/acceptance properties of the CDs layer. Specifically, five distinct electron/energy transfer states of the CDs-decorated TiO_2 nanotubes (CDs/TNTs) system can be dynamically tuned by different pulse-bias treatments. To investigate the photoelectrochemical and electron/energy transfer process in the CDs/TNTs system, a series of photoelectrochemical experiments was performed in a two-electrode cell system with Pt as the counter electrode. The mechanism of the present pulse-bias tunable opto-electronic conversion logic phenomenon in the CDs/TNTs system as follows. When the negative bias applied to the CDs/TNTs electrode was increased from −1, −2, to −3 V, the positively charged CDs would be attracted increasingly closer to TNTs, leading to gradual reduction in interface impedance. At 0 and −1 V pulse-bias treatment, or high and medium interface impedance, photo-excited electrons of TNTs would favor to transfer to the Pt electrode, and those of CDs would favor to return to a low-energy state. And then the up-converted PL of CDs would excite the TiO_2, leading to higher IPCEmax (opto-electronic conversion). At −2 V pulse-bias treatment, or low interface impedance, the symbiotic effect of up-converted PL of CDs and high electron transfer ability between TNTs and CDs would lead to the highest $IPCE_{max}$ (opto-electronic conversion). While at −3 V pulse-bias treatment, the interface impedance between TNTs and CDs would be lowest, and recombination of photo-induced electrons and holes of TNTs through TiO_2-CDs-TiO_2 pathway be highest, thus leading to least up-converted PL and lowest opto-electronic conversion. The present pulse-bias treatment method would enable the opto-electronic conversion logic phenomenon in the CDs/TNTs nanosystem. Application of different pulse-bias treatment can generate high, medium or low opto-electronic conversion efficiency. The IPCE of electrode increased from 13.5 to 19 % after −1 V pulse-bias treatment, and from 13.5 to 24 % after −2 V pulse-bias treatment, but decreased from 13.5 to 2.5 % after −3 V pulse-bias treatment. This phenomenon represents a simple logic function, which yields a medium output at no bias (or low positive pulse bias), a high output at −2 V pulse bias, and a low output at −3 V pulse bias. Significantly, the programmable character (based on different applied bias potential) of the present system is extremely versatile and facile. This bias-mediated electron/energy transfer process at the

CDs/TiO_2 interface may provide a new opportunity not only for the manipulation of opto-electronic conversion process, but also for the development of opto-electronic conversion devices and photocatalyst design.

4.6 CDs Stabilized Gold Nanoparticles Catalyst System [183]

General speaking, the adsorption effect should be taken into the consideration firstly to improve the catalytic activity and stability, which is the precondition and crucial process in nanocatalyst design of catalytic reactions. Au nanoparticles (AuNPs), with their excellent electronic, magnetic, optical, thermal, catalytic, and biochemical properties, [184–187] have been intensively studied in many fields (such as sensors, biolabeling, drug delivery, and photonics), especially as novel catalysts for catalysis [185–192]. However, the catalytic chemistry of AuNPs still face some challenges, such as, the catalytic activity needs to be improved and AuNPs are still unstable in the reaction process. It was expected to enhance the adsorption capacity and stability of AuNPs based nanocatalysts to improve the catalytic efficiency. CDs with abundant functional groups (–OH, –COOH, C=O) in surface were especially designed to enhance the adsorption capacity and the catalytic activity of AuNPs. The CDs stabilized gold nanoparticles (AuNPs-CDs) were greenly synthesized by one-step reduction $HAuCl_4$ with CDs used as both reductant and stabilizer under visible light irradiation. The resulted AuNPs-CDs possess high catalytic activity toward 4-nitrophenol reduction. Further detailed adsorption kinetics data indicated that the present adsorption systems follow predominantly the second-order rate model and CDs capped on the surface of AuNPs also enhanced the adsorption capacity and the catalytic activity of AuNPs. CDs have been successfully used in photocatalyst design and optical sensor due to their excellent chemical and optical properties [193, 194]. From the view of catalyst design, the novel advantages of CDs used in catalysis are as follows: First, CDs have suited sizes (below 10 nm) for catalysis system [40]. Second, CDs have excellent photo-chemical properties (photoinduced electron transfer and redox properties, luminescence and so on) [195]. Third, the surface of CDs is adjustable with overall oxygen contents ranging from 5 to 50 wt%, depending upon the experimental conditions used [196]. Finally, the functional groups (–OH, –COOH, etc.) in surface of CDs are helpful for improving the adsorption of reactants [197–200]. Therefore, CDs are good candidates for nanocatalyst design. CDs were specially designed to enhance the adsorption capacity and catalysis activity of AuNPs. The abundant functional groups (–OH, –COOH, C=O) in the surface of CDs not only can reduce $HAuCl_4$ to AuNPs but also can enhance the stability of AuNPs in a water-based environment [201–204]. The resultant AuNPs-CDs show excellent hydrophilic, high stability and good biocompatible. Adsorption kinetics manifests the present AuNPs-CDs sorption systems follow predominantly the second-order rate model. The synthesized AuNPs-CDs also can be used as an efficient catalyst for toxic pollutant 4-NP totally reduction to 4-AP (conversion, 100 %). A good linear correlation of $\ln(C_i/C_0)$ versus time was obtained and a kinetic rate constant of AuNPs-CDs was determined to be 0.68 min^{-1}. The functional groups (–OH, –COOH, –C=O) in surface of CDs are good for enhancing the adsorption capacity and then the catalytic activity of AuNPs. In present system, the results obviously indicated CDs capped on the surface of AuNPs enhanced the adsorption and stability of AuNPs, which accorded with the highly catalytic activity of AuNPs-CDs. The equilibrium of sorption was evaluated by two well-known models of Langmuir and Freundlich isotherm. The Langmuir isotherm model is the most widely used two-parameter equation, which assumes monolayer adsorption onto a surface containing a limited number of adsorption sites with no transmigration of the adsorbate in the plane of the surface. The Freundlich isotherm in the other hand takes heterogeneous systems into account and is not restricted to the formation of the monolayer. The Langmuir isotherm with correlation coefficient of 0.995 and 0.997 represents a better fit of experimental data than Freundlich model with correlation coefficient of 0.982 and 0.975, for CDs and AuNPs-CDs, respectively. It indicates that monolayer adsorption of 4-NP takes place on the homogeneous surface of AuNPs-CDs adsorbents. Moreover, the adsorption is favorable and rather irreversible. Based on the results of adsorption and reduction rate, it suggests that CDs help in adsorbing 4-nitrophenote ions closer to the surface of AuNPs, AuNPs subsequently help in facilitating the reduction of 4-NP by lowering the activation energy of the reaction and play the role of catalyst. Adsorption kinetics manifests that the present AuNPs-CDs sorption systems follow predominantly the second-order rate model.

4.7 Photocatalyst Design Based on CDs and Semiconductor Nanoparticles [40, 168, 169]

CDs/TiO_2 Photocatalysts [40]. As one of the most popular photocatalysts, TiO_2 has been used in the removal of organic pollutants and in the generation of H_2 through water splitting. However, a major drawback in its photocatalytic efficiency resides in its ineffective utilisation of visible light as the irradiation source. Because the bandgap of bulk TiO_2 lies in the UV region (3.0–3.2 eV), only less than 5 % of sunlight is utilised by TiO_2. TiO_2–CD nanocomposites are able to completely degrade MB (50 mg mL^{-1}) within 25 min under visible light irradiation, where only <5 % of MB is degraded when pure TiO_2 is used as the photocatalyst. Apart from harvesting visible light and converting it to shorter wavelength light through up-conversion, which in turn excites TiO_2 to form electron–hole pairs, it is believed that the CDs in the nanocomposites facilitate the transfer of electrons from TiO_2 and the electrons can be shuttled freely along the conducting paths of the CDs, allowing charge separation, stabilisation and hindering recombination, thereby generating long-lived holes on the TiO_2 surface. The longer-lived holes then account for the much enhanced photocatalytic activity of the TiO_2–CD nanocomposites. Likewise, similar behavior was observed with SiO_2–CD nanocomposites in the photocatalytic degradation of MB36 and selective hydrocarbon oxidation. Nonetheless, more work is needed to improve the lifespan of the above-mentioned photocatalysts before they can be employed in practical scenarios.

In addition to being used as photocatalysts, CDs have been capturing the attention of researchers as potential photosensitizers in solar cells [205]. For example, a CD–RhB–TiO_2 system showed that the CDs effectively bridge the RhB molecules to the TiO2 substrate by acting as a one-way electron transfer intermediary. Comparing to the RhB–TiO_2 system, the presence of CDs significantly enhanced the photoelectric conversion efficiency by as much as seven times. In another report, a CD/TiO_2 electrode was employed in a solar cell. The photocurrent density was 2.7 times larger than that of pristine TiO_2 electrode under visible light illumination. Enhanced performance of a solar cell was also obtained when N-CDs were used as photosensitizers. Although the photo-to-electricity conversion efficiency of the above-mentioned solar cells is far from satisfactory, these findings definitely encourage more research in the application of CDs in photovoltaic devices and photocatalyst design [205].

CDs/Ag_3PO_4 Photocatalysts [168]. Ag_3PO_4 is slightly soluble in aqueous solution, which greatly reduces its structural stability. During the photocatalytic process, the transformation of Ag^+ into Ag is usually accompanied, which results in the photocorrosion of Ag_3PO_4 in the absence of electron acceptors. Furthermore, along with the photocatalytic reaction, the appearance of the by-products, black metallic Ag particles would inevitably prevent visible light absorption of Ag_3PO_4, which decrease the photocatalytic activity. CDs display abundant photo-physical properties. Especially, the strong size and excitation wavelength dependent PL behaviors further enhance their photocatalytic properties. Considering such remarkable properties of CDs and the limitations of the Ag_3PO_4 photocatalytic system, the combination of CDs and Ag_3PO_4 may be regarded as an ideal strategy to construct the stable and efficient complex photocatalytic system (such as, CDs/Ag_3PO_4 and CDs/Ag/Ag_3PO_4 photocatalysts). The key roles of CDs in the complex photocatalysts were soundly investigated. The CDs layer on the surface of Ag_3PO_4 and Ag/Ag_3PO_4 particles can effectively protect Ag_3PO_4 from dissolution in aqueous solution. The unique photoinduced electron transfer properties of CDs can make Ag_3PO_4 avoid photocorrosion. The up-converted PL property of CDs make the CDs/Ag_3PO_4 and CDs/Ag/Ag_3PO_4 complex systems effectively utilize the full spectrum of sunlight to greatly enhance the photocatalytic activity. Moreover, CDs can act as an electron reservoir to hinder the electron/hole pairs' recombination probability. More interestingly, the existence of Ag accompanied by SPR can further enhance the utilization of sunlight and formation of electron/hole pairs. The synergistic effects of CDs and Ag lead to the highest photocatalytic activity of CDs/Ag/Ag_3PO_4 complex photocatalyst.

CDs/Cu_2O Photocatalysts [169]. So far, near infrared (NIR) and IR light (total called (N)IR) of sunlight (account for 53 %) are still not fully utilized by present photocatalytic system, which has become gradually an urgent challenge for chemistry and materials science. To effectively utilize the (N)IR light, an ideal photocatalytic system should be effectively activated by (N)IR light. As discussed above, the TiO_2/CDs and SiO_2/CDs complex systems were demonstrated to show excellent photocatalytic ability in visible light, while they are still impuissant in (N) IR wavelength zone. In light of its relative narrow band gap (about 2.2 eV), high photostability, and low toxicity, Cu_2O was a potential candidate for designing the CDs based (N)IR light sensitive photocatalyst. It was reported that the CDs/Cu_2O photocatalytic system could harness the (N)IR light to enhance the photocatalytic activity based on the collective effect of superior light reflecting ability of the Cu_2O protruding nanostructures and the up-converted PL property of CDs. When the CDs/Cu_2O composite photocatalyst is illuminated, the protruding nanostructures allow the multiple reflections of (N)IR light among the vacant space between these protruding particles, which can make better use of the source light and therefore offering an improved photocatalytic activity. On the other hand, CDs can absorb (N)IR light (>700 nm), and then emit shorter wavelength light (390–564 nm) as a result of up–conversion, which in turn further excites Cu_2O to form electron/hole (e^-/h^+) pairs. The electron/hole pairs then react with the adsorbed oxidants/reducers (usually O_2/OH^-) to produce active oxygen radicals (e.g. $^{\cdot}O_2$, $^{\cdot}OH$), which subsequently cause the degradation of organic dye (MB). Significantly, when CDs are attached on the surface of Cu_2O, the relative position of CDs band edge permits the transfer of electrons from the Cu_2O surface, allowing the charge separation, stabilization, and then hindering e^-/h^+ pairs' recombination. The electrons can be shuttled freely along the conducting network of CDs, and the longer-lived holes on the Cu_2O then account for the higher activity of the composite photocatalyst. In addition, for organic pollution degradation, the p–p interaction between conjugated structure of CDs and benzene ring of MB is beneficial to the enrichment of MB on the surface of CDs/Cu_2O composite.

4.8 CDs/NiFe Layered Double Hydroxide Composite Electrocatalyst [206]

The oxygen evolution reaction (OER) is kinetically slow owning to its multistep proton-coupled electron transfer process, so the electrolysis must require a relative high potential than thermodynamic potential for water-splitting. 10 RuO_2 and IrO_2 are regarded as the most active OER catalysts, but the lacking of Ru and Ir makes it impractical to use the metals on a large scale [207–209]. Layered double hydroxides (LDHs), as a family of layered anionic materials, have attracted considerable attention because of their flexible structures and chemical versatility, which have been exploited as a fruitful source of materials with applications in electrochemistry, magnetism, catalysis and chemical sensing [210–213]. NiFe-LDH possesses a layered and relatively open structure, which makes it much easier for rapid diffusion of reactants/products, and even fast proton-coupled electron transfer process in water oxidation reaction. But the poor electrical conductivity restricts its massive applications in electrocatalysis. Although other carbon-related materials (such as carbon nanotubes (CNTs), graphene) have been reported to composite with LDHs for electrocatalysis applications, their poor charge transport efficiencies and complicated functional groups design are still weaknesses [214–216]. While, CDs known as a novel class of nanocarbon with abundant functional groups on the surface which are needed for nucleating and anchoring the pristine nanocrystals with CDs solidly to achieve intense electrostatic interaction (or covalent attachment), have attracted considerable attention because of their unique physical or chemical properties. The CDs/NiFe-LDH complex exhibits high electrocatalytic activity (with overpotential ~235 mV in 1 M KOH at a current density of 10 mA cm^{-2}) and stability for oxygen evolution. This almost exceeded those of any previous Ni–Fe compounds and even was comparable to that of the lowest overpotential reported in Ni-Fe catalysts, ~230 mV at 10 mA cm^{-2} for electrodeposited Ni–Fe films [169, 207, 217]. Typically, nickel acetate and iron nitrate (with a molar ratio of Ni/Fe = 5) were hydrolyzed in a mixed N,N-dimethylformamide (DMF) and CDs aqueous solution at 85 °C for 4 h. Followed by redispersing the intermediate product in a DMF/H_2O mixed solvent and through a solvothermal treatment at 120 °C for 12 h, then a second solvothermal step at 160 °C for 2 h. The solvothermal treatment may lead to the crystallization of NiFe-LDH nanoplates and formation of the nanocomposites. The high electrocatalytic property was primarily attributed to the NiFe-LDH phase, and further enhanced by strong associating the LDH with CDs, which possess small size, excellent conductivity, rapid electron transfer, and electron reservoir properties. Specifically, the small size of CDs may provide large specific surface area for more convenient elecocatalytic reaction. And the rapid electron transfers from CDs to NiFe-LDH on the surface could further improve the electrocatalytic activities. The surface functional groups on CDs (such as, carboxy C=O) make it easier for the formation of CDs/NiFe-LDH composites solidly due to the strong electrostatic interactions between NiFe-LDH and CDs (or generating new covalent bond (like C–O–Ni or C–O–Fe)). Anyway, through synergistic effect between NiFe-LDH and CDs afforded by direct integrating the Ni-Fe LDH nanoplates with the surface functional groups on CDs contributed to the optimal OER activity of the CDs/NiFe-LDH composite catalysts. Given the diversity and versatility of structural design of the present CDs/NiFe-LDH composite system, the combination of CDs and NiFe-LDH nanoplates as a superb electrocatalyst may provide a new approach to high-efficiency CDs-related electrocatalysts design for applications towards new energy sources, green chemistry, and environmental issues.

4.9 CDs for Porous Co, N-Codoped Carbon Electrocatalyst Design [218]

Although materials based on Pt are believed to be the most effective electrocatalysts for ORR, they still suffer from the high cost and poor supply of Pt, promoting the research on highly active catalysts [219–222]. Numerous studies have focused on the preparation of low-cost Co based nitrogen-containing materials as electrocatalysts for ORR, in which the heat-treatment of macrocyclic compounds under the inert atmosphere was found to be an effective method for the preparation of such Co/N/C catalysts [223–225]. For example, Chang et al. studied the effect of the structures of pyrolyzed cobalt-based macrocyclic compounds on ORR and proposed that the pyrolyzed cobalt-corrin compounds (py-Co-Corrin/C) exhibited the highest activity for ORR [226]. Zagal et al. [227] studied the application of Vitamin B12 (VB12) absorbed on graphite electrode for ORR. Liang et al. [228] investigated the mesoporous catalyst fabricated from VB12 and silica nanoparticles exhibited a remarkable activity for ORR in acid medium with half-wave potential of 0.79 V versus reversible hydrogen electrode (RHE). Wang et al. reported a serious of applications of pyrolyzed VB12 as non-precious metal catalyst for ORR in polymer electrolyte fuel cells (PEFC) and microbial fuel cells (MFC) [229]. Although these Co, N-codoped materials were demonstrated to be potential non-precious catalysts for ORR, there are still underlying problems yet to be dealt with urgently: such as, the finely control and optimization of Co and N concentrations, which is closely related to the catalytic ability optimization; the construction of porous structure; whether the Co and N concentrations affect the electrocatalytic activity for ORR synergistically. The solutions of above issues are of great importance and will effectively promote widespread application of Co/N/C catalysts [218]. With the aim of constructing a material with large surface area and well crystalline structure in favor of electron transportation, CDs could be regarded as the promising building blocks as a result of their tiny size, highly crystalline structure (they are fragments of graphite) and functionalized surface. Taking advantages of CDs, the pyrolysis of VB12 and CDs could lead to the porous structure and well controlled concentrations of Co and N, which are beneficial for the enhancement of electrocatalytic activity of catalyst. In a typical experiments, CDs and VB12 were utilized as raw materials to fabricate Co/N/C catalysts [218]. The porous cobalt, nitrogen-codoped carbon materials (Co/N/C) were synthesized by a facile one-step pyrolysis of VB12 and CDs. Varying the initial mass ratio of CDs and VB12 leads to controllable concentrations of Co (0–3.88 %) and N (0–5.88 %) after pyrolysis. The obtained Co/N/C was evaluated by ORR in both alkaline and acid media. Particularly, the Co/N/C with 1.12 % Co and 2.92 % N prepared at 700 °C exhibited the best catalytic ability for ORR with cathodic peak at −0.165 and 0.185 V (vs. SCE) in 0.1 M KOH and 0.1 M $HClO_4$ solution, respectively, which are comparable to that of Pt/C (20 %). The obtain catalyst also showed long-term stability and high methanol tolerance, which outperforms commercial Pt/C (20 %). According to the rotating disk electrode (RDE) and rotating ring-disk electrode (RRDE) results of Co1.12/N2.92/C-700 catalyst, the numbers of electron-transferred for the ORR were calculated to be 3.6, which indicated a four-electron pathway. Our results further demonstrated that the Co1.12/N2.92/C-700 catalyst showed superior methanol tolerance and durability than the commercial Pt/C (20 %) in alkaline and acid media. With the increase of Co concentration, the peak potential shifts to more positive and reach a maximum at the optimum Co concentration of 1.12 %, then shifts negatively. For the effect of N concentration on the ORR peak potential, the peak potential reaches a maximum at N concentration of 2.92 %. On the other hand, with the increase of Co doping concentration, the current density increases and reaches the maximum value of 1.16 mA cm^{-2} at 1.12 % for Co then drops with the increase of Co concentration. While, the highest current density is obtained when the doping concentration of N is 2.92 %. More or less N concentration will lead to the decrease of the current density. Co/N/C with 1.12 % Co and 2.92 % N exhibits the highest current density in acid medium. These results indicate that the Co and N concentrations exhibit a synergistic effect on the catalytic ability of Co/N/C system. Although, at present stage, no more information was provided to confirm the most optimized Co, N doping concentrations for ORR catalytic ability of Co/N/C system, the present results still demonstrate a primary rule for electrocatalyst design and component contents optimization to achieve more efficient catalyst: The effects of Co and N doping concentrations on the catalytic ability of Co/N/C are synergistic. With the increase of Co and N concentrations, the catalytic ability of Co/N/C increases to reach a maximum at the concentration of 1.12 % for Co and 2.92 % for N, respectively, then decreases with higher doping concentration of Co and N (Co > 1.12 and N > 2.92). The stability of Co1.12/N2.92/C-700 was investigated by chronoamperometric measurements in O_2 saturated 0.1 M KOH and 0.1 M $HClO_4$ solution. The continuous oxygen reduction reaction causes only a slight loss (7 %) of the current for Co1.12/N2.92/C-700 catalyst after 4000s of reaction under alkaline condition. As a comparison, the current loss on Pt/C (20 %) is as high as 26 % after the same reaction duration. The excellent ability of Co1.12/N2.92/C-700 was attributed to the existence of clusters with excellent crystallinity and porous structure which resulted in formation of large interface and surface. These factors facilitated the transportation of electrons. The numbers of electron-transferred for the ORR were calculated to be around 3.6, which indicated the four-electron pathway.

5 Outlook [1–5]

In this chapter, we have described the recent advances in the research on C-dots, focusing on their synthesis, properties, and applications in catalysis. The examples given above are taken from the literature published until December 2014. We believe that many other reactions will be studied using CDs nanocatalysts and some of them will undoubtedly give new examples of catalysis by CDs.

A variety of synthesis techniques already exist for producing CDs of different characteristics. The PL and optical properties of C-dots are both interesting and intriguing, constituting a rich and hot research topic. The general strategy for the adoption of environmentally benign CDs as photocatalysts in synthetic chemistry represents an attractive approach in the development of green chemistry, which may eventually lessen the burden of energy consumption, product clean-up and waste disposal. The immediate goal in this emerging area should be geared toward the discovery of photochemical solutions for increasingly ambitious synthetic goals. The long-term goals should be to improve efficiency and synthetic utility and to eventually perform chemical synthesis under sunlight. Compared to other applications of CDs, there have been fewer studies in the usability of CDs as electrocatalysts for ORR and OER. In-depth theoretical and experimental studies are needed to delicately design CD-based electrocatalysts with desirable electrocatalytic activity and long-term operation stability. The combination of CD doping and CD-based nanocomposites with other nanomaterials may open up new avenues to systematically study the effect of structural parameters and chemical compositions on the catalytic performance of the electrocatalysts, thus leading to fundamental insights and practical applications. The potential of CDs in the storage and transport of electrons impacted by light has yet to be exploited fully.

In the future, we expect the advent of more facile and robust synthetic routes and creative applications to better realize the potential of the increasingly important CDs materials. New properties and how to subtly tune these properties, such as fresh phosphorescence and debatable up-converted PL, are also challengeable for its nubilous luminescent mechanism. The amorphous to nearly crystalline internal structure, nonquantitative surface structure and virtual size polydispersion may block the clarification of the luminescence mechanism. This problem will be solved through the accurate synthesis, careful analysis, and intelligent consideration. CDs stand to have a huge impact in biotechnological and environmental applications because of their potential as nontoxic alternates to traditional heavy-metal-based QDs. In addition, the unique photoinduced electron transfer ability, as well as excellent light harvesting capability, make CDs an exceptional candidate for photocatalytic and photovoltaic applications. Carefully designed C-dots composites have the potential to expand the capabilities of next-generation energy-storage and photovoltaic devices, photocatalysts and sensors. By surface and band gap modification of CDs via functionalization or semiconductors, we can expect to design novel catalysts from CDs for green chemistry and energy issues.

References

1. S.N. Baker, G.A. Baker, Luminescent carbon nanodots: emergent nanolights. Angew. Chem. Int. Ed. **49**, 6726–6744 (2010). doi:10.1002/anie.200906623
2. H. Li, Z. Kang, Y. Liu, S.T. Lee, Carbon nanodots: synthesis, properties and applications. J. Mater. Chem. **22**, 24230 (2012). doi:10.1039/C2JM34690G

3. L.L. Li, G.H. Wu, G.H. Yang, J. Peng, J.W. Zhao, J.J. Zhu, Focusing on luminescent graphene quantum dots: current status and future perspectives. Nanoscale **5**, 4015 (2013). doi:10.1039/C3NR33849E
4. Y.F. Wang, A.G. Hu, Carbon quantum dots: synthesis, properties and applications. J. Mater. Chem. C **2**, 6921 (2014). doi:10.1039/C4TC00988F
5. S.Y. Lim, W. Shen, Z.Q. Gao, Carbon quantum dots and their applications. Chem. Soc. Rev. **44**, 362 (2015). doi:10.1039/C4CS00269E
6. J. Sun, S. Yang, Z. Wang, H. Shen, T. Xu, L. Sun, H. Li, W. Chen, X. Jiang, G. Ding, Z. Kang, X. Xie, M. Jiang, Ultra-high quantum yield of graphene quantum dots: aromatic-nitrogen doping and photoluminescence mechanism. Part. Part. Syst. Charact. (2014). doi:10.1002/ppsc.201400189
7. H. Li, R. Liu, S. Lian, Y. Liu, H. Huang, Z. Kang, Near-infrared light controlled photocatalytic activity of carbon quantum dots for highly selective oxidation reaction. Nanoscale **5**, 3289–3297 (2013). doi:10.1039/C3NR00092C
8. G. Palmisano, E. Garcia-Lopez, G. Marci, V. Loddo, S. Yurdakal, V. Augugliaro, L. Palmisano, Advances in selective conversions by heterogeneous photocatalysis. Chem. Commun. **46**, 7074 (2010). doi:10.1039/C0CC02087G
9. M. Zhang, Q. Wang, C. Chen, L. Zang, W. Ma, J. Zhao, Oxygen atom transfer in the photocatalytic oxidation of alcohols by TiO_2: oxygen isotope studies. Angew. Chem. Int. Ed. **48**, 6081 (2009). doi:10.1002/anie.200900322
10. K. Kalyanasundaram, Photophysics, photochemistry and solar energy conversion with tris (bipyridyl)ruthenium(II) and its analogues. Coord. Chem. Re. **46**, 159–244 (1982). doi:10.1016/0010-8545(82)85003-0
11. B. Liu, F. Wu, N.-S. Deng, UV-light induced photodegradation of 17α-ethynylestradiol in aqueous solutions. J. Hazard. Mater. **98**, 311 (2003). doi:10.1016/S0304-3894(02)00321-7
12. Y. Zhang, J.L. Zhou, B. Ning, Photodegradation of estrone and 17β-estradiol in water. Water Res. **41**, 19 (2007). doi:10.1061/j.watres.2006.09.020
13. A.S. Kumar, T. Ye, T. Takami, B.-C. Yu, A.K. Flatt, J.M. Tour, P.S. Weiss, Reversible photo-switching of single azobenzene molecules in controlled nanoscale environments. Nano Lett. **8**, 1644 (2008). doi:10.1021/nl1080323+
14. M. Zhang, C.C. Chen, W.H. Ma, J.C. Zhao, Visible-Light-induced aerobic oxidation of alcohols in a coupled photocatalytic system of dye-sensitized TiO_2 and TEMPO. Angew. Chem. Int. Ed. **47**, 9730 (2008). doi:10.1002/ange.200803630
15. N. Zhang, S.Q. Liu, X.Z. Fu, Y.-J. Xu, A simple strategy for fabrication of "Plum-Pudding" Type Pd@CeO_2 semiconductor nanocomposite as a visible-light-driven photocatalyst for selective oxidation. J. Phys. Chem. C **115**, 22901–22909 (2011). doi:10.1021/jp205821b
16. S. Higashimoto, N. Kitao, N. Yoshida, T. Sakura, M. Azuma, H. Ohue, Y. Sakata, Selective photocatalytic oxidation of benzyl alcohol and its derivatives into corresponding aldehydes by molecular oxygen on titanium dioxide under visible light irradiation. J. Catal. **266**, 279–285 (2009). doi:10.1016/j.jcat.2009.06.018
17. Q. Wang, M. Zhang, C.C. Chen, W.H. Ma, J.C. Zhao, Photocatalytic aerobic oxidation of alcohols on TiO_2: the acceleration effect of a Brønsted acid. Angew. Chem. Int. Ed. **49**, 7976 (2010). doi:10.1002/anie.201001533
18. Z.-R. Tang, Y.H. Zhang, Y.-J. Xu, Tuning the optical property and photocatalytic performance of titanate nanotube toward selective oxidation of alcohols under ambient conditions. ACS Appl. Mater. Interfaces, **4**, 1512 (2012). doi:10.1021/am3001852
19. Z.H. Kang, E.B. Wang, B.D. Mao, Z.M. Su, L. Gao, L. Niu, H.Y. Shan, L. Xu, Heterogeneous hydroxylation catalyzed by multi-walled carbon nanotubes at low temperature. Appl. Catal. A **299**, 212 (2006). doi:10.1016/j.apcata.2005.10.038
20. D.R. Dreyer, H.-P. Jia, C.W. Bielawski, Graphene oxide: a convenient carbocatalyst for facilitating oxidation and hydration reactions. Angew. Chem. Int. Ed. **49**, 6813 (2010). doi:10.1002/ange.201002160

21. J. Zhang, X. Liu, R. Blume, A. Zhang, R. Schlogl, D.S. Su, Surface-modified carbon nanotubes catalyze oxidative dehydrogenation of n-butane. Science **322**, 73–77 (2008). doi:10.1126/science.1161916
22. J. Pyun, Graphene oxide as catalyst: application of carbon materials beyond nanotechnology. Angew. Chem. Int. Ed. **50**, 46 (2011). doi:10.1002/anie.201003897
23. Z.H. Kang, C.H.A. Tsang, N.-B. Wong, Z.D. Zhang, S.T. Lee, Silicon quantum dots: a general photocatalyst for reduction, decomposition, and selective oxidation reactions. J. Am. Chem. Soc. **129**, 12090 (2007). doi:10.1021/ja075184x
24. H. Li, R. Liu, W. Kong, J. Liu, Y. Liu, L. Zhou, X. Zhang, S.-T. Lee, Z. Kang, Carbon quantum dots with photo-generated proton property as efficient visible light controlled acid catalyst. Nanoscale **6**, 867–873 (2014). doi:10.1039/C3NR03996J
25. M. Kitano, K. Nakajima, J.N. Kondo, S. Hayashi, M. Hara, In vitro reconstitution of mycobacterial ergothioneine biosynthesis. J. Am. Chem. Soc. **132**, 6632–6633 (2010). doi:10.1021/ja101721e
26. I. Čorić, B. List, Asymmetric spiroacetalization catalysed by confined Brønsted acids. Nature **483**, 315 (2012). doi:10.1038/nature10932
27. H.Z. Liu, T. Jiang, B.X. Han, S.G. Liang, Y.X. Zhou, Selective phenol hydrogenation to cyclohexanone over a dual supported Pd-Lewis acid catalyst. Science **326**, 1250 (2009). doi:10.1126/science.1179713
28. J.H. Clark, Solid acids for green chemistry. Acc. Chem. Res. **35**, 791 (2002). doi:10.1021/ar010072a
29. A. Corma, From microporous to mesoporous molecular sieve materials and their use in catalysis. Chem. Rev. **97**, 2373 (1997). doi:10.1021/cr960406n
30. P. Chen, Solid acid catalysts: stain and shine. Nat. Chem. **3**, 839 (2011). doi:10.1038/nchem.1183
31. R.A. Sheldon, R.S. Downing, Heterogeneous catalytic transformations for environmentally friendly production. Appl. Catal. A **189**, 163–183 (1999). doi:10.1016/S0926-860X(99)00274-4
32. T. Okuhara, Water-Tolerant solid acid catalysts. Chem. Rev. **102**, 3641 (2002). doi:10.1021/cr0103569
33. S.K. Yang, W.P. Cai, H.W. Zhang, H.B. Zeng, Y. Lei, A general strategy for fabricating unique carbide nanostructures with excitation wavelength-dependent light emissions. J. Phys. Chem. C **15**, 7279–7284 (2011). doi:10.1021/jp111873k
34. D.A. Nicewicz, D.W.C. MacMillan, Merging photoredox catalysis with organocatalysis: the direct asymmetric alkylation of aldehydes. Science **322**, 77 (2008). doi:10.1126/science.1161976
35. M.A. Ischay, M.E. Anzovino, J. Du, T.P. Yoon, Efficient visible light photocatalysis of [2+2] enone cycloadditions. J. Am. Chem. Soc. **130**, 12886 (2008). doi:10.1021/ja805387f
36. J.M.R. Narayanam, J.W. Tucker, C.R. Stephenson, Electron-transfer photoredox catalysis: development of a tin-free reductive dehalogenation reaction. J. Am. Chem. Soc. **131**, 8756 (2009). doi:10.1021/ja9033582
37. F.Z. Su, S.C. Mathew, G. Lipner, X.Z. Fu, M. Antonietti, S. Blechert, X.C. Wang, mpg-C3N4-catalyzed selective oxidation of alcohols Using O_2 and visible light. J. Am. Chem. Soc. **132**, 16299 (2010). doi:10.1021/ja102866p
38. K. Nakajima, M. Hara, Amorphous carbon with SO_3H Groups as a solid Brønsted acid catalyst. ACS Catal. **2**, 1296 (2012). doi:10.1021/cs300103k
39. F.J. Liu, J. Sun, L.F. Zhu, X.J. Meng, C.Z. Qi, F.-S. Xiao, Sulfated graphene as an efficient solid catalyst for acid-catalyzed liquid reactions. J. Mater. Chem. **22**, 5495 (2012). doi:10.1039/C2JM16608A
40. H.T. Li, X.D. He, Z.H. Kang, H. Huang, Y. Liu, J.L. Liu, S.Y. Lian, C.H.A. Tsang, X.B. Yang, S.T. Lee, Water-soluble fluorescent carbon quantum dots and catalyst design. Angew. Chem. Int. Ed. **49**, 4430–4434 (2010). doi:10.1002/anie.200906154

41. Y.Z. Han, H. Huang, H.C. Zhang, Y. Liu, X. Han, R.H. Liu, H.T. Li, Z.H. Kang, Carbon quantum dots with photoenhanced hydrogen-bond catalytic activity in aldol condensations. ACS Catal. **4**(3), 781–787 (2014). doi:10.1021/cs401118x
42. R.R. Knowles, E.N. Jacobsen, Attractive noncovalent interactions in asymmetric catalysis: links between enzymes and small molecule catalysts. Proc. Natl. Acad. Sci. U.S.A. **107**, 20678–20685 (2010). doi:10.1073/pnas.1006402107
43. P.R. Carlier, Threading the needle: Mimicking natural toroidal catalysts. Angew. Chem. Int. Ed. **43**(20), 2602–2605 (2004). doi:10.1002/anie.200301731
44. P.I. Dalko, L. Moisan, In the golden age of organocatalysis. Angew. Chem. Int. Ed. **43**, 5138–5175 (2004). doi:10.1002/anie.200400650
45. T. Akiyama, J. Itoh, K. Yokota, K. Fuchibe, Enantioselective Mannich-type reaction catalyzed by a Chiral Brønsted acid. Angew. Chem. Int. Ed. **43**, 1566–1568 (2004). doi:10.1002/anie.200353240
46. V.B. Gondi, M. Gravel, V.H. Rawal, Hydrogen bond catalyzed enantioselective vinylogous Mukaiyama aldol reaction. Org. Lett. **7**, 5657–5660 (2005). doi:10.1021/ol052301p
47. M.S. Sigman, P. Vachal, E.N. Jacobsen, A general catalyst for the asymmetric Strecker reaction. Angew. Chem. Int. Ed. **39**, 1279–1281 (2000). doi:10.1002/(SICI)1521-3773(20000403)39:7<1279:AID-ANIE1279>3.0.CO;2-U
48. T. Okino, Y. Hoashi, Y. Takemoto, Enantioselective Michael reaction of malonates to nitroolefins catalyzed by bifunctional organocatalysts. J. Am. Chem. Soc. **125**, 12672–12673 (2003). doi:10.1021/ja036972z
49. B.M. Nugent, R.A. Yoder, J.N. Johnston, Chiral proton catalysis: a catalytic enantioselective direct Aza-Henry reaction. J. Am. Chem. Soc. **126**, 3418–3419 (2004). doi:10.1021/ja031906i
50. D. Uraguchi, M. Terada, Chiral Brønsted acid-catalyzed direct Mannich reactions via electrophilic activation. J. Am. Chem. Soc. **126**, 5356–5367 (2004). doi:10.1021/ja0491533
51. Y. Huang, A.K. Unni, A.N. Thadani, V.H. Rawal, Hydrogen bonding: single enantiomers from a chiral-alcohol catalyst. Nature **424**, 146 (2003). doi:10.1038/424146a
52. A.K. Unni, N. Takenaka, H. Yamamoto, V.H. Rawal, Axially chiral biaryl diols catalyze highly enantioselective hetero-Diels–Alder reactions through hydrogen bonding. J. Am. Chem. Soc. **127**, 1336–1337 (2005). doi:10.1021/ja044076x
53. Y. Hoashi, T. Okino, Y. Takemoto, Enantioselective Michael addition to α, β-unsaturated Imides catalyzed by a bifunctional organocatalyst. Angew. Chem. Int. Ed. **44**(26), 4032–4035 (2005). doi:10.1002/anie.200500459
54. S. Rajaram, M.S. Sigman, Design of hydrogen bond catalysts based on a modular oxazoline template: application to an enantioselective hetero Diels–Alder reaction. Org. Lett. **7**, 5473–5475 (2005). doi:10.1021/ol052300x
55. M.S. Sigman, P. Vachal, E.N. Jacobsen, A general catalyst for the asymmetric Strecker reaction. Angew. Chem. Int. Ed. **39**(7), 1279–1281 (2000). doi:10.1002/(SICI)1521-3773(20000403)39:7<1279:AID-ANIE1279>3.0.CO;2-U
56. A.G. Wenzel, E.N. Jacobsen, Asymmetric catalytic Mannich reactions catalyzed by urea derivatives: enantioselective synthesis of β-Aryl-β-amino acids. J. Am. Chem. Soc. **124**, 12964–12965 (2002). doi:10.1021/ja028353g
57. T. Akiyama, J. Itoh, K. Yokota, K. Fuchibe, Enantioselective Mannich-type reaction catalyzed by a Chiral Brønsted acid. Angew. Chem. Int. Ed. **43**(12), 1566–1568 (2004). doi:10.1002/anie.200353240
58. J.D. McGilvra, A.K. Unni, K. Modi, V.H. Rawal, Highly diastereo- and enantioselective Mukaiyama Aldol reactions catalyzed by hydrogen bonding. Angew. Chem. Int. Ed. **45**, 6130–6133 (2006). doi:10.1002/ange.200601638
59. A.G. Doyle, E.N. Jacobsen, Small-molecule H-bond donors in asymmetric catalysis. Chem. Rev. **107**, 5713–5743 (2007). doi:10.1021/cr068373r
60. T.R. Kelly, P. Meghani, V.S. Ekkundi, Diels-alder reactions: rate acceleration promoted by a biphenylenediol. Tetrahedron Lett. **31**(24), 3381–3384 (1990). doi:10.1016/S0040-4039(00)97402-1

61. N. Momiyama, H. Yamamoto, Brønsted acid catalysis of achiral enamine for regio- and enantioselective nitroso aldol synthesis. J. Am. Chem. Soc. **127**, 1080–1081 (2005). doi:10.1021/ja0444637
62. F. Niu, J.Y. Wu, L.S. Zhang, P. Li, J.F. Zhu, Z.Y. Wu, C.R. Wang, W.-G. Song, Hydroxyl group rich C_{60} fullerenol: an excellent hydrogen bond catalyst with superb activity, selectivity, and stability. ACS Catal. **1**, 1158–1161 (2011). doi:10.1021/cs200317d
63. N.T. McDougal, S.E. Schaus, Asymmetric Morita–Baylis–Hillman reactions catalyzed by Chiral Brønsted acids. J. Am. Chem. Soc. **125**, 12094–12095 (2003). doi:10.1021/ja037705w
64. F. Niu, C.C. Liu, Z.M. Cui, J. Zhai, L. Jiang, W.-G. Song, Promotion of organic reactions by interfacial hydrogen bonds on hydroxyl group rich nano-solids. Chem. Commun. **24**, 2803–2805 (2008). doi:10.1039/B801361F
65. F. Niu, J. Zhai, L. Jiang, W.-G. Song, Light induced activity switch in interfacial hydrogen-bond catalysis with photo sensitive metal oxides. Chem. Commun. **31**, 4738–4740 (2009). doi:10.1039/B908834B
66. F. Niu, L. Zhang, S.Z. Luo, W.-G. Song, Room temperature aldol reactions using magnetic $Fe_3O_4@Fe(OH)_3$ composite microspheres in hydrogen bond catalysis. Chem. Commun. **46**, 1109–1111 (2010). doi:10.1039/B920009F
67. L. Cao, S. Sahu, P. Anilkumar, C.E. Bunker, J. Xu, K.A. Shiral Fernando, P. Wang, E.A. Guliants, K.N. Tackett, Y.-P. Sun, Carbon nanoparticles as visible-light photocatalysts for efficient CO_2 conversion and beyond. J. Am. Chem. Soc. **133**, 4754–4757 (2011). doi:10.1021/ja200804h
68. P. Usubharatana, D. McMartin, A. Veawab, P. Tontiwachwuthikul, Photocatalytic process for CO_2 emission reduction from industrial flue gas streams. Ind. Eng. Chem. Res. **45**, 2558–2568 (2006). doi:10.1021/ie0505763
69. S.C. Roy, O.K. Varghese, M. Paulose, C.A. Grimes, Toward solar fuels: photocatalytic conversion of carbon dioxide to hydrocarbons. ACS Nano **4**, 1259–1278 (2010). doi:10.1021/nn9015423
70. X. Chen, S. Shen, L. Guo, S.S. Mao, Semiconductor-based photocatalytic hydrogen generation. Chem. Rev. **110**, 6503–6570 (2010). doi:10.1021/cr1001645
71. K. Maeda, M. Eguchi, W.J. Youngblood, T.E. Mallouk, Niobium oxide nanoscrolls as building blocks for dye-sensitized hydrogen production from water under visible light irradiation. Chem. Mater. **20**, 6770 (2008). doi:10.1021/cm801807b
72. X. Wang, L. Cao, F. Lu, M.J. Meziani, H. Li, G. Qi, B. Zhou, B.A. Harruff, Y.-P. Sun, Photoinduced electron transfers with carbon dots. Chem. Commun. 3774–3776 (2009). doi:10.1039/B906252A
73. X. Wang, L. Cao, S.-T. Yang, F. Lu, M.J. Meziani, L. Tian, K.W. Sun, M.A. Bloodgood, Y.-P. Sun, Bandgap-like strong fluorescence in functionalized carbon nanoparticles. Angew. Chem. Int. Ed. **49**, 5310–5314 (2010). doi:10.1002/ange.201000982
74. Q.H. Zhang, W.D. Han, Y.J. Hong, J.G. Yu, Photocatalytic reduction of CO_2 with H_2O on Pt-loaded TiO_2 catalyst. Catal. Today **148**, 335–340 (2009). doi:10.1016/j.cattod.2009.07.081
75. M.A. Fox, M.T. Dulay, Heterogeneous photocatalysis. Chem. Rev. **93**, 341–357 (1993). doi:10.1021/cr00017a016
76. P. Pathak, M.J. Meziani, L. Castillo, Y.-P. Sun, Metal-coated nanoscale TiO_2 catalysts for enhanced CO_2 photoreduction. Green Chem. **7**, 667–670 (2005). doi:10.1039/B507103H
77. H. Yan, J. Yang, G. Ma, G. Wu, X. Zong, Z. Lei, J. Shi, C. Li, Effect of post-calcination on photocatalytic activity of $(Ga_{1-x}Zn_x)(N_{1-x}O_x)$ solid solution for overall water splitting under visible light. J. Catal. **254**, 198–204 (2008). doi:10.1016/j.jcat.2007.12.009
78. K. Maeda, K. Domen, Photocatalytic water splitting: recent progress and future challenges. J. Phys. Chem. Lett. **1**, 2655–2661 (2010). doi:10.1021/jz1007966
79. T.-F. Yeh, C.-Y. Teng, S.-J. Chen, H. Teng, Nitrogen-doped graphene oxide quantum dots as photocatalysts for overall water-splitting under visible light illumination. Adv. Mater. **26**, 3297–3303 (2014). doi:10.1002/adma.201305299

80. T.F. Yeh, S.J. Chen, C.S. Yeh, H. Teng, Tuning the electronic structure of graphite oxide through ammonia treatment for photocatalytic generation of H_2 and O_2 from water splitting. J. Phys. Chem. C **117**, 6516–6524 (2013). doi:10.1021/jp312613r
81. G. Eda, Y.Y. Lin, C. Mattevi, H. Yamaguchi, H.A. Chen, I.S. Chen, C.W. Chen, M. Chhowalla, Blue photoluminescence from chemically derived graphene oxide. Adv. Mater. **22**, 505–509 (2010). doi:10.1002/adma.200901996
82. H. Tetsuka, R. Asahi, A. Nagoya, K. Okamoto, I. Tajima, R. Ohta, A. Okamoto, Optically tunable amino-functionalized graphene quantum dots. Adv. Mater. **24**, 5333–5338 (2012). doi:10.1002/adma.201201930
83. S.H. Jin, D.H. Kim, G.H. Jun, S.H. Hong, S. Jeon, Tuning the photoluminescence of graphene quantum dots through the charge transfer effect of functional groups. ACS Nano **7**, 1239–1245 (2013). doi:10.1021/nn304675g
84. L.A. Ponomarenko, F. Schedin, M.I. Katsnelson, R. Yang, E.W. Hill, K.S. Novoselov, Chaotic Dirac billiard in graphene quantum dots. Science **320**, 356–358 (2008). doi:10.1126/science.1154663
85. Y.W. Son, M.L. Cohen, S.G. Louie, Energy gaps in graphene nanoribbons. Phys. Rev. Lett. **97**, 216803 (2006). doi:10.1103/PhysRevLett.97.216803
86. X. Li, X. Wang, L. Zhang, S. Lee, H. Dai, Chemically derived, ultrasmooth graphene nanoribbon semiconductors. Science **319**, 1229–1232 (2008). doi:10.1126/science.1150878
87. S.B. Yang, X.L. Feng, X.C. Wang, K. Müllen, Graphene-based carbon nitride nanosheets as efficient metal-free electrocatalysts for oxygen reduction reactions. Angew. Chem. Int. Ed. **50**, 5339–5343 (2011). doi:10.1002/anie.201100170
88. Y. Li, Y. Zhao, H.H. Cheng, Y. Hu, G.Q. Shi, L.M. Dai, L.T. Qu, Nitrogen-doped graphene quantum dots with oxygen-rich functional groups. J. Am. Chem. Soc. **134**, 15–18 (2012). doi:10.1021/ja206030c
89. R. Yan, H. Wu, Q. Zheng, J.Y. Wang, J.L. Huang, K.J. Ding, Q.G. Guo, J.Z. Wang, Graphene quantum dots cut from graphene flakes: high electrocatalytic activity for oxygen reduction and low cytotoxicity. RSC Adv. **4**, 23097–23106 (2014). doi:10.1039/C4RA02336F
90. Y. Liu, P.Y. Wu, Graphene quantum dot hybrids as efficient metal-free electrocatalyst for the oxygen reduction reaction. ACS Appl. Mater. Interfaces **5**, 3362–3369 (2013). doi:10.1021/am400415t
91. W.A. Saidi, Oxygen reduction electrocatalysis using N-doped graphene quantum-dots. J. Phys. Chem. Lett. **4**, 4160–4165 (2013). doi:10.1021/jz402090d
92. C.Z. Zhu, J.F. Zhai, S.J. Dong, Bifunctional fluorescent carbon nanodots: green synthesis via soy milk and application as metal-free electrocatalysts for oxygen reduction. Chem. Commun. **48**, 9367–9369 (2012). doi:10.1039/C2CC33844K
93. L. Zhou, J. Liu, X. Zhang, R. Liu, H. Huang, Y. Liu, Z. Kang, Template-free fabrication of mesoporous carbons from carbon quantum dots and their catalytic application to the selective oxidation of hydrocarbons. Nanoscale **6**, 5831–5837 (2014). doi:10.1039/C4NR00716F
94. G.A. Ozin, Nanochemistry: synthesis in diminishing dimensions. Adv. Mater. **4**, 612–649 (1992). doi:10.1002/adma.19920041003
95. A. Mehdi, C. Reye, R. Corriu, From molecular chemistry to hybrid nanomaterials. Design and functionalization. Chem. Soc. Rev. **40**, 563–574 (2011). doi:10.1039/B920516K
96. D.H. Zhang, C. Zhou, Z.H. Sun, L.-Z. Wu, C.-H. Tung, T.R. Zhang, Magnetically recyclable nanocatalysts (MRNCs): a versatile integration of high catalytic activity and facile recovery. Nanoscale **4**, 6244–6255 (2012). doi:10.1039/C2NR31929B
97. R.H. Baughman, A.A. Zakhidov, W.A. de Heer, Carbon nanotubes—the route toward applications. Science **297**, 787–792 (2002). doi:10.1126/science.1060928
98. Y.W. Zhu, S. Murali, W.W. Cai, X.S. Li, J.W. Suk, J.R. Potts, R.S. Ruoff, Graphene and graphene oxide: synthesis, properties, and applications. Adv. Mater. **22**, 3906–3924 (2010). doi:10.1002/adma.201001068

99. G. Williams, B. Seger, P.V. Kamat, TiO_2-graphene nanocomposites. UV-assisted photocatalytic reduction of graphene oxide. ACS Nano **2**, 1487–1491 (2008). doi:10.1021/nn800251f
100. T. Rueckes, K. Kim, E. Joselevich, G.Y. Tseng, C.-L. Cheung, C.M. Lieber, Carbon nanotube-based nonvolatile random access memory of molecular computing. Science **289**, 94–97 (2000). doi:10.1126/science.289.5476.94
101. B.Z. Fang, H.S. Zhou, I. Honma, Ordered porous carbon with tailored pore size for electrochemical hydrogen storage application. J. Phys. Chem. B **110**, 4875–4880 (2006). doi:10.1021/jp056063r
102. H.L. Wang, Q.M. Gao, J. Hu, High hydrogen storage capacity of porous carbons prepared by using activated carbon. J. Am. Chem. Soc. **131**, 7016–7022 (2009). doi:10.1021/ja8083225
103. M. Hartmann, A. Vinu, G. Chandrasekar, Adsorption of Vitamin E on mesoporous carbon molecular sieves. Chem. Mater. **17**, 829–833 (2005). doi:10.1021/cm048564f
104. J.-S. Yu, S. Kang, S.B. Yoon, G. Chai, Fabrication of ordered uniform porous carbon networks and their application to a catalyst supporter. J. Am. Chem. Soc. **124**, 9382–9383 (2002). doi:10.1021/ja0203972
105. S.L. Zhang, L. Chen, S.X. Zhou, D.Y. Zhao, L.M. Wu, Facile synthesis of hierarchically ordered porous carbon via in situ self-assembly of colloidal polymer and silica spheres and its use as a catalyst support. Chem. Mater. **22**, 3433–3440 (2010). doi:10.1021/cm1002274
106. F.B. Su, J.H. Zeng, X.Y. Bao, Y.S. Yu, J.Y. Lee, X.S. Zhao, Preparation and characterization of highly ordered graphitic mesoporous carbon as a Pt catalyst support for direct methanol fuel cells. Chem. Mater. **17**, 3960–3967 (2005). doi:10.1021/cm0502222
107. K.X. Wang, Y.G. Wang, Y.R. Wang, E. Hosono, H.S. Zhou, Mesoporous carbon nanofibers for supercapacitor application. J. Phys. Chem. C **113**, 1093–1097 (2008). doi:10.1021/jp807463u
108. S. Oh, K. Kim, Synthesis of a new mesoporous carbon and its application to electrochemical double-layer capacitors. Chem. Commun. 2177–2178 (1999). doi:10.1039/A906872D
109. T. Kyotani, Z.X. Ma, A. Tomita, Template synthesis of novel porous carbons using various types of zeolites. Carbon **41**, 1451–1459 (2003). doi:10.1016/S0008-6223(03)00090-3
110. Y.D. Xia, R. Mokaya, Ordered mesoporous carbon monoliths: CVD nanocasting and hydrogen storage properties. J. Phys. Chem. C **111**, 10035–10039 (2007). doi:10.1021/jp071936y
111. R.H. Liu, H. Huang, H.T. Li, Y. Liu, J. Zhong, Y.Y. Li, S. Zhang, Z.H. Kang, Metal nanoparticle/carbon quantum dot composite as a photocatalyst for high-efficiency cyclohexane oxidation. ACS Catalysis **1**, 328–336 (2014). doi:10.1021/cs400913h
112. H. Hattori, Y. Ide, S. Ogo, K. Inumaru, M. Sadakane, Sano, efficient and selective photocatalytic cyclohexane oxidation on a layered titanate modified with iron oxide under sunlight and CO_2 atmosphere. T. ACS Catal. **2**, 1910–1915 (2012). doi:10.1021/cs300339f
113. C.H. Wang, L.F. Chen, Z.W. Qi, One-pot synthesis of gold nanoparticles embedded in silica for cyclohexane oxidation. Catal. Sci. Technol. **3**, 1123–1128 (2013). doi:10.1039/C2CY20692G
114. D. Bonnet, T. Ireland, E. Fache, J.-P. Simonato, Innovative direct synthesis of adipic acid by air oxidation of cyclohexane. Green Chem. **8**, 556–559 (2006). doi:10.1039/B600903D
115. M.S. Hamdy, A. Ramanathan, T. Maschmeyer, U. Hanefeld, J.C. Jansen, Co-TUD-1: a ketone-selective catalyst for cyclohexane oxidation. Chem. Eur. J. **12**, 1782–1789 (2006). doi:10.1002/chem.200500166
116. U. Pillai, R.E. Sahle-Demessie, A highly efficient oxidation of cyclohexane over Au/ZSM-5 molecular sieve catalyst with oxygen as oxidant. Chem. Commun. **0**, 2142–2143 (2002). doi:10.1039/B315098D
117. F.A. Chavez, C.V. Nguyen, M.M. Olmstead, P.K. Mascharak, Synthesis, properties, and structure of a stable cobalt(iii) alkyl peroxide complex and its role in the oxidation of cyclohexane. Inorg. Chem. **35**, 6282–6291 (1996). doi:10.1021/ic960500m

118. P. Comba, M. Maurer, P. Vadivelu, Oxidation of cyclohexane by high-valent iron bispidine complexes: tetradentate versus pentadentate ligands. Inorg. Chem. **48**, 10389–10396 (2009). doi:10.1021/ic901702s
119. A.S. Goldstein, R.H. Beer, R.S. Drago, Catalytic oxidation of hydrocarbons with O_2 or H_2O_2 using a sterically hindered ruthenium complex. J. Am. Chem. Soc. **116**, 2424–2429 (1994). doi:10.1021/ja00085a023
120. S.-M. Yiu, W.-L. Man, T.-C. Lau, Efficient catalytic oxidation of alkanes by Lewis acid/$[Os^{VI}(N)Cl_4]^-$ using peroxides as terminal oxidants. Evidence for a metal-based active intermediate. J. Am. Chem. Soc. **130**, 10821–10827 (2008). doi:10.1021/ja802625e
121. V.I. Pârvulescu, D. Dumitriu, G. Poncelet, Hydrocarbons oxidation with hydrogen peroxide over germanic faujasites catalysts. Catal. A-Chem. **140**, 91–105 (1999). doi:10.1016/S1381-1169(98)00221-0
122. E.V. Spinacé, H.O. Pastore, U. Schuchardt, Cyclohexane oxidation catalyzed by titanium silicalite(TS-1): overoxidation and comparison with other oxidation systems. J. Catal. **157**, 631–635 (1995). doi:10.1006/jcat.1995.1328
123. G.M. Lü, D. Ji, G. Qian, Y.X. Qi, X.L. Wang, J.S. Suo, Gold nanoparticles in mesoporous materials showing catalytic selective oxidation cyclohexane using oxygen. Appl. Catal. A **280**, 175–180 (2005). doi:10.1016/j.apcata.2004.10.034
124. R. Raja, G. Sankar, J.M. Thomas, Powerful redox molecular sieve catalysts for the selective oxidation of cyclohexane in air. J. Am. Chem. Soc. **121**, 11926–11927 (1999). doi:10.1021/ja9935052
125. Y. Yuan, H.B. Ji, Y.X. Chen, Y. Han, X.F. Song, Y.B. She, R.G. Zhong, Oxidation of cyclohexane to adipic acid using Fe-porphyrin as a biomimetic catalyst. Org. Process Res. Dev. **8**, 418–420 (2004). doi:10.1021/op049974s
126. B. Sarkar, P. Prajapati, R. Tiwari, R. Tiwari, S. Ghosh, S.S. Acharyya, C. Pendem, R.K. Singha, L.N.S. Konathala, J. Kumar, T. Sasaki, R. Bal, Room temperature selective oxidation of cyclohexane over Cu-nanoclusters supported on nanocrystalline Cr_2O_3. Green Chem. **14**, 2600–2606 (2012). doi:10.1039/C2GC35658A
127. A. Wittstock, V. Zielasek, J. Biener, C.M. Friend, M. Bäumer, Nanoporous gold catalysts for selective gas-phase oxidative coupling of methanol at low temperature. Science **327**, 319–322 (2010). doi:10.1126/science.1183591
128. J.L. Gu, Y.L. Gu, Y. Huang, S.P. Elangovan, Y.S. Li, W.R. Zhao, I. Toshio, Y. Yamazaki, J.L. Shi, Highly dispersed copper species within SBA-15 introduced by the hydrophobic core of a surfactant micelle as a carrier and their enhanced catalytic activity for cyclohexane oxidation. J. Phys. Chem. C **115**, 21211–21217 (2011). doi:10.1021/jp206132a
129. K.-I. Shimizu, Y. Murata, A. Satsuma, Dicopper(II)–Dioxygen complexes in Y zeolite for selective catalytic oxidation of cyclohexane under photoirradiation. J. Phys. Chem. C **111**, 19043–19051 (2007). doi:10.1021/jp0767821
130. S.S. Shankar, A. Rai, A. Ahmad, M. Sastry, Rapid synthesis of Au, Ag and bimetallic Au-shell nanoparticles using Neem. J. Colloid Interf. Sci. **275**, 496–502 (2004). doi:10.1016/j.jcis.2004.03.003
131. J. Liu, H.C. Zhang, D. Tang, X. Zhang, L.K. Yan, Y.Z. Han, H. Huang, Y. Liu, Z.H. Kang, Carbon quantum dot/silver nanoparticle/polyoxometalate composites as photocatalysts for overall water splitting in visible light. ChemCatChem **6**, 2634–2641 (2014). doi:10.1002/cctc.201402227
132. A. Fujishima, K. Honda, Electrochemical photolysis of water at a semiconductor electrode. Nature **238**, 37–38 (1972). doi:10.1038/238037a0
133. A. Iwase, Y.H. Ng, Y. Ishiguro, A. Kudo, R. Amal, Reduced graphene oxide as a solid-state electron mediator in Z-scheme photocatalytic water splitting under visible light. J. Am. Chem. Soc. **133**, 11054–11057 (2011). doi:10.1021/ja203296z
134. H.G. Kim, P.H. Borse, J.S. Jang, C.W. Ahn, E.D. Jeong, J.S. Lee, Engineered nanorod perovskite film photocatalysts to harvest visible light. Adv. Mater. **23**, 2088–2092 (2011). doi:10.1002/adma.201004171

135. A. Mukherji, R. Marschall, A. Tanksale, C. Sun, S.C. Smith, G.Q. Lu, L. Wang, N-doped $CsTaWO_6$ as a new photocatalyst for hydrogen production from water splitting under solar irradiation. Adv. Funct. Mater. **21**, 126–132 (2011). doi:10.1002/adfm.201000591
136. Y. Liu, H. Ming, Z. Ma, H. Huang, S.Y. Lian, H.T. Li, X.D. He, H. Yu, K.M. Pan, Z.H. Kang, Nanoporous TiO_2 spheres with narrow pore size distribution and improved visible light photocatalytic abilities. Chem. Commun. **47**, 8025–8027 (2011). doi:10.1039/C1CC12557E
137. K. Maeda et al., Photocatalytic overall water splitting promoted by two different cocatalysts for hydrogen and oxygen evolution under visible light. Angew. Chem. **122**, 4190–4193 (2010). doi:10.1002/ange.201001259
138. H. Zhang, G. Chen, X. He, Y. Li, Electronic structure and water splitting under visible-light irradiation of Zn and Ag co-doped $In(OH)_yS_z$ photocatalysts. Int. J. Hydrogen Energy **37**, 5532–5539 (2012). doi:10.1016/j.ijhydene.2011.12.155
139. A. Kudo, Y. Miseki, Heterogeneous photocatalyst materials for water splitting. Chem. Soc. Rev. **38**, 253–278 (2009). doi:10.1039/B800489G
140. K. Iwashina, A. Kudo, Rh-Doped $SrTiO_3$ photocatalyst electrode showing cathodic photocurrent for water splitting under visible-light irradiation. J. Am. Chem. Soc. **133**, 13272–13275 (2011). doi:10.1021/ja2050315
141. R. Abe, M. Higashi, K. Domen, Facile fabrication of an efficient oxynitride TaON photoanode for overall water splitting into H_2 and O_2 under visible light irradiation. J. Am. Chem. Soc. **132**, 11828–11829 (2010). doi:10.1021/ja1016552
142. G.R. Bamwenda, T. Uesigi, Y. Abe, K. Sayama, H. Arakawa, The photocatalytic oxidation of water to O_2 over pure CeO_2, WO_3, and TiO_2 using Fe^{3+} and Ce^{4+} as electron acceptors. Appl. Catal. A-Gen. **205**, 117–128 (2001). doi:10.1016/S0926-860X(00)00549-4
143. T. Ohno, L. Bai, T. Hisatomi, K. Maeda, K. Domen, Photocatalytic water splitting using modified GaN:ZnO solid solution under visible light: long-time operation and regeneration of activity. J. Am. Chem. Soc. **134**, 8254–8259 (2012). doi:10.1021/ja302479f
144. N. Arai, N. Saito, H. Nishiyama, Y. Shimodaira, H. Kohayashi, Y. Inoue, K. Sato, Photocatalytic activity for overall water splitting of RuO_2-loaded $Y_xIn_{2-x}O_3$ ($x = 0.9$–1.5). J. Phys. Chem. C **112**, 5000–5005 (2008). doi:10.1021/jp709629t
145. S.H. Hwang, C. Kim, J. Jang, SnO_2 nanoparticle embedded TiO_2 nanofibers-highly efficient photocatalyst for the degradation of rhodamine B. Catal. Commun. **12**, 1037–1041 (2011). doi:10.1016/j.catcom.2011.02.024
146. J. Sato, N. Saito, Y. Yamada, K. Maeda, T. Takata, J.N. Kondo, M. Hara, H. Kobayashi, K. Domen, Y. Inoue, RuO_2-loaded β-Ge_3N_4 as a non-oxide photocatalyst for overall water splitting. J. Am. Chem. Soc. **127**, 4150–4151 (2005). doi:10.1021/ja042973v
147. K. Maeda, K. Teramura, D. Lu, T. Takata, N. Saito, Y. Inoue, K. Domen, Photocatalyst releasing hydrogen from water-enhancing catalytic performance holds promise for hydrogen production by water splitting in sunlight. Nature **440**, 295-295 (2006). doi:10.1038/440295a
148. D. Wang, R. Li, J. Zhu, J. Shi, J. Han, X. Zong, C. Li, Photocatalytic water oxidation on $BiVO_4$ with the electrocatalyst as an oxidation cocatalyst: essential relations between electrocatalyst and photocatalyst. J. Phys. Chem. C **116**, 5082–5089 (2012). doi:10.1021/jp210584b
149. D.K. Zhong, S. Choi, D.R. Gamelin, Near-complete suppression of surface recombination in solar photoelectrolysis by "Co-Pi" catalyst-modified W:$BiVO_4$. J. Am. Chem. Soc. **133**, 18370–18377 (2011). doi:10.1021/ja207348x
150. S.K. Pilli, T.E. Furtak, L.D. Brown, T.G. Deutsch, J.A. Turner, A.M. Herring, Cobalt-phosphate (Co-Pi) catalyst modified Mo-doped $BiVO_4$ photoelectrodes for solar water oxidation. Energy Environ. Sci. **4**, 5028–5034 (2011). doi:10.1039/C1EE02444B
151. J.A. Seabold, K.S. Choi, Effect of a cobalt-based oxygen evolution catalyst on the stability and the selectivity of photo-oxidation reactions of a WO_3 photoanode. Chem. Mater. **23**, 1105–1112 (2011). doi:10.1021/cm1019469

152. K. Sayama, K. Mukasa, R. Abe, Y. Abe, H. Arakawa, A new photocatalytic water splitting system under visible light irradiation mimicking a Z-scheme mechanism in photosynthesis. J. Photochem. Photobiol. A **148**, 71–77 (2002). doi:10.1016/S1010-6030(02)00070-9
153. F.M. Toma et al., Efficient water oxidation at carbon nanotube–polyoxometalate electrocatalytic interfaces. Nat. Chem. **2**, 826–831 (2010). doi:10.1038/nchem.761
154. Z.Y. Zhang, Q.P. Lin, S.T. Zheng, X.H. Bu, P.Y. Feng, A novel sandwich-type polyoxometalate compound with visible-light photocatalytic H_2 evolution activity. Chem. Commun. **47**, 3918–3920 (2011). doi:10.1039/C0CC04697C
155. X. Liu, Y. Li, S. Peng, G. Lu, S. Li, Photocatalytic hydrogen evolution under visible light irradiation by the polyoxometalate α-[AlSiW11(H2O)O39]5–Eosin Y system. Int. J. Hydrogen Energy **37**, 12150–12157 (2012). doi:10.1016/j.ijhydene.2012.06.028
156. A. Sartorel, P. Miró, E. Salvadori, S. Romain, M. Carraro, G. Scorrano, M.D. Valentin, A. Llobet, C. Bo, M. Bonchio, Water oxidation at a tetraruthenate core stabilized by polyoxometalate ligands: experimental and computational evidence to trace the competent intermediates. J. Am. Chem. Soc. **131**, 16051–16053 (2009). doi:10.1021/ja905067u
157. B. Nohra et al., Polyoxometalate-based metal organic frameworks (POMOFs): structural trends, energetics, and high electrocatalytic efficiency for hydrogen evolution reaction. J. Am. Chem. Soc. **133**, 13363–13374 (2011). doi:10.1021/ja201165c
158. A. Sartorel, M. Carraro, G. Scorrano, R.D. Zorzi, S. Geremia, N.D. McDaniel, S. Bernhard, M. Bonchio, Polyoxometalate embedding of a Tetraruthenium(IV)-oxo-core by template-directed metalation of $[\gamma\text{-SiW}_{10}O_{36}]^{8-}$: a totally inorganic oxygen-evolving catalyst. J. Am. Chem. Soc. **130**, 5006–5007 (2008). doi:10.1021/ja077837f
159. P.V. Kamat, Photophysical, photochemical and photocatalytic aspects of metal nanoparticles. J. Phys. Chem. B **106**, 7729–7744 (2002). doi:10.1021/jp0209289
160. S. Hara, M. Yoshimizu, S. Tanigawa, L. Ni, B. Ohtani, H. Irie, Hydrogen and oxygen evolution photocatalysts synthesized from strontium titanate by controlled doping and their performance in two-step overall water splitting under visible light. J. Phys. Chem. C **116**, 17458–17463 (2012). doi:10.1021/jp306315r
161. C. Costa-Coquelard, S. Sorgues, L. Ruhlmann, Photocatalysis with polyoxometalates associated to porphyrins under visible light: an application of charge transfer in electrostatic complexes. J. Phys. Chem. A **114**, 6394–6400 (2010). doi:10.1021/jp101261n
162. S. Kim, J. Yeo, W. Choi, Simultaneous conversion of dye and hexavalent chromium in visible light-illuminated aqueous solution of polyoxometalate as an electron transfer catalyst. Appl. Catal. B **84**, 148–155 (2008). doi:10.1016/j.apcatb.2008.03.012
163. P. Wang, B.B. Huang, X.Y. Qin, X.Y. Zhang, Y. Dai, J.Y. Wei, M.-H. Whangbo, Ag@AgCl: a highly efficient and stable photocatalyst active under visible light. Angew. Chem. **120**, 8049–8051 (2008). doi:10.1002/anie.200802483
164. M.R. Hoffmann, S.T. Martin, W. Choi, W. Bahnemann, Environmental applications of semiconductor photocatalysis. Chem. Rev. **95**, 69–96 (1995). doi:10.1021/cr00033a004
165. A. Müller, F. Peters, M.T. Pope, D. Gatteschi, Polyoxometalates: very large clusters nanoscale magnets. Chem. Rev. **98**, 239–272 (1998). doi:10.1021/cr9603946
166. J.T. Rhule, C.L. Hill, D.A. Judd, R.F. Schinazi, Polyoxometalates in medicine. Chem. Rev. **98**, 327–357 (1998). doi:10.1021/cr960396q
167. E. Coronado, C.J. Gómez-García, Polyoxometalate-based molecular materials. Chem. Rev. **98**, 273–296 (1998). doi:10.1021/cr970471c
168. H. Zhang, H. Huang, H. Ming, H. Li, L. Zhang, Y. Liu, Z.H. Kang, Carbon quantum Dots/Ag_3PO_4 complex photocatalysts with enhanced photocatalytic activity and stability under visible light. J. Mater. Chem. **22**, 10501–10506 (2012). doi:10.1039/C2JM30703K
169. H.T. Li, R.H. Liu, Y. Liu, H. Huang, H. Yu, H. Ming, S.Y. Lian, S.T. Lee, Z.H. Kang, Carbon quantum dots/Cu_2O composites with protruding nanostructures and their highly efficient (near) infrared photocatalytic behavior. J. Mater. Chem. **22**, 17470–17475 (2012). doi:10.1039/C2JM32827E

170. X. Zhang, F. Wang, H. Huang, H. Li, X. Han, Y. Liu, Z. Kang, Carbon quantum dot sensitized TiO_2 nanotube arrays for photoelectrochemical hydrogen generation under visible light†. Nanoscale **5**, 2274–2278 (2013). doi:10.1039/C3NR34142A
171. S. Nishimura, N. Abrams, B.A. Lewis, L.I. Halaoui, T.E. Mallouk, K.D. Benkstein, J. van de Lagemaat, A.J. Frank, Standing wave enhancement of red absorbance and photocurrent in dye-sensitized titanium dioxide photoelectrodes coupled to photonic crystals. J. Am. Chem. Soc. **125**, 6306–6310 (2003). doi:10.1021/ja034650p
172. L.I. Halaoui, N.M. Abrams, T.E. Mallouk, Increasing the conversion efficiency of dye-sensitized TiO_2 photoelectrochemical cells by coupling to photonic crystals. J. Phys. Chem. B **109**, 6334–6342 (2005). doi:10.1021/jp044228a
173. W.J. Youngblood, S.H.A. Lee, K. Maeda, T.E. Mallouk, Visible light water splitting using dye-sensitized oxide semiconductors. Acc. Chem. Res. **42**, 1966–1972 (2009). doi:10.1021/ar9002398
174. W.J. Youngblood, S.H.A. Lee, Y. Kobayashi, E.A. Hernandez-Pagan, P.G. Hoertz, T.A. Moore, A.L. Moore, D. Gust, T.E. Mallouk, Photoassisted overall water splitting in a visible light-absorbing dye-sensitized photoelectrochemical cell. J. Am. Chem. Soc. **131**, 926–927 (2009). doi:10.1021/ja809108y
175. L. Li, L.L. Duan, Y.H. Xu, M. Gorlov, A. Hagfeldt, L.C. Sun, A photoelectrochemical device for visible light driven water splitting by a molecular ruthenium catalyst assembled on dye-sensitized nanostructured TiO_2. Chem. Commun. **46**, 7307–7309 (2010). doi:10.1039/C0CC01828G
176. G. Calogero, G.D. Marco, S. Caramori, S. Cazzanti, R. Argazzi, C.A. Bignozzi, Natural dye senstizers for photoelectrochemical cells. Energy Environ. Sci. **2**, 1162–1172 (2009). doi:10.1039/B913248C
177. X.-F. Gao, H.-B. Li, W.-T. Sun, Q. Chen, F.-Q. Tang, L.-M. Peng, CdTe quantum dots-sensitized TiO_2 nanotube array photoelectrodes. J. Phys. Chem. C **113**, 7531–7535 (2009). doi:10.1021/jp810727n
178. W.T. Sun, Y. Yu, H.Y. Pan, X.F. Gao, Q. Chen, L.M. Peng, CdS quantum dots sensitized TiO_2 nanotube-array photoelectrodes. J. Am. Chem. Soc. **130**, 1124–1125 (2008). doi:10.1021/ja0777741
179. G.M. Wang, X.Y. Yang, F. Qian, J.Z. Zhang, Y. Li, Double-sided CdS and CdSe quantum dot co-sensitized ZnO nanowire arrays for photoelectrochemical hydrogen generation. Nano Lett. **10**, 1088–1092 (2010). doi:10.1021/nl100250z
180. J. Hensel, G.M. Wang, Y. Li, J.Z. Zhang, Synergistic effect of CdSe quantum dot sensitization and nitrogen doping of TiO_2 nanostructures for photoelectrochemical solar hydrogen generation. Nano Lett. **10**, 478–483 (2010). doi:10.1021/nl903217w
181. J.H. Bang, P.V. Kamat, Solar cells by design: photoelectrochemistry of TiO_2 nanorod arrays decorated with CdSe. Adv. Funct. Mater. **2010**, 20 (1970). doi:10.1002/adfm.200902234
182. F. Wang, Y. Zhang, Y. Liu, X. Wang, M. Shen, S.-T. Lee, Z. Kang, Opto-electronic conversion logic behaviour through dynamic modulation of electron/energy transfer states at the TiO_2–carbon quantum dot interface. Nanoscale **5**, 1831–1835 (2013). doi:10.1039/C3NR33985H
183. R. Liu, J. Liu, W. Kong, H. Huang, X. Han, X. Zhang, Y. Liu, Z. Kang, Adsorption dominant catalytic activity of a carbon dots stabilized gold nanoparticles system. Dalton Trans. **43**, 10920–10929 (2014). doi:10.1039/C4DT00630E
184. H.B. Xia, S. Bai, J. Hartmann, D. Wang, Synthesis of monodisperse quasi-spherical gold nanoparticles in water via silver(I)-assisted citrate reduction. Langmuir **26**, 3585–3589 (2010). doi:10.1021/la902987w
185. B.T. Teranishi, I. Kiyokawa, M. Miyake, Synthesis of monodisperse gold nanoparticles using linear polymers as protective agents. Adv. Mater. **10**, 596–599 (1998). doi:10.1002/(SICI)1521-4095(199805)10:8<596:AID-ADMA596>3.0.CO;2-Y

186. J.S. Oh, L.N. Dang, S.W. Yoon, P.C. Lee, D.O. Kim, K.J. Kim, J.D. Nam, Amine-functionalized polyglycidyl methacrylate microsphere as a unified template for the synthesis of gold nanoparticles and single-crystal gold plates. Macromol. Rapid Commun. **34**, 504–510 (2013). doi:10.1002/marc.201200713
187. J. Liu, P.-Y. Gu, N.-J. Li, L.-H. Wang, C.-Y. Zhang, Q.-F. Xu, J.-M. Lu, Preparation of a polymer containing indole groups by RAFT polymerization and one-phase synthesis of AuNPs-polymer nanocomposites. J. Appl. Polym. Sci. **129**, 2913–2921 (2013). doi:10.1002/app.39026
188. J. Wagner, J.M. Köhler, Continuous synthesis of gold nanoparticles in a microreactor. Nano Lett. **5**, 685–691 (2005). doi:10.1021/nl050097t
189. Y.-C. Yeh, B. Creran, V.M. Rotello, Gold nanoparticles: preparation, properties, and applications in bionanotechnology. Nanoscale **4**, 1871–1880 (2012). doi:10.1021/cm049532v
190. H. Hiramatsu, F.E. Osterloh, A simple large-scale synthesis of nearly monodisperse gold and silver nanoparticles with adjustable sizes and with exchangeable surfactants. Chem. Mater. **16**, 2509–2511 (2004). doi:10.1021/cm049532v
191. L. Dykmana, N. Khlebtsov, Gold nanoparticles in biomedical applications: recent advances and perspectives. Chem. Soc. Rev. **41**, 2256–2282 (2012). doi:10.1039/C1CS15166E
192. M. Schulz-Dobrick, K.V. Sarathy, M. Jansen, Surfactant-free synthesis and functionalization of gold nanoparticles. J. Am Chem. Sco. **127**, 12816–12817 (2005). doi:10.1021/ja054734t
193. S.N. Baker, G.A. Baker, Luminescent carbon nanodots: emergent nanolights. Angew. Chem. Int. Ed. **49**, 6726–6744 (2010). doi:10.1002/anie.200906623
194. S.C. Ray, A. Saha, N.R. Jana, R. Sarkar, Fluorescent carbon nanoparticles: synthesis, characterization, and bioimaging application. J. Phys. Chem. C **113**, 18546–18551 (2009). doi:10.1021/jp905912n
195. H.T. Li, Z.H. Kang, Y. Liu, S.-T. Lee, Carbon nanodots: synthesis, properties and applications. J. Mater. Chem. **22**, 24230–24253 (2012). doi:10.1039/C2JM34690G
196. H. Wu, X. Huang, M.M. Gao, X.P. Liao, B. Shi, Polyphenol-grafted collagen fiber as reductant and stabilizer for one-step synthesis of size-controlled gold nanoparticles and their catalytic application to 4-nitrophenol reduction. Green Chem. **13**, 651–658 (2011). doi:10.1039/C0GC00843E
197. S.M. Henrichs, S.F. Sugai, Adsorption of amino acids and glucose by sediments of Resurrection Bay, Alaska, USA: functional group effects. Geochim. Cosmochim. Acta **57**, 823–835 (1993). doi:10.1016/0016-7037(93)90171-R
198. L.R. Radovic, I.F. Silva, J.I. Ume, J.A. Menéndez, C.A. Leon, Y. Leon, A.W. Scaroni, An experimental and theoretical study of the adsorption of aromatics possessing electron-withdrawing and electron-donating functional groups by chemically modified activated carbons. Carbon **35**, 1339–1348 (1997). doi:10.1016/S0008-6223(97)00072-9
199. C.H. Tessmer, R.D. Vidic, L.J. Uranowski, Impact of oxygen-containing surface functional groups on activated carbon adsorption of phenols. Environ. Sci. Technol. **31**, 1872–1878 (1997). doi:10.1021/es960474r
200. E.C. Cho, L. Au, Q. Zhang, Y.N. Xia, The effects of size, shape, and surface functional group of gold nanostructures on their adsorption and internalization by cells. Small **6**, 517–522 (2010). doi:10.1002/smll.200901622
201. G. Muralidharan, L. Subramanian, S.K. Nallamuthu, V. Santhanam, S. Kumar, Effect of reagent addition rate and temperature on synthesis of gold nanoparticles in microemulsion route. Ind. Eng. Chem. Res. **50**, 8786–8791 (2011). doi:10.1021/ie2002507
202. T. Yang, Z. Li, L. Wang, C.L. Guo, Y.J. Sun, Synthesis, characterization, and self-assembly of protein lysozyme monolayer-stabilized gold nanoparticles. Langmuir **23**, 10533–10538 (2007). doi:10.1021/la701649z
203. C. Jayaseelan, R. Ramkumar, A.A. Rahuman, P. Perumal, Green synthesis of gold nanoparticles using seed aqueous extract of *Abelmoschus esculentus* and its antifungal activity. Ind. Crop. Prod. **45**, 423–429 (2013). doi:10.1016/j.indcrop.2012.12.019

204. S.K. Das, C. Dickinson, F. Lafir, D.F. Brougham, E. Marsili, Synthesis, characterization and catalytic activity of gold nanoparticles biosynthesized with *Rhizopus oryzae* protein extract. Green Chem. **14**, 1322–1334 (2012). doi:10.1039/C2GC16676C
205. H.T. Li, X.D. He, Z.H. Kang, H. Huang, Y. Liu, J.L. Liu, S.Y. Lian, C.C.A. Tsang, X.B. Yang, S.T. Lee, Water-soluble fluorescent carbon quantum dots and catalyst design. Angew. Chem. Int. Ed. **49**, 4430–4434 (2010). doi:10.1002/anie.200906154
206. D. Tang, J. Liu, X. Wu, R. Liu, X. Han, Y. Han, H. Huang, Y. Liu, Z. Kang, Carbon quantum dot/NiFe layered double-hydroxide composite as a highly efficient electrocatalyst for water oxidation. ACS Appl. Mater. Interfaces **6**, 7918–7925 (2014). doi:10.1021/am501256x
207. L. Trotochaud, J.K. Ranney, K.N. Williams, S.W. Boettcher, Solution-cast metal oxide thin film electrocatalysts for oxygen evolution. J. Am. Chem. Soc. **134**, 17253–17261 (2012). doi:10.1021/ja307507a
208. Y. Lee, J. Suntivich, K.J. May, E.E. Perry, Y. Shao-Horn, Synthesis and activities of rutile IrO_2 and RuO_2 nanoparticles for oxygen evolution in acid and alkaline solutions. J. Phys. Chem. Lett. **3**, 399–404 (2012). doi:10.1021/jz2016507
209. M. Fekete, R.K. Hocking, S.L.Y. Chang, C. Italiano, A.F. Patti, F. Arena, L. Spiccia, Highly active screen-printed electrocatalysts for water oxidation based on β-manganese oxide. Energy Environ. Sci. **6**, 2222–2232 (2013). doi:10.1039/C3EE40429C
210. G. Abellán, E. Coronado, C. Martí-Gastaldo, E. Pinilla-Cienfuegos, A. Ribera, Hexagonal nanosheets from the exfoliation of Ni^{2+}-Fe^{3+} LDHs: a route towards layered multifunctional materials. J. Mater. Chem. **20**, 7451–7455 (2010). doi:10.1039/C0JM01447H
211. D.G. Evans, X. Duan, Preparation of layered double hydroxides and their applications as additives in polymers, as precursors to magnetic materials and in biology and medicine. Chem. Commun. 485–496 (2006). doi:10.1039/B510313B
212. G.A. Caravaggio, C. Detellier, Z. Wronski, Synthesis, stability and electrochemical properties of NiAl and NiV layered double hydroxides. J. Mater. Chem. **11**, 912–921 (2001). doi:10.1039/B004542J
213. G. Abellán, F. Busolo, E. Coronado, C. Martí-Gastaldo, A. Ribera, Hybrid magnetic multilayers by intercalation of Cu(II) phthalocyanine in LDH hosts. J. Phys. Chem. C **116**, 15756–15764 (2012). doi:10.1021/jp303537v
214. O.C. Compton, S.T. Nguyen, Graphene oxide, highly reduced graphene oxide, and graphene: versatile building blocks for carbon-based materials. Small **6**, 711–723 (2010). doi:10.1002/smll.200901934
215. Y. Zhang, N. Zhang, Z.-R. Tang, Y.-J. Xu, Graphene transforms wide band gap ZnS to a visible light photocatalyst. The new role of graphene as a macromolecular photosensitizer. ACS Nano **6**, 9777–9789 (2012). doi:10.1021/nn304154s
216. X. Huang, Z. Yin, S. Wu, X. Qi, Q. He, Q. Zhang, Q. Yan, F. Boey, H. Zhang, Graphene-based materials: synthesis, characterization, properties, and applications. Small **7**, 1876–1902 (2011). doi:10.1002/smll.201002009
217. D. Tang, H. Zhang, H. Huang, R. Liu, Y. Han, Y. Liu, C. Tong, Z. Kang, Carbon quantum dots enhance the photocatalytic performance of $BiVO_4$ with different exposed facets. Dalton Trans. **42**, 6285–6289 (2013). doi:10.1039/c3dt50567g
218. H. Li, R. Liu, Y. Liu, H. Huang, H. Yu, H. Ming, S. Lian, S.-T. Lee, Z. Kang, Carbon quantum dots/Cu_2O composites with protruding nanostructures and their highly efficient (near) infrared photocatalytic behavior. J. Mater. Chem. **22**, 17470–17475 (2012). doi:10.1039/C2JM32827E
219. Y.M. Yang, J. Liu, Y.Z. Han, H. Huang, N.Y. Liu, Y. Liu, Z.H. Kang, Porous cobalt, nitrogen-codoped carbon nanostructures from carbon quantum dots and VB12 and their catalytic properties for oxygen reduction. Phys. Chem. Chem. Phys. **16**, 25350–25357 (2014). doi:10.1039/C4CP04119D
220. N. Alexeyeva, K. Tammeveski, Electroreduction of oxygen on gold nanoparticle/PDDA-MWCNT nanocomposites in acid solution. Anal. Chim. Acta **618**, 140–146 (2008). doi:10.1016/j.aca.2008.04.056

221. Z. Jin, H. Nie, Z. Yang, J. Zhang, Z. Liu, X. Xu, S. Huang, Metal-free selenium doped carbon nanotube/graphene networks as a synergistically improved cathode catalyst for oxygen reduction reaction. Nanoscale **4**, 6455–6460 (2012). doi:10.1039/C2NR31858J
222. X. Ma, H. Meng, M. Cai, P.K. Shen, Bimetallic carbide nanocomposite enhanced Pt catalyst with high activity and stability for the oxygen reduction reaction. J. Am. Chem. Soc. **134**, 1954–1957 (2012). doi:10.1021/ja2093053
223. X. Xia, F. Zhang, X. Zhang, P. Liang, X. Huang, B.E. Logan, Use of pyrolyzed iron ethylenediaminetetraacetic acid modified activated carbon as air-cathode catalyst in microbial fuel cells. ACS Appl. Mater. Interfaces **5**, 7862–7866 (2013). doi:10.1021/am4018225
224. Z.-S. Wu, L. Chen, J. Liu, K. Parvez, H. Liang, J. Shu, H. Sachdev, R. Graf, X. Feng, K. Müllen, High-performance electrocatalysts for oxygen reduction derived from cobalt porphyrin-based conjugated mesoporous polymers. Adv. Mater. **26**, 1450 (2014). doi:10.1002/adma.201304147
225. B. Piela, T.S. Olson, P. Atanassov, P. Zelenay, Highly methanol-tolerant non-precious metal cathode catalysts for direct methanol fuel cell. Electrochim. Acta **55**, 7615–7621 (2010). doi:10.1016/j.electacta.2009.11.085
226. S.-T. Chang, H.-C. Huang, H.-C. Wang, H.-C. Hsu, J.-F. Lee, C.-H. Wang, Effects of structures of pyrolyzed corrin, corrole and porphyrin on oxygen reduction reaction. Int. J. Hydrogen Energy **39**, 934–941 (2014). doi:10.1016/j.ijhydene.2013.10.082
227. J.H. Zagal, M. Páez, C. Páez, Reactivity of immobilized cobalt phthalocyanines for the electroreduction of molecular oxygen in terms of molecular hardness. J. Electroanal. Chem. **489**(1), 96–100 (2000). doi:10.1016/S0022-0728(00)00209-6
228. H.-W. Liang, W. Wei, Z.-S. Wu, X.L. Feng, K. Müllen, Mesoporous metal–nitrogen-doped carbon electrocatalysts for highly efficient oxygen reduction reaction. J. Am. Chem. Soc. **135**, 16002–16005 (2013). doi:10.1021/ja407552k
229. S.-T. Chang, C.-H. Wang, H.-Y. Du, H.-C. Hsu, C.-M. Kang, C.-C. Chen, J.C.S. Wu, S.-C. Yen, W.-F. Huang, L.-C. Chen, M.C. Linf, K.-H. Chen, Vitalizing fuel cells with vitamins: pyrolyzed vitamin B_{12} as a non-precious catalyst for enhanced oxygen reduction reaction of polymer electrolyte fuel cells. Energy Environ. Sci. **5**, 5305–5314 (2012). doi:10.1039/C1EE01962G

Diamond Nanostructures and Nanoparticles: Electrochemical Properties and Applications

Nianjun Yang and Xin Jiang

Abstract Macro-sized diamond films have been widely applied as the electrode for electrochemical and electroanalytical applications. Due to the non-uniform doping in diamond, boundary effects, and the varied ratios of graphite to diamond, only averaged electrochemical signals are detected over the full electrode. The studies of diamond electrochemistry at the nanoscale are thus highly required. In this chapter we overview recent progress and achievements about electrochemical properties and applications of diamond nanostructures and nanoparticles. After a brief introduction of the formation of these nanostructures and nanoparticles, electrochemical behavior of diamond nanostructures (e.g., diamond nanotexures, nanowires, networks, etc.) and nanoparticles (undoped, doped nanoparticles) in the presence/absence of redox probes is summarized. Their electroanalytical (e.g., electrochemical, biochemical sensing, etc.) and electrochemical (e.g., energy storage using capacitors and batteries, electrocatalysis, etc.) applications are shown. Diamond nanoelectrode array is introduced and highlighted as a promising tool to investigate diamond electrochemistry at the nanoscale as well.

Keywords Electrochemistry · Electroanalysis · Diamond · Nanowires · Nanoparticles · Nanoelectrodes and arrays

1 Introduction

Iwaki et al. in 1983 [1] and later Pleskov et al. in 1987 [2] introduced diamond as an electrode material. Since then boron-doped diamond has been recognized as one of the best electrode materials. It shows numerous unique physical and chemical

N. Yang (✉) · X. Jiang (✉)
Institute of Materials Engineering, University of Siegen,
57076 Siegen, Germany
e-mail: nianjun.yang@uni-siegen.de

X. Jiang
e-mail: xin.jiang@uni-siegen.de

N. Yang et al. (eds.), *Carbon Nanoparticles and Nanostructures*,
Carbon Nanostructures, DOI 10.1007/978-3-319-28782-9_9

properties [3–5], such as (i) high chemical stability in harsh environments and/or at high voltage/current densities; (ii) weak or no surface bio-fouling; (iii) biocompatibility; (iv) low and stable capacitive currents in aqueous and non-aqueous solutions; and (v) a wide electrochemical potential window. Moreover, the diamond surface can be terminated with hydrogen, hydroxyl, and oxygen, which allow tuning the electronic properties of the solid/electrolyte interface with respect to energy alignments of interacting levels. Furthermore, diamond is ultra-hard (50–150 GaP) and various diamond nanostructures can be formed [6–9].

Many successful reports have been thus shown about the use of planar macroscopic diamond electrodes for electrochemistry, bio electrochemistry, electroanalysis, electrocatalysis, and environmental related applications [3–6]. However due to the non-uniform doping in diamond, boundary effects, and the varied ratios of graphite to diamond, in most studies only averaged electrochemical signals are detected over the full electrode. To solve this problem, the investigation of diamond electrochemistry at the nanoscale, in other words, electrochemistry of diamond nanostructures and nanoparticles, is highly needed. In this way the difference of electric and electrochemical properties from their bulk electrodes might be detected. The effect of the size, shape, and composition of diamond nanostructures and nanoparticles on their electrochemical properties can be clarified [10–12]. These studies will play significant role in their applications for electrochemical energy storage and conversion, electrocatalysis, electrochemical sensor development, and related fields.

We therefore summarize in this chapter recent progress and achievements about diamond electrochemistry using diamond nanostructures (e.g., nanotextures, nanowires) [13–56], diamond nanoparticles (un-doped and doped nanoparticles) [57–128], and diamond nanoelectrode arrays [129–132]. The applications of these nanostructures and nanoparticles for electroanalytical (e.g., electrochemical, biochemical sensing) and electrochemical applications (e.g., energy storage with capacitors and batteries, electrocatalysis, etc.) [133–141] are shown. Diamond nanoelectrode array is introduced and highlighted as a promising tool to investigate diamond electrochemistry at the nanoscale. After comparing the results published in literature, we close this chapter with a conclusion about the future and tendency of electrochemistry using diamond nanostructures and nanoparticles.

2 Diamond Nanostructures

2.1 *Fabrication Approaches*

Diamond nanostructures, including nanotextures (also called nanograss, nanowinkles, nanocone, etc.) with dimensions of typically a few nanometers, nanowires (also called nanoneedles, nanoforests, etc.) with lengths of a few micrometers, and networks (also called porous films, etc.) with pore sizes from few nanometers to

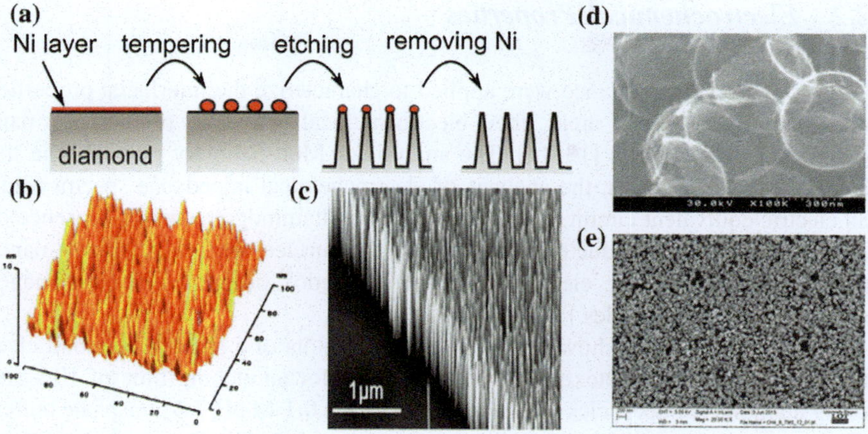

Fig. 1 **a** Schematic illustration of forming diamond nanowires using a top-down approach and Ni nanoparticles as the etching mask; **b** AFM tapping-mode image of diamond nanotextures; SEM images of **c** diamond nanowires, **d** diamond foam and **e** diamond networks [137]

micrometers have been fabricated using top-down, bottom-up, or template-free approaches. In the top-down approach, diamond is etched away with reactive ions in a plasma using an etching (hard) mask. Figure 1a shows schematically such an approach where nickel nanoparticles are applied as the etching mask. The morphology (e.g., size, length, density, etc.) of the resulting nanostructures are determined mainly by the etching mask (e.g., nature, size, shape, etc.) and etching conditions (e.g., temperature, gas, pressure, time, etc.). Various mask materials have been thus applied, including Al, SiO_2, Au, Ni, Mo, polymers, oxides, nitrides, and diamond nanoparticles, etc. [6–9]. For example, the etching mask of 5–10 nm diamond nanoparticles was applied to form diamond nanotextures [13]. With an etching time of 10 s the textures (1–5 nm in diameters, 11 nm in distance) were produced. One typical AFM tapping-mode image of such nanotextures is shown in Fig. 1b. If nickel nanoparticles were applied as the etching mask, the formed diamond nanowires [14] have a height of 1200 (±200) nm, a width of 35 (±5) nm, and a density of $\sim 10^{10}$ cm^{-2}, as shown in the SEM image in Fig. 1c. In a bottom-up approach, diamond nanostructures are produced by the overgrowth of other nanostructures. Up to the morphology of the templates, various diamond nanostructures have been formed [6–9]. The templates from silicon nanowires, carbon nanotubes (CNTs), SiO_2 spheres, etc. have been employed. Figure 1d shows one SEM image of diamond foam fabricated with the overgrowth of SiO_2 sphere. Figure 1e gives the SEM image of a diamond network [48], synthesized with a template-free approach, namely through selective and wet-chemical removal of silicon carbide from a diamond/SiC composite film with a mixture of HF and HNO_3.

2.2 Electrochemical Properties

Voltammetry and impedance were applied to characterize the interfacial properties (e.g., the real electrode areas, their electrochemical activities) of diamond nanotextures and nanowires [13–15]. The analysis of Mott-Schottky plots [13] in the absence of redox probes, the analysis of electrochemical impedance spectra using an electric equivalent circuit [15] and the cyclic voltammograms in the presence of redox probes were conducted to determine the surface areas of diamond nanostructures as well as the electrode processes of redox probes on diamond nanostructures based electrodes [13–15].

For example, Fig. 2 shows the Mott-Schottky plots of a smooth diamond electrode and diamond nanotextures based electrode (for an etching time of 10 s and using diamond nanoparticles as the etching mask) in 0.1 M pH 7.4 phosphate buffer. The measurements were done at a fixed frequency of 1.0 kHz. The built-in potential of diamond nanotextures, calculated by use of Mott-Schottky equation, was 1.6 V. Its surface area was 2.1 times higher than that of a smooth diamond [13].

Figure 3a shows the cyclic voltammograms of a smooth diamond electrode and diamond nanowires based electrode (using a top-down approach and nickel nanoparticles as the etching mask) in 0.1 M KCl. The surface area of diamond nanowires based electrode, calculated from the capacitive current, is 10-times larger than that of a smooth diamond electrode [14]. On such an electrode the electrode process of $Fe(CN)_6^{3-/4-}$ was quasi-reversible and diffusion-controlled, as shown in Fig. 3b. The calculated electrode active area of diamond nanowires based electrode is however only 1.5 times larger than that of a smooth diamond. Similar result, namely the electrode area calculated from the capacitive current is much larger than that from the Faradaic current of redox probes, was obtained on diamond networks [48]. This is because the faradaic current of ferri-/ferro- cyanide is dominated by their diffusion lengths of analytes, which are typically in the range

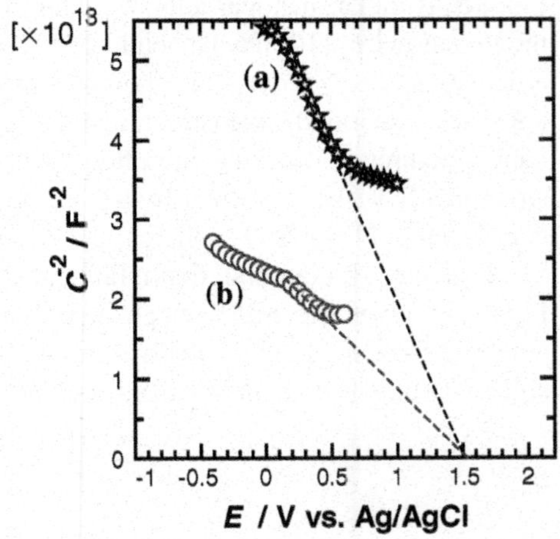

Fig. 2 Mott-Schottky plots of **a** a smooth diamond and (**b**) diamond nanotextures for an etching time of 10 s in 0.1 M pH 7.4 phosphate buffer at a fixed frequency of 1.0 kHz [13]

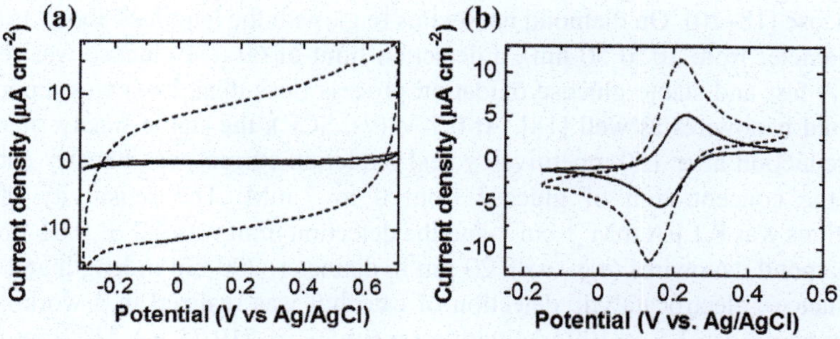

Fig. 3 Cyclic voltammograms of (solid lines) a smooth diamond electrode and (dashed lines) diamond nanowires based electrode in **a** 0.1 M KCl at a scan rate of 0.050 V s^{-1} and in **b** 1.0 mM $Fe(CN)_6^{3-/4-}$ + 0.1 M KCl at a scan rate of 0.1 V s^{-1} [14]

of tens of micrometers. Since the distance in between nanowires or diamond pores is much smaller than the value of diffusion lengths of analytes, diamond nanowires or diamond networks based electrode actually behaves as a macro-sized electrode [14].

2.3 *Electrochemical Applications*

Due to improved reactive sites, promoted electrocatalytic activities, and accelerated electron transfer rates, diamond nanotextures/nanowires are promising electrodes to improve the efficiency, sensitivity, selectivity and reproducibility of biomedical and chemical sensors [16–39]. Due to their wide electrochemical potential windows, enhanced surface areas, as well as high chemical stability, diamond nanostructures have been used for energy storage and related applications [42–56].

2.3.1 Electroanalytical Applications

Diamond nanostructures have been widely applied for electroanalytical applications [16–55], such as for electrochemical sensing of dopamine in the presence of ascorbic acid and uric acid [16, 17], for non-enzymatic monitoring of glucose [18–20], for the detection and immobilization of biomolecules (e.g., tryptophan [21, 22], tyrosine [22], catechol [23], and cytochrome c [24–26]), for enhanced electron transfer of shewanella loihica PV-4 [27], and for the construction of toxicity sensor by use of shewanella loihica PV-4 planktonic cells as the recognition element in bioelectrochemical systems [28].

For example, N-doped diamond nanowires showed excellent electrocatalytic activity towards the oxidation of ascorbic acid, dopamine, and uric acid. The electrocatalytic activity results from the increased sp^2 graphitic phase and the nanowire-like structure [17]. Diamond nanowires allowed non-enzymatic oxidation of glucose [18–20]. On diamond nanowires (e.g., with the length of about 3 μm and the diameter from 10 to 50 nm), a detection limit of 60 μM glucose was reported [19]. A fast and stable glucose oxidation process (less than 20 s) was reported on diamond nanowires as well [18]. At 0.7 V (vs. SCE), the stable steady-state (only 8 % reduction after 150 repetitive cycles) oxidation current was linearly enhanced with the concentration of glucose from 0 to 7 mM. The sensitivity of these detections was 8.1 $\mu A\ mM^{-1}\ cm^{-2}$ and the detection limit was 0.2 ± 0.01 μM [18]. On diamond nanowires (e.g., with 20 nm in diameter, 200 nm in length and 50 nm in distance), electrocatalytic detection of catechol was realized in a working concentration range of 5 to 100 μM with a sensitivity of 719.71 $\mu A\ M^{-1}\ cm^{-2}$ and a detection limit of 1.3 μM [23]. Differential pulse voltammetric detection of tryptophan on diamond nanowires was reported with a detection limit of 5×10^{-7} M [21]. Simultaneous detection of tryptophan and tyrosine was conducted successfully as well by differential pulse voltammetry on diamond nanowires when the amount ratio of tryptophan to tyrosine was less than 0.5 [22].

Direct electrochemistry of cytochrome c was realized on the OH-terminated diamond nanotextrues [24–26]. As shown in Fig. 4a, diamond nanotextures act as molecular traps, leading to a more efficient electron transfer process. The surface coverage of cytochrome c on diamond nanotextures was evaluated to be $4.2 \times 10^{12}\ cm^{-2}$ and its electron transfer rate constant was $(1.43 \pm 0.05)\ s^{-1}$, higher than some of reported values [24–26]. The enhancement of electron transfer rate results from the electrostatic and hydrophobic interaction of cytochrome c with the OH-terminated diamond. Electrocatalytic reactions towards oxygen reduction as well as AFM tapping and scratching experiments in buffer [29–31] further confirmed that the electrostatic interaction controls coarse orientation of cytochrome c while hydrophobic interaction assists in the formation of the electron transfer complex, as schematically shown in Fig. 4b. Diamond nanograss showed an enhanced electron transfer from outer membrane c-type cytochromes of shewanella loihica PV-4 to the electrode [27]. Using such a shewanella loihica PV-4 planktonic cell as the recognition element in bioelectrochemical systems (BES), a toxicity sensor based on the electrochemical and impedance detection of tobramycin was proposed [28].

Decorated diamond nanotextures/nanowires with nanoparticles (e.g., from nickel [19, 32, 33] and platinum [34]), with nitrophenyl [35–37], and with carboxylic

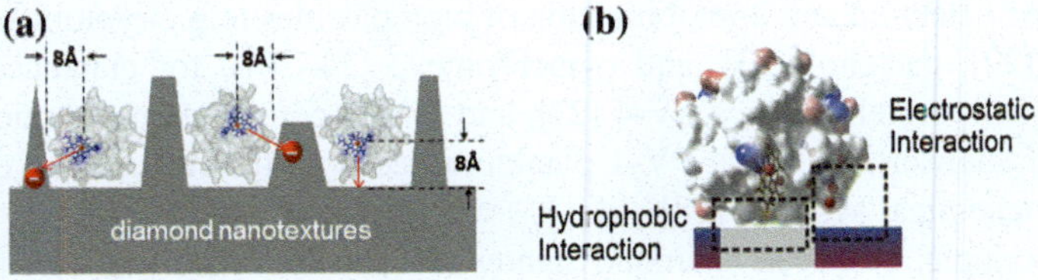

Fig. 4 Schematic demonstration of **a** trapping of cytochrome c using diamond nanotextures for an efficient electron transfer process and **b** the interaction of cytochrome c with the OH-terminated diamond surface [26, 30]

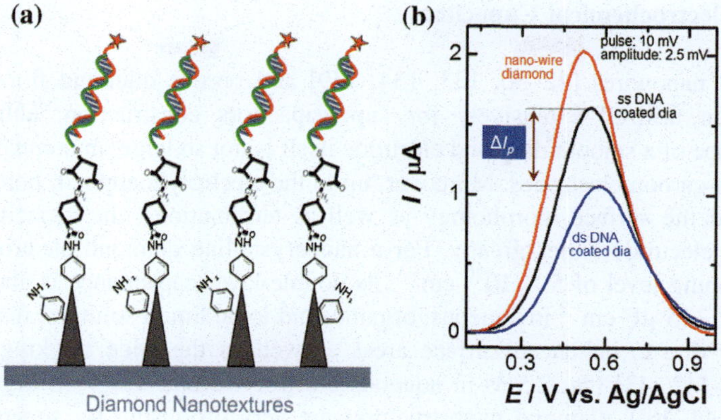

Fig. 5 **a** Schematic plots of the tip-functionalization of diamond nanotextures for DNA immobilization and **b** differential pulse voltammetric detection of DNA hybridization on diamond nanotexures using negatively charged redox mediators [35, 36]

acid-terminated poly(pyrrole) [38] were applied to construct electrochemical [19], DNA biosensors [35–37, 39], and immunosensors [32, 33, 38].

For example, tip coating of diamond nanowires with nickel nanoparticles improved the detection limit for glucose detection to 10 μM [19]. Non-enzymatic oxidation of glucose and electrocatalytic oxidation of methanol have been reported on diamond nanotextures decorated with platinum nanoparticles [34]. Diamond nanotextures were functionalized via electrochemical grafting at −0.05 V (vs. Ag/Ag^{+}) for 4 s with 1.0 mM diazonium salts. The density of phenyl molecules at diamond nanotextures was about 2×10^{13} cm^{-2} [36]. Constant-current mode scanning tunneling microscope (STM) conducted before and after grafting with nitrophenyl on diamond textures indicated the preferential bonding of nitrophenyl to the tips of wires [13]. Such tip-functionalized diamond nanotextures were applied to immobilize DNA with aid of SSMCC chemistry, as shown in Fig. 5a [35, 36]. Electrochemical detection of DNA hybridization on such an interface was realized by use of negatively charged redox indicators [35, 36]. Figure 5b shows the differential pulse voltammograms of $Fe(CN)_6^{3-/4-}$ on diamond nanotextures based electrode before and after functionalized with single-strand DNA (ss DNA) and double-strand DNA (ds DNA). The difference of peak currents of redox indicators on these interfaces was applied to detect DNA hybridization. A detection limit of about 2 pM was realized on 0.03 cm^2 sensor area over 30 hybridization/denaturation cycles. The discrimination of single-base mismatched complementary DNA was achieved [35–37, 39].

Furthermore, several diamond nanowires based immunosensors have been reported [33, 34, 38]. For instance, biotinylated anti-IgG was specifically linked to nickel particles. The charger transfer resistance, detected from electrochemical impedance spectroscopy, was linear with IgG concentration in the range of 0.3–400 ng mL^{-1}. The detection limit of IgG was found to be 0.3 ng mL^{-1} [33].

2.3.2 Electrochemical Capacitors

Diamond nanowires [42–50, 133, 134, 139] and porous diamond films [51–53] have been utilized extensively for supercapacitors construction, although the capacitance of a smooth diamond electrode itself is not so huge, in comparison with other non-carbon electrodes. Moreover, up to the electrolyte applied, boron-doping level, and the surface morphology as well as terminations, the capacitance of a diamond electrode varies greatly. For a nanocrystalline diamond electrode with a boron-doping level of 5×10^{20} cm^{-3}, its double-layer capacitance is about 3.6–7, 14–20, 11–15 μF cm^{-2} in aqueous, organic, and ionic liquid solution, respectively. However due to enhanced surface areas as well as the wide working potential windows [42, 43] (ca. 2.5 V in aqueous electrolytes and 7.3 V in organic electrolytes [44–46]), diamond nanostructures are very promising for supercapacitors construction. Taking diamond networks as an example, a diamond network with a porosity from 15 to 68 % led to hundreds of times enhancement of the surface areas than that of flat diamonds (e.g., 490-fold for a 3 μm thick diamond network).

Electric double layer capacitors (EDLCs) based on diamond honeycomb nanostructures showed a capacitance of 3910 μF cm^{-2} and 666 μF cm^{-2} in aqueous and organic solution, respectively [44–46]. Diamond foam based EDLCs attained specific capacitances of 598 and 436 μF cm^{-2} in aqueous and organic solutions, respectively. A high power density of 807 W cm^{-3} was achieved, which touched the best power performance of electrolytic capacitors [47]. In 0.1 M H_2SO_4, the double layer capacitance of a diamond network was calculated to be 13.7 F g^{-1} or 17.3 F cm^{-3} at a scan rate of 0.1 V s^{-1} [48]. Figure 6a compares the capacitive behaviour of a smooth diamond electrode with two diamond networks in 0.1 M H_2SO_4 [48]. Silicon nanowires coated with a thin diamond film (240 nm in thickness) was employed for EDLCs application. The capacitance of such an EDLC was 105 μF cm^{-2} in a mixture

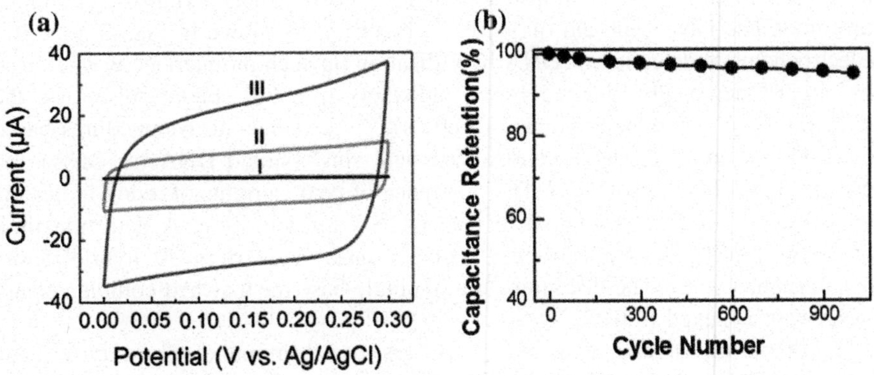

Fig. 6 Cyclic voltammograms of **a** flat diamond film (I), diamond network-1 (II) and diamond network-2 (III) in 0.1 M H_2SO_4 at a scan rate of 0.1 V s^{-1}; **b** Capacitance retention of a diamond EDLC as a function of charge/discharge cycles, tested in 1.0 M Na_2SO_4 at a scan rate of 0.1 V s^{-1} [43, 48]

of propylene carbonate with 1-methyl-1-propylpyrrolidinium bis(trifluoromethylsulfonyl)imide, one room temperature ionic liquid. A high energy density of 84 μJ cm^{-2} and power density of 0.94 mW cm^{-2} and good stability (retention stability of 93.3 % after 10,000 cycles at a scan rate of 5 V s^{-1}) were achieved [49]. Figure 6b presents a very recent retention stability test result for a diamond EDLC in 1.0 M Na_2SO_4 [43]. Only 5 % reduction of the capacitance is seen. This is due to high chemical stability of diamond electrodes in any kind of media as well as at different working potentials. On TiO_2 nanotubes coated with boron-doped diamond (200–500 nm in thickness), the specific capacitance is dependent on the boron concentration. In 0.1 M $NaNO_3$, the specific capacitance of 2.10, 4.79, and 7.46 mF cm^{-2} was obtained at a scan rate of 0.01 V s^{-1} for a [B]/[C] ratio of 2000, 5000 and 10,000, respectively. The substantial improvement of electrochemical performance and the excellent rate capability was explained with the synergistic effect of TiO_2 treatment in CH_4:H_2 plasma and the high electrical conductivity of boron-doped diamond layers [50]. Recently, "diamond paper" showed in aqueous electrolyte a capacitance of 0.688 mF cm^{-2} per layer, or 0.645 F g^{-1}. The specific power of these diamond based supercapacitors reached 1×10^5 W kg^{-1} [134].

Other porous diamond films/membranes [51–53] have been employed for supercapacitor applications. For example, a two-step thermal treatment method was developed for the fabrication of porous conductive boron-doped diamond films [51]. The sizes of the pores were from several tens to several hundred nanometer. Such a porous membrane exhibited a double-layer capacitance of ca. 140 μF cm^{-2} in an aqueous electrolyte, estimated from cyclic voltammetry and galvanostatic measurements [52]. Porous boron-doped diamond films overgrew on CNTs, showed c.a. 450 times greater electroactive areas and double-layer capacitance values than those for the equivalent flat boron-doped diamond electrodes [51]. Another porous diamond, based on the overgrowth of a highly porous polypyrrole scaffold with a thin boron-doped diamond film, exhibited remarkable electrochemical properties, including a large double layer capacitance up to 3 mF cm^{-2} in aqueous $LiClO_4$ and a low electrochemical impedance [53].

Diamond nanowires were coated as well with nickel hydroxide [133] to construct diamond pseudocapacitors. Although much higher capacitances of diamond pseudocapacitors than those of diamond EDLCs were obtained, the big shortcoming of diamond pseudocapacitors is the poor retention stability of their capacitances. This results from low stability of pseudo or redox-active species on diamond electrode surface (e.g., these metal oxides are easily stripped from diamond during the charging/discharging processes) [43, 133].

2.3.3 Other Applications

Diamond nanostructures have been proposed as well for other electrochemical applications [40, 54–56, 135, 138]. For example, porous diamond membrane was applied as a filter to separate differently charged 5-carboxyfluorescein and substance P under different biases [55]. Diamond nanowires was also applied to enhance the intensity of electrogenerated chemiluminescence (ECL) of a ruthenium tris(2,2′) bipyridyl/tripropylamine system [40]. Owing to their superior properties (e.g., large surface areas, improved electrocatalytic activities, and accelerated electron transfer rates), the enhancement of ECL intensity was attributed mainly to the highly facile oxidation of tripropylamine on diamond nanowires. This study revealed an effective method for the ultrasensitive detection of ruthenium tris(2,2′)bipyridyl. It will help to increase the efficiency of immunoassays and DNA analysis based on ruthenium tris (2,2′)bipyridyl electrogenerated chemiluminescence. Diamond nanowires have been utilized as well for electrocatalytic applications [135, 138]. For example, diamond nanowires coated with Pt have been applied for electrocatalytic hydrogen evolution [135]. Very recently, electrochemical CO_2 reduction has been achieved on diamond coated silicon nanowires, where doped diamond acts as a metal-free electrocatalyst [138]. Recently, diamond foams composed of hollow spheres of polycrystalline boron-doped diamond have been chemically modified with two donor–acceptor type molecular dyes and further utilized as electrode materials for p-type dye-sensitized solar cells in an aqueous electrolyte solution containing methyl viologen as a redox mediator [141].

3 Diamond Nanoparticles

3.1 Synthesis Methods

3.1.1 Un-doped Diamond Nanoparticles

Un-doped diamond nanoparticles with size down to 20 nm have been produced via ball-milling of micro-sized high-pressure-high-temperature (HPHT) diamond films. In this way, diamond nanoparticles less than 10 nm were seldomly obtained. With the aid of bead-assisted sonic disintegration, the production of diamond nanoparticles 70–80 nm in diameter was realized by milling polycrystalline chemical-vapor-deposited (CVD) diamond films. These nanoparticles have faceted shapes with sharp edges, which correspond to fractured crystallographic planes. Using such a method, few spherical nanoparticles were found. Another more widely applied approach to produce spherical diamond nanoparticles these days is to use dynamic processes from molecules of explosives and different graphite precursors [5, 57], including the direct transformation of graphite by an external shock wave, the detonation of graphite mixed with explosive, and the detonation of high energy explosive. Diamond nanoparticles synthesized from these detonation methods have a core of sp^3 diamond with a size of 4–5 nm and a shell of a mixture of sp^2 and sp^3 carbon as well as oxygen based functional groups (e.g., carboxylic acids, esters, lactones, etc.) [5, 57].

3.1.2 Doped Diamond Nanoparticles

Doped diamond particles, mainly boron-doped diamond particles, have been obtained mainly via the overgrowth of insulating diamond particles with boron-doped diamond, solid-state diffusion, and milling [119, 120]. For example, boron-doped diamond powder was synthesized under HPHT using B-doped graphite intercalation compositions as carbon sources [121]. Recently, Kruger et al. [123] applied a multistep milling process followed by purification and surface oxidation to produce 10–60 nm boron-doped diamond nanoparticles. The starting material used for such a milling process was heavily boron-doped diamond films.

3.2 Electrochemical Properties

To investigate electrochemistry of diamond nanoparticles, diamond nanoparticles are always treated and/or cleaned. The applied processes include acid boiling, oxidation by Osswald method, plasma treatment, thermal annealing, etc. For example, surface treatment of diamond nanoparticles by heating in air and in a hydrogen flow results in oxygenated (O-) and hydrogenated (H-) nanoparticles, respectively. The acid boiling removes effectively the responses associated with sp^2 carbon impurities. On these diamond nanoparticles based film electrodes, a potential independent capacitive signal has been obtained and checked by recording their cyclic voltammograms in solutions without any redox species [67].

Diamond nanoparticles based film electrodes have been fabricated with various techniques [129], such as drop coating from their ethanol suspensions, smearing a mineral oil paste of diamond nanoparticles, grinding diamond powders into the tip of a Pt wire sealed in a small pipette, electrophoretically deposition, co-deposition, non-contact microprinting, and layer-by-layer self-assembly by a high pressure/high-temperature methodology, etc.

3.2.1 Un-doped Diamond Nanoparticles

Diamond nanoparticles, one form of un-doped diamond, showed different electrochemical properties from un-doped bulk diamond. They are surprisingly electrochemically active. Their electrochemical activities depend greatly on the nature/type (e.g., detonation, HPHT particles), surface terminals, as well as the pH value of the solutions.

Detonation Diamond Nanoparticles

Novoselova et al. [58] studied for the first time the redox activity of diamond powders based film electrode in aqueous electrolytes. Redox couples of

$[Fe(CN)_6]^{3-/4-}$ and $Ce^{3+/4+}$ were applied as the probes. The voltammetric response of $Fe(CN)_6^{3-/4-}$ was however superimposed on a linearly sloping background and with small peak currents. The reduction currents were two times larger than the oxidation currents. In the case of the $Ce^{3+/4+}$ redox couple, additional current responses were noted, besides the expected current peaks. Zang et al. [59] used cavity electrodes to investigate redox activities of diamond nanoparticles. Stable background currents in KCl electrolyte over a wide potential range (−1.2 to 2.0 V vs. Ag/AgCl) and a quasi-reversible reversible electrode reaction for the $Fe(CN)_6^{3-/4-}$ couple with the electrode reaction rate constant of 2.87×10^{-3} cm s^{-1} were obtained. The recorded AC impedance spectra were consistent with those obtained on a porous electrode [59].

Holt and her colleague [60–65] made extensive investigation on the electrochemistry of detonation diamond nanoparticles. Differential pulse voltammetry of electrode-immobilised layers of diamond nanoparticles in the absence of solution redox species revealed oxidation and reduction peaks, resulting from direct electron transfer reactions of diamond nanoparticles themselves [60]. Moreover, the presence of detonation diamond nanoparticles on the (diamond) electrode modified the cyclic voltammetric response of $Fe(CN)_6^{3-/4-}$ and $IrCl_6^{3-/2-}$ when the scan rate was slow and the concentration of redox couples was low. For example, enhancements of oxidation currents were noted at potentials where the oxidation of the surface of diamond nanoparticles started. The enhancements of reduction currents were likewise observed where diamond nanoparticle was reducible [60]. Attenuated total reflectance infrared spectroscopy was then used to monitor spectral features of the surface of diamond nanoparticles [61]. Aqueous $IrCl_6^{2-}$ was added in these studies. They found that electron transfer between the surface of diamond nanoparticle and the solution redox species results in the oxidation of 8.5 % of surface alcohol groups, with concomitant formation of unsaturated ketone or quinone-like moieties [61]. Scanning electrochemical microscopy (SECM) [62] was applied as well to investigate the redox behavior of detonation diamond nanoparticles based film electrode (Fig. 7). Different collection modes and various redox mediators were

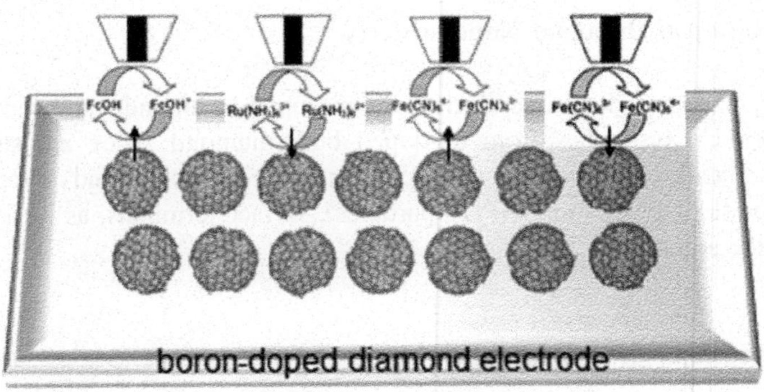

Fig. 7 Schematic illustration of SECM investigation of redox activities of diamond nanoparticles based film electrode with different redox probes and collection modes [62]

used to estimate quantitatively heterogeneous rate constants and overpotentials of redox probes. On such an electrode, extremely sluggish kinetics were found for all redox couples, but the reduction of $Fe(CN)_6^{3-}$ was found to be especially slow when compared to the oxidation of $Fe(CN)_6^{4-}$. Overall, the rate constants were about 10 times faster at diamond nanoparticles based film electrode than that on the boron-doped diamond film [62]. Supported with in-situ infrared (IR) experimental results, they stated that electron transfer at the diamond nanoparticle surface takes place at similar sites as on the boron-doped diamond film. But the reaction sites are present at higher relative concentrations due to the higher surface to bulk atom ratio of the nanoparticles. The modification of an electrode with an immobilised layer of diamond nanoparticles was found later to enhance significantly the redox currents for reversible oxidation of ferrocene methanol. Current enhancement is dependence of the diameter of diamond nanoparticles, with enhancement increasing in the order 1000 nm < 250 nm < 100 nm < 10 nm < 5 nm [64].

To explain such the 'molecule-like' redox behaviour of diamond nanoparticles, Holt et al. proposed a so-called feedback mechanism [60–64]. Figure 8 shows schematically such a mechanism. A self-conducted oxidation and reduction process via surface states at specific potentials triggers the redox activities of diamond nanoparticles. The magnitude of current enhancement depends on the standard potential of the redox couple relative to those of the surface states of diamond nanoparticles. Provided that at the nanoscale surface properties of diamond nanoparticles dominate over those of the bulk, electron transfer occurs between these essentially insulating particles and a redox species in solution or an underlying electrode. The occurrence of reversible reduction of diamond nanoparticles via electron injection into available surface states at well-defined reduction potentials is thus speculated. In this process, diamond nanoparticles act as a source and a sink of electrons for the promotion of solution redox reactions [67]. They then concluded that the electrochemical activity of un-doped diamond nanoparticles is attributed to unsaturated bonding at the diamond nanoparticle surface [60]. Electron transfer occurs between electroactive species generated at the underlying electrode during voltammetry and the immobilized diamond nanoparticles in the interfacial region, leading to regeneration of the starting species and hence enhancement in currents.

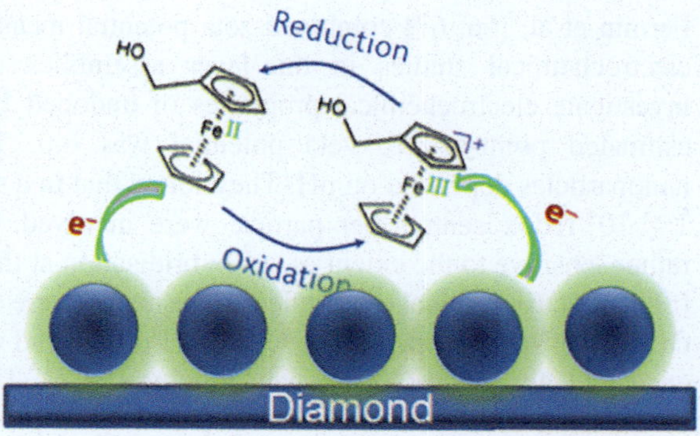

Fig. 8 Schematic plot of a catalytic feedback mechanism: catalytic reactions of ferrocene methanol on diamond nanoparticles based film electrode [60]

Moreover, electrochemical activity of detonation diamond nanoparticles (5 nm in diameter) based film electrode was tunable via surface functionalizations and graphitization degree of diamond nanoparticles. For example, both redox reactions of $Fe(CN)_6^{3-/4-}$ and $Fe^{3+/2+}$ were quasi-reversible on the pristine diamond nanoparticles based film electrodes. The O-terminated diamond nanoparticles exhibited the greatest electrochemical activity for the redox couples $Ru(NH_3)_6^{3+/2+}$ and $Fe(CN)_6^{4-/3-}$ while the H-terminated diamond nanoparticles the least. After fluorination of diamond nanoparticles, the electrode reactions of $Fe(CN)_6^{3-/4-}$ became slower, while the amino modification accelerated the electron transfer process of $Fe(CN)_6^{3-/4-}$ anions but slowed the redox reaction of $Fe^{3+/2+}$ cations [69]. Annealing of diamond nanoparticles in vacuum led to the variation of electrochemical activities of diamond nanoparticles. The electron transfer rate of the $Fe(CN)_6{}^{3-/4-}$ redox couple in aqueous solutions decreased with an increase of the annealing temperature. Re-annealing in air restored the original electrochemical activity. This is because vacuum annealing below 850 °C removed parts of the oxygen-containing surface functionalities from the surface of diamond nanoparticles and produced more sp^2 carbon atoms in the shell. When the annealing temperature was at 900–1100 °C, more serious graphitization produced a continuous fullerenic shell wrapped around a diamond core, which had a high conductivity and electrochemical activity [68].

Furthermore, electrochemical activity of detonation diamond nanoparticles based film electrode was found to be affected by pH values of the solution, namely the by solution proton concentration [60, 62]. In the solutions with pH value of 4 and 5, well-defined peaks in the potential range of −0.1 to 0.5 V (vs. Ag/AgCl) were seen from differential pulse voltammograms. As the solution pH increased, they became much smaller in magnitude and far less resolved [60, 62]. This is because the electrochemical response of diamond nanoparticles based film electrode resulted from the oxidation and reduction of surface states of diamond nanoparticles. The potentials of these surface states are however pH-dependent; moreover they are able to interact with solution redox species [60, 62].

HPHT Diamond Nanoparticles

Fermin et al. [66, 67] combined zeta potential measurements in the solutions and electrochemical studies in thin-layer assemblies of diamond nanoparticles to investigate electrochemical properties of undoped HPHT diamond particles. The estimated point-of-zero zeta potential was 6.6. The zeta potentials of these nanoparticles depended on pH. They found that in a single electron transfer process 1×10^4 redox centres per particle were involved. Electrochemical signals were rather sensitive to the extent of sp^2 hybridisation at the surface of diamond powders [66]. Electrochemical field-effect transistors were employed to investigate the charge transport properties of O- and H-terminated diamond particles in the presence and absence of metal nanostructures [67]. The assembly of H-terminated diamond particle was characterized by a charging process at a potential above 0.1 V (vs. Ag/AgCl). The responses were found to be associated with hole-injection into the valence band edge, which is shifted to approximately −4.75 eV (vs. vacuum) upon hydrogenation. The position of the valence band edge as well as hole number density at the H-terminated diamond particle surface varied as a function of the applied potential [70]. Through the discussion in terms of the electrochemical formation of charge carriers in the diamond particles, percolation theory, and charge screening at the double layer, Fermin et al. suggested that charge transport on un-doped diamond particles is not only determined by the intrinsic surface conductivity of individual diamond particles, but also by particle-to-particle charge transfer [71]. The latter contribution effectively controls the assembly conductivity in the presence of an electrolyte solution as the difference between hydrogenated and oxygenated particles vanishes. The conductivity in the presence of metal nanoparticles is mainly determined by the metal volume fraction, while diamond surface termination and the presence of electrolyte solutions exert only minor effects [71].

3.2.2 Doped Diamond Nanoparticles

To check electrochemical activities of boron-doped diamond particles, redox probes of $Fe(CN)_6^{3-/4-}$ and $Ru(NH_3)_6^{3+/2+}$ and a polytetrafluoroethylene binder were used. These particles were produced via microwave plasma-assisted CVD growth of a thin boron-doped layer on insulating diamond powders (8–12 μm in diameter). At scan rates between 10 and 500 mV s^{-1}, the peak difference of the anodic wave from the cathodic wave was high (in the range 0.14–0.35 V vs. Ag/AgCl) for both redox probes, suggesting significant ohmic resistance within the powder/binder electrode [119]. Later the same technique was applied to overgrow small-sized diamond powders (100 and 500 nm in diameter). Both powders had increased conductivities. Well-defined electrochemical responses were obtained on these powders based film electrode for the redox reactions of $Fe(CN)_6^{3-/4-}$, $Ir(Cl)_6^{2-/3-}$, and $Fe^{2+/3+}$, in comparable to typical responses shown on the high-quality, boron-doped nanocrystalline diamond thin-film [120]. Later the electrochemical characteristics of boron-doped diamond powders based film electrodes were investigated by measuring the cyclic voltammetric curves and AC impedance spectra [121]. The powders were synthesized under HPHT using B-doped graphite intercalation compositions as carbon sources. For the $Fe(CN)_6^{3-/4-}$ redox couple, the electrode reaction process is reversible or quasi-reversible at the scan rates of 0.01–1.0 V s^{-1}. At the low scan rates the linear relation between peak current and square root of scan rate indicates that the electrode process is a diffusion-controlled mass-transport process. The electrochemical behavior is similar to a planar electrode. With an increase of the scan rate the electrode process is controlled by the mass transport plus kinetic process. AC impedance spectra exhibited the porous structure characteristic of boron-doped diamond powders based film electrode [121]. The boron-doped diamond nanoparticles made by solid-state diffusion method showed a lower capacitance but a higher conductivity than undoped diamond nanoparticles [122]. In the potential range of −0.3 to 1.8 V (vs. SCE), a featureless voltammetric response was obtained. Recently high quality boron-doped diamond nanoparticles with a size of 10–60 nm and a boron concentration of approximately 2.3×10^{21} cm^{-3} have been produced by Kruger et al. [123]. However the electrochemistry of those nanoparticles has not been reported yet.

3.3 Electrochemical Applications

3.3.1 Un-doped Diamond Nanoparticles

Electroanalytical Applications

Due to the features of giant specific surface areas and large numbers of surface defects as well as the cluster structure, detonation diamond nanoparticles have increased electrical conductivities. As a novel type of electrode materials, they have been employed frequently for electrochemical and biochemical sensing applications [72–86].

Many electrochemical sensors based on diamond nanoparticles have been reported, including the sensors for the detection of azathioprine [72], epinephrine and uric acid in the presence of ascorbic acid [73], nitrite [74], tryptophan, and 5-hydroxytryptophan [75]. The matrix of diamond and silver nanoparticles were applied for electrochemical monitoring of thioridazine [76] and hydrogen peroxide [77]. For example, on an electrode based on a chitosan matrix and the mixture of nanographite and diamond nanoparticles, electrocatalytic detection of azathioprine was realized in a concentration range from 0.2 to 100 μM with a detection limit of 65 nM [72]. The same electrode was applied successfully to detect epinephrine (0.01–10 μM) and uric acid (0.01–60 μM) in the presence of ascorbic acid. The detection limit was 3 nM for both epinephrine and uric acid [73]. Voltammetric monitoring of 30 nM tryptophan and 6 nM 5-hydroxytryptophan was shown to be possible on diamond nanoparticles based film electrode [75]. By decorating diamond nanoparticles with silver nanoparticles, voltammetric determination of thioridazine was achieved in the concentration range of 0.08–100 μM with a detection limit of 0.01 μM [76]. Synergistic effect of two kinds of nanoparticles was proposed to demonstrate the satisfactory electrochemical activity [76]. Such a matrix was applied to fabricate a non-enzymatic hydrogen peroxide sensor [77]. On such a sensor, hydrogen peroxide was detected in the range of 0.1–34.0 μM with a detection limit of 0.01 μM and a sensitivity of 1.59×10^6 $\mu A\ M^{-1}$ [77]. TiO_2 nanoparticles coated diamond nanoparticles exhibited higher electrochemical activity than the pristine diamond nanoparticles, especially higher catalytic ability towards the oxidation of nitrite anions. A detection limit of 0.55 μM and a linear range of 0.05–1.0 mM for the detection of nitrite ions were achieved [78].

Diamond nanoparticles based electrochemical biosensors have been reported. Glucose oxide, cytochrome c, hemoglobin, horseradish peroxidase, alcohol dehydrogenase, and lactate oxidase have been immobilized on diamond nanoparticles for the detection of glucose [79–83], and alcohols [84], lactate [85], respectively. For example, alcohol dehydrogenase (ADH) has been adsorbed on oxidized diamond nanoparticles. The adsorption of the non-covalently immobilized ADH, estimated with Langmuir isotherms, was dependent on pH values of the solutions. A higher packing density was achieved at the isoelectric point of ADH. Its relative activity was retained up to 70 % under optimum pH conditions. An ethanol bioelectrochemical cell and an alcohol biosensor were then proposed [84]. On the diamond nanoparticles and polyaniline based electrode, direct electrochemistry of cytochrome c was achieved, leading to electrocatalytic detection of nitrite ions in a concentration range from 0.5 μM to 3 mM with a detection limit of 0.16 μM [81]. Diamond nanoparticles and porous poly(aniline)–poly(2-acrylamido 2-methyl propane sulfonic acid) network based sponges were prepared to entrap horseradish peroxidase [83]. On such a matrix, electrocatalytic reduction of hydrogen peroxide was realized in a concentration range of 1–45 mM with a rapid response time of 5 s, a high sensitivity of 129.6 μA M^{-1} and a low detection limit of 59 μM [83]. An electrochemical biosensor based on diamond nanoparticles was proposed as well for lactate determination. The workable concentration range was from 50 μM to 0.7 mM, the sensitivity was 4.0 μA mM^{-1}, and the detection limit was 15 μM [85]. Antibody immobilization was reported on diamond nanoparticles seeded inter-digitated electrodes (IDEs). Such an impedance biosensor improved the overall detection sensitivity, namely the resistance to charge transfer. The sensor performance was better than those based on gold or ITO electrodes. When sensing bacteria from 106 cfu mL^{-1} *E. coli* O157:H7, the resistance to charge transfer at the IDEs decreased by 38.8 %, which is nearly 1.5 times better than that reported previously using redox probes. Further in the case of 108 cfu mL^{-1} *E. coli* O157:H7, the charge transfer resistance changed by 46 % [86].

Electrocatalysts

Different kinds of diamond nanoparticles (e.g., surface graphitized diamond nanoparticles with a diamond core covered by a graphitic carbon shell, bucky diamond nanoparticles with a nanoscale diamond core surrounded by a fullerene shell, and graphene coated diamond nanoparticles) have been used as the support to load catalysts (e.g., Pt [87–93], Ni [94, 95], Ti [96], Pd [97], Pt/Ni [98], Pt/Ru [99–102], Sn/Pb [103], Pt/Eu [91, 99], and metal oxides [104–106]). The mostly investigated electrocatalytic reactions include electrocatalytic oxidation of methanol/formic acid/CO as well as oxygen reduction reaction. To load these catalysts on diamond nanoparticles, numerous approaches have been developed, including electrodeposition [88, 92], chemical reduction [93], and microwave-assisted polyol synthesis [91, 99]. For example, electrodeposition of Pt nanoparticles on un-doped diamond nanoparticles (5–100 nm in diameter) were conducted in 1.1 mM chloroplatinic acid solution. The electrodeposited Pt nanoparticles were well-dispersed on the facet surfaces of diamond nanoparticles [88, 107]. Pt and Ru nanoparticles were chemically deposited on un-doped and boron-doped diamond nanoparticles through the use of $NaBH_4$ as reducing agent and sodium dodecyl benzene sulfonate as a surfactant [93]. Microwave-assisted reduction method [91, 98] has been used for the preparation of catalysts of Pt [91], Pt/Ni [98], and Pt/Ru [102] on the surface of diamond nanoparticles. A microwave heating polyol method was used to prepare the Pt/Ru electrocatalyst on the surface of un-doped diamond nanoparticles. It was found that Ru was partly dissolved in the face-centered cubic Pt lattice. The Pt/Ru nanoparticles were small and uniform with the size of 2–4 nm, and highly dispersed on the surface of diamond nanoparticles [102]. A simple ultrasonic treatment in the presence of diamond powders prior to electrodeposition improved spatial distribution and a higher Pt dispersion over the electrode [99–102]. A two-step method was reported to modify diamond nanoparticles with Pt and TiO_2 nanoparticles, namely first by a microwave hydrolysis step, and then electrodeposition of Pt nanoparticles [104–106].

For the first time Fermin et al. [97] utilized Pd and HPHT diamond nanoparticles (500 nm in diameter) based electrocatalysts for the electrochemical stripping of CO and oxidation of formic acid in the acid solutions. Later the same group studied the electrocatalytic reactivity of Pt nanoparticles coated HPHT diamond particles towards the oxidation of adsorbed CO, methanol, and formic acid with differential electrochemical mass spectrometry. Diamond nanoparticles with different surface terminations were used for these electrocatalytic oxidation reactions, leading to different oxidation mechanisms [89]. Towards the ability of the electrocatalytic oxidation of methanol, diamond nanoparticles with smaller diameters (e.g., 5 nm) exhibited better electrocatalytic activity than bigger ones (e.g., 100 nm) after the surface of diamond nanoparticles were coated with electrodeposited Pt nanoparticles [88, 107]. The application of these Pt modified diamond electrodes in the electrochemical oxidation of hydrogen peroxide was demonstrated [92]. The ink-paste method was used to prepare the membrane electrode assembled with Pt and Pt/Ru modified un-doped and boron-doped diamond nanoparticle catalytic systems [100]. Their performances were examined in a direct methanol fuel cell system [100]. The Pt/Ru catalyst exhibited higher activity and stability for methanol electrooxidation reaction than individual Pt catalyst [102]. The investigation of electrocatalytic reduction of oxygen on metal catalysts coated diamond nanoparticles has been conducted further using cyclic voltammetry, chronoamperometry and linear sweep voltammetry [98]. The Pt/Ni catalysts exhibited better electrocatalytic activities than the Pt catalysts either for methanol oxidation reaction or for oxygen reduction reaction [98]. In the acid medium the Pt/TiO_2 catalyst system possessed higher electrocatalytic activity for methanol oxidation reaction compared with the individual Pt catalyst [96]. TiN coated diamond nanoparticles showed higher catalytic activity and better stability in methanol oxidation and oxygen reduction reactions compared with the individual catalysts on carbon and on diamond nanoparticles [106].

Energy Storage

Diamond nanoparticles have been employed as the electrode material for energy storage, such as for electrochemical capacitors [108–113, 136], lithium batteries [114, 115], and dye-sensitive solar cells [116]. For supercapacitors, diamond nanoparticles are always thermally annealed at the temperatures above 1000 °C. Due to the generation of carbon onions, the energy is possible to be stored under a high current density and a high capacitance.

For example, Gogotsi et al. investigated and compared in organic and aqueous electrolytes the performance of EDLCs based on carbon onions, diamond nanoparticles, carbon black and multi-walled carbon nanotubes [108]. Different methods were applied, including galvanostatic cycling, electrochemical impedance spectroscopy and cyclic voltammetry. To construct pseudocapacitors, the surface of diamond nanoparticles and carbon onions was coated with a layer of phosphomolybdate [111] or polyanilline [112, 113] (produced via electropolymerization in a cavity electrode or a chemical oxidation approach [113]). Carbon onions and phosphomolybdate based pseudocapacitance exhibited a 20 % increase in the capacitance (up to 600 mF cm^{-2} at 5 V s^{-1}) [111]. Due to the porous network structure [113], the pseudocapacitance based on diamond nanoparticles and polyaniline (with the weight ratios of 3–28 %) increased to 640 F g^{-1} in 1.0 M H_2SO_4. This capacitance was 3–4 times higher than that of the activated carbons and more than 15 times higher than that of diamond nanoparticles and carbon onions. Moreover the charge-discharge characteristics were stable for 10,000 cycles [113].

Some experiments to use diamond nanoparticles for lithium batteries were reported as well [114, 115]. For example, a volumetric capacity of less than 23 mA h cm^{-3} has been shown, although it was much lower than 450–700 mA h cm^{-3} offered by state-of-the-art high-density cathodes used in commercial Li-ion batteries [115]. The composite system of polyaniline and diamond nanoparticles prepared via electrochemical polymerization techniques was applied toward the iodine/iodide redox couple for the construction of dye-sensitised solar cells [116].

Other Applications

Diamond nanoparticles have been applied as corrosion inhibition [117], and as the metal-free catalysts for oxidant- and steam-free dehydrogenation [118, 140]. For example, the nanocomposite of polyaniline and diamond nanoparticles showed an ohmic junction and wide potential values, independent of redox characteristics of both polyaniline and diamond nanoparticles. This is due to its chain conformation and electronic properties (achieved by the interaction of the free electron pairs of the nitrogen atoms in the polyaniline with a charged molecule on the surface of diamond nanoparticles), leading to excellent corrosion inhibitor characteristics [117].

3.3.2 Doped Diamond Particles

Boron-doped diamond powders based film electrodes were shown to be dimensionally stable at 1.4 V (vs. Ag/AgCl) for 1 h in 0.5 M H_2SO_4 at 80 °C. They are thus corrosion-resistant during anodic polarization. In contrast, glassy carbon powders polarized under identical conditions underwent significant microstructural degradation and corrosion [124]. Electrocatalytic oxidation of methanol was tested on the composite based on boron-doped diamond particles (500 nm and 5 μm in diameter) and metallic oxides [125]. By coating boron-doped diamond nanoparticles with $Ni(OH)_2$, a non-enzymatic glucose sensor was constructed. The detection limit for glucose detection was 1.2 μM [126]. Recently, boron-doped diamond nanoparticles have been electrostatically self-assembled on carbon nanotubes. Such a 3D network showed a low electron transfer resistance but a large effective surface area, resulted in an improved electrochemical performance in glucose detection [128]. The construction of electrochemical capacitors using RuO_2 coated diamond powders has been reported as well [127].

4 Diamond Nanoelectrode Arrays

Small-dimensional electrodes (e.g., nanoelectrodes) offer various benefits over planar macroscopic electrodes [10, 142, 143], such as reduced Ohmic resistance, enhanced mass transport, decreased charging currents, decreased deleterious effects of solution resistance, and high possibility for fast voltammetric measurements. However, single nanoelectrode only generates a small current that is relatively difficult to detect with conventional electrochemical setups. This has been circumvented by fabricating nanoelectrode arrays or ensembles that operate in parallel. They amplify the signal of individual nanoelectrodes but do not lose their beneficial characteristics. If diamond nanoelectrode arrays or ensembles are applied for sensor applications, their performances with respect to the sensitivity, detection limit, life time, and reproducibility, will be highly improved [10, 130, 142, 143]. This is because boron-doped diamond is one of the most appropriate and optimized material for the fabrication of these arrays and ensembles [6, 137]. Moreover, since a macroscopic diamond electrode shows a higher degree of inhomogeneity with respect to boron-doping level and termination effects due to its macroscopic dimensions, one would thus expect a homogenized behavior on a diamond nanoelectrode array (NEA) or nanoelectrode ensemble (NEE) [131, 132]. This is due to the small grains of diamond films as well as a more effective termination of these small electrochemical active areas.

4.1 Production Procedures

NEAs and NEEs were fabricated using E-beam lithography and nanosphere lithography, respectively [131]. The following is the fabrication steps of NEAs [131]. On a 200 nm thin boron-doped nanocrystalline diamond film, a 200 nm thick SiO_2 is deposited. This oxide layer is structured using E-beam lithography with subsequent nickel deposition and SF_6 etching of SiO_2. In the next step, metal contacts are deposited using photolithography to allow electrical contact for electrochemical characterization. In the crucial step, a 140 nm thin insulating nanocrystalline diamond film is grown on the part of the boron-doped nanocrystalline diamond layer that is exposed to the CVD plasma and not protected by SiO_2 islands. With the removal of SiO_2 in hydrofluoric acid, the arrays of recessed boron-doped nanocrystalline diamond nanoelectrodes surrounded by insulating diamond (NEAs) are obtained. In the NEAs we fabricated the nanoelectrodes are distributed in a hexagonal order, having a well-defined radius of 250 nm and a distance of 10 μm next to other nanoelectrodes and an electrode density of 11×10^5 cm^{-2}. One SEM image of such a NEA is shown in Fig. 9a.

Nanosphere lithography was developed to fabricate NEEs [131]. Initially, a photolithography step is used to deposit metal contacts on diamond. Thereafter, the sample is immersed in a solution of SiO_2 spheres. The next step involves the growth of insulating diamond around the above mentioned spheres. Insulating diamond selectively grows on the area exposed to the plasma. After the removal of SiO_2 spheres in hydrofluoric acid, nanoelectrodes having a concave shape are fabricated. These nanoelectrodes thus have the same size as the diameter of SiO_2 spheres. Since the concentration of the SiO_2 solution is directly correlated to the density of spheres on the diamond surface as well as to the average distance of neighboring spheres, the density of those nanoelectrodes are controllable. The size of the nanoelectrodes can be well-defined by selecting market-available SO_2 spheres as

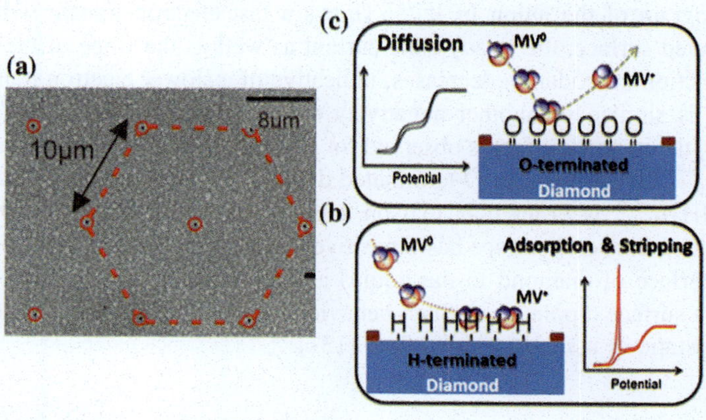

Fig. 9 **a** SEM image of a diamond NEA; Schematic plots of volumetric behavior of methyl viologen on **b** H- and **c** O-terminated diamond NEAs, respectively [131, 132]

required. For example, to obtain sigmoidal voltammograms, we chose a SiO_2 concentration of 9.55×10^8 cm^{-3}, corresponding to a surface density of 9.7×10^5 cm^{-2} and an average distance of neighboring spheres of $\sim$10 μm. The size of the nanoelectrodes in a NEE is about 175 nm. The density of nanoelectrodes is about 8.5×10^5 cm^{-2} [131].

4.2 Electrochemical Properties

Cyclic voltammetry of $Fe(CN)_6^{3-/4-}$, $Ru(NH_3)_6^{2+/3+}$ and $IrCl_6^{2-/3-}$ were conducted on the NEA and NEE in 0.1 M KCl solution. The scan rates were varied from few mV s^{-1} up to 10 V s^{-1}. At small scan rates (e.g., 20 mV s^{-1} for the NEA and 1 mV s^{-1} for the NEE), the voltammograms have mixed shapes, indicating partially overlapping diffusion hemispheres. Increasing the scan rates leads to typical steady-state sigmoidal voltammograms on both electrodes. The change is more distinct on the NEA than on the NEE [131]. Impedance was performed on the NEA and NEE in 0.1 M KCl at open circuit potentials. The redox couple of $Fe(CN)_6^{3-/4-}$ (1.0 mM) was added. Their impedance spectra have similar characteristics, exhibiting a large semicircle in the high-frequency regime and at low frequencies a transition to linear diffusion with unity slope (particularly observable for the NEA) [131]. A semicircle at high frequency regime is due to a three-dimensional hemispherical diffusion on the diamond NEA and NEE [144–146]. The transition at low frequencies represents the regime of overlapping diffusion hemispheres. These behaviors are similar with those obtained from voltammetry [131].

Moreover, the voltammetric response of $Ru(NH_3)_6^{2+/3+}$ and $IrCl_6^{2-/3-}$ on a diamond NEA show the dependence of surface termination on the charge of the analytes [131]. Please note that on planar macroscopic diamond electrodes both analytes show no dependence of electron transfer rate constants on the surface termination of diamond electrodes. On the H-terminated diamond NEAs, the voltammogram of the anion $IrCl_6^{2-/3-}$ shows a fast electron transfer while at the O-terminated surface, the steady-state current as well as the slope of the transition from reduction to oxidation decreases, indicative of a slower electron transfer. This tendency is similar for another negatively charged redox couple of $Fe(CN)_6^{3-/4-}$. However, the opposite effect is observed for the positively charged redox molecules $Ru(NH_3)_6^{2+/3+}$. That is, on an O-terminated diamond NEA, the electron transfer rate for $Ru(NH_3)_6^{2+/3+}$ is faster than that on a H-terminated surface. It is known that H-terminated diamond surface has a positive surface dipole layer ("positive" refers to the interface of diamond to the liquid) and the O-terminated surface results in a negative surface dipole layer. Such behaviors are therefore probably due to either an electrostatic or a site blocking effect [131].

4.3 Electrochemical Applications

Diamond NEAs have been applied for electrochemical sensing [130] and the investigation of surface-sensitive adsorption phenomena [132]. The adsorption of neutral methyl viologen (MV^0) was used as a model system. Diffusion-controlled processes manifest themselves as sigmoidal-shaped voltammograms on O-terminated diamond NEAs, whereas adsorption-controlled processes result in peaks in the voltammogram for H-terminated diamond NEAs. The change in the shapes of these voltammograms is due to the drastic changes that occur in the diffusion profiles during the transition. It alters from hemispherical diffusion on the O-terminated surface to thin-layer electrochemistry upon the adsorption on the H-terminated surface. In this way the de-convolution of diffusion-controlled current from adsorption-controlled current was conducted. By analysing anodic stripping process at high scan rates, the deposition of amorphous MV^0 was approved on H-terminated diamond NEAs. These results are schematically shown in Fig. 9b, c for H- and O-terminated diamond NEAs, respectively [132]. The types and the concentration of the buffer solutions were changed to alter the interaction of MV^0 with H-terminated diamond NEAs [132]. Increasing urea concentrations leads to the same impact on the adsorption of MV^0 as guanidine, which weakens hydrophobic interaction. This effect of ions on the interaction of MV^0 and the hydrophobic diamond surface is correlated with the Hofmeister series [132]. Subsequently, the adsorption of MV^0 on H-terminated diamond NEAs is controlled by hydrophobic interaction [132]. Therefore diamond NEA is ideal for the study of adsorption phenomena at the liquid-solid interface in voltammetry [132].

5 Summary and Outlook

Electrochemistry using diamond nanostructures, nanoparticles, nanoelectrodes, in other words, diamond nanoelectrochemistry, has been paid much attention in the fields of electrochemical sensors, energy, and electrocatalysts during the past years. In such electrochemical systems, nanostructured diamond (e.g., textures, wires, pores, nanoelectrodes, etc.) and diamond nanoparticles were used as the working electrode, instead of macro-sized diamond bulk electrode. From fundamental aspects of diamond nanoelectrochemistry, future activities should focus on the effect of surface termination of diamond at the nanoscale, diamond-to-graphite ratios, surface defects, and morphology (e.g., size, shape, etc.) effects of diamond nanostructures on their electrochemical properties in the absence and presence of redox probes. Coating these diamond nanostructures and nanoparticles with stable and electroactive modifiers is important for their applications for sensors, energy storage and conversion, and catalytic reactions. Doping diamond nanostructures and nanoparticles with dopants such as N and other atoms will widen their applications such as for electrocalaytic reactions (e.g., oxygen reduction/evolution reaction, CO_2 reduction reaction, etc.). Combination of diamond nanostructures with diamond nanoparticles, or the formation of their hybrid nanocomposites, for example, diamond nanowires based diamond nanoelectrodes, diamond nanoparticles coated diamond nanoelectrodes, will take full advantages of diamond as well as nanoelectrochemistry, leading to more novel concepts and applications [147, 148]. For example, on such hybrid nanocomposites, the investigation of diamond electrochemistry and electroanalysis at the nanoscale (e.g., capacitive current at a single nanowire, single molecule detection, etc.), which is hard to be realized on other electrode materials, will be feasible.

In conclusion, progress and achievements on electrochemistry using diamond nanostructures, nanoparticles and nanoelectrodes are summarized. Through the input from material scientists, chemists, physics, and engineers, more and nicer results in the fields of electrochemical properties and applications of diamond nanostructures and nanoparticles will be obtained in coming years. By showing and comparing the results published in literature, we believe this chapter will help the readers to know more how electrochemistry of diamond nanostructures and nanoparticles started as well as where and how it goes in future.

Acknowledgements The authors thank the financial support from German Research Foundation (DFG) under the project (YA344/1-1).

References

1. M. Iwaki, S. Sato, K. Takahashi, H. Sakairi, Electrical conductivity of nitrogen and argon implanted diamond. Nucl. Instrum. Methods Phys. Res. **209–210**, 1129 (1983). doi:10.1016/0167-5087(83)90930-4
2. Y.V. Pleskov, A.Y. Sakharova, M.D. Krotova, L.L. Bouilov, B.V. Spitsyn, Photoelectrochemical properties of semiconductor diamond. J. Electroanal. Chem. **228**, 19 (1987). doi:10.1016/0022-0728(87)80093-1
3. A. Fujishima, Y. Einaga, T.N. Rao, D.A. Tryk (eds.), *Diamond Electrochemistry* (Elsevier, Tokyo, 2005)
4. R.L. McCreery, Advanced carbon electrode materials for molecular electrochemistry. Chem. Rev. **108**, 2646–2687 (2008). doi:10.1021/cr068076m
5. E. Brillas, C.A. Martinez-Huitle (eds.), *Synthetic Diamond Films: Preparation, Electrochemistry, Characterization, and Applications* (Wiley, 2011)
6. N. Yang (ed.), *Novel Aspects of Diamond* (Springer, 2014)
7. N. Yang, W. Smirnov, C.E. Nebel, Diamond nanotextures: technologies, properties, and electrochemical applications, M. Chehimi, J. Pinson (eds.), *Applied Surface Chemistry of Nanomaterials* (NOVA Publisher, 2013), pp 33–54
8. Y. Yu, L. Wu, J. Zhi, Diamond nanowires: fabrication, structure, properties, and applications. Angew. Chem. Int. Ed. **53**(2014), 14326–14351 (2013). doi:10.1002/anie.10803
9. S. Szunerits, Y. Coffinier, R. Boukherroub, Diamond nanowires: a novel platform for electrochemistry and matrix-free mass spectrometry. Sensors **15**, 12573–12593 (2015). doi:10.3390/s150612573
10. R.W. Murray, Nanoelectrochemistry: metal nanoparticles, nanoelectrodes, and nanopores. Chem. Rev. **108**, 2688–2720 (2008). doi:10.1021/cr068077e
11. J.D. Wadhawan, R.G. Compton (eds.), *Electrochemistry*, vol. 11, (RSC Publisher, 2012)

12. M.V. Mirkin, S. Amemiya (eds.), *Nanoelectrochemistry*, (CRC Press, 2015)
13. N. Yang, H. Uetsuka, E. Osawa, C.E. Nebel, Vertically aligned nanowires from boron-doped diamond. Nano Lett. **8**, 3572–3576 (2008). doi:10.1021/nl801136h
14. W. Smirnov, A. Kriele, N. Yang, C.E. Nebel, Aligned diamond nano-wires: fabrication and characterisation for advanced applications in bio and electrochemistry. Diam. Relat. Mater. **18**, 186–189 (2009). doi:10.1016/j.diamond.2009.09.001
15. Y.S. Zou, Y. Yang, Y.L. Zhou, Z.X. Li, H. Yang, B. He, I. Bello, W.J. Zhang, Surface nanostructuring of boron-doped diamond films and their electrochemical performance. J. Nanosci. Nanotech. **11**, 7914–7919 (2011). doi:10.1016/j.diamond.2003.10.066
16. M. Wei, C. Terashima, M. Lv, A. Fujishima, Z.-Z. Gu, Boron-doped diamond nanograss array for electrochemical sensors. Chem. Commun. 3624–3626 (2009). doi:10.1039/B903284C
17. J. Shalini, K.J. Sankaran, C.L. Dong, C.Y. Lee, N.H. Tai, I.N. Lin, In situ detection of dopamine using nitrogen incorporated diamond nanowire electrode. Nanoscale **5**, 1159–1167 (2013). doi:10.1039/C2NR32939E
18. D. Luo, L. Wu, J. Zhi, Fabrication of boron-doped diamond nanorod forest electrodes and their application in nonenzymatic amperometric glucose sensing. ACS Nano **3**, 2121–2128 (2009). doi:10.1021/nn9003154
19. Q. Wang, P. Subramanian, M. Li, W.S. Yeap, K. Haenen, Y. Coffinier, R. Boukherroub, S. Szunerits, Non-enzymatic glucose sensing on long and short diamond nanowires electrodes. Electrochem. Commun. **34**, 286–290 (2013). doi:10.1016/j.elecom.2013.07.014
20. N. Yang, W. Smirnov, C.E. Nebel, Three-dimensional electrochemical reactions on tip-coated diamond nanowires with nickel nanoparticles. Electrochem. Commun. **27**, 89–91 (2013). doi:10.1016/j.elecom.2012.10.044
21. S. Szunerits, Y. Coffinier, E. Galopin, J. Brenner, R. Boukherroub, Preparation of boron-doped diamond nanowires and their application for sensitive electrochemical detection of tryptophan. Electrochem. Commun. **12**, 438–441 (2010). doi:10.1016/j.elecom.2010.01.014
22. Q. Wang, A. Vasilescu, P. Subramanian, A. Vezeanu, V. Andrei, Y. Coffinier, M. Li, R. Boukherroub, S. Szunerits, Simultaneous electrochemical detection of tryptophan and tyrosine using boron-doped diamond and diamond nanowires electrodes. Electrochem. Commun. **35**, 84–87 (2013). doi:10.1016/j.elecom.2013.08.010
23. M. Lv, M. Wei, F. Rong, C. Terashima, A. Fujishima, Z.-Z. Gu, Electrochemical detection of catechol based on as-grown and nanograss array boron-doped diamond electrodes. Electroanalysis **22**, 199–203 (2010). doi:10.1002/elan.200900296
24. N. Yang, R. Hoffmann, W. Smirnov, A. Kriele, C.E. Nebel, Interface properties of cytochrome c on nano-textured diamond surface. Diam. Relat. Mater. **20**, 269–273 (2011). doi:10.1016/j.diamond.2010.12.012
25. N. Yang, W. Smirnov, A. Kriele, R. Hoffmann, C.E. Nebel, Nano-textured surface for enhanced protein redox activity. Phys. Status Solidi A **207**, 2069–2072 (2010). doi:10.1002/pssa.201000085
26. N. Yang, R. Hoffmann, W. Smirnov, A. Kriele, C.E. Nebel, Direct electrochemistry of cytochrome c on nano-textured diamond surface. Electrochem. Commun. **12**, 1218–1221 (2010). doi:10.1016/j.elecom.2010.06.023
27. W. Wu, L. Bai, X. Lin, Z. Tang, Z.-Z. Gu, Nanograss array boron-doped diamond electrode for enhanced electron transfer from Shewanella loihica PV-4. Electrochem. Commun. **13**, 872–874 (2011). doi:10.1016/j.elecom.2011.05.025
28. W. Wu, Z.-Z. Gu, X. Liu, L. Bai, Z. Tang, Nanograss array boron-doped diamond electrode for toxicity sensor with Shewanella loihica PV-4 in bioelectrochemical systems. Sens. Lett. **12**, 191–196 (2014). doi:10.1166/sl.2014.3272
29. R. Hoffmann, A. Kriele, S. Kopta, W. Smirnov, N. Yang, C.E. Nebel, Intentional adsorption of cytochrome c to diamond. Phys. Status Solidi A **207**, 2073–2077 (2010). doi:10.1002/pssa.201000043

30. R. Hoffmann, A. Kriele, H. Obloh, N. Tokuda, W. Smirnov, N. Yang, C.E. Nebel, The creation of a biomimetic interface between boron-doped diamond and immobilized proteins. Biomaterials **30**, 7325–7332 (2011). doi:10.1016/j.biomaterials.2011.06.052
31. R. Hoffmann, H. Obloh, N. Tokuda, N. Yang, C.E. Nebel, Fractional surface termination of diamond by electrochemical oxidation. Langmuir **28**, 47–50 (2012). doi:10.1021/la2039366
32. P. Subramanian, J. Foord, D. Steinmueller, Y. Coffinier, R. Boukherroub, S. Szunerits, Diamond nanowires decorated with metallic nanoparticles: a novel electrical interface for the immobilization of histidinylated biomolecules. Electrochim. Acta **110**, 4–8 (2013). doi:10.1016/j.electacta.2012.11.010
33. P. Subramanian, A. Motorina, W.S. Yeap, K. Haenen, Y. Coffinier, V. Zaitsev, J. Niedziolka-Jonsson, R. Boukherroub, S. Szunerits, Impedimetric immunosensor based on diamond nanowires decorated with nickel nanoparticles. Analyst **139**, 1726–1731 (2014). doi:10.1039/C3AN02045B
34. I. Shpilevaya, W. Smirnov, S. Hirsz, N. Yang, C.E. Nebel, J.S. Foord, Nanostructured diamond decorated with Pt particles: preparation and electrochemistry. RSC Adv. **4**, 531–537 (2014). doi:10.1039/C3RA43763A
35. N. Yang, H. Uetsuka, E. Osawa, C.E. Nebel, Vertically aligned diamond nanowires for DNA sensing. Angew. Chem. Int. Ed. **47**, 5183–5185 (2008). doi:10.1002/anie.200801706
36. N. Yang, H. Uetsuka, C.E. Nebel, Biofunctionalization of vertically aligned diamond nanowires. Adv. Funct. Mater. **19**, 887–893 (2009). doi:10.1002/adfm.200990018
37. N. Yang, H. Uetsuka, O.A. Williams, E. Osawa, N. Tokuda, C.E. Nebel, Vertically aligned diamond nanowires: Fabrication, characterization, and application for DNA sensing. Phys. Stat. Sol. A **206**, 2048–2056 (2009). doi:10.1002/pssa.200982222
38. P. Subramanian, I. Mazurenko, Y. Coffinier, Y. Zaitsev, R. Boukherroub, S. Szunerits, Diamond nanowires modified with poly[3-(pyrrolyl)carboxylic acid] for the immobilization of histidine-tagged peptides. Analyst **139**, 4343–4349 (2014). doi:10.1039/C4AN00146J
39. C.E. Nebel, N. Yang, H. Uetsuka, E. Osawa, N. Tokuda, O. Williams, Diamond nano-wires, a new approach towards next generation electrochemical gene sensor platforms. Diam. Relat. Mater. **18**, 910–917 (2009). doi:10.1016/j.diamond.2008.11.024
40. Y. Yang, J.-W. Oh, Y.-R. Kim, C. Terashima, A. Fujishima, J.S. Kim, H. Kim, Enhanced electrogenerated chemiluminescence of a ruthenium tris(2,2′)bipyridyl/tripropylamine system on a boron-doped diamond nanograss array. Chem. Commun. **46**, 5793–5795 (2010). doi:10.1039/C0CC00773K
41. N. Yang, W. Smirnov, C.E. Nebel, Fabrication, properties and electrochemical applications of diamond nanostructures. MRS Proceedings **1511**, mrsf12-1511-ee07-01. doi:10.1557/opl.2012.1661
42. V.D. van Wyk, P.G.L. Baker, T. Waryo, E.I. Iwuoha, C. O'Sullivan, Electrochemical evaluation of a novel boron doped diamond (BDD) material for application as potential electrochemical capacitor. Anal. Lett. **44**, 2005–2018 (2011). doi:10.1080/00032719.2010.539735
43. S. Yu, N. Yang, H. Zhuang, J. Meyer, S. Mandal, O.A. Williams, I. Lilge, H. Schönherr, X. Jiang, Electrochemical supercapacitors from diamond. J. Phys. Chem. C **33**, 18918–18926 (2015). doi:10.1021/acs.jpcc.5b04719
44. K. Honda, T.N. Rao, D.A. Tryk, A. Fujishima, M. Watanabe, K. Yasui, H. Masuda, Electrochemical characterization of the nanoporous honeycomb diamond electrode as an electrical double-layer capacitor. J. Electrochem. Soc. **147**, 659–664 (2000). doi:10.1149/1.1393249
45. K. Honda, T.N. Rao, D.A. Tryk, A. Fujishima, M. Watanabe, K. Yasui, H. Masuda, Impedance characteristics of the nanoporous honeycomb diamond electrodes for electrical double-layer capacitor applications. J. Electrochem. Soc. **148**, A668–A679 (2001). doi:10.1149/1.1373450
46. M. Yoshimura, K. Honda, R. Uchikado, T. Kondo, T.N. Rao, D.A. Tryk, A. Fujishima, Y. Sakamoto, K. Yasui, H. Masuda, Electrochemical characterization of nanoporous

honeycomb diamond electrodes in non-aqueous electrolytes. Diam. Relat. Mater. **10**, 620–626 (2001). doi:10.1016/S0925-9635(00)00381-2

47. F. Gao, M. Wolfer, C.E. Nebel, Highly porous diamond foam as a thin-film micro-supercapacitor material. Carbon **80**, 833–840 (2014). doi:10.1016/j.carbon.2014.09.007
48. H. Zhuang, N. Yang, H. Fu, L. Zhang, C. Wang, N. Huang, X. Jiang, Diamond network: template-free fabrication and properties. ACS Appl. Mater. Interfaces **7**, 5384–5390 (2015). doi:10.1021/am508851r
49. F. Gao, G. Lewes-Malandrakis, M. Wolfer, W. Müller-Sebert, P. Gentile, D. Aradilla, T. Schubert, C.E. Nebel, Diamond-coated silicon wires for supercapacitor applications in ionic liquids. Diam. Relat. Mater. **51**, 1–6 (2015). doi:10.1016/j.diamond.2014.10.009
50. K. Siuzdak, R. Bogdanowicz, M. Sawczak M. Sobaszek, Enhanced capacitance of composite TiO_2 nanotube/boron-doped diamond electrodes studied by impedance spectroscopy. Nanoscale **7**, 551–558 (2015). doi:10.1039/C4NR04417G
51. H. Zanin, P.W. May, D.J. Fermin, D. Plana, S.M.C. Vieira, W.I. Milne, E.J. Corat, Porous boron-doped diamond/carbon nanotube electrodes. ACS Appl. Mater. Interfaces **6**, 990–995 (2014). doi:10.1021/am4044344
52. T. Kondo, Y. Kodama, S. Ikezoe, K. Yajima, T. Aikawa, M. Yuasa, Porous boron-doped diamond electrodes fabricated via two-step thermal treatment. Carbon **77**, 783–789 (2014). doi:10.1016/j.carbon.2014.05.082
53. C. Hebert, E. Scorsone, M. Mermoux, P. Bergonzo, Porous diamond with high electrochemical performance. Carbon **90**, 102–109 (2015). doi:10.1016/j.carbon.2015.04.016
54. H. Kato, J. Hees, R. Hoffmann, M. Wolfer, N. Yang, S. Yamasaki, C.E. Nebel, Diamond foam electrodes for electrochemical applications. Electrochem. Commun. **33**, 88–91 (2013). doi:10.1016/j.elecom.2013.04.028
55. F. Gao, C. Giese, G. Lewes-Malandrakis, C.E. Nebel, Porous diamond membrane fabricated by templated growth for electrochemical separation processes, *2015 ECS Meeting Abstract, 213-A*
56. S. Ruffinatto, H.A. Girard, F. Becher, J.C. Arnault, D. Tromson, P. Bergonzo, Diamond porous membranes: A material toward analytical chemistry. Diam. Relat. Mater. **55**, 123–130 (2015). doi:10.1016/j.diamond.2015.03.008
57. O.A. Williams (ed.), *Nanodiamond* (RSC Publisher, 2014)
58. I.A. Novoselova, E.N. Fedoryshena, E.V. Panov, A.A. Bochechka, L.A. Romanko, Electrochemical properties of compacts of nano-and microdisperse diamond powders in aqueous electrolytes. Phys. Solid State **46**, 748–750 (2004). doi:10.1134/1.1711465
59. J.B. Zang, Y.H. Wang, S.Z. Zhoa, L.Y. Bian, J. Lu, Electrochemical properties of nanodiamond powder electrodes. Diam. Relat. Mater. **16**, 16–20 (2007). doi:10.1016/j.diamond.2006.03.010
60. K.B. holt, Undoped diamond nanoparticles: origins of surface redox chemistry. Phys. Chem. Chem. Phys. **12**, 2048–2058 (2010). doi:10.1039/B920075D
61. J. Scholz, A.J. McQuillan, K.B. Holt, Redox transformations at nanodiamond surfaces revealed by in situ infrared spectroscopy. Chem. Commu. **47**, 12140–12142 (2011). doi:10.1039/C1CC14961J
62. K.B. Holt, C. Ziegler, J. Zang, J. Hu, J.S. Foord, Scanning electrochemical microscopy studies of redox processes at undoped nanodiamond surfaces. J. Phys. Chem. C **113**, 2761–2770 (2009). doi:10.1021/jp8038384
63. K.B. Holt, D.J. Caruana, E.J. Millan-Barrios, Electrochemistry of undoped diamond nanoparticles: accessing surface redox states. J. Am. Chem. Soc. **131**, 11272–11273 (2009). doi:10.1021/ja902216n
64. K.B. Holt, C. Ziegler, D.J. Caruana, J. Zang, E.J. Millan-Barrios, J. Hu, J.S. Foord, Redox properties of undoped 5 nm diamond nanoparticles. Phys. Chem. Chem. Phys. **10**, 303–310 (2008). doi:10.1039/B711049
65. A.T.S. Varley, M. Hirani, G. Harrison, K.B. Holt, Nanodiamond surface redox chemistry: influence of physicochemical properties on catalytic processes. Faraday Discuss. **172**, 349–364 (2014). doi:10.1039/C4FD00041B

66. W. hongthani, D.J. Fermin, Layer-by-layer assembly and redox properties of undoped HPHT diamond particles. Diam. Relat. Mater. **19**, 680–684 (2010). doi:10.1016/j.diamond.2010.01.039
67. D. Plana, J.J.L. Humphrey, K.A. Bradley, V. Celorrio, D.J. Fermin, Charge transport across high surface area metal/diamond nanostructured composites. ACS Appl. Mater. Interfaces **5**, 2985–2990 (2013). doi:10.1021/am302397p
68. J. Zang, Y. Wang, L. Bian, J. Zhang, F. Meng, Y. Zhao, S. Ren, X. Qu, Surface modification and electrochemical behaviour of undoped nanodiamonds. Electrochim. Acta **72**, 68–73 (2012). doi:10.1016/j.electacta.2012.03.169
69. Y. Wang, H. Huang, J. Zang, F. Meng, L. Dong, J. Su, Electrochemical behavior of fluorinated and aminated nanodiamond. Int. J. Electrochem. Sci. **7**, 6807–6815 (2012)
70. W. Hongthani, N.A. Fox, D.J. Fermin, Electrochemical properties of two dimensional assemblies of insulating diamond particles. Langmuir **27**, 5112–5118 (2011). doi:10.1021/la1045833
71. D. Plana, J.J.L. Humphrey, K.A. Bradley, V. Celorrio, D.J. Fermin, Charge transport across high surface area metal/diamond nanostructured composites. ACS Appl. Mater. Interfaces **5**, 2985–2990 (2013). doi:10.1021/am302397p
72. S. Shahrokhian, M. Ghalkhani, Glassy carbon electrodes modified with a film of nanodiamond–graphite/chitosan: application to the highly sensitive electrochemical determination of azathioprine. Electrochim. Acta **55**, 3621–3627 (2010). doi:10.1016/j.electacta.2010.01.099
73. S. Shahrokhian, M. Khafaji, Application of pyrolytic graphite modified with nano-diamond/graphite film for simultaneous voltammetric determination of epinephrine and uric acid in the presence of ascorbic acid. Electrochim. Acta **55**, 9090–9096 (2010). doi:10.1016/j.electacta.2010.08.043
74. L.H. Chen, J.B. Zang, Y.H. Wang, L.Y. Bian, Electrochemical oxidation of nitrite on nanodiamond powder electrode. Electrochim. Acta **53**, 3442–3445 (2008). doi:10.1016/j.electacta.2007.12.023
75. S. Shahrokhian, M. Bayat, Pyrolytic graphite electrode modified with a thin film of a graphite/diamond nano-mixture for highly sensitive voltammetric determination of tryptophan and 5-hydroxytryptophan. Microchim. Acta **174**, 361–366 (2011). doi:10.1007/s00604-011-0631-2
76. S. Shahrokhian, N.H. Nassab, Nanodiamond decorated with silver nanoparticles as a sensitive film modifier in a jeweled electrochemical sensor: application to voltammetric determination of thioridazine. Electroanalysis **25**, 417–425 (2013). doi:10.1002/elan.201200339
77. B. Habibi, M. Jahanbakhshi, Sensitive determination of hydrogen peroxide based on a novel nonenzymatic electrochemical sensor: silver nanoparticles decorated on nanodiamonds. J. Iran. Chem. Soc. **12**, 1431–1438 (2015). doi:10.1007/s13738-015-0611-2
78. L.Y. Bian, Y.H. Wang, J. Lu, J.B. Zang, Synthesis and electrochemical properties of TiO_2/nanodiamond nanocomposite. Diam. Relat. Mater. **19**, 1178–1182 (2010). doi:10.1016/j.diamond.2010.05.007
79. W. Zhao, J.J. Xu, Q.Q. Qiu, H.Y. Chen, Nanocrystalline diamond modified gold electrode for glucose biosensing. Biosens. Bioelectron. **22**, 649–655 (2006). doi:10.1016/j.bios.2006.01.026
80. M. Briones, E. Casero, M.D. Petit-Dominguez, M.A. Ruiz, A.M. Parra-Alfambra, F. Pariente, E. Lorenzo, L. Vazquez, Diamond nanoparticles based biosensors for efficient glucose and lactate determination. Biosens. Bioelectron. **68**, 521–528 (2015). doi:10.1016/j.bios.2015.01.044
81. A.I. Gopalan, K.-P. Lee, S. Komathi, Bioelectrocatalytic determination of nitrite ions based on polyaniline grafted nanodiamond. Biosens. Bioelectron. **26**, 1638–1643 (2010). doi:10.1016/j.bios.2010.08.042

82. J.-T. Zhu, C.-G. Shi, J.-J. Xu, H.-Y. Chen, Direct electrochemistry and electrocatalysis of hemoglobin on undoped nanocrystalline diamond modified glassy carbon electrode. Bioelectrochemistry **71**, 243–248 (2007). doi:10.1016/j.bioelechem.2007.07.002
83. A.I. Gopalan, S. Komathi, G.S. Anand, K.P. Lee, Nanodiamond based sponges with entrapped enzyme: a novel electrochemical probe for hydrogen peroxide. Biosens. Bioelectron. **46**, 136–141 (2013). doi:10.1016/j.bios.2013.02.036
84. E. Nicolau, J. Mendez, J.J. Fonseca, K. Griebenow, C.R. Cabrera, Bioelectrochemistry of non-covalent immobilized alcohol dehydrogenase on oxidized diamond nanoparticles. Bioelectrochemistry **85**, 1–6 (2012). doi:10.1016/j.bioelechem.2011.11.002
85. M. Briones, E. Casero, M.D. Petit-Dominguez, M.A. Ruiz, A.M. Parra-Alfambra, F. Pariente, E. Lorenzo, L. Vazquez, Diamond nanoparticles based biosensors for efficient glucose and lactate determination. Biosens. Bioelectron. **68**, 521–528 (2015). doi:10.1016/j.bios.2015.01.044
86. W.L. Zhang, K. Patel, A. Schexnider, S. Banu, A.D. Radadia, Nanostructuring of biosensing electrodes with nanodiamonds for antibody immobilization. ACS Nano **8**, 1419–1428 (2014). doi:10.1021/nn405240g
87. L.Y. Bian, Y.H. Wang, J.B. Zang, F.W. Meng, Y.L. Zhao, Detonation-synthesized nanodiamond as a stable support of Pt electrocatalyst for methanol electrooxidation. Int. J. Electrochem. Sci. **7**, 7295–7303 (2012)
88. L. Bian, Y. Wang, J. Zang, J. Yu, H. Huang, Electrodeposition of Pt nanoparticles on undoped nanodiamond powder for methanol oxidation electrocatalysts. J. Electroanal. Chem. **644**, 85–88 (2010). doi:10.1016/j.jelechem.2010.04.001
89. V. Celorrio, D. Plana, J. Florez-Montano, M.G. Montes de Oca, A. Moore, M.J. Lazaro, E. Pastor, D.J. Fermin, Methanol oxidation at diamond-supported Pt nanoparticles: effect of the diamond surface termination. J. Phys. Chem. C **117**, 21735–21742 (2013). doi:10.1021/jp4039804
90. J. Zang, Y. Wang, L. Bian, J. Zhang, F. Meng, Y. Zhao, R. Lu, X. Qu, S. Ren, Graphene growth on nanodiamond as a support for a Pt electrocatalyst in methanol electro-oxidation. Carbon **50**, 3032–3038 (2012). doi:10.1016/j.carbon.2012.02.089
91. J. Zang, Y. Wang, L. Bian, J. Zhang, F. Meng, Y. Zhao, X. Qu, S. Ren, Bucky diamond produced by annealing nanodiamond as a support of Pt electrocatalyst for methanol electrooxidation. Int. J. Hydrogen Energy **37**, 6349–6355 (2012). doi:10.1016/j.ijhydene.2012.01.034
92. J. Hu, X. Lu, J.S. Foord, Nanodiamond pretreatment for the modification of diamond electrodes by platinum nanoparticles. Electrochem. Comm. **12**, 676–679 (2010). doi:10.1016/j.elecom.2010.03.004
93. L. La-Torre-Riveros, R. Guzman-Blas, A.E. Mendez-Torres, M. Prelas, D.A. Tryk, C.R. Cabrera, Diamond nanoparticles as a support for Pt and PtRu catalysts for direct methanol fuel cells. ACS Appl. Mater. Interface **4**, 1134–1147 (2012). doi:10.1021/am2018628
94. E.A. Levashov, P.V. Vakaev, E.I. Zamulaeva, A.E. Kudryashov, V.V. Kurbatkina, D.V. Shtansky, A.A. Voevodin, A. Sanz, Disperse-strengthening by nanoparticles advanced tribological coatings and electrode materials for their deposition. Surf. Coat. Technol. **201**, 6176–6181 (2007). doi:10.1016/j.surfcoat.2006.08.134
95. L.-N. Tsai, G.-R. Shen, Y.-T. Cheng, W. Hsu, Performance improvement of an electrothermal microactuator fabricated using Ni-diamond nanocomposite. J. Microelectromech. Syst. **15**, 149–158(2006). doi:10.1109/JMEMS.2005.863737
96. Y. Wang, Y. Zhao, R. Lu, L. Dong, J. Zang, J. Lu, X. Xu, Nano titania modified nanodiamonds as stable electrocatalyst supports for direct methanol fuel cells. J. Electrochem. Soc. **162**, F211–F215 (2015). doi:10.1149/2.0051503jes
97. A. Moore, V. Celorrio, M.M. de Oca, D. Plana, W. Hongthani, M.J. Lazaro, D.J. Fermin, Insulating diamond particles as substrate for Pd electrocatalysts. Chem. Commun. **47**, 7656–7658 (2011). doi:10.1039/c1cc12387d

98. Y. Wang, J. Zang, L. Dong, H. Pan, Y. Yuan, Y. Wang, Graphitized nanodiamond supporting PtNi alloy as stable anodic and cathodic electrocatalysts for direct methanol fuel cell. Electrochim. Acta **113**, 583–590 (2013). doi:10.1016/j.electacta.2013.09.091
99. X. Lu, J.-P. Hu, J.S. Foord, Q. Wang, Electrochemical deposition of Pt-Ru on diamond electrodes for the electrooxidation of methanol. J. Electroanal. Chem. **654**, 38–43 (2011). doi:10.1016/j.jelechem.2011.01.034
100. L. La-Torre-Riveros, R. Guzman-Blas, A.E. Mendez-Torres, M. Prelas, D.A. Tryk, C.R. Cabrera, Diamond nanoparticles as a support for Pt and Pt-Ru catalysts for direct methanol fuel cells. ACS Appl. Mater. Interfaces **4**, 1134–1147 (2012). doi:10.1021/am2018628
101. L. La-Torre-Riveros, E. Abel-Tatis, A.E. Mendez-Torres, D.A. Tryk, M. Prelas, C.R. Cabrera, Synthesis of platinum and platinum-ruthenium-modified diamond nanoparticles. J. Nanopart. Res. **13**, 2997–3009 (2011). doi:10.1007/s11051-010-0196-8
102. R. Lu, J. Zang, Y. Wang, Y. Zhao, Microwave synthesis and properties of nanodiamond supported PtRu electrocatalyst for methanol oxidation. Electrochim. Acta **60**, 329–333 (2012). doi:10.1016/j.electacta.2011.11.068
103. T. Fujimura, V.Y. Dolmatov, G.K. Burkat, E.A. Orlova, M.V. Veretennikova, Electrochemical codeposition of Sn–Pb–metal alloy along with detonation synthesis nanodiamonds. Diam. Relat. Mater. **13**, 2226–2229 (2004). doi:10.1016/j.diamond.2004.06.009
104. G.R. Salazar-Banda, K.I.B. Eguiluz, L.A. Avaca, Boron-doped diamond powder as catalyst support for fuel cell applications. Electrochem. Commun. **9**, 59–64 (2006). doi:10.1016/j.elecom.2006.08.038
105. L.Y. Bian, Y.H. Wang, J. Lu, J.B. Zang, Synthesis and electrochemical properties of TiO_2/nanodiamond nanocomposite. Diam. Relat. Mater. **19**, 1178–1182 (2010). doi:10.1016/j.diamond.2010.05.007
106. Y. Zhao, Y. Wang, L. Dong, J. Huang, J. Zang, J. Lu, X. Xu, Core-shell structural nanodiamond@TiN supported Pt nanoparticles as a highly efficient and stable electrocatalyst for direct methanol fuel cells. Electrochim. Acta **148**, 8–14 (2014). doi:10.1016/j.electacta.2014.10.024
107. L.Y. Bian, Y.H. Wang, J.B. Zang, F.W. Meng, Y.L. Zhao, Microwave synthesis and characterization of Pt nanoparticles supported on undoped nanodiamond for methanol electrooxidation. Int. J. Hydrogen Energy **37**, 1220–1225 (2012). doi:10.1016/j.ijhydene.2011.09.118
108. C. Portet, G. Yushin, Y. Gogotsi, Electrochemical performance of carbon onions, nanodiamonds, carbon black and multiwalled nanotubes in electrical double layer capacitors. Carbon **45**, 2511–2518 (2007). doi:10.1016/j.carbon.2007.08.024
109. C. Portet, J. Chmiola, Y. Gogotsi, S. Park, K. Lian, Electrochemical characterizations of carbon nanomaterials by the cavity microelectrode technique. Electrochim. Acta **53**, 7675–7680 (2008). doi:10.1016/j.electacta.2008.05.019
110. Y.Q. Sun, Q. Wu, Y.X. Xu, H. Bai, C. Li, G.G. Shi, Highly conductive and flexible mesoporous graphitic films prepared by graphitizing the composites of graphene oxide and nanodiamond. J. Mater. Chem. **21**, 7154–7160 (2011). doi:10.1039/C0JM04434B
111. S. Park, K. Lian, Y. Gogotsi, Pseudocapacitive behavior of carbon nanoparticles modified by phosphomolybdic acid. J. Electrochem. Soc. **156**, A921–A926 (2009). doi:10.1149/1.3223964
112. J. Zang, Y. Wang, X. Zhao, G. Xin, S. Sun, X. Qu, S. Ren, Electrochemical synthesis of polyaniline on nanodiamond powder. Int. J. Electrochem. Sci. **7**, 1677–1687 (2012)
113. I. Kovalenko, D.G. Bucknall, G. Yushin, Detonation nanodiamond and onion-like-carbon-embedded polyaniline for supercapacitors. Adv. Func. Mater. **20**, 3979–3986 (2010). doi:10.1002/adfm.201000906
114. A. Kausar, R. Ashraf, M. Siddiq, Polymer/nanodiamond composites in Li-ion batteries: a review. Polymer-Plast Technol. **53**, 550–563 (2014). doi:10.1080/03602559.2013.854386

115. W. Gu, N. Peters, G. Yushin, Functionalized carbon onions, detonation nanodiamond and mesoporous carbon as cathodes in li-ion electrochemical energy storage devices. Carbon **53**, 292–301 (2013). doi:10.1016/j.carbon.2012.10.061
116. E. Tamburri, S. Orlanducci, V. Guglielmotti, G. Reina, M. Rossi, M.L. Terranova, Engineering detonation nanodiamond–polyaniline composites by electrochemical routes: structural features and functional characterizations. Polymer **52**, 5001–5008 (2011). doi:10.1016/j.polymer.2011.09.003
117. H. Gomez, M.K. Ram, F. Alvi, E. Stefanakos, A. Kumar, Novel synthesis, characterization, and corrosion inhibition properties of nanodiamond-polyaniline films. J. Phys. Chem. C **114**, 18797–18804 (2010). doi:10.1021/jp106379e
118. Z. Zhao, Y. Dai, Nanodiamond/carbon nitride hybrid nanoarchitecture as an efficient metal-free catalyst for oxidant- and steam-free dehydrogenation. J. Mater. Chem. A **2**, 13442–13451 (2014). doi:10.1039/C4TA02282C
119. A.E. Fischer, G.M. Swain, Preparation and characterization of boron-doped diamond powder: a possible dimensionally stable electrocatalyst support material. J. Electrochem. Soc. **152**, B369–B375 (2005). doi:10.1149/1.1984367
120. A. Ay, V.M. Swope, G.M. Swain, The physicochemical and electrochemical properties of 100 and 500 nm diameter diamond powders coated with boron-doped nanocrystalline diamond. J. Electrochem. Soc. **155**, B1013–B1022 (2005). doi:10.1149/1.2958308
121. J. Zang, Y. Wang, H. Huang, W. Tang, Electrochemical behavior of high-pressure synthetic boron doped diamond powder electrodes. Electrochim. Acta **52**, 4398–4402 (2007). doi:10.1016/j.electacta.2006.12.028
122. L. Cunci, C.R. Cabrera, Preparation and electrochemistry of boron-doped diamond nanoparticles on glassy carbon electrodes. Electrochem. Solid-State Lett. **14**, K17–K19 (2011). doi:10.1149/1.3532943
123. S. Heyer, W. Janssen, S. Turner, Y.-G. Lu, W.S. Yeap, J. Verbeeck, K. Haenen, A. Krueger, Toward deep blue nano hope diamonds: heavily boron-doped diamond nanoparticles. ACS Nano **8**, 5757–5764 (2014). doi:10.1021/nn500573x
124. A.A.V.M. Swope, G.M. Swain, The Physicochemical and electrochemical properties of 100 and 500 nm diameter diamond powders coated with boron-doped nanocrystalline diamond. J. Electrochem. Soc. **155**, B1013–B1022 (2005). doi:10.1149/1.2958308
125. G.R. Salazar-Banda, K.I.B. Eguiluz, L.A. Avaca, Boron-doped diamond powder as catalyst support for fuel cell applications. Electrochem. Comm. **9**, 59–64 (2007). doi:10.1016/j.elecom.2006.08.038
126. C.Y. Ko, J.H. Huang, S. Raina, W.P. Kang, A high performance non-enzymatic glucose sensor based on nickel hydroxide modified nitrogen-incorporated nanodiamonds. Analyst **138**, 3201–3208 (2013). doi:10.1039/C3AN36679K
127. N. Spătaru, X. Zhang, T. Spătaru, D.A. Tryk, A. Fujishima, Anodic Deposition of RuO x•nH2O at conductive diamond films and conductive diamond powder for electrochemical capacitors. J. Electrochem. Soc. **155**, D73–D77 (2008). doi:10.1149/1.2804379
128. S.K. Lee, M.J. Song, J.H. Kim, T.S. Kan, Y.K. Lim, J.P. Ahn, D.S. Lim, 3D-networked carbon nanotube/diamond core-shell nanowires for enhanced electrochemical performance. NPG Asia Mater. **6**, e115 (2014). doi:10.1038/am.2014.50
129. C. Dincer, E. Laubender, J. Hees, C.E. Nebel, G. Urban, J. Heinze, SECM detection of single boron doped diamond nanodes and nanoelectrode arrays using phase-operated shear force technique. Electrochem. Commun. **24**, 123–127 (2012). doi:10.1016/j.elecom.2012.08.005
130. C. Dincer, R. Ktaich, E. Lauberder, J.J. Hees, J. Kieninger, C.E. Nebel, J. Heinze, G.A. Urabn, Nanocrystalline boron-doped diamond nanoelectrode arrays for ultrasensitive dopamine detection. Electrochim. Acta. **185**, 101–106 (2015). doi:10.1016/j.electacta.2015.10.113
131. J. Hees, R. Hoffmann, A. Kriele, W. Smirnov, H. Obloh, K. Glorer, B. Raynor, R. Driad, N. Yang, O.A. Williams, C.E. Nebel, Nanocrystalline diamond nanoelectrode arrays and ensembles. ACS Nano **5**, 339–3346 (2011). doi:10.1021/nn2005409

132. J. Hees, R. Hoffmann, N. Yang, C.E. Nebel, Diamond nanoelectrode arrays for the detection of surface sensitive adsorption. Chem. Euro. J. **19**, 11287–11292 (2013). doi:10.1002/chem.201301763
133. F. Gao, C.E. Nebel, Diamond nanowire forest decorated with nickel hydroxide as a pseudocapacitive material for fast charging-discharging. Phys. Status. Solidi. A. **212**, 2533–2538 (2015). doi:10.1002/pssa.201532131
134. F. Gao, C.E. Nebel, Diamond-based supercapacitors: realization and properties. ACS Appl. Mater. Interfaces (2016). doi:10.1021/acsami.5b07027
135. F. Gao, R. Thomann, C.E. Nebel, Aligned Pt-diamond core-shell nanowires for electrochemical catalysis. Electrochem. Commun. **50**, 32–35 (2015). doi:10.1016/j.elecom.2014.11.006
136. Q. Wang, N. Plylahan, M.V. Shelke, R.R. Devarapalli, M. Li, P. Subramanian, T. Djenizian, R. Boukherroub, S. Szunerits, Nanodiamond particles/reduced graphene oxide composites as efficient supercapacitor electrodes. Carbon **68**, 175–184 (2014). doi:10.1016/j.carbon.2013.10.077
137. N. Yang, J.S. Foord, X. Jiang, Diamond electrochemistry at the nanoscale: A review. Carbon **99**, 90–110 (2016). doi:10.1016/j.carbon.2015.11.061
138. Y. Liu, S. Chen, X. Quan, H. Yu, Efficient electrochemical reduction of carbon dioxide to acetate on nitrogen-doped nanodiamond. J. Am. Chem. Soc. **137**, 11631–11636 (2015). doi:10.1021/jacs.5b02975
139. D. Aradilla, F. Gao, G. Lewes-Malandrakis, W. Muller-Sebert, D. Gaboriau, P. Gentile, B. Iliev, T. Schubert, S. Sadki, G. Bidan, C.E. Nebel. A step forward into hierarchically nanostructured materials for high performance micro-supercapacitors: Diamond-coated SiNW electrodes in protic ionic liquid electrolyte. Electrochem. Commun. **63**, 34–38 (2016). doi:10.1016/j.elecom.2015.12.008
140. T.T. Thanh, H. Ba, L. Truong-Phuoc, J.-M. Nhut, O. Ersen, D. Begin, I. Janowska, D.L. Nguyen, P. Granger, C. Pham-Huu, A few-layer graphene–graphene oxide composite containing nanodiamonds as metal-free catalysts. J. Mater. Chem. A. **2**, 11349–11357 (2014). doi:10.1039/C4TA01307G
141. H. Krysova, L. Kavan, Z.V. Zivcova, W.S. Yeap, P. Verstappen, W. Maes, K. Haenen, F. Gao, C.E. Nebel, Dye-sensitization of boron-doped diamond foam: champion photoelectrochemical performance of diamond electrodes under solar light illumination. RSC Adv. **5**, 81069–81077 (2015). doi:10.1039/C5RA12413A
142. D.W.M. Arrigan, Nanoelectrodes, nanoelectrode arrays and their applications. Analyst **129**, 1157–1165 (2004). doi:10.1039/B415395M
143. R.G. Compton, G.G. Wildgoose, N.V. Rees, I. Streeter, R. Baron, Design, fabrication, characterisation and application of nanoelectrode arrays. Chem. Phys. Lett. **459**, 1–17 (2008). doi:10.1016/j.cplett.2008.03.095
144. J. Guo, E. Lindner, Cyclic voltammograms at coplanar and shallow recessed microdisk electrode arrays: guidelines for design and experiment. Anal. Chem. **81**, 130–138 (2009). doi:10.1021/ac801592j
145. M. Fleischmann, S. Pons, J. Daschbach, The ac impedance of spherical, cylindrical, disk, and ring microelectrodes. J. Electroanal. Chem. **317**, 1–26 (1991). doi:10.1016/0022-0728(91)85001-6
146. M. Fleischmann, S. Pons, The behavior of microdisk and microring electrodes. Mass transport to the disk in the unsteady state: the ac response. J. Electroanal. Chem. **250**, 277–283 (1988). doi:10.1016/0022-0728(88)85169-6
147. N. Yang, S.R. Waldvogel, X. Jiang, Electrochemistry of carbon dioxide on carbon electrodes. ACS Appl. Mater. Interfaces (2016). doi:10.1021/acsami.5b09825
148. N. Yang, G.M. Swain, X. Jiang, Nanocarbon electrochemistry and electroanalysis: current status and future perspectives. Electroanalysis **28**, 27–34 (2016). doi:10.1002/elan.201500577

Carbon-Based Nanostructures for Matrix-Free Mass Spectrometry

Yannick Coffinier, Rabah Boukherroub and Sabine Szunerits

Abstract Matrix-assisted laser desorption/ionization mass spectrometry (MALDI-MS) has become a widespread analytical tool for peptides, proteins and most other biomolecules. However, due to a competitive desorption of parasitic ions from the matrix, it is difficult to detect low molecular weight compounds (<700 Da). To enable desorption/ionization of small molecules, techniques operating in absence of an organic matrix were developed. These techniques known as surface assisted laser desorption/ionization mass spectrometry (SALDI-MS) rely on the use of nanostructured surfaces as laser desorption/ionization-assisted material. As compared to traditional MALDI-MS, SALDI-MS offers several advantages such as the ability to detect small molecules (<700 Da), easy sample preparation, low noise background, high salt tolerance and fast data collection. Carbon-based interfaces such as carbon-like graphite, carbon nanotubes, fullerenes or amorphous carbon have been employed as SALDI substrates for the detection of small macromolecules such as synthetic polymers and biomolecules. While the drawback of fullerenes and their derivatives is the general limited sensitivity, carbon nanotubes, which exhibit high sensitivities, are hardly soluble in aqueous solutions, limiting their use in bioanalytical applications. More recently, diamond-like carbon (DLC) and diamond nanowires have been successfully introduced as SALDI interfaces. This chapter summarizes recent developments in the use of carbon-based materials for SALDI-MS. A particular emphasis will be put on the use of diamond nanowires as novel SALDI substrates.

Keywords Surface assisted laser desorption/ionization mass spectrometry (SALDI-MS) · Carbon-based nanostructures · Diamond nanowires · Small molecules · Biomolecules

---

Y. Coffinier · R. Boukherroub · S. Szunerits (✉)
Institut d'Electronique, de Microélectronique et de Nanotechnologie (IEMN, UMR 8520), Cité Scientifique, Avenue Poincaré, B.P. 60069, 59652 Villeneuve d'Ascq, France
e-mail: sabine.szunerits@iri.univ-lille1.fr

N. Yang et al. (eds.), *Carbon Nanoparticles and Nanostructures*,
Carbon Nanostructures, DOI 10.1007/978-3-319-28782-9_10

1 Introduction

Mass spectrometry (MS) is widely accepted as a 'gold-standard' method for identifying chemicals or biological products. It is nowadays applied in highly diversified domains like those directly or indirectly tied to healthcare or regulation-driven demand, such as drug development, diagnostics, food and environmental safety testing. Among the different methodologies, matrix-assisted laser desorption/ionization mass spectrometry (MALDI-MS), first introduced in 1988 by Hillenkampf and Karas [1], constitutes one of the soft ionization techniques that provides the nondestructive vaporization and ionization of analytes using UV-absorbing organic matrices. The utility of MALDI-MS for protein and peptide analysis lies in its ability to provide high accurate molecular-weight information on intact molecules. Acquiring optimum MALDI data depends not only on the functional and structural properties of the analyte itself, but also largely on the choice of suitable matrices. As the matrix is the medium by which the analyte is transported to the gaseous phase and provides the conditions that make ionization possible, the matrix and the sample-matrix preparation procedure greatly influence the quality of MALDI mass spectra. Organic matrices such as 2,5-dihydroxylbenzoic acid, α-cyano-4-hydroxycinnamic acid or 3,5-dimethoxy-4-hydroxycinnamic acid are commonly used. However, the choice of the matrix remains an empirical issue. Although MALDI-MS has been successfully used to analyze large molecules [2–4], it has rarely been applied to low-molecular weight compounds (<700 Da) as a large number of matrix ions appear in the low-mass range.

Using inorganic species as the assisting material in MALDI-MS is an alternative approach to avoid the problems arising in conventional MALDI analysis using organic compounds as matrix systems. Tanaka et al. [5] investigated a suspension of cobalt nanoparticles of 30 nm in size mixed with glycerol as inorganic matrix for the laser desorption/ionization of small analytes. Since then, several alternative inorganic matrices have been proposed as assisting materials [6] in a process that was named surface assisted laser desorption/ionization mass spectrometry (SALDI-MS) by Sunner et al. [7]. The basic principle of SALDI-MS analysis is schematically outlined in Fig. 1. Two different approaches are currently used. One, a true matrix-free approach, is based on the deposition of the analyte solution on a nanostructured surface, previously deposited on MALDI-plate (Fig. 1a). The other approach uses a mixture of the analyte together with SALDI nanostructures for analysis (Fig. 1b). There are indeed several advantages associated with inorganic matrix systems for MALDI-MS analysis, such as the elimination of co-crystallization of the sample with an organic matrix, limiting formation of hot spots. Moreover, the absence of peaks from the matrix in the low mass range encountered in classical MALDI-MS allows increasing the background of the spectra and lowering the sensitivity of detection of small molecules.

One of the most notable developments in the use of solid interfaces as matrix for LDI-MS has certainly been desorption/ionization on porous silicon [8] (DIOS) for the analysis of low molecular mass compounds. Porous silicon substrates are easily

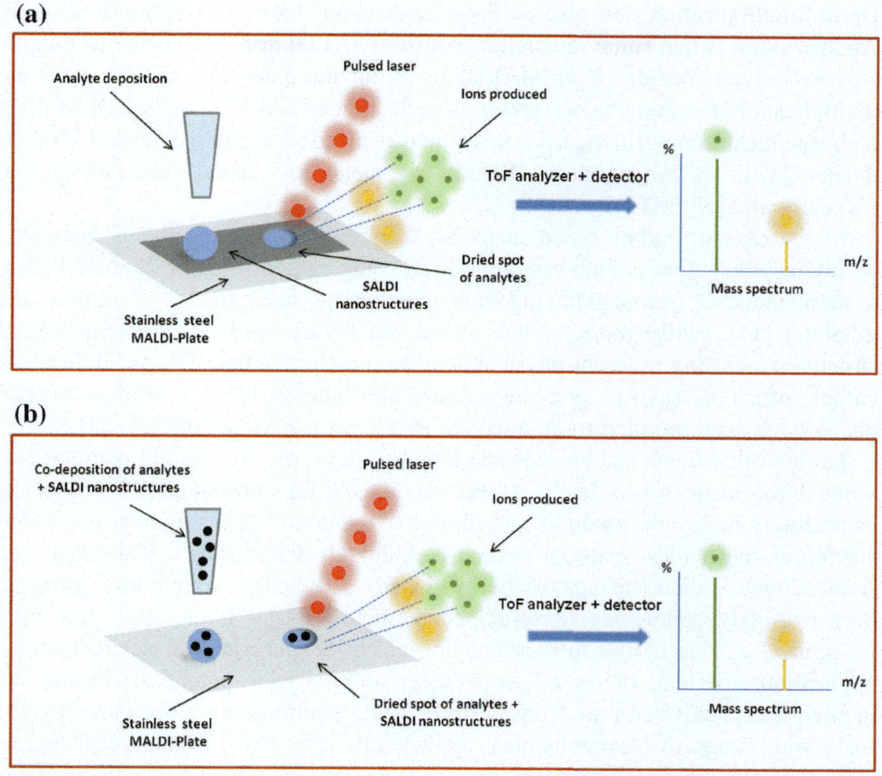

Fig. 1 Schematic presentation of the principle of surface assisted laser desorption/ionization mass spectrometry (SALDI-MS): LDI-MS process for analyte detection with analyte deposition on SALDI based nanostructures (**a**), and co-deposition of analyte and SALDI-based material (**b**)

obtained by electrochemical anodization of crystalline silicon in HF-based solutions. Using the same mass spectrometer as for MALDI (i.e. N_2 laser $\lambda = 327$ nm and time-of-flight (TOF) analyzer), an optimal performance for molecules less than 3 kDa was achieved [9]. It is believed that the low thermal conductivity of porous silicon confines heat close to the surface, thus increasing locally the temperature and resulting in sensitivities exceeding those of classical MALDI-MS [10]. Furthermore, by using silicon nanowires rather than porous silicon allowed to decrease the laser irradiation energy required for analyte desorption, resulting in an extremely gentle ionization process [11, 12] for SALDI-MS.

Along with nanostructured silicon, other semiconductors with good UV absorbance are in addition promising candidates for SALDI-MS [6]. The large number of nanomaterials, together with different preparation techniques, make classification into well-defined LDI categories and distinguishing it from MALDI somewhat difficult. This situation is not helped by the large number of different acronyms that are currently used for SALDI-MS, many of which describe similar or essentially the same methodologies. For instance, PALDI is referred as Particles-Assisted Laser

Desorption/Ionization, but also as Polymer-Assisted Laser Desorption/Ionization. Another point is that some researchers consider SALDI as a matrix-free technique, whereas others consider it as MALDI using an inorganic matrix. To avoid the multiplication of acronyms, we prefer classification of SALDI-based matrices upon their chemical composition. Like this, three main groups emerge as SALDI candidates [13]: (i) semiconductor-based, (ii) metal and metal-oxide based and (iii) carbon-based SALDI structures.

In the case of carbon-based matrices, this classification scheme is, however, highly insufficient, as carbon can range from being a pure insulator (diamond) over a semiconductor (semiconducting nanotubes) to a metal (metal nanotubes and graphite) [14]. Furthermore, carbon atoms can be arranged into different crystal structures revealing different physical and chemical properties. The sp^2 hybridized carbon structures such as graphite, carbon nanotubes (CNTs), fullerenes or graphene have been reported to be effective matrices for SALDI-MS [15–23].

On the other hand, sp^3 hybridized carbon in the form of diamond particles and wires have also shown lately to be useful SALDI substrates [24]. With its exceedingly high bulk modulus and hardness, diamond has historically been considered as the hardest material available. Although diamond is a wide band gap semiconductor, diamond can be easily doped, particularly with boron atoms to prepare highly conductive boron-doped diamond (BDD) films. BDD has been recognized as one of the most promising electrode material for electrochemical applications, because of its unique features such as wide potential window and negligible capacitive current. Consequently, BDD electrodes have been investigated for a wide range of electrochemical applications [25, 26]. The analytical performance of BDD-based biosensors was considerably improved by increasing the surface area through structuring BDD interfaces into nanowires [27–30]. Luo et al. [31] showed that diamond nanoforest electrodes are electrochemically active towards glucose sensing under basic conditions. The importance of the diamond nanowires' length for the non-enzymatic detection of glucose was recently underlined by our group [32]. The electrochemical activity of 3 μm long diamond nanowires is 3.5 times larger compared to as-grown planar BDD electrode and 2.4 times higher than 1 μm long diamond nanowires resulting in increased sensitivity to glucose [32]. Such diamond nanowires proved to be also an excellent electrical interface for the detection of tryptophan and/or tyrosine, two aromatic acids that are important precursors for adrenaline, dopamine or melatonine [32–34]. Next to these examples where the intrinsic properties of diamond nanowires were exploited for sensitive and selective sensing, polymer [35] and metallic nanoparticles coated diamond nanowires [30, 36, 37] have shown additional interest for electrochemical sensing.

While the use of diamond nanowires electrodes for electrochemical sensing has been recently reviewed [38], the intent of this article is to make the reader more familiar with recent developments for the preparation of diamond nanostructures and the use of such structures as LDI-MS surfaces. To put the work on diamond nanowires into perspective, first, the current state of the art on carbon-based nanomaterials such as graphite and carbon nanotubes (CNTs), for the detection of analytes by LDI-MS will be described. Thereafter, the synthetic routes for diamond nanowires and their use as SALDI-MS substrates will be presented around some pertinent examples.

2 Carbon-Based Materials for SALDI

Carbon-based nanomaterials can been considered ideal SALDI-MS matrices. They provide high-affinity binding sites for peptides and proteins by virtue of their chemical functionalities where hydrophobic and hydrophilic forces are taking place. To minimize probable contamination from these materials due to ionization, extended surface modification strategies can be adopted to decrease the tendency to fly off into the instrument when a laser pulse was applied. This chemical diversity together with remarkable charge mobility and universal optical absorption properties make carbon-based structures the focus for SALDI (Table 1).

Table 1 Carbon-based materials for LDI-MS

Carbon material	Analysis	Comments	References
Colloidal graphite	Fatty acids, flavonoids (negative mode, fmolar range)		[47]
Colloidal graphite	Peptides and proteins		[7]
Colloidal graphite aerosols	Different small molecules (fatty acid, carbohydrates, flavonoids, phospholipids, etc.)	In isopropyl alcohol	[45–47]
Activated carbon	Peptides and proteins	Deposited on TLC plates	[39]
Graphite plate	Polymers, fatty acid		[40–42]
Pencil line	Peptides	On silica gel matrix	[44]
Pyrolitic graphite sheet (PGS)	Environmental related compounds (e.g. bisphenol A, 4-choloraniline, etc.)		[48]
CNTs	Peptides and proteins profiling	Derivatized with iminodiacetic acid (IDA), and loaded with copper	[49]
CNTs-citric acid	selective capture of positively charged species such as cytochrome c	Citric acid-treated	[50]
Oxidized CNTs	Amino acids, small molecules, oligosaccharides		[22, 23, 53]
CNTS	Amino acids		[54]
CNTs	Herbal components		[59, 60]

(continued)

Table 1 (continued)

Carbon material	Analysis	Comments	References
Oxidized CNTs	Organotin, organoarsenic, organophosphate, organomercury, alkyl phenol, anilines, atrazine derivatives, naphthol derivatives, chlormequat chloride, diquat, paraquat, humic acid, and decabromodiphenyl oxide		[61]
Oxidized CNTs	Amino acids, Jatrorrhizine		[62]
SWNTs	Peptides	$(CH_2)_2COOH$ modified	[21]
Fullerenes	Proteins		[56]
Fullerenes	Proteins	Hexa(sulfonbutyl) modified	[15]
Fullerenes	N-terminal sulfonated peptides	N-dimethyl pyrrolidinium iodide modified	[57]
Fullerenes	Peptides	$(CH_2)_2COOH$ modified	[21]
Fullerenes	Proteins	Modified	[58]
Diamond particles	Albumin, myoglobin, cytochrome C	Presence of HCCA	[63]
Nanocrystalline diamond (NCD)	Protein profiling	Activation by glycidyl methacrylate and introduction of IDA-Cu^{2+} complex	[64]
Diamond nanocrystals (100 nm diam.)	Solid phase extraction of DNA and MS detection	Poly-lysine coating on carboxylated diamond nanocrystals	[65]
Diamond nanowires	Peptides, verapamil (zeptomole range of detection), cortisone	Octadecyl-terminated	[24]

2.1 *Graphite*

Graphite particles were the first carbon-based materials introduced as SALDI-MS matrix [7]. Laser desorption/ionization time-of-flight (TOF) mass spectra of peptides and proteins, as well as low molecular weight analytes, could be obtained by using a pulsed UV laser (337 nm) to irradiate mixtures of 2–150 μm graphite particles and solutions of the analytes in glycerol (Fig. 2a). At optimum conditions, mass spectra with signal intensities as good as those in conventional MALDI with very few background peaks, even at low mass were obtained. For example, Fig. 2a shows the low-mass region of a graphite SALDI spectrum of glycerol. The ion at m/z = 93 is protonated glycerol. This ion and the fragments at m/z = 75, 57, 45, 29 and 19 are all prominent in the spectrum. The other fragments are due to alkali salt impurities and alkali ions/glycerol adducts. A series of peaks due to carbon clusters,

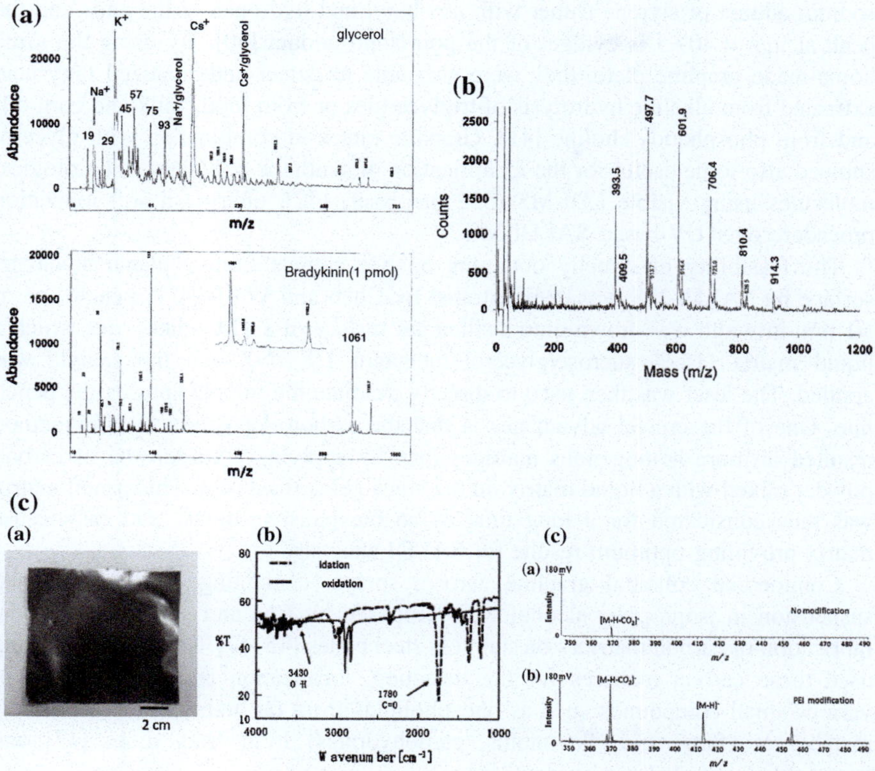

Fig. 2 **a** Graphite SALDI of glycerol and Bradykinin (reprint with permission [7]). **b** Graphite plate assisted SALDI-MS analysis of low-mass polystyrene by depositing 1 μL (1 mg/mL) of the polymer sample on the graphite plate without additional matrix (reprint with permission from [40]). **c** Image of pyrolitic graphite sheet (PGS) (*a*) together with FT-IR ATR spectrum and (*c*) Mass spectrum of PFPA using non-modified and PEI-modified PGS in negative ion mode. The sample concentration was 0.1 mg/ml (reprint with permission from [48])

$C_nH_m{}^+$ with n = 1–25 and beyond, and m = 0–4 are observed, being low in intensity. The technique was also suitable for the analysis of biomolecules such as bradykinin in glycerol (1 pmol).

The same group further developed activated carbon particles deposited on thin-layer chromatography (TLC) plates and used them for SALDI-MS analysis [39]. Indeed, one of the advantages of graphite is that it can be used either in solution (graphite/liquid mixture) or as dry graphite (without liquid). Kim et al. [40] investigated, for example, a graphite plate as substrate and matrix for the SALDI-MS detection of polystyrene and poly(methylsilsesquioxane) (PMSSQ) materials derived from methyltriethoxysilane (MTES) in the range of 100–1000 Da. Figure 2b shows a graphite plate SALDI mass spectrum of a polystyrene standard in the low mass range. The styrene oligomers were separated by 104.3 mass units, related to $[(C_6H_5)CHCH_2]$ repeat units. The peak at m/z = 393.5 is the sodium adduct of styrene trimer with *tert*-butyl and hydrogen at the ends and the peak at m/z = 409.5 is evident of the potassium adduct [40]. By using the same home-made graphite plate, they were also able to detect underivatized fatty acid extracted from alkaline hydrolysis of triglycerides or from food (milk and corn oil) and from phosphatidylcholine [41]. Graphite plates in combination with glycerol showed also to be useful for the identification of synthetic polymers and biological molecules using visible LDI-MS (532 nm laser) [42], offering a soft ionization procedure over UV-based SALDI-MS.

The feasibility of directly obtaining analyte signals from a planar silica gel surface by SALDI-MS was demonstrated by Chen and Wu [43]. A pencil line of ≈1 mm in width was drawn onto a silica gel surface of a TLC plate, over which a liquid matrix (15 % sucrose/glycerol–methanol 1:1 v/v) with the analyte was applied. The laser was then used to directly irradiate the sample spot on the pencil line. One of the crucial advantages is that the carbon deposition by this method resulted in more homogenous matrices than by applying a suspension of carbon powder mixed with a liquid matrix on the silica gel surface [44]. This pencil matrix was thus considered for a long time as an inexpensive, quick, and easy-to-use matrix providing optimum results for SALDI analysis.

Commercial colloidal graphite aerosol sprays, containing colloidal graphite suspension in isopropylic alcohol were proposed by Cha and co-workers for the fabrication of carbon matrices on stainless steel plates [45–47]. The group of Zhang used these carbon matrices for the detection, localization and identification of various small compounds such as phospholipids from rat brain tissue (slice) [45], metabolites (fatty acid, flavonoïds, carbohydrates) from fruit slices [47] and flavonoïds from plant tissue wax [46].

Highly oriented graphite film known under the name Pyrolitic Graphite Sheet (PGS) has been employed by Kawasaki to analyze low-mass analytes by SALDI-MS [48]. PGS is a synthetic, uniform and highly oriented graphite polymer film that provides thermal management/heat-sinking in limited spaces (Fig. 2c). It shows high thermal conductivity (600–1500 Wm^{-1} K), being twice that of copper and ten times of ordinary graphite. The large heat conduction anisotropy of PGS allowed the rapid heating of the PGS surface by the laser radiation, which is crucial for efficient LDI-MS. Various environmentally related compounds (e.g. bisphenol A, 4-chloroaniline, perfluorooctanoic acid (PFOA), etc.) were detected using this approach [48]. The influence of surface modification of PGS on the signal intensity for perfluorooctanoic acid (PFOA) analysis was in addition demonstrated (Fig. 2c). The signal intensity of PFOA detected on PGS modified with polyethyleneimine (PEI) showed a ten-fold increase over that obtained from desorption/ionization on porous silicon (DIOS), the reference SALDI surface. PFOA could be quantified using SALDI and a wide linear dynamic range response from 20 to 1000 ppb was observed [48]. The advantages of PGS as SALDI matrix are therefore linked to the large film morphology of the PGS, that low volume of analyte solution can be directly deposited onto PGS, resulting in good reproducibility of intra and inter-sample spots, making quantitative analysis possible. Surface modification of the PGS is in addition easy and various functionalized surfaces to concentrate target analytes can be obtained. All these features make PGS a suitable material for the LDI-MS of small molecules.

2.2 Carbon Nanotubes (CNTs)

The development of reliable methods for the fabrication of carbon nanotubes (CNTs) has enabled the use of this material in SALDI-MS analysis [14, 18, 49, 50]. Wang and co-workers showed that CNTs of 250 nm in diameter and ca. 60 μm in length, which can disperse very well in solution, can be employed as SALDI matrix [51] (Fig. 3a). The same group reported a year later, that adding a citrate buffer provides an extra proton-source and also reduced the intensities of sodium adduct ions of analytes during SALDI-MS analysis [52]. As the surface of CNTs can be easily modified in numerous ways, they formed anionic CNTs through immersion into citric acid and used the treated CNTs as affinity probes for the selective capture of positively charged species [50]. The potential of citrate-treated CNTs was demonstrated through the detection of very low concentrations of insulin. The pI value of insulin is 5.3 and thus it has a net positive charge in solutions at pH 4. Figure 3a displays the SALDI-MS spectra of insulin solution (10^{-8} M) at pH 4 and pH 10. The protonated pseudomolecular ion of insulin (MH^+) at m/z = 5734 dominates the spectrum at pH 4, while no insulin peak was detected at pH 10 [50]. Cytochrome c with a molecular weight of 1–12 kDa is the largest molecule detected by this technique.

Oxidized CNTs have been widely tested as matrix substance for the analysis of small molecules by SALDI-MS. Oxidized CNTs allowed for example the analysis of mixtures of amino acids containing Asn, Glu, His, and Phe [22]. As shown in Fig. 3b, mass spectra of these amino acids using CNTs and oxidized CNTs as matrix substrates displayed well-separated peaks as $[M + H]^+$, $[M + Na]^+$ and $[M + K]^+$. Intensity and S/N ratios were considerably improved for the oxidized CNTs matrix compared to simple CNTs due to the removal of many impurities during the oxidation procedure together with its better dispersibility in water (Fig. 3a). Pan et al. [53] developed in addition an approach where CNTs served both as the adsorbent in solid-phase extraction and as matrix for LDI-MS analysis for small molecules (β-Arg, N-Alpha-Benzoyl-L-Arginine ester ethyl (BAEE), propranolol, and Leu-Tyr) in solution. The detection limit for analytes could be improved 10- to 100-fold with this approach. Subsequently, twenty common amino acids have been analyzed by LDI-MS by Zhang et al. [54]. The mass spectra of a mixture of seven amino acids, including Phe, Trp, Tyr, Leu, Ile, Val, and Met, showed good signals with small intermolecular interferences. Leu and Ile were not differentiated because of similar molecular weights, whereas a detection limit of 6 pmol was achieved [54].

Next to CNTs, single-walled carbon nanotubes (SWNTs) modified with $-(CH_2)_2COOH$ groups were proposed by Ugarov et al. for the analysis of peptide solutions by SALDI-MS [21]. The authors showed that the position of acidic

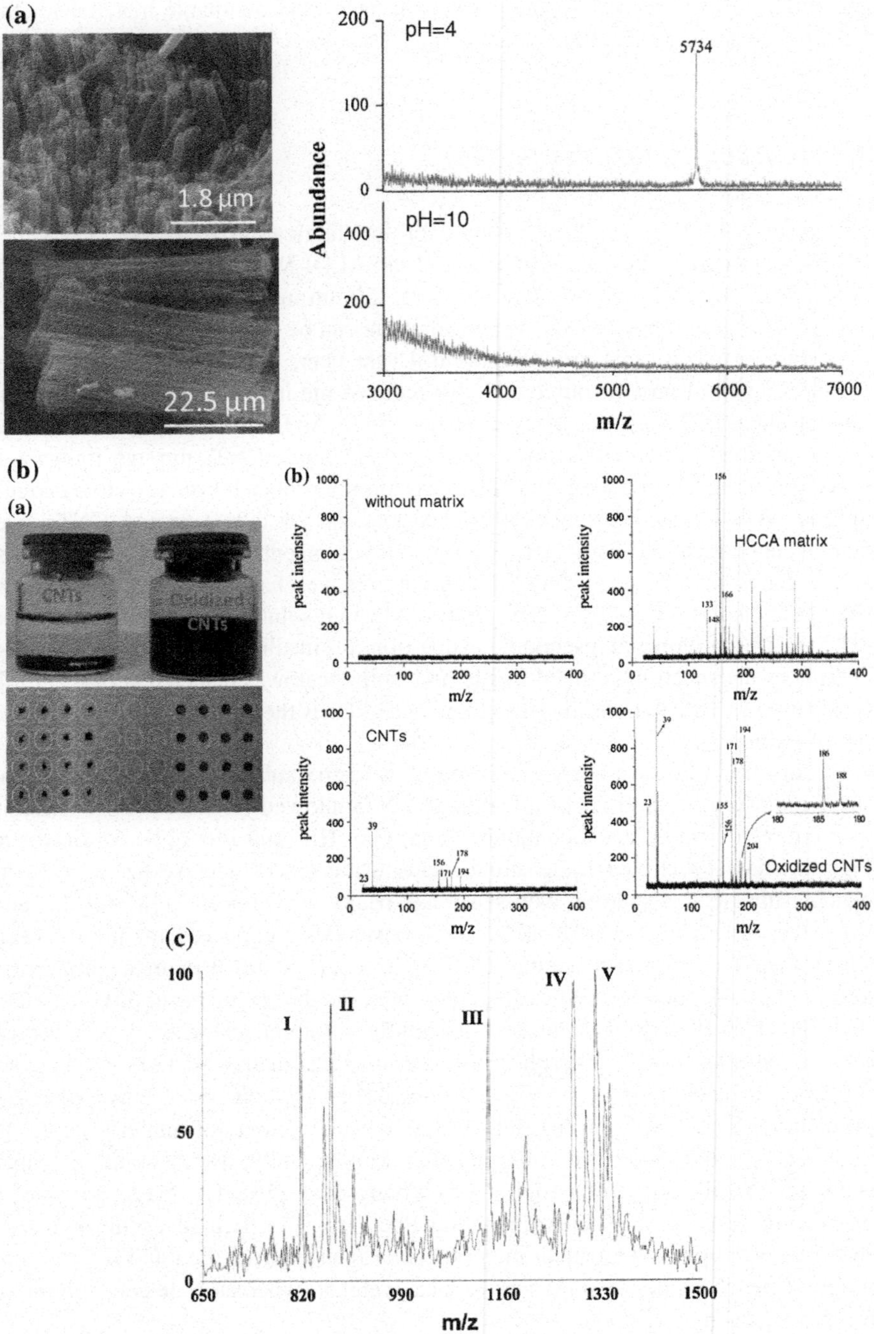
(a)
1.8 µm
22.5 µm
pH=4
5734
pH=10
Abundance
m/z
(b)
(a)
CNTs
Oxidized CNTs
(b)
without matrix
HCCA matrix
CNTs
Oxidized CNTs
peak intensity
m/z
(c)
I
II
III
IV
V
m/z

◀ **Fig. 3** **a** SEM images of CNTs and SALDI-MS spectra of insulin obtained using citrate-treated CNTs as affinity probes to trap insulin from sample solutions containing 10^{-8} M insulin at pH 4 and pH 10 (reprint with permission from [50]). **b** (*a*) Comparison of solubility of CNTs and oxidized CNTs in water and matrix layers of CNTs and oxidized CNTs, (*b*) Mass spectra for 4 amino acid mixture solutions containing Asn, Clu, His and Phe in the LDI mode without matrix, and with HCCA matrix, CNTs and oxidized CNTs (reprint with permission from [22]). **c** SALDI spectrum of 4 peptides at femtomolar loading. *I* RRPYIL (81,800 amu), *II* dynorphon 1–7 YGGFLRR (868.0 amu), *III* dynorphin 1–9 YGGFLRRIR (1138.4 amu); and *IV* goosefish angiotensin I NRVYVHPFHL (1281.5 amu). The concentration of each peptide was 25 fmol/μL (water solution) (reprint with permission from [21])

moieties on the walls of nanotubes is essential for SALDI analysis. Indeed, CNTs with the functional groups attached to the ends do not work as LDI-MS matrices, which means that the attachment to the walls is critical for desorption/ionization. Four peptides (RRPYIL (818.0 amu), dynorphin 1–7 YGGFLRR (868.0 amu), dynorphin 1–9 YGGFLRRIR (1138.4 amu), and goosefish angiotensin I NRVYVHPFHL (1281.5 amu) were efficiently detected with this system at femtomole concentrations [21] as seen in Fig. 3c.

2.3 Fullerenes

The discovery of buckminster fullerene (C_{60}) together with its subsequent large-scale synthesis have initiated a growing research field in the area of biochemistry and life sciences using this material [55]. Hopwood was one of the first considering C_{60} as LDI matrix [56]. In that case, the analytes were deposited on a pre-deposited thin film (10 nm) of fullerenes and measured the positive ions generated with a nitrogen laser. Although some proteins, such as Cytochrome C (12 kDa) and bovine serum albumin (BSA, 66 kDa), are detected by this method, the sensitivity is still lower than with organic acids as matrix. The reason for this lower sensitivity may be the uneven dispersion of polar analytes in the non-polar fullerene matrix.

To overcome this problem, a water-soluble fullerene derivative, hexa(sulfonbutyl)fullerene $C_{60}\left[(CH_2)SO_3{}^-\right]_6$ (HSBF) was synthesized [15]. HSBF consisted of a C_{60} cage, covalently bonded with six negatively charged sulphonate arms (Fig. 4a). This matrix was used as an ion-parting reagent to precipitate analytes in a mixture, depending upon the charge, structure, and hydrophobic character of each analyte [15]. As HSBF shows strong optical absorbance in the UV range, no additional organic matrix was required to conduct SALDI-MS analysis. The affinity of various proteins to HSBF was studied by the authors [15]. For example, digestion of myoglobin with trypsin generated peptides that are positively charged and the peptides in abundant quantity suppress the signals of some important trace peptides, which were quite evident on precipitation by HSBF, due to strong affinity of the fullerene derivative to lysine ends of these peptides (Fig. 4a).

Other fullerene modifications such as the *N*-dimethyl pyrrolidinium iodide modified fullerene allowed the enrichment of N-terminal sulfonated peptides [57]. In parallel to the work on SWNTs, Ugarov demonstrated the interest of acid-terminated fullerenes [C_{60} $(CH_2)_2COOH)_n$] as matrix material for LDI-MS analysis of a solution of 4 peptides [21]. Due to the presence of acidic groups, no additional components are needed for decreasing the pH and perform an effective ionization. Vallant et al. [58] showed that fullerene-based materials with different chemical moieties can be successfully employed for selective enrichment and identification of low mass serum constituents. Dioctadecyl methano fullerenes, fullerenoacetic acid and iminoacetic acid (IDA)-fullerenes were prepared (Fig. 4b) and subjected to a comprehensive characterization study including protein binding properties and capacity. Indeed, in this study a new strategy was elaborated enabling both screening of biological samples using SALDI and identification of serum constituents in a broad mass range (m/z = 1000–30,000) after elution of bound analytes as seen in Fig. 4b. It has been shown that diverse functionalities result in characteristic human serum patterns in terms of signal intensity as well as the number of detectable masses (Fig. 4b). As depicted in Figs. 4b, the carboxylated fullerenes provided a number of signals, whereas the more hydrophobic dioactadecyl methano fullerene was characterized by a total absence of expedient

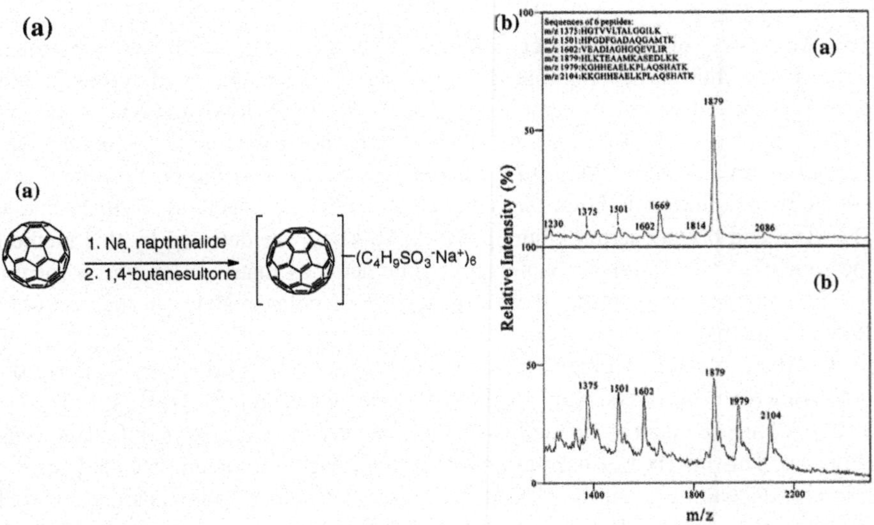

Fig. 4 **a** (*a*) Synthetic procedure of the sodium salt water soluble fullerene derivative hexa (sulfonbutyl)fullerene $C_{60}[(CH_2)SO_3^-]_6$ (HSBF), (*b*) SALDI spectrum of the trypsin-digested myoglobulin solution (1 mM) with particle LDI mass spectrum; **b** (*a*) different fullerene derivatives used for LDI-MS, (*b*) representation of low molecular mass working flow of human serum enriched on derivatized fullerenes: (*A*–*C*) fullereneacetic acid, (*D*–*F*) dioctadecyl methanofullerene: *A* + *D* MALDI spectra, *B* + *C* MALDI before MC-separation, *C* + *F* MALDI after liquid chromatography separation and fractionation (reprint with permission from [58])

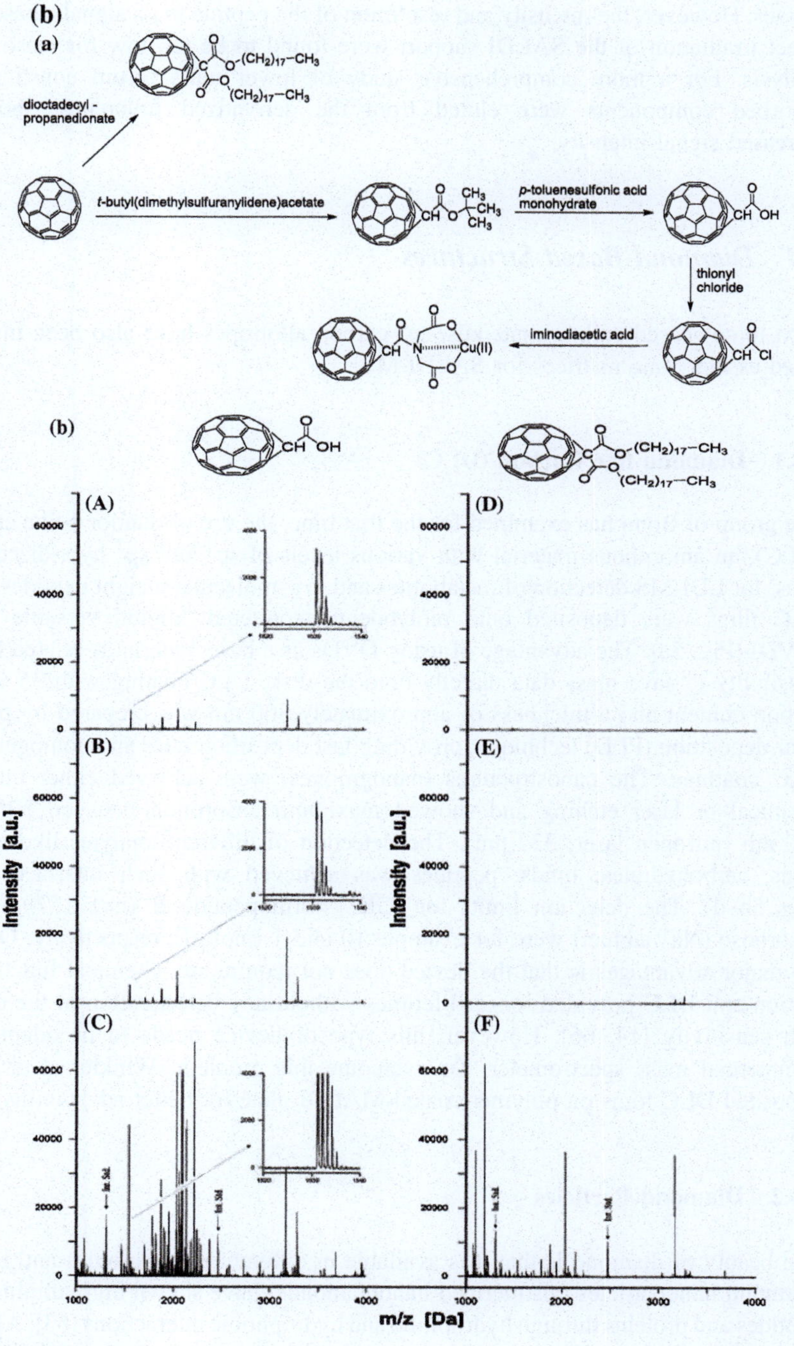

Fig. 4 (continued)

masses. However, the intensity and resolution of the peptide mass signals, based on direct irradiation of the SALDI support were found to be too low for reasonable analysis. For a more comprehensive study of lower mass serum constituents, adsorbed components were eluted from the derivatized fullerenes, showing increased signal intensity.

2.4 *Diamond-Based Structures*

Next to sp^2-based carbon materials, sp^3 carbon allotropes have also been investigated as inorganic matrices for SALDI-MS.

2.4.1 Diamond-like Carbon (DLC)

The group of Bonn has examined for the first time, the use of diamond-like carbon (DLC), an amorphous material with various levels of sp^3 and sp^2 hybridized carbons, for LDI-MS detection of metabolites and low-molecular weight peptides [14]. DLC films were deposited onto molybdenum-sputtered "digital versatile disc" (DVD) (Fig. 5a). The advantage of using DVDs as a base material is related to the possibility to save mass data directly onto the disk. DLC coating with 35 % sp^3 carbon content and a thickness of approximately 300 nm was prepared by pulsed laser deposition (PLD) technique, providing fast deposition rates and homogeneous DLC coatings. The nanostructures (nanogrooves) were achieved either through chemical or laser etching and showed maximum absorption between 305 and 330 nm (nitrogen laser, 337 nm). The detection of different analytes like amino acids, carbohydrates, lipids, peptides was achieved with such interfaces [14] (Fig. 5b–d). The detection limits for [Glu^1]-fibrinopeptide B (m/z 1570.6) and L-sorbose (Na^+ adduct) were for example 10 and 1 fmol/μL, respectively. One of the major advantages is that the device does not require any chemical functionalization and DLC provided longer lifetimes without any deterioration in the detection sensitivity [14, 66]. However, this type of device needs to be adapted to commercial mass spectrometer. To overcome this problem, Winkler et al. have deposited DLC films on polymer coated-MALDI-plate for better adaptability [67].

2.4.2 Diamond Particles

The beauty of diamond is that it is available in different morphologies and forms. Diamond nanoparticles (also termed nanodiamonds) have shown high affinities for peptides and proteins through hydrophilic and hydrophobic interactions [63]. Affinity allows biological molecules in dilute solutions to be captured easily by diamond and directly measured by matrix-assisted laser desorption/ionization (MALDI) time-of flight (TOF) mass spectrometry (MS) with α-cyano-4-hydroxycinnamic acid (HCCA)

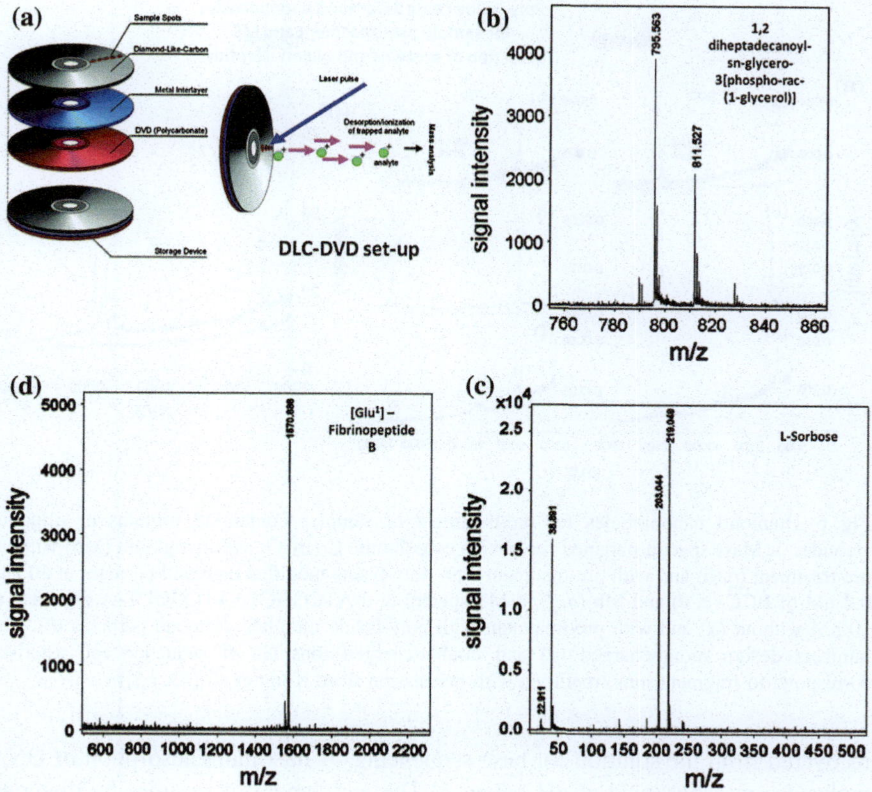

Fig. 5 DLC DVD disk set-up (**a**) developed by Najam-ul-Haq et al. for the LDI-MS detection of lipid (**b**), carbohydrate (**c**) and peptide (**d**) (reprint with permission from Najam-ul-haq et al. [14])

matrix. From 50-fold diluted human blood serum, the spectral quality was greatly improved in comparison to sample without the pre-treatment with diamond particles, with an enhancement in detection sensitivity by more than 2 orders of magnitude. Thanks to diamond particles, proteins such as cytochrome c, myoglobin, and albumin were detected at 100 pM from a 1-mL solution (Fig. 6a).

Diamond has also been utilized in solid-phase extraction (SPE), where the isolation, concentration, purification, digestion, and analysis of DNA were carried out in complex protein solutions and cell lysates [65]. Poly(L-lysine) coated carboxylated/oxidized diamond nanocrystals (100 nm in diameter) were prepared and used to isolate, concentrate, purify, and digest DNA oligonucleotides in one microcentrifuge tube for matrix-assisted laser desorption/ionization (MALDI) time of-flight (TOF) mass spectrometry. It was demonstrated that using diamond nanocrystals as a solid-phase extraction support not only permits concentration of oligonucleotides in highly diluted solutions, but also facilitates separation of oligonucleotides from proteins in heavily contaminated solutions. In addition, enzymatic digestions can be conducted on particles, and the digests can be easily

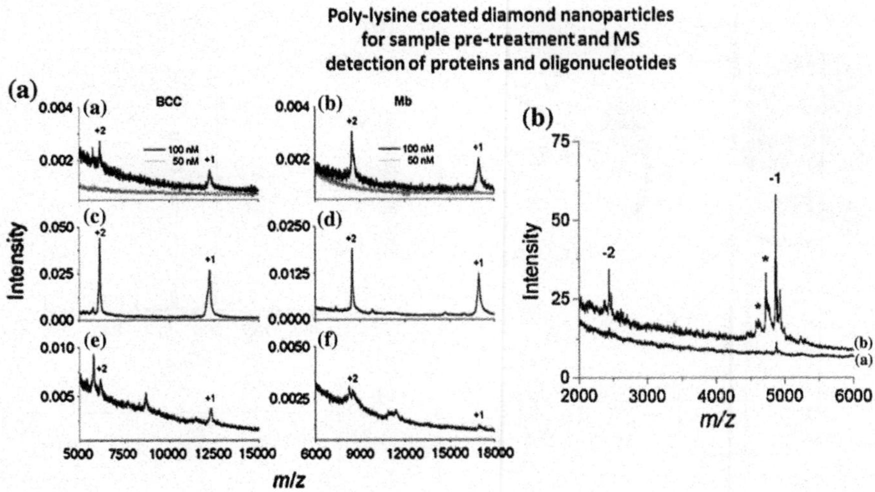

Fig. 6 Diamond nanoparticles for pre-treatment of samples containing proteins or oligonucleotides. **a** Mass spectra obtained for bovine cytochrome C (BCC) and myoglobin (Mb) without pre-treatment (*a*, *b*) and with pre-treatment with poly-lysine modified diamond particles at 50 and 0.2 nM of BCC (*c*, *e*) and Mb (*d*, *f*). **b** MS spectra of d(ATCGGCTAATCGGCTA) sequence at 10 nM without (*a*) and with pre-treatment with poly-lysine modified diamond particles (*b*). The numbers design mono-charged (*1*) and double-charged ions (*2*) of sequence and asterisks correspond to fragment ions (modified with permission from Kong et al. [63, 65])

recovered from the solution for base sequencing. A maximum adsorption of DNA molecules was observed at pH below 3. The enrichment of samples by diamond particles enhanced the ion signals detected for a 500-µL solution at 5 nM. Both singly and doubly charged ions of oligonucleotides can be clearly seen at m/z 4880 and 2440, respectively. Concentrations as low as 100 pM were measured (Fig. 6b).

2.4.3 Nanocrystalline Diamond (NCD)

Nanocrystalline diamond (NCD) can also be chemically modified as shown by Najam-ul-Haq et al. [64] using glycidyl methacrylate (GMA) under ultraviolet (UV) light. This modification was followed by the introduction of iminodiacetic acid (IDA), a chelating agent or C_{18} alkyl chain via the reaction with octadecylamine. IDA is a chelating agent, and when loaded with copper ions permits to create immobilized metal affinity chromatographic (IMAC) supports for improving sensitivity, selectivity, and capacity of detection whereas C_{18} acted as reversed phase surface [64]. Figure 7 depicts the chemical modification routes of the NCD films (IDA-Cu^{2+} and C_{18}) and the different mass spectra obtained on both surfaces for peptides and proteins profiling from serum.

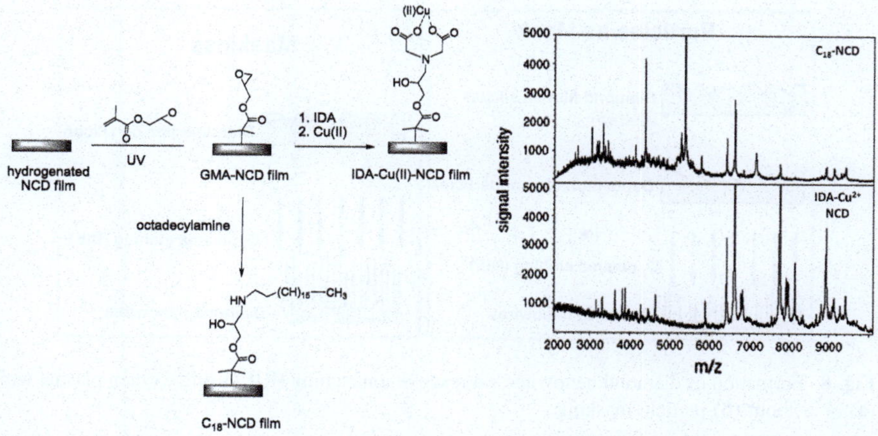

Fig. 7 NCD chemical modifications by IDA-Cu^{2+} or octadecylamine for protein and peptide mass spectrometry profiling from serum in presence of α-cyano-4-hydroxycinnamic acid (HCCA) (reprint with permission from [64])

2.4.4 Diamond Nanowires

Diamond nanowires are among the relatively new and very promising materials for chemical and biochemical sensing. The first attempt to prepare diamond nanowires dates back to 1968 using a radiation heating unit, developed from a super-high pressure Xenon lamp [68]. It was however only around the beginning of 21th century that further attempts for the synthesis of diamond nanostructures using bottom-up and top-down approaches were undertaken. While the bottom-up methods [69–74] are in general more efficient, since the growth of a thick diamond layer is time consuming and its post-processing to form nanostructures can be somehow cumbersome, only the top-down approaches are currently used for the formation of diamond nanostructures for SALDI applications [24, 75].

Top-down approaches are mainly based on the etching of CVD grown diamond films using arrays of nanoparticles seeded onto the diamond films as masks (Fig. 8a). Aluminum dots [76], SiO_2 particles [77], gold nanoparticles [78] and diamond nanoparticles [27, 29, 79, 80] showed to be useful etching masks. The choice of the material of the seeding particle is mainly governed by the possibility of depositing a uniform layer onto the CVD diamond films, the ease of forming well dispersed particle solutions, and the selectivity of etching and its rate for diamond versus metal [71]. Gold nanoparticles were found to be the most suitable seeding masks than aluminum oxide and SiO_2 nanoparticles as they are easier to disperse, resulting in single nanoparticles on the surface, requiring no further processing step.

Top-down approaches without the use of any mask have been proposed recently for the preparation of diamond nanowires [24, 75, 81–83]. Such methods have the intrinsic advantage of being simple and straightforward not requiring complicated processing steps such as mask deposition or template removal.

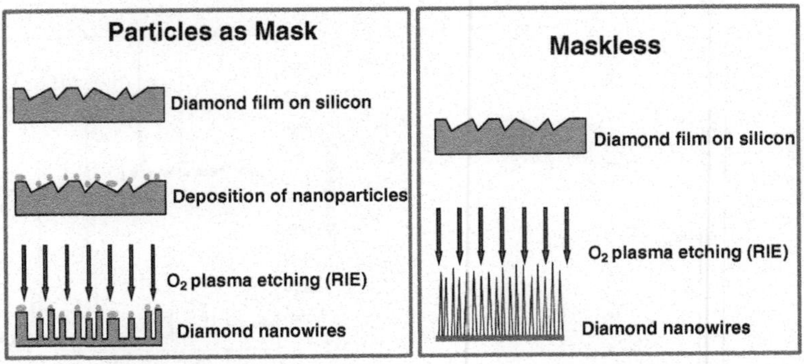

Fig. 8 Formation of diamond nanowires via reactive ion etching (RIE) using oxygen plasma with (**a**) or without (**b**) particle masking

We demonstrated that BDD nanowires can be formed from highly-doped polycrystalline diamond thin films using RIE with an oxygen plasma [24, 32, 75, 82]. Figure 9 shows some of the characterization of the formed diamond nanowires. X-ray photoelectron spectroscopy (XPS) analysis of the chemical composition of the diamond nanowires revealed that next to C_{1s} at 285 eV and O_{1s} at 532 eV, additional peaks at 402, 104 and 169 eV due to N_{1s}, Si_{2p} and S_{2p} are also present (Fig. 9a). The latter elements are believed to originate from surface contamination during the RIE process. Indeed, deposition of a SiO_x shell around the BDD NWs was confirmed by HR-TEM (Fig. 9d) and EDX analysis (Fig. 9b). SiO_x was most likely deposited during the etching process due to sputtering of the substrate holder or the silicon wafer onto which the diamond film was deposited. A similar behavior was observed by Baik et al. when Mo sample holder was used [83]. The SiO_x deposits can be easily removed in HF aqueous solutions as confirmed by XPS and EDX analysis (Fig. 9a, b). A change in nanowires morphology was in addition observed after HF treatment with sharper wires being formed thereafter (Fig. 9c).

Depending on the diamond film thickness, the resulting nanowires' dimensions can vary between 0.6 and 1.4 ± 0.1 μm in length and exhibit constant tip and base radius of $r_{tip} = 10 \pm 5$ nm and $r_{base} = 40 \pm 5$ nm, respectively. The wires are about 60–140 times longer than aligned diamond nanowires prepared using diamond nanoparticles as a hard mask [24, 32, 33, 75, 82]. We have also demonstrated that boron atoms are not essentially involved in the nanowires formation mechanism since undoped and doped diamond substrates led to the formation of diamond nanowires [24]. However, we have shown that the BDD film doping level can have an influence on nanowires' density (Fig. 10).

The use of diamond nanowires as a matrix-free LDI substrate for the D/I of peptides and small molecules, and their analysis by mass spectrometry with a very high sensitivity has been shown by us lately [24]. To minimize droplet spreading, the nanowires were chemically functionalized with octadecyltrichlorosilane (OTS) and then UV/ozone treated to reach a final water contact angle of 120 °C (Fig. 11a) [75]. The sub-bandgap absorption under UV laser irradiation and the heat

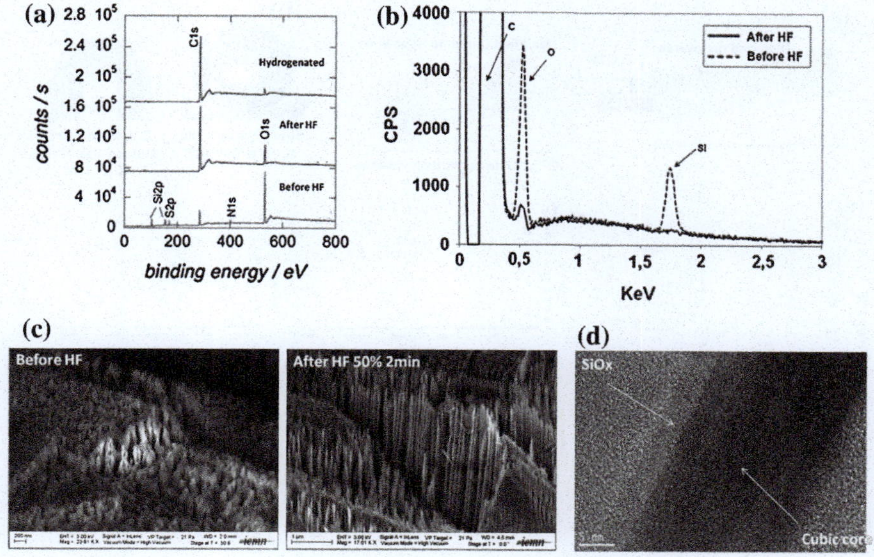

Fig. 9 **a** XPS spectra recorded on the as-prepared diamond nanowires, after HF treatment and after hydrogenation; **b** EDX spectra of the nanowires before and after HF treatment; **c** SEM images of the nanowires before and after HF treatment; **d** HR-TEM image of the as-prepared diamond nanowires [75]

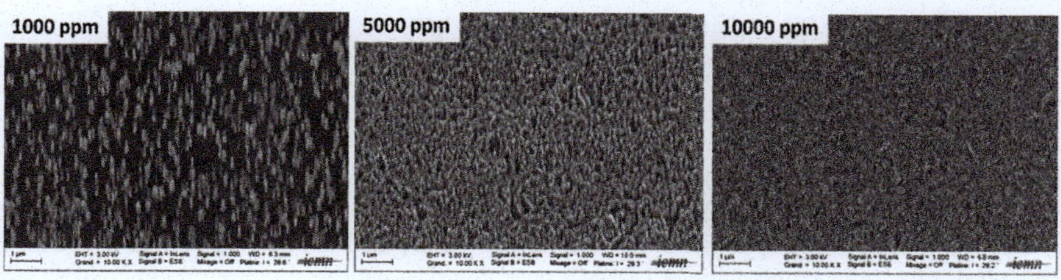

Fig. 10 Densities of diamond nanowires obtained from diamond films with different boron concentrations during the growth of BDD films (1000, 5000 and 10,000 ppm). Diamond nanowires were synthesized using O_2 plasma during 30 min at a power of 350 W, flow rate of 20 sccm O_2, pressure chamber of 150 mT (unpublished results)

confinement inside the nanowires allowed LDI-MS of various compounds, most likely via a thermal mechanism. A detection limit of 200 zeptomole for verapamil was achieved (Fig. 11b). In this study, we have observed that, for the realization of LDI-MS on such diamond surface, nanostructuration of the NCD film is required (formation of diamond nanowires). The investigated diamond nanowires, undoped (UDD) or boron-doped (BDD), display antireflective properties permitting photons absorption at $\lambda = 337$ nm (Fig. 11c, d). However, only the BDD nanowires have shown the best results in terms of LDI-MS efficiency [24].

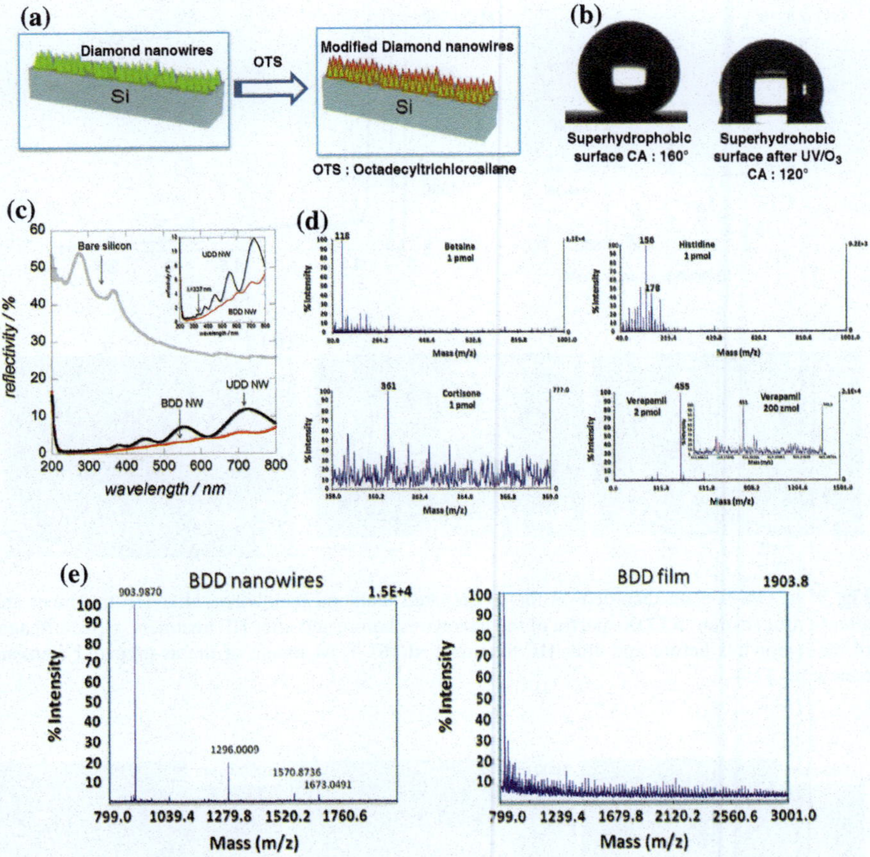

Fig. 11 Diamond nanowires for SALDI: **a** OTS modified diamond nanowires, **b** contact angle measurement before and after UV/O_3 treatment, **c** reflectivity of undoped diamond nanowires (UDD NW) and boron doped nanowires (BDD NW) compared to silicon wafer, **d** LDI-MS on BDD NW of various compounds, **e** LDI-MS detection of a peptide mixture on either BDD NW and BDD film (reprint with permission from Coffinier et al. [24])

3 Conclusion

In conclusion, different carbon nanostructures in the form of films and particle suspensions have shown to work as SALDI matrices for small molecules, peptides, proteins and other analytes. At present, most of the examples reported in the literature are based on sp^2 hybridized carbon materials. One of the drawbacks of such carbon materials as SALDI interface is the contamination of the ion source. CNTs, when used as 2D films, have the tendency to fly off into the instrument when laser pulses are applied. AB Sciex sent a note in which they specifically do not recommend the use of CNTs as matrix in any of their mass spectrometers. Attempts to overcome this limitation were undertaken such as the addition of polyurethane adhesives to CNT matrixes [84]. The other possibility is based on the use of different matrixes.

Due to the rapid advancements made over the last years in the possibility to from sp^3 based structures in the form of diamond films, diamond nanoparticles and diamond nanowires, this material is currently considered in more detail for SALDI based applications. While there are currently only a handful of reports on the use of diamond for matrix-free mass spectrometry analysis, the encouraging results are hoped to trigger further interest of these structures for mass spectrometry based identification and quantification. A full and detailed understanding of the importance of doping level, wire length and other physical properties is needed to foster further developments in this direction. Diamond nanostructures display several advantages as SALDI substrates such as high surface area, availability (diamond nanoparticles), and ease of preparation (diamond nanowires). Moreover, the existence of numerous possibilities for integrating surface functionalities [26, 85] will allow to design chips for specific capture of analytes and their subsequent MS analysis with high sensitivity. The robust surface chemistry offered by diamond and its exceptional mechanical properties are a real asset for the reusability of the chips. Finally, the non wetting properties (superhydrophobicity) of chemically modified diamond nanowires open new opportunities for their integration in lab-on-chip devices for MS analysis [86–88]. Although diamond nanostructures are highly suitable for SALDI-MS analysis, there are some points that need special attention to make these substrates of prime interest in this field. The main hurdle is connected with the price of these substrates. Even though BDD thin films are commercially available, the price is still quite high for instance compared to silicon substrate, the starting material for the preparation of porous silicon and silicon nanowires commonly used as SALDI substrates. This can be lowered by the reutilization of the substrates. Also the preparation of BDD films of controlled doping type and level is restricted to specialized laboratories and as such limits access to these materials. However, given the promising results obtained with BDD nanowires and exceptional properties of this material, it is likely that these limitations will be overcome easily and thus make the future of diamond nanostructures for SALDI analysis an interesting challenge.

Acknowledgement The authors gratefully acknowledge financial support from the Centre National de la Recherche Scientifique (CNRS), the Université Lille 1 and the Nord Pas de Calais region.

References

1. M. Karas, F. Hillenkamp, Laser desorption ionization of proteins with molecular masses exceeding 10,000 daltons. Anal. Chem. **60**, 259–280 (1988). doi:10.1021/ac00171a028
2. S.D. Hanton, Mass spectrometry of polymers and polymer surfaces. Chem. Rev. **101**, 527–569 (2001). doi:10.1021/cr9901081

3. R. Knochenmuss, R. Zenobi, MALDI ionization: the role of in-plume processes. Chem. Rev. **103**, 441–452 (2003). doi:10.1021/cr0103773
4. L. Li, *MALDI Mass Spectrometry for Synthetic Polymer Analysis* (*Chemical Analysis: A Series of Monographs on Analytical Chemistry and Its Applications*) (Wiley-VCH, 2009). ISBN: 978–0-471-77579-9
5. K. Tanaka, H. Waki, Y. Ido, S. Akita, Y. Yoshida, T. Yoshida, Protein and polymer analyses up to m/z 100,000 by laser ionization time-of-flight mass spectrometry. Rapid Commun. Mass Spectrom. 151–153 (1988). doi:10.1002/rcm.1290020802
6. R. Arakawa, H. Kawasaki, Functionalized nanoparticles and nanostructured surfaces for surface-assisted laser desorption/ionization mass spectrometry. Anal. Sci. **26**, 1229 (2010). doi:10.2116/analsci.26.1229
7. J. Sunner, E. Dratz, Y.C. Chen, Graphite surface-assisted laser desorption/ionization time-of-flight mass spectrometry of peptides and proteins from liquid solutions. Anal. Chem. **67**, 4335–4342 (1995). doi:10.1021/ac00119a021
8. J. Wei, J.M. Buriak, G. Siuzdak, Desorption–ionization mass spectrometry on porous silicon. Nature **399**, 243–246 (1999). doi:10.1038/20400
9. J.J. Thomas, Z. Shen, J.E. Crowell, M.G. Finn, G. Siuzdak, Desorption/ionization on silicon (DIOS): a diverse mass spectrometry platform for protein characterization. Proc. Natl. Acad. Sci. USA **98**, 4932–4937 (2001). doi:10.1073/pnas.081069298
10. S.A. Trauger, E.P. Go, Z. Shen, J.V. Apon, B.J. Compton, E.S.P. Bouvier, M.G. Finn, G. Siuzdak, High sensitivity and analyte capture with desorption/ionization mass spectrometry on silylated porous silicon. Anal. Chem. **76**, 4484–4489 (2004). doi:10.1021/ac049657j
11. G. Piret, H. Drobecq, Y. Coffinier, O. Melnyk, R. Boukherroub, Matrix-free laser desorption/ionization mass spectrometry on silicon nanowire arrays prepared by chemical etching of crystalline silicon. Langmuir **26**(2), 1354–1361 (2010). doi:10.1021/la902266x
12. E.P. Go, J.V. Apon, G. Luo, A. Saghatelian, R.H. Daniels, V. Sahi, R. Dubrow, B.F. Cravatt, A. Vertes, G. Siuzdak, Desorption/ionization on silicon nanowires. Anal. Chem. **77**, 1641–1646 (2005). doi:10.1021/ac048460o
13. K.P. Law, J.R. Larkin, Recent advances in SALDI-MS techniques and their chemical and bioanalytical applications. Anal. Bioanal. Chem. **399**, 2597–2622 (2011). doi:10.1007/s00216-010-4063-3
14. M. Najam-ul-haq, M. Rainer, Z. Szabo, R. Vallant, C.W. Huck, G.K. Bonn, Role of carbon nano-materials in the analysis of biological materials by laser desorption/ionization-mass spectrometry. J. Biochem. Biophys. Methods **70**, 319–328 (2007). doi:10.1016/j.jbbm.2006.11.004
15. J.T. Shiea, J.P. Huang, C.F. Teng, J.Y. Jeng, L.Y. Wang, L.Y. Chiang, Use of a water-soluble fullerene derivative as precipitating reagent and matrix-assisted laser desorption/ionization matrix to selectively detect charged species in aqueous solutions. Anal. Chem. **75**, 3587–3595 (2003). doi:10.1021/ac020750m
16. M.J. Dale, R. Knochenmuss, R. Zenobi, Graphite/liquid mixed matrices for laser desorption/ionization mass spectrometry. Anal. Chem. **68**, 3321–3329 (1996). doi:10.1021/ac960558i
17. S. Zumbuhl, R. Knochenmuss, S. Wulfert, F. Dubois, M.J. Dale, R. Zenobi, A graphite-assisted laser desorption/ionization study of light-induced aging in triterpene dammar and mastic varnishes. Anal. Chem. **70**, 707–715 (1998). doi:10.1021/ac970574v
18. S. Xu, Y. Li, H. Zou, J. Qiu, Z. Guo, B. Guo, Carbon nanotubes as assisted matrix for laser desorption/ionization time-of-flight mass spectrometry. Anal. Chem. **75**, 6191–6195 (2003). doi:10.1021/ac0345695
19. X. Zhou, Y. Wei, Q. He, F. Boey, Q. Zhang, H. Zhang, Reduced graphene oxide films used as matrix of MALDI-TOF-MS for detection of octachlorodibenzo-p-dioxin. Chem. Commun. **46**, 6974–6976 (2010). doi:10.1039/C0CC01681K
20. Y.-K. Kim, H.-K. Na, S.-J. Kwack, S.-R. Ryoo, Y. Lee, S. Hong, S. Hong, Y. Jeong, D.-H. Min, Synergistic effect of graphene oxide/MWCNT films in laser desorption/ionization mass

spectrometry of small molecules and tissue imaging. ACS Nano **5**, 4550–4561 (2011). doi:10.1021/nn200245v

21. M.V. Ugarov, T. Egan, D.V. Khabashesku, J.A. Schultz, H. Peng, V.N. Khabashesku, H. Furutani, K.S. Prather, H.W.J. Wang, S.N. Jackson, A.S. Woods, MALDI matrices for biomolecular analysis based on functionalized carbon nanomaterials. Anal. Chem. **76**, 6734–6742 (2004). doi:10.1021/ac049192x
22. C. Pan, S. Xu, L. Hu, X. Su, J. Ou, H. Zou, Z. Guo, Y. Zhang, B. Guo, Using oxidized carbon nanotubes as matrix for analysis of small molecules by MALDI-TOF MS. J. Am. Soc. Mass Spectrom. **16**, 883–892 (2005). doi:10.1016/j.jasms.2005.03.009
23. S.F. Ren, Y.L. Guo, Oxidized carbon nanotubes as matrix for matrix-assisted laser desorption/ionization time-of-flight mass spectrometric analysis of biomolecules. Rapid Commun. Mass Spectrom. **19**, 255–260 (2005). doi:10.1002/rcm.1779
24. Y. Coffinier, S. Szunerits, H. Drobecq, O. Melnyk, R. Boukherroub, Diamond nanowires for highly sensitive matrix-free mass spectrometry analysis of small molecules. Nanoscale **4**, 231–238 (2012). doi:10.1039/C1NR11274K
25. Y. Yu, L. Wu, J. Zhi, Diamond nanowires: fabrication, structure, properties, and applications. Angew. Chem. Int. Ed. **53**, 14326–14351 (2014). doi:10.1002/anie.201310803
26. S. Szunerits, R. Boukherroub, Different strategies for chemical functionalization of diamond surfaces. J. Solid-State Electrochem. **12**, 1205–1218 (2008). doi:10.1007/s10008-007-0473-3
27. N. Yang, H. Uetsuka, E. Osawa, C.E. Nebel, Vertically aligned diamond nanowires for DNA sensing. Angew. Chem. Int. Ed. **47**, 5183–5185 (2008). doi:10.1002/anie.200801706
28. H. Uetsuka, D. Shin, N. Tokuda, K. Saeki, C.E. Nebel, Electrochemical grafting of boron-doped single-crystalline chemical vapor deposition diamond with nitrophenyl molecules. Langmuir **23**, 3466–3472 (2007). doi:10.1021/la063241e
29. C.E. Nebel, N. Yang, H. Uetsuka, E. Osawa, N. Tokuda, O. Williams, Diamond nano-wires, a new approach towards next generation electrochemical gene sensor platforms. Diamond Relat. Mater. **18**, 910 (2009). doi:10.1016/j.diamond.2008.11.024
30. N. Yang, W. Smirnov, C.E. Nebel, Three-dimensional electrochemical reactions on tip-coated diamond nanowires with nickel nanoparticles. Electrochem. Commun. **27**, 89–91 (2013). doi:10.1016/j.elecom.2012.10.044
31. D. Luo, L. Wu, J. Zhi, Fabrication of boron-doped diamond nanorod forest electrodes and their application in nonenzymatic amperometric glucose sensing. ACS Nano **3**, 2121–2128 (2009). doi:10.1021/nn9003154
32. Q. Wang, P. Subramanian, M. Li, W.S. Yeap, K. Haenen, Y. Coffinier, R. Boukherroub, S. Szunerits, Non-enzymatic glucose sensing on long and short diamond nanowires electrodes. Electrochem. Commun. **34**, 286–290 (2013). doi:10.1016/j.elecom.2013.07.014
33. Q. Wang, A. Vasilescu, P. Subramanian, V. Andrei, Y. Coffinier, M. Li, R. Boukherroub, S. Szunerits, Simultaneous electrochemical detection of tryptophan and tyrosine using boron-doped diamond and diamond nanowires electrodes. Electrochem. Commun. **35**, 84–87 (2013). doi:10.1016/j.elecom.2013.08.010
34. S. Szunerits, Y. Coffinier, E. Galopin, J. Brenner, R. Boukherroub, Preparation of boron-doped diamond nanowires and their application for sensitive electrochemical detection of tryptophan. Electrochem. Commun. **12**, 438 (2010). doi:10.1016/j.elecom.2010.01.014
35. P. Subramanian, I. Mazurenko, V. Zaitsev, Y. Coffinier, R. Boukherroub, S. Szunerits, Diamond nanowires modified with poly[3-(pyrrolyl)carboxylic acid] for the immobilization of histidine-tagged peptides. Analyst **139**, 4343–4349 (2014). doi:10.1039/c4an00146j
36. P. Subramanian, J. Foord, D. Steinmueller, Y. Coffinier, R. Boukherroub, S. Szunerits, Diamond nanowires decorated with metallic nanoparticles: a novel electrical interface for the immobilization of histidinylated biomolecules. Electrochim. Acta **110**, 4–8 (2013). doi:10.1016/j.electacta.2012.11.010
37. P. Subramanian, A. Motorina, W.S. Yeap, K. Haenen, Y. Coffinier, V. Zaitsev, J. Niedziolka-Jonsson, R. Boukherroub, S. Szunerits, Impedimetric immunosensor based on diamond nanowires decorated with nickel nanoparticles. Analyst **139**, 1726–1731 (2014). doi:10.1039/c3an02045b

38. S. Szunerits, Y. Coffinier, R. Boukherroub, Diamond nanowires: a recent success story for biosensing. In: *Nanosensor Technology*. Springer Series on Chemical Sensors and Biosensors (Springer, Heidelberg, 2015) (in print)
39. Y.C. Chen, J. Shiea, J. Sunner, Thin-layer chromatography–mass spectrometry using activated carbon, surface-assisted laser desorption/ionization. J. Chromatogr. A **826**, 77–86 (1998). doi:10.1016/S0021-9673(98)00726-2
40. H.J. Kim, J.K. Lee, S.J. Park, H.W. Ro, D.Y. Yoo, D.Y. Yoon, Observation of low molecular weight poly(methylsilsesquioxane)s by graphite plate laser desorption/ionization time-of-flight mass spectrometry. Anal. Chem. **72**, 5673–5678 (2000). doi:10.1021/ac0003899
41. K.H. Park, H.J. Kim, Analysis of fatty acid by graphite plate laser desorption/ionization time of flight mass spectrometry. Rapid Commun. Mass Spectrom. **15**, 1494–1499 (2001). doi:10.1002/rcm.387
42. J. Kim, K. Paek, W. Kang, Visible surface-assisted laser desorption/ ionization mass spectrometry of small macromolecules deposited on the graphite plate. Bull. Korean Chem. Soc. **23**, 315–319 (2002). doi:10.1002/pmic.200401023
43. Y.C. Chen, J.Y. Wu, Analysis of small organics on planar silica surfaces using surface-assisted laser desorption/ionization mass spectrometry. Rapid Commun. Mass Spectrom. **20**, 1899–1903 (2001). doi:10.1002/rcm.451
44. C. Black, C. Poile, J. Langley, J. Herniman, The use of pencil lead as a matrix and calibrant for matrix-assisted laser desorption/ionisation. Rapid Commun. Mass Spectrom. **20**, 1053–1060 (2006). doi:10.1002/rcm.2408
45. S. Cha, E.S. Yeung, Colloidal graphite-assisted laser desorption/ionization mass spectrometry and MSn of small molecules. 1. Imaging of cerebosides directly from rat brain tissue. Anal. Chem. **79**, 2373–2385 (2007). doi:10.1021/ac062251h
46. S. Cha, H. Zhang, H.I. Ilarsaln, Z.S. Wurtele, L. Brachova, B.J. Nikolau, E.S. Yeung, Direct profiling and imaging of plant metabolites in intact tissues by using colloidal graphite-assisted laser desorption ionization mass spectrometry. Plant. J. **55**, 348–360 (2008). doi:10.1111/j.1365-313X.2008.03507
47. H. Zhang, S. Cha, E.S. Yeung, Colloidal graphite-assisted laser desorption/ionization MS and MS n of small molecules. 2. Direct profiling and MS imaging of small metabolites from fruits. Anal. Chem. **79**, 6575–6584 (2007). doi:10.1021/ac0706170
48. H. Kawasaki, T. Takahashi, F. Fujimori, O. Okumura, W. Watanabe, M. Matsumura, T. Takemine, T. Nakano, R. Arakawa, Functionalized pyrolytic highly oriented graphite polymer film for surface-assisted laser desorption/ ionization mass spectrometry in environmental analysis. Rapid Commun. Mass Spectrom. **23**, 3323–3332 (2009). doi:10.1002/rcm.4254
49. M. Najam-ul-haq, M. Rainer, T. Schwarzenauer, C.W. Huck, G.K. Bonn, Chemically modified carbon nanotubes as material enhanced laser desorption ionisation (MELDI) material in protein profiling. Anal. Chim. Acta **561**, 32–39 (2006). doi:10.1016/j.aca.2006.01.012
50. W.Y. Chen, L.S. Wang, H.T. Chiu, Y.C. Chen, C.Y. Lee, Carbon nanotubes as affinity probes for peptides and proteins in MALDI MS analysis. J. Am. Soc. Mass Spectrom. **15**, 629–635 (2004). doi:10.1016/j.jasms.2004.08.001
51. L.-S. Wang, C.-Y. Lee, H.-T. Chiu, New nanotube synthesis strategy—application of sodium nanotubes formed inside anodic aluminium oxide as a reactive template. Chem. Commun. **15**, 1964–1965 (2003). doi:10.1039/B305610D
52. C.-T. Chen, Y.-C. Chen, Desorption/ionization mass spectrometry on nanocrystalline titania sol–gel-deposited films. Rapid Commun. Mass Spectrom. **18**, 1956–1964 (2004). doi:10.1002/rcm.1572
53. C. Pan, S. Xu, H. Zou, Z. Guo, Y. Zhang, B. Guo, Carbon nanotubes as adsorbent of solid-phase extraction and matrix for laser desorption/ ionization mass spectrometry. J. Am. Soc. Mass Spectrom. **16**, 263–270 (2005). doi:10.1016/j.jasms.2004.11.005
54. J. Zhang, H.Y. Wang, Y.L. Guo, Amino acids analysis by MALDI mass spectrometry using carbon nanotube as matrix. Chin. J. Chem. **23**, 185–189 (2005)

55. E. Nakamura, H. Isobe, Functionalized fullerenes in water. The first 10 years of their chemistry, biology, and nanoscience. Acc. Chem. Res. **36**, 807–815 (2003). doi:10.1021/ar030027y
56. F.G. Hopwood, L. Michalak, D.S. Alderdice, K.J. Fisher, G.D. Willet, C_{60}-assisted laser desorption/ionization mass spectrometry in the analysis of phospho tungstic acid. Rapid Commun. Mass Spectrom. **8**, 881–885 (1994). doi:10.1002/rcm.1290081105
57. Y.H. Lee, J.W. Shin, S. Ryu, S.W. Lee, C.H. Lee, K. Lee, Enrichment of N-terminal sulfonated peptides by water-soluble fullerene derivative and its applications to highly efficient proteomics. Anal. Chim. Acta 556, 140–144 (2006). doi:10.1016/j.aca.2005.06.060
58. R.M. Vallant, Z. Szabo, L. Trojer, M. Najam-ul-Haq, M. Rainer, C.W. Huck, R. Bakry, G.K. Bonn, A new analytical approach for the determination of low mass serum constituents employing fullerene derivatives for selective enrichment. J. Proteome Res. **6**, 44–53 (2007). doi:10.1021/pr060347m
59. X. Chen, L. Hu, X. Su, L. Kong, M. Ye, H. Zou, Separation and detection of compounds in Honeysuckle by integration of ion-exchange chromatography fractionation with reversed-phase liquid chromatography-atmospheric pressure chemical ionization mass spectrometer and matrix-assisted laser desorption/ionization time-of-flight mass spectrometry analysis. J. Pharm. Biomed. Anal. **40**, 559–570 (2006). doi:10.1016/j.jpba.2005.07.043
60. X. Chen, L. Kong, X. Su, C. Pan, M. Ye, H. Zou, Integration of ion-exchange chromatography fractionation with reversed-phase liquid chromatography atmospheric pressure chemical ionization mass spectrometer and matrix assisted laser desorption/ionization time-of-flight mass spectrometry for isolation and identification of compounds in *Psoralea corylifolia*. J. Chromatogr. A **1089**, 87–100 (2005). doi:10.1016/j.chroma.2005.06.067
61. L. Hu, S. Xu, C. Pan, C. Yuan, H. Zou, G. Jiang, Matrix-assisted laser desorption/ionization time-of-flight mass spectrometry with a matrix of carbon nanotubes for the analysis of low-mass compounds in environmental samples. Environ. Sci. Technol. **39**, 8442–8447 (2005). doi:10.1021/es0508572
62. M. Rainer, M.N. Quershi, G.K. Bonn, Matrix-free and material-enhanced laser desorption/ionization mass spectrometry for the analysis of low molecular weight compounds. Anal. Bioanal. Chem. **400**, 2281–2288 (2011). doi:10.1007/s00216-010-4138-1
63. X.L. Kong, L.C.L. Huang, C.M. Hsu, W.H. Chen, C.C. Han, H.C. Chang, High affinity capture of proteins by diamond nanoparticles for mass spectrometric analysis. Anal. Chem. **77**, 259–265 (2005). doi:10.1021/ac048971a
64. M. Najam-ul-Haq, M. Rainer, C.W. Huck, G. Stecher, I. Feuerstein, D. Steinmueller, G.K. Bonn, Chemically modified nano crystalline diamond layer as material enhanced laser desorption ionisation (MELDI) surface in protein profiling. Curr. Nanosci. **2**, 1–7 (2006). doi:10.2174/157341306775473836
65. X.L. Kong, L.C.L. Huang, S.C.V. Liau, C.C. Han, H.C. Chang, Polylysine-coated diamond nanocrystals for MALDI-TOF mass analysis of DNA oligonucleotides. Anal. Chem. **77**, 4273–4277 (2005). doi:10.1021/ac050213c
66. L. Sage, Femtomolar sensitivity with matrix-free LDI MS. Anal. Chem. **80**, 5515–5523 (2008). doi:10.1021/ac801668w
67. W. Winkler, W. Balika, P. Hausberger, H. Kraushaar, G. Allmaier, Diamond-like diamond coated polymer-based target in microscope slide format for MALDI mass spectrometry. J. Mass Spectrom. **45**, 566–569 (2010). doi:10.1002/jms.1744
68. B.V. Derjaguin, D.V. Fedoseev, V.M. Lukyanovich, B.V. Spitzin, V.A. Ryabov, A.V. Lavrentyev, Filamentary diamond crystals. J. Cryst. Growth **2**, 380–384 (1968). doi:10.1016/0022-0248(68)90033-X
69. N. Shang, P. Papakonstantinou, P. Wang, A. Zakharov, U. Palnitkar, I.N. Lin, M. Chu, A. Stamboulis, Self-assembled growth, microstructure, and field-emission high-performance of ultrathin diamond nanorods. ACS Nano **3**, 1032–1038 (2009). doi:10.1021/nn900167p
70. C.-H. Hsu, J. Xu, Diamond nanowire—a challenge from extremes. Nanoscale **4**, 5293 (2012). doi:10.1039/c2nr31260c

71. B.J.M. Hausmann, M. Khan, Y. Zhang, T.M. Bainec, K. Martinick, M. McCutcheon, P. Hemmer, M. Loncar, Fabrication of diamond nanowires for quantum information processing applications. Diamond Relat. Mater. **19**, 621–629 (2010). doi:10.1016/j.diamond.2010.01.011
72. H. Masuda, M. Watanaba, K. Yasui, D. Tryk, T. Rao, A. Fujishima, Fabrication of a nanostructured diamond honeycomb film. Adv. Mater. **12**, 444–447 (2000). doi:10.1002/(SICI)1521-4095(200003)12:63.3.CO;2-B
73. T.M. Babinec, B.J.M. Hausmann, M. Khan, Y. Zhang, J.R. Maze, P.R. Hemmer, M. Loncar, A diamond nanowire single-photon source. Nat. Nanotechnol. **5**, 195–199 (2010). doi:10.1038/nnano.2010.6
74. H. Masuda, T. Yanagishita, K. Yasui, K. Nishio, I. Yagi, N. Rao, A. Fujishima, Synthesis of well-aligned diamond nanocylinders. Adv. Mater. **13**, 247 (2001). doi:10.1002/1521-4095(200102)13:4<247:AID-ADMA247>3.0.CO;2-H
75. Y. Coffinier, E. Galopin, S. Szunerits, R. Boukherroub, Preparation of superhydrophobic and oleophobic diamond nanograss array. J. Mater. Chem. **20**, 10671–10675 (2010). doi:10.1039/C0JM01296C
76. Y. Ando, Y. Nishibayashi, A. Sawaben, 'Nano-rods' of single crystalline diamond. Diamond Relat. Mater. **13**, 633 (2004). doi:10.1016/j.diamond.2003.10.066
77. S. Okuyama, S.I. Matsushita, A. Fujishima, Periodic submicrocylinder diamond surfaces using two-dimensional fine particle arrays. Langmuir **18**, 8282–8287 (2002). doi:10.1021/la011107i
78. Y.S. Zou, T. Yang, W.J. Zhang, Y.M. Chong, B. He, I. Bello, S.T. Lee, Fabrication of diamond nanopillar and their arrays. Appl. Phys. Lett. **92**, 053105 (2008). doi:10.1063/1.2841822
79. N. Yang, H. Uetsuka, E. Osawa, C.E. Nebel, Vertically aligned nanowires from boron-doped diamond. Nano Lett. **8**, 3572–3576 (2008). doi:10.1021/nl801136h
80. W. Smirnov, A. Kriele, N. Yang, C.F. Nebel, Aligned diamond nano-wires: fabrication and characterisation for advanced applications in bio and electrochemistry. Diamond Relat. Mater. **18**, 186–189 (2009). doi:10.1016/j.diamond.2009.09.001
81. M. Wei, C. Terashima, M. Lv, A. Fujishima, Z.-Z. Gu, Boron-doped diamond nanograss array for electrochemical sensors. Chem. Commun. 3624 (2009). doi:10.1039/B903284C
82. S. Szunerits, Y. Coffinier, E. Galopin, J. Brenner, R. Boukherroub, Preparation of boron-doped diamond nanowires and their application for sensitive electrochemical detection of tryptophan. Electrochem. Commun. **12**, 438–441 (2010). doi:10.1016/j.elecom.2010.01.014
83. E.-S. Baik, Y.-J. Baik, D. Jeaon, Aligned diamond nanowhiskers. J. Mater. Res. **15**, 923 (2000). doi:10.1557/JMR.2000.0131
84. S.F. Ren, L. Zhang, Z.H. Cheng, Y.L. Guo, Immobilized carbon nanotubes as matrix for MALDI-TOF-MS analysis: applications to neutral small carbohydrates. J. Am. Soc. Mass Spectrom. **16**, 333–339 (2005). doi:10.1016/j.jasms.2004.11.017
85. S. Szunerits, C.E. Nebel, R.J. Hamers, Surface functionalization and biological applications of CVD diamond. MRS Bull. **309**(6), 517–524 (2014). doi:10.1557/mrs.2014.99
86. F. Lapierre, G. Piret, H. Drobecq, O. Melnyk, Y. Coffinier, V. Thomy, R. Boukherroub, High sensitive matrix-free mass spectrometry analysis of peptides using silicon nanowires-based digital microfluidic device. Lab Chip **11**, 1620–1628 (2011). doi:10.1039/c0lc00716a
87. M. Jönsson-Niedziolka, F. Lapierre, Y. Coffinier, S.J. Parry, F. Zoueshtiagh, T. Foat, V. Thomy, R. Boukherroub, EWOD driven cleaning of bioparticles on hydrophobic and superhydrophobic surfaces. Lab Chip **11**, 490–496 (2011). doi:10.1039/c0lc00203h
88. F. Lapierre, M. Harnois, Y. Coffinier, R. Boukherroub, V. Thomy, Split and flow: reconfigurable capillary connection for digital microfluidic systems. Lab Chip **14**, 3589–3593 (2014). doi:10.1039/c4lc00650j

The manufacturer's authorised representative in the EU is Springer Nature Customer Service Centre GmbH, Europaplatz 3, 69115 Heidelberg, Germany. If you have any concerns regarding our products, please contact ProductSafety@springernature.com

Printed and bound by CPI Group (UK) Ltd, Croydon, CR0 4YY
30/09/2026
02225435-0007